Nanobiotechnology

This book covers topics related to drug delivery, biomaterials, drug design, formulation development, nanoscience, and nanotechnology. It describes the fundamental concepts in nanotechnology and their different applications in biotechnology to solve engineering challenges and generate new areas of technological development. Nanobiotechnology: Applications of Nanomaterials in Biotechnology, Medicine and Healthcare covers vast application areas that include medical science, material science, pharmaceutical science, and environmental science. Section 1 presents recent research updates on the different nanomaterials, which are promising in different medical and biotechnological applications. Applications of nanomaterials as bone replacement orthopedic implants have revolutionized the treatment of orthopedic surgery. Nanostructured polymeric materials have gained immense research attention as therapeutic carriers for the precise delivery of drugs at targeted sites. Nanocellulose is recognized as a promising green nanomaterial due to its renewability and abundance in nature. Scientific topics on the most recent scientific and technological advances and applications of different nanostructured materials are presented in this section. Section 2 focuses on the novel synthesis methods that are used extensively and promising for large-scale production of inorganic and nanostructured materials. Section 3 covers the applications of nanotools in the treatment of different diseases, including cancers and genetic diseases. The increasing use of nanotechnology will bring changes in the manufacturing processes of nanomaterials. The applications of nanomaterials in the field of medical imaging and molecular detection are presented in section 4. This book will be useful for students, researchers, scientists, academicians, and industrial manufacturers to understand the importance and applicability of nanomaterials in the field of biotechnology and medical science.

Nanobiotechnology

Applications of Nanomaterials in Biotechnology, Medicine and Healthcare

Edited by
Tridib Kumar Bhowmick
Kalyan Gayen
Sunil K. Maity

CRC Press is an imprint of the
Taylor & Francis Group, an **informa** business

Designed cover image: © Tridib Kumar Bhowmick

First edition published 2024
by CRC Press
2385 NW Executive Center Drive, Suite 320, Boca Raton FL 33431

and by CRC Press
4 Park Square, Milton Park, Abingdon, Oxon, OX14 4RN

CRC Press is an imprint of Taylor & Francis Group, LLC

ISBN: 978-1-032-30387-1 (hbk)
ISBN: 978-1-032-30529-5 (pbk)
ISBN: 978-1-003-30558-3 (ebk)

DOI: 10.1201/9781003305583

Typeset in Times New Roman
by MPS Limited, Dehradun

Contents

Preface

Nanotechnology is recognized as the driving force for a new industrial revolution and generated tremendous excitement in the vast research fields. Nanotechnology enables controlled manipulation of the materials at 1–100 nm length scale. In this operational space, a plethora of unique size-dependent physical and chemical properties have evolved from the materials suitable for commercial exploration. The application of nanotechnology to address medical problems or solve biological issues is commonly known as nanobiotechnology. This emerging field has opened the possibilities of manipulating and controlling matters at the sub-cellular level with nanometer precision. Recent advancements in biotechnology, genomics, proteomics, genetic engineering, material science, and medical science are combined with nanotechnology, and their combination led to the evolution of a new generation of tools that have a tremendous impact on biotechnological research and medical applications. Nanobiotechnology is a rapidly growing and challenging multidisciplinary field with wide applicability, including biology and medical sciences.

This book has four different sections: 1) Materials at the nanoscale, 2) Novel synthesis of nanomaterials for biotechnological applications, 3) Nanobiomedicine: Tool for effective and safe therapy, and 4) Nanomaterials in diagnostics imaging and therapy. Section 1 presents recent research updates on various nanomaterials, which are promising in different medical and biotechnological applications. Applications of nanomaterials as bone replacement orthopedic implants have revolutionized orthopedic surgery. Nanostructured polymeric materials have gained immense research attention as therapeutic carriers for the precise delivery of drugs at targeted sites. Nanocellulose is recognized as a promising green nanomaterial due to its renewability and abundance in nature. The recent scientific and technological advances and applications of different nanostructured materials are presented in this section. Section 2 focuses on novel synthesis methods that are used extensively for large-scale production of inorganic and nanostructured materials. Section 3 covers applications of nanotools in the treatment of different diseases like cancers and genetic diseases. Nanotechnology brings changes in nanomaterial manufacturing processes. The applications of nanomaterials in medical imaging and molecular detection are presented in section 4.

In this book, every effort has been made to portray fascinating areas of nanobiotechnology to attract students, faculties, scientists, and engineers working in the industry, persons in regulatory agencies, and researchers working at the molecular or atomic level with unique skills. We hope that this book will stimulate research efforts, increase collaborative research between industry and academic disciplines, and motivate readers toward the advancement of nanobiotechnology.

About the Editors

Tridib Kumar Bhowmick, Assistant Professor, Department of Bioengineering, National Institute of Technology Agartala, completed his Ph.D. from Department of Chemical Engineering, Indian Institute of Technology Bombay, India, in 2009. Currently, Dr. Bhowmick is involved to explore north east region of India, representing a biodiversity hotspot with endemic flora and fauna, to identify promising microalgae strains as resources for alternative renewable energy production. During his PhD, he worked on Indian traditional medicine. He was a Postdoctoral Research Fellow at the Institute for Bioscience and Biotechnology Research, University of Maryland, College Park, Maryland, from 2008 to 2013. His research on nanomaterials and targeted delivery of therapeutic molecules has made a number of publications in reputed international journals.

Dr. Kalyan Gayen is working as an academician in the Department of Chemical Engineering, NIT Agartala, India. Previously, he worked as an Assistant Professor for two years in the Department of Chemical Engineering at IIT Gandhinagar, India, and also worked as a Post-Doctoral Fellow at University of California Santabarbara, USA, for two years. He has completed his PhD in the field of Chemical Engineering from IIT Bombay, India, and master degree from IIT Kharagpur, India. His current research is primarily focused on the microalgae-based biofuels and bioproducts, value-added chemicals from lignocellulosic biomass, fermentation technology, and systems biology. He has sucessfully executed a number of external funding projects. Dr. Kalyan Gayen guided five PhD students and published more than 50 international journal articles, fourteen book chapters, and two edited books.

Dr. Sunil K. Maity is currently working as a Professor in the Department of Chemical Engineering, Indian Institute of Technology Hyderabad, India. Before this, he served about two and half years as an Assistant Professor at the National Institute of Technology Rourkela, India. He received his Ph.D. degree from the Indian Institute of Technology Kharagpur. His current research interests are as follows: 1) biorefinery for biofuels and chemicals, 2) heterogeneous catalysis and chemical reaction engineering, and 3) techno-economic analysis using Aspen Plus and pinch technology. He is currently working on hydrodeoxygenation of vegetable oils to produce green diesel, steam and oxidative steam reforming, production of gasoline, butylene, and aromatics from bio-butanol, sustainable aviation fuel production via hydroxyalkylation-alkylation reaction, and catalytic conversion of butanediols to value-added chemicals. He has executed several externally funded and consultancy projects. He published two edited books, nine book chapters, forty-seven journal articles, and several national and international conferences.

Contributors

Mohammed Jamir Ahemad
Institute of Advanced Composite Materials
Korea Institute of Science and Technology
Jeollabuk-do, Republic of Korea

Tarun Kanti Bandyopadhyay
Department of Bioengineering
National Institute of Technology Agartala
Agartala, India

Tridib Kumar Bhowmick
Department of Bioengineering
National Institute of Technology Agartala
Agartala, India

Bhabatush Biswas
Department of Bioengineering
National Institute of Technology Agartala
Agartala, India

Swati Biswas
Department of Pharmacy
Birla Institute of Technology and Science-Pilani
Hyderabad Campus, Shamirpet
Hyderabad, India

Kaberi Chatterjee
Department of Pharmaceutical Sciences
and Technology
Birla Institute of Technology
Mesra, Ranchi-835215, India

Pranjal P. Das
Centre for the Environment
Indian Institute of Technology Guwahati
Guwahati-781039, India

Subrata Das
Applied Chemistry lab, Dept of Chemistry
National Institute of Technology (NIT) Patna
Ashok Rajpath, Patna, India

Tushar Das
Applied Chemistry lab, Dept of Chemistry
National Institute of Technology (NIT) Patna
Ashok Rajpath, Patna, India

Nimeet Desai
Department of Biomedical Engineering
Indian Institute of Technology Hyderabad
Hyderabad, India

Poonam Dogra
Department of Biomedical Engineering
Indian Institute of Technology Hyderabad
Hyderabad, India

Prangan Duarah
Centre for the Environment
Indian Institute of Technology Guwahati
Guwahati, India

Kalyani Eswar
Department of Biomedical Engineering
Indian Institute of Technology Hyderabad
Hyderabad, India

Kalyan Gayen
Department of Chemical Engineering
National Institute of Technology Agartala
Agartala, India

Bhagyashri Gaykwad
Department of Chemical Engineering
Indian Institute of Technology
Gandhinagar, India

Aparajita Ghosh
Department of Pharmacy
Birla Institute of Technology and Science-Pilani
Hyderabad Campus, Shamirpet,
Hyderabad, India

Balaram Ghosh
Department of Pharmacy
Birla Institute of Technology and Science-Pilani
Hyderabad Campus, Shamirpet
Hyderabad, India

Jyotsnendu Giri
Department of Biomedical Engineering
Indian Institute of Technology Hyderabad
Hyderabad, India

Ambati Himaja
Department of Pharmacy
Birla Institute of Technology and Science-Pilani
Hyderabad Campus, Shamirpet
Hyderabad, India

Farhana Islam
Single Molecule Biophysics Lab
Chemical Sciences Division
Saha Institute of Nuclear Physics
1/AF Bidhannagar, Kolkata, India

Keerti Jain
Department of Pharmaceutics
National Institute of Pharmaceutical Education and Research (NIPER)-Raebareli
Lucknow, India

Vineet Kumar Jain
Department of Pharmaceutics
Delhi Pharmaceutical Sciences and Research University (DPSRU)
New Delhi, India

Vruddhi Jani
Department of Chemical Engineering
Indian Institute of Technology
Gandhinagar, India

Kabeer Jasuja
Department of Chemical Engineering
Indian Institute of Technology
Gandhinagar, India

Rohit Joshi
Department of Chemical Engineering
IIT Bombay
Mumbai, India

Onkar Kulkarni
Department of Pharmacy
Birla Institute of Technology and Science-Pilani
Hyderabad Campus, Shamirpet
Hyderabad, India

Sunil K. Maity
Department of Chemical Engineering
Indian Institute of Technology Hyderabad
Hyderabad, India

Abhijit Majumder
Center for Research in Nano Technology and Research
Department of Chemical Engineering IIT Bombay
Mumbai, India

Suresh Mamidia
Department of Chemical Engineering
National Institute of Technology Agartala
Jirania, Agartala, India

Papiya Mitra Mazumder
Department of Pharmaceutical Sciences and Technology
Birla Institute of Technology
Mesra, Ranchi, India

Padmaja P. Mishra
Single Molecule Biophysics Lab
Chemical Sciences Division
Saha Institute of Nuclear Physics
1/AF Bidhannagar, Kolkata, India

Suchetan Pal
Department of Chemistry
Indian Institute of Technology-Bhilai
Raipur, India

Shreya Pande
Department of Biomedical Engineering
Indian Institute of Technology Hyderabad
Hyderabad, India

Iason Papademetriou
Application lab, Internal Research and Development
MaxCyte
Rockville, MD, 20850, USA

Manisha Patel
Department of Pharmaceutics
National Institute of Pharmaceutical Education and Research (NIPER)-Raebareli
Lucknow, India

Parth Patel
Department of Pharmaceutics
National Institute of Pharmaceutical Education and Research (NIPER)-Raebareli
Lucknow, India

Paloma Patra
Department of Biomedical Engineering
Indian Institute of Technology Hyderabad
Hyderabad, India

Harvinder Popli
Department of Pharmaceutics
Delhi Pharmaceutical Sciences and Research University (DPSRU)
New Delhi, India

Mihir K. Purkait
Department of Chemical Engineering, Centre for the Environment
Indian Institute of Technology Guwahati
Guwahati, India

Shubham Raj
Applied Chemistry lab, Dept of Chemistry
National Institute of Technology (NIT) Patna
Ashok Rajpath, Patna, India

P. S. Rajalakshmi
Department of Biomedical Engineering
Indian Institute of Technology Hyderabad
Hyderabad, India

Tatini Rakshit
Department of Chemistry
Shiv Nadar IoE
Greater Noida, India

Aravind Kumar Rengan
Department of Biomedical Engineering
Indian Institute of Technology Hyderabad
Hyderabad, India

Subhadip Roy
Department of Chemistry
The ICFAI University Tripura
Mohanpur, Agartala, India

Rajdeep Saha
Department of Pharmaceutical Sciences and Technology
Birla Institute of Technology
Mesra, Ranchi, India

Biswatrish Sarkar
Department of Pharmaceutical Sciences and Technology
Birla Institute of Technology
Mesra, Ranchi, India

Shobha Shukla
Center for Research in Nano Technology and Research
Department of Metallurgical Engineering and Material Science
IIT Bombay
Mumbai, India

Tejas Suryawanshi
Center for Research in Nano Technology and Research
IIT Bombay
Mumbai, India

Anshul Rasyotra
Department of Chemical Engineering
Indian Institute of Technology
Gandhinagar, India

Anupma Thakur
Department of Chemical Engineering
Indian Institute of Technology
Gandhinagar, India

1 Importance of Nanomaterials for Biological Applications

Bhabatush Biswas, Kalyan Gayen, Sunil K. Maity, Tarun Kanti Bandyopadhyay, and Tridib Kumar Bhowmick

1.1 INTRODUCTION

The advancement of nanoscience and nanotechnology has broadened the scope to create novel materials with unique properties. Nanomaterials display outstanding physical, optical, electronic, magnetic, and chemical properties due to their smaller size. The application of nanotechnology has shown marked improvement in various science and technology disciplines, such as information technology, energy, environmental science, agriculture, food technology, medicine, biomedical, etc. Currently, varieties of nanomaterials are under investigation for biomedical applications. Nanotechnology offers significant advantages in biomedical applications from the beginning of the 21st century. Nanomaterials are now actively used in bioimaging, biosensor, medical diagnostics, medical implants, targeted therapeutic delivery, hyperthermia for cancer treatment, etc. Researchers working on nanomedicines are taking advantage of nanoparticles to precisely deliver drugs, vaccines, and genes to the affected organs. The nanoparticles, which dramatically improve immunogenicity and increase the duration of immunity, are quite promising for universal vaccine development (Deng and Wang 2018). Research in regenerative medicine has progressed significantly using nanomaterials in bone and neural tissue engineering and finding remedies for several medical problems, including spinal cord injuries, providing scaffolds for cell growth, and growing organs for transplants (Kumar et al. 2017; Sawah et al. 2022). Nanoparticles have been explored intensively for the potential treatment of cancer and other diseases to minimize the risk of damaging healthy cells (Akbarzadeh-Khiavi et al. 2022; Giordo et al. 2022). Medical tools expanded their reach in diagnostics and imaging using nanomaterials, enabling early disease detection and cure for different diseases (Arshad et al. 2022; George Kerry et al. 2021). Applications of nanomaterials are portrayed in Figure 1.1.

Traditional knowledge of medicine was based on indigenous theories, beliefs, and experiences transferred from our ancestors. In traditional medicines, a number of medicinal preparations use particulate materials as ingredients. For instance, the traditional medicinal system, *Ayurveda*, has been originated in India and documented in 1000 BC (Yano 1975b, a). In herbo-mineral medicinal preparations, particulate materials (originated from metals, minerals, and animal bones) combined with herbs and juices are used as ingredients. *Bhasma* is a traditional herbo-mineral medicinal preparation that belongs to the *Ayurveda* and *Siddha* systems. Animal sources, such as coral (*Praval Bhasma*), diamond (*Hiraka bhasma*), conch shell (*Sankha Bhasma*), peacock feather (*Mayurpuchha Bhasma*), etc., are used for preparing *Ayurvedic* medicines (Dash et al. 2022; Kotrannavar et al. 2013; Mishra et al. 2014; Rajoria et al. 2021). Several metals, such as iron (*Louha Bhasma*), gold (*Swarna Bhasma*), copper (*Tamra Bhasma*), zinc (*Jasada Bhasma*), silver (*Roupya Bhasma*), mercury (*Kajjali*), etc., are also used in *Ayurvedic* and *Siddha* preparations (Chaudhary 2011; Dash et al. 2022;

DOI: 10.1201/9781003305583-1

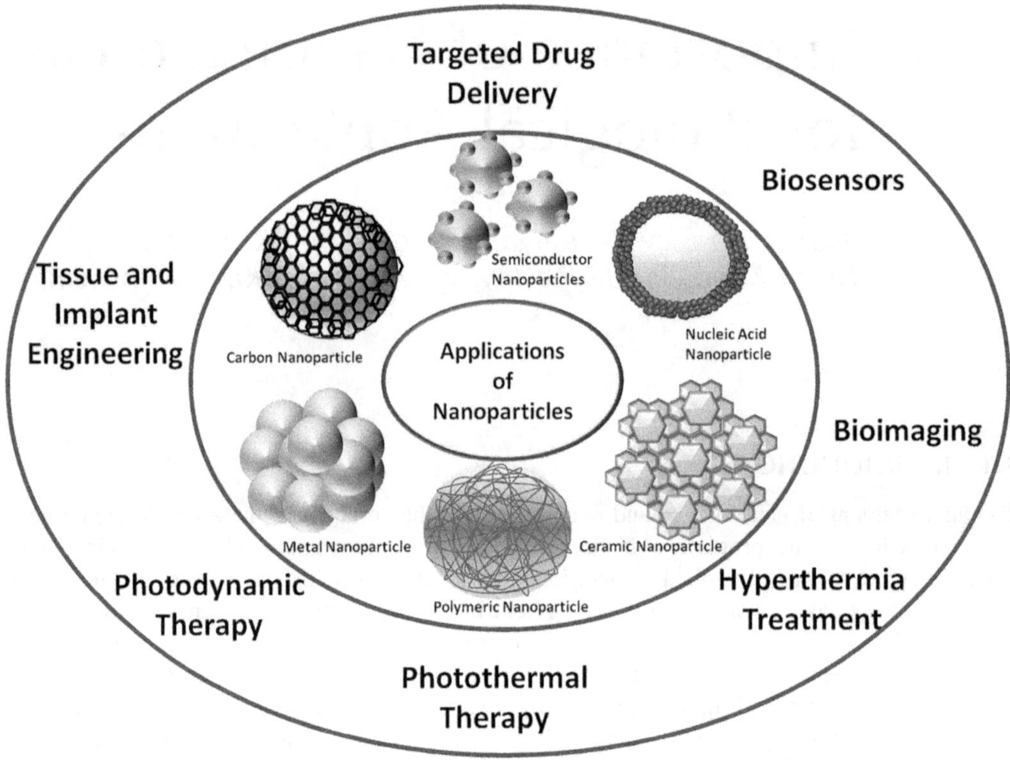

FIGURE 1.1 Varieties of nanomaterials applied in different biomedical problems.

Pandit et al. 1999; Sarkar and Das Mukhopadhyay 2021; Srikanth et al. 2018; Umrani and Paknikar 2015). Minerals like mica are used in the preparation of *Abhrak Bhasma* (Balkrishna et al. 2021). A wide range of therapeutic uses, potency at extremely small doses, and absence of undesirable taste are the advantages of *Bhasmas*. The studies revealed that the presence of particulate matter with nanometer size is responsible for their biological activities in these *Ayurvedic/Siddha* formulations (Al-Ansari et al. 2021; Kantak et al. 2019; Rasheed et al. 2014). Currently, nanotechnology is applied to develop different Traditional Chinese Medicine (TCM) to overcome the obstacles of pharmaceutical dosage forms and broaden the applications (Huang et al. 2015; Wei et al. 2022). The traditional formulation of the *Unani* system of medicine, which was developed in the middle east of Asia, is currently under investigation to combine with nanotechnology to find better treatment options for different diseases (Khoobchandani et al. 2020).

Nowadays, it is well-recognized that nanoparticles have unique properties that are different from their bulk counterpart. This feature is often found to be advantageous for biomedical applications. Diverse carbon-based allotropes, such as amorphous carbon, carbon nano-tube (CNT), carbon quantum dots (CD), graphene oxide (GO), fullerene, etc., attracted tremendous attention for exploring their unique properties for biomedical applications. The ease of functionalization of GO with different groups or molecules, therapeutic agents, and fluorescence dyes makes them smart biomaterials for various biomedical applications, such as *in vivo* imaging, carriers for drug delivery, diagnostics, etc. (Bhardwaj et al. 2021; Bolibok et al. 2021). On the other hand, CD exhibits excellent photoluminescence properties due to quantum confinement and is an attractive nanomaterial for detecting and imaging molecules (Xiang et al. 2022; Zhang et al. 2023). Based on the above brief background, this chapter presents an overview of traditional particulate medicines that are used in the past to treat numerous diseases. Specifically, this chapter highlights the biomedical applications of various nanomaterials (carbon-based, ceramic-based, metal-based, polymer-based, etc.). The use of nucleic acid nanomaterials in the biomedical fields is also briefly covered in this chapter.

1.2 TRADITIONAL MEDICINE IN ASIA: APPLICATION OF PARTICULATE MEDICINE

Nanotechnology emerged as an innovative scientific technique of the 21st century that can be combined with traditional medicines for better efficacy to improve human health. Several particle-based medicinal formulations exist in different traditional medicinal systems (e.g., *Ayurveda*, *Siddha*, and *Unani* Traditional Chinese medicine). *Bhasmas* are unique *Ayurvedic* preparations of metals and minerals combined with juices of herbs and fruits to treat various chronic ailments. The metals that are converted into *Bhasmas* are gold, iron, zinc, mercury, silver, arsenic, copper, tin, potassium, sodium, magnesium, manganese, etc. (Pal and Gurjar). Traditional preparation of *Bhasmas* is long with multiple steps, such as *Shodhana* (surface purification), *Marana* (incineration), *Puta* (controlled heat treatment), and *Bhavana* (trituration with juices). These steps are followed strictly under the guidance of an *Ayurvedic* expert to obtain the final *Bhasma* in fine powder form. *Mandura bhasma* is an iron-based preparation used as a therapeutic in anemia, jaundice, poor digestion, edema, skin diseases, etc. Mulik and Jha (2011) showed the presence of crystalline iron silicate (200–300nm size) in *Mandura bhasma* using X-ray diffraction and scanning electron microscopy. Another medicinal preparation, Jasada Bhasma, is made from metallic zinc and used to treat diabetes and age-related eye diseases and as a diuretic. The study showed the presence of non-stoichiometric hexagonal zinc oxide particles of 15–20nm (Bhowmick et al. 2009). Rasheed et al. tested *Trivanga Bhasma* (containing a mixture of three metals lead, tin, and zinc) on Winstar rats in a dose-dependent manner. Results showed that treatment of *Trivanga Bhasma* with 200–400mg/kg dose caused a dose-dependent increment in urine volume and reduced the fasting blood glucose level after 15 days of treatment. This study showed the efficacy of the particulate *Trivanga Bhasma* formulation (~60nm) in treating diabetic disease (Rasheed et al. 2014).

Another traditional medicine, Unani, uses particulate matter to treat different diseases. For example, a poly-herbo-mineral formulation Safoof-e-Pathar phori is used as an anti-urolithiatic and antioxidant agent. Ahmad et al. showed that treatment of Safoof-e-Pathar phori formulation (700–1,000mg/kg) significantly lowered the creatinine and urea in the blood, decreased urinary calcium and lipid peroxidation after 21 days of treatment of urolithiatic rats (Ahmad et al. 2021). The advantages of particulate-based formulations for the delivery of phytochemicals are numerous. Fenugreek seeds (*Trigonella foenum-graecum*) or methi is used in different *Ayurvedic/Unani* formulations for controlling diabetes and lowering the cholesterol levels in the blood. Mushtaq and Hannan used one *Unani* formulation, *Kushta* (Calx), to improve the solubility of hydrophobic plant materials in the aqueous phase and increase the bioavailability of the formulation.

Sparganii Rhizoma belongs to Traditional Chinese Medicine (TCM), which is used to treat hyperacidity, dysmenorrhea, bruises, menstruation disorder, gynecologic oncology, etc. Aluminum-rich polysaccharide is extracted from Sparganii Rhizoma (SpaTa) and exists in nanoparticulate form (~152nm) with a regular physical morphology. Wu et al. used SpaTa for diagnostic imaging and endocrine therapy in cancer. This formulation exhibited notable NIR fluorescence in the MCF-7 xenografted mouse model. The same model was also employed to analyze *in vivo* therapeutic targeting of SpaTA. Quantitative immunohistochemistry (IHC) analysis showed that anti-cancer apoptotic protein expression in tumor-grafted mice significantly improved the apoptotic effect. The study indicated the potential use of SpaTa as an anti-cancer agent (Wu et al. 2020).

Bai-Hu-Tang (BHT) is another herbal decoction used as an antipyretic agent that belongs to the TCM. This decoction aggregate is formed from *Anemarrhena asphodeloides Bunge, Glycyrrhizae*, Japonica rice, and Gypsum. Ping et al. elucidated the anti-pyritic effect of the nanoaggregates (approx. 100nm) in the rats. The induced fever in rats was treated with BHT (20mL/kg) and compared with positive control (aspirin administered –3.20mg/kg). The results showed that secretion of plasma cytokines (IL-1, IL-2, IL-6, IL-10, TNF-α, and IFN-γ) in the rats treated with BHT was significantly decreased, comparable to the positive control and indicated the anti-pyritic effect (Ping et al. 2020). Traditional medicines have been used in the form of particulates for the treatment of several diseases. Even though the nano size of metal particulates and herbal extracts were not known, they were researched in modern times, and their relatable nano range sizes and biomedical applications were explored.

1.3 CARBON-BASED NANOPLATFORMS FOR BIOLOGICAL APPLICATIONS

Carbon-based nanomaterials are extensively studied for biomedical applications due to their unique chemical and physical properties, including thermal, mechanical, electrical, optical, and structural diversity. A variety of carbon-based nanomaterials, such as carbon nanotubes (CNT), graphene oxide (GO), quantum dots (QDs), carbon nanotubes (CNT), and carbon nano onions (CNOs), etc., are investigated for phototherapy, imaging of the cells, drug delivery, aiding in the healing process, etc. (Maiti et al. 2019). Wang et al. had developed mesoporous carbon nanospheres (OMCN), for targeted therapy of ovarian cancer (Wang et al. 2022). The nanocarriers (size 131.6 to 198.3nm) were loaded with anti-cancer drug doxorubicin and small peptides, which act as targeting moieties of interleukin-6 receptors. Results showed that synthesized multi-functional OMCN have a high anti-cancer drug storage capacity and excellent photothermal effects, triggered by the pH and near-infrared laser irradiation. *In vitro* analysis demonstrated that the ovarian tumor cell line transcytosis process mediated by IL-6 receptors helped the uptake of nanocarriers in the cells. The combined chemo and thermo treatment with the OMCN particles on the ovarian cancer cells exhibited excellent anti-tumor activity and holds promises for future development.

Hollow mesoporous carbon dots (CDs) were investigated as a carrier for NIR-absorbing molecule (polydopamine) to increase photothermal effects at low power. Lu et al. showed that hollow mesoporous carbon (size ~5nm) has a high specific surface area with a high loading capacity (~41%) of the carrier (Lu et al. 2021). The anti-cancer drug (doxorubicin hydrochloride)-loaded CDs were tested on tumor-bearing Balb/c mice and exhibited anti-tumor activity at a low power irradiation (0.75W/cm^2). CDs have excellent fluorescence and luminescence properties with a high quantum yield, which is explored for efficient cellular imaging. Doping of CDs with cerium (Ce-CDs) has some extra advantages, such as faster wound healing, quick cell proliferation, and protection against oxidative cell damage due to the excellent antioxidant properties of cerium. Zhang et al. synthesized biocompatible Ce-CDs (size ~2–4nm) and tested their wound-healing properties in rats. Small Ce-CDs particles can penetrate cell membranes efficiently, thereby arresting oxidative cell damage *in vitro* (L929 cells). Histopathological analysis of wounded skin tissues of rats showed that Ce-CDs promoted the effect of fibroplasia and angiogenesis, leading to an accelerated wound-healing process (Zhang et al. 2021). Another carbon-based nanoplatform (loaded with urokinase plasminogen activators) was investigated for targeted thrombolytic therapy rats (Zhang et al. 2021). A carbonized metal-organic framework (Fe) (size ~ 100nm) was synthesized to deliver urokinase plasminogen activators (thrombolytic agent) in rats induced with acute femoral vein thrombosis. This nanoplatform (51.4% loading efficiency) showed magnetothermal and NIR-mediated photothermal properties, which were explored for targeting and thrombolytic therapy in rats (Zhang et al. 2021).

TABLE 1.1
Biomedical Applications of Different Carbon-Based Nanomaterials

Carbon Nanomaterials	Drug	Disease/Model System	Ref.
Graphene Quantum Dots	Doxorubicin	Human squamous carcinoma cells (HSC-3)	(Li et al. 2022)
4-hydroxyphenyl methacrylate -Carbon nano onions (PMPMA-CNOs) nanocomposite	Doxorubicin	Human fibroblast cells	(Mamidi et al. 2020)
Multi-walled CNTs	Doxorubicin	Human breast cancer cells	(Thakur et al. 2022)
Single-walled CNTs	Doxorubicin	Human breast cancer cells (MCF-7)	(Yang et al. 2020)

Other carbon-based nanomaterials, such as CNTs, graphene oxide (GO), and Carbon Nano Onions (CNO), have unique properties, making them attractive for a wide range of biomedical applications (Table 1.1). One atomic layer thick graphene oxide sheet has excellent dispersity properties in the aqueous medium. The presence of oxygen on the surface and electron-rich carbon as a matrix made them suitable for functionalization (Mouhat et al. 2020). Li et al. have shown that surface-functionalized GO with the anti-cancer drug doxorubicin is useful for treating oral squamous cell carcinoma (Li et al. 2022). The release of doxorubicin in human squamous cancer cells (HSC-3) from graphene oxide surface was pH dependent and prolonged the half-life of the drug. Mamidi et al. synthesized nanocomposite material with 4-hydroxyphenyl methacrylate and CNOs and used it as a thermo-sensitive carrier platform for drug (doxorubicin) release (Mamidi et al. 2020). This platform showed excellent biocompatibility with the human fibroblast cells and released 95% of the drugs at 43°C.

The sustained release of drugs from nanocomposite platforms over 15 days demonstrated promising technology for the delivery of therapeutics. Thakur et al. used multi-walled CNTs to deliver doxorubicin in breast cancer cells (Thakur et al. 2022). The large available surface of the multi-walled CNTs was coated with lysine for increasing loading capacity. Glycosylation of CNTs ensures the enhanced uptake and sustained release (over 120 hrs) of drugs at pH 5 in cancer cells. Yang et al. loaded doxorubicin molecules on functionalized single-walled CNTs conjugated with poly(ethylene glycol) and polyethylenimine for delivery of drug in human breast cancer cells (MCF-7) (Yang et al. 2020). Results showed a higher affinity of breast carcinoma cells for drug-loaded CNTs, higher uptake in cells, controlled drug release at acidic pH, and better anti-tumor effects (in vitro). These CNTs-based substrates have shown the potential to develop nanoscale technology to improve cancer treatment through sustained delivery of anti-cancer drug molecules at tumor sites.

1.4 CERAMIC NANOSTRUCTURES FOR BIOLOGICAL APPLICATIONS

Ceramics biomaterials are biocompatible and preferred in various biomedical applications, such as replacement, repair, and regeneration of tissues, including bone and teeth. Nanostructured ceramics exhibit better mechanical resistance, hardness, fracture toughness, and corrosion resistance than bulk ceramics (Paramita et al. 2021; Reddy et al. 2012). Several ceramics, such as zirconia, alumina, and hydroxyapatite, are gaining popularity for biomedical and biological applications (Filip et al. 2022). A class of ceramics has been named bioceramics due to their enhanced biocompatibility. Professor Larry Hench developed bioactive glass ceramics

(Bioglass 45S5) based on silica and mixed with other elements (e.g., Ca, Zn, etc.) to mimic bone tissue and regenerate new bone in the fractured region (Hench 2006). Bone is a rigid organ with adaptive tissue composed of inorganic phase, collagen, and water. Medical intervention in bone tissue is often required during damage induced by traumatic injury, bone tumor, etc. Nanostructured ceramics can mimic the dimensions of constituent components of bone tissues more accurately and enhance cellular adhesion, increasing biocompatibility. Bigham et al. synthesized nano-sized (10–20nm) Gehlenite ($Ca_2Al_2SiO_7$) ceramic particles and coated the surface with polycaprolactone (PCL)-Forsterite (Mg_2SiO_4) to prepare a nanocomposite for their applications in bone regeneration (Bigham et al. 2020). Synthesized Gehlenite scaffold material was found to be highly porous, and the surface fabricated with PCL-Forsterite nanocomposite improved the scaffold's brittleness and biological properties. *In vitro,* the bioactivity of nanocomposite was estimated by immersion of scaffolds into simulated body fluid (SBF) for 21 days at different temperatures. The degradation properties of scaffolds were evaluated in phosphate buffer saline (PBS, pH = 7.4) and citric acid buffer (pH = 3). Results showed that nanocomposite materials made with PCL (6wt%)-Forsterite (11wt%) enhanced the regeneration of bone tissue dramatically (Bigham et al. 2020).

The effect of silica-based nanocomposites (silica calcium phosphate) on cell differentiation and bone regeneration was further evaluated in bone marrow mesenchymal stromal cells (MSCs) (Altwaim et al. 2021). The nanocomposites with MSCs showed that cells had a better attachment and spread across the calcium phosphate layer over the scaffold surface. The biocompatibility testing of nanocomposites showed that the proliferation rate and viability of MSCs were increased over seven days. Rat bone marrow stromal cells were used for *in vivo* implantation. Histological examination showed well-organized fibrous connective tissue infiltrating the graft and healing bone defects with new bone tissues (Altwaim et al. 2021). Paramita et al. studied nano-sized zinc-doped bioglass for bone tissue repair (Antoniac et al. 2021). In this study, nano-bioglass ceramics (338–708nm) was synthesized by sol-gel method using SiO_2, CaO, and P_2O_5 (55:40:5 Mol %). It was found that incorporating Zn in nanocomposite enhanced osteoblast cellular activity and antibacterial activity. Nano-sized zinc-doped bioglass material was found promising for application in bone tissue engineering. Further, a nanostructured novel composite scaffold composed of collagen-hydroxyapatite doped with magnesium was reported for their application in improving biodegradation rate and hard tissue regeneration (Antoniac et al. 2021). The degradation rate of composite materials was investigated in simulated body fluid, and it was found that the presence of magnesium in nanocomposite slows the degradation rate. Generally, nano-sized bioglass as implant material enhances the dissolution rates of soluble ions from nanocomposite surfaces in body fluids. During this process, concentrations of soluble ions, e.g., silica, calcium, phosphorous, sodium, magnesium, zinc, etc., increase in the local environment. These ions combine with body fluid ions, forming calcium phosphate layers on the bioglass surface. This new surface layer results in new bond formation with bone tissue cells and enhances the bioactivity of bioglass materials.

1.5 METAL-BASED NANOMATERIALS FOR BIOMEDICAL APPLICATIONS

Nanotechnology helped to evolve a wide range of metal-based nano-sized materials. However, their distinctive mechanical, electrical, optical, thermal, magnetic, and chemical properties are size dependent. Engineered metal-based nano-sized particles have broad applications in pharmaceutical and biomedical fields, such as sunscreen lotion, diagnostics kits, implant materials, contrast-enhancing agents in magnetic resonance imaging (MRI), drug delivery carriers, and many more. Metal and metal-based nano-sized materials are classified based on their composition. Pure metal-based nanomaterials include copper, magnesium, iron, zinc, nickel, titanium, silver, gold, etc. The most common biomedical applications of nickel-titanium (NiTi) alloy are vascular stents, staples, catheter guide wires, bone fracture fixtures, orthodontic wires, endodontic files, and reamers. The NiTi alloy-coated surface increases the resistance against corrosion and adhesion with endothelial cells (Zhao et al. 2012). NiTi alloys composite in nano-hydroxyapatite exhibits a high hardness value, for which it can be exploited as biomedical implant materials (Akmal et al. 2016). NiTi alloys are considered elastic and highly biocompatible compared to other Ni alloys. Rokaya et al. showed that nanocomposite (graphene oxide/silver nanoparticle) coated surfaces improve mechanical properties (Berkovich hardness and Young's modulus) and reduce friction coefficient (Rokaya et al. 2019). The coated NiTi alloys are anticipated to be a useful material for biomedical applications (Rokaya et al. 2019).

Other categories of metal-based nanomaterials are metal oxides, metal sulfides, metal/metal oxide-doped nanomaterials, and metal-organic frameworks. Different metal oxide nanoparticles, such as TiO_2, ZnO, CuO, Ag_2O, and MnO_2, have antibacterial properties and are used widely in the biomedical field (Fathy Abo-Elmahasen et al. 2022; Liu et al. 2021; Raj et al. 2018; Subhan et al. 2014; Wu et al. 2019; Zhang et al. 2018). Nano-sized TiO_2 and ZrO_2 are especially used as dental and orthopedic implants due to their high mechanical strength and excellent biocompatibility (Frandsen et al. 2013). The biocompatibility, excellent mechanical strength, and thermal properties of nano-sized alumina (Al_2O_3) make them useful as bone implants (Wei et al. 2019). Metal sulfide nano-sized materials, such as CdS, CuS, MnS, etc., have cytotoxicity and are used in magnetic resonance imaging-guided photothermal therapy for cancer treatment (Chen et al. 2019). Nano-sized noble metals, such as Au, Pt, and Ag, exhibit localized surface plasmon resonance arising from charge density oscillation inside metallic nano-sized particles. The nontoxic nature of Au nanoparticles with outstanding localized surface plasmon resonance properties has been explored for early detection of amyloid β protein in the brain responsible for Alzheimer's disease (Nair et al. 2020).

Semiconductors nanomaterials have specific electrical properties, whose electrical conductivity lies between conductor and insulator. Examples of semiconductors nanomaterials and their applications have been depicted in Table 1.2. Semiconductors dominate various aspects of the biomedical industry, allowing the market to categorize based on their use, such as medical equipment, portable monitoring systems, clinical diagnostic devices, and medical imaging. Semiconductor nanomaterials are also categorized by types of sensing, such as high sensitivity, gesture, proximity, and movement sensing. Jiao et al. demonstrated the applications of silver indium sulfide (AIS) quantum dots (QDs) as bioimaging agents (Jiao et al. 2020). The AIS-QDs

TABLE 1.2
Biomedical Applications of Different Semiconductor Nanomaterials

Semiconductor Nanomaterial	Biomedical Applications	Ref.
Mn-doped ZnS quantum dots	Phosphorescent immunoassay for quantification of Prostate Specific Antigen (PSA) in prostate cancer cells.	(GarcÃa-CortÃ©s et al. 2017)
Poly(vinyl pyrrolidone) (PVP)-modified Bi_2WO_6 nanoplates	Tumor Radiosensitization and Radiocatalysis	(Zang et al. 2019)
ZnO@Gd_2O_3 nanostructures	Used as good contrast agents, useful for magnetic resonance imaging.	(Babayevska et al. 2017)
Quasi-2D In_2O_3 thin film	Biosensor for the detection of glucose concentration at femtomolar levels.	(Chen et al. 2017)
Mg-doped ZnO	Biosensor	(Fu Lin et al. 2020)
Silicon- and Germanium-based semiconductor	Temporary biomedical implants, environmentally degradable sensors	(Kang et al. 2015)

(2.5–3nm) were co-ligated with electron-rich L-glutathione to achieve enhanced photoluminescence quantum yield. The *in vitro* cell imaging experiments were carried out using human HeLa cells. The AIS-QDs treated cells were imaged with a confocal microscope, and 37% emission of quantum yield was found. These biocompatible quantum dots are useful for real-time bioimaging (Jiao et al. 2020).

The wide-gap semiconductors oxides, such as ZnO (zinc oxide), are the other semiconductor-based nanomaterials and exhibit unique optical and luminescent properties. The photoluminescence spectrum emitted by ZnO has two different wavelengths (UV and visible), which is very important for integrated optoelectronics and biological fluorescence labeling. Babayevska et al. synthesized $ZnO.Gd_2O_3$ hybrid nanomaterials (40–50nm) and functionalized with folic acid and anti-cancer drug doxorubicin. The $ZnO.Gd_2O_3$ nanomaterial exhibited as a promising contrast-enhancing agent in the magnetic resonance imaging technique. Further, these $ZnO.Gd_2O_3$ nanostructures functionalized with doxorubicin were evaluated for pH-controlled drug release studies. Around 30% of doxorubicin was released in the first 20 hours at pH 7.4, and almost 60% of doxorubicin was released in PBS at pH 5.3 (Babayevska et al. 2017). This doxorubicin functionalized $ZnO.Gd_2O_3$ nanomaterials could be used as a good contrast agent and anti-cancer drug delivery vehicles.

1.6 POLYMERIC NANOSTRUCTURES FOR BIOMEDICAL APPLICATIONS

Polymeric nanomaterials are nowadays an attractive choice for possible applications in the biomedical field. Polymeric materials extracted from natural sources, such as chitosan, gelatin, cellulose, and alginate, are preferable in bone tissue engineering due to their minimal immunogenic response. Synthetic polymers, such as polycaprolactone (PCL), poly(L-lactide) (PLA), and poly(lactic acid-co-glycolic acid) (PLGA), are also preferred for the delivery of therapeutics. The advantages of using these synthetic nano-sized polymers as therapeutic cargo are biodegradability and quick elimination of degraded metabolites from the body. Functionalization of the nano-sized polymeric surface has been explored for various biomedical applications (Raspantini et al. 2021). Raspantini et al. explored the delivery of the anti-cancer drug docetaxel with the help of a nano-sized copolymer of polycaprolactone for treating prostrate cancer (Raspantini et al. 2021). Biological assays demonstrated that anti-cancer drug-encapsulated polymeric nanoparticles successfully internalized inside prostate cancer cell line (PC-3 cell), causing cell death at 0.476 ± 0.092nM (EC_{50}) concentration. *In vivo* studies in a mouse model demonstrated a significant reduction in tumor volume compared to control groups (Raspantini et al. 2021).

The use of polymeric nanomaterials as scaffolds for regenerating the nervous system and neuronal tissue engineering is growing rapidly. The novel polymeric nanomaterials that promote stem cell adhesion, aligned migration, and differentiation are under current research focus (Bal et al. 2021; Cheng and Kisaalita 2010). Polo et al. synthesized a biodegradable scaffold with a copolymer of L-lactide and ε-caprolactone and functionalized the surface with polydopamine-coated graphene oxide (Polo et al. 2020). The study demonstrated the generation of both immature neurons and supporting glial cells over a polydopamine-coated functionalized copolymer scaffold *in vitro*. This system could reestablish spatially oriented neural precursor cell connectivity, constituting a promising tool for future cellular therapy, including nerve tissue regeneration (Polo et al. 2020).

Polymer-based nanoplatforms were also explored for fluorescence imaging and photothermal photodynamic therapy. Poly[2,6-(4,4-bis-(2-ethylhexyl)-4H-cyclopenta [2,1-b;3,4-b]dithiophene)-alt-4,7(2,1,3-benzothiadiazole) are one of the examples of conjugated polymers. Nagy-Simon et al. studied the photothermal stability and efficiency of this conjugated polymer with irradiation of near-infrared (NIR) laser for bioimaging and phototherapy purposes (Nagy-Simon et al. 2021).

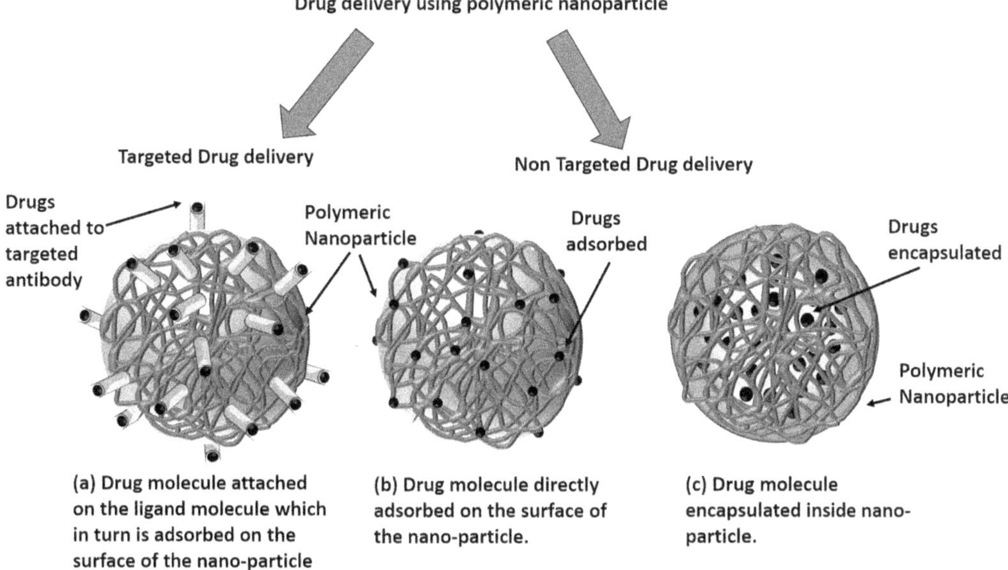

FIGURE 1.2 Use of polymeric nanoparticles for delivery of therapeutic molecules. 1) Ligand attached drug molecules adsorbed on the nanoparticle surface. 2) Drug molecules directly adsorbed on the nanoparticle surface. 3) Drug molecules encapsulated inside the nanoparticle.

This conjugated polymer in aqueous dispersion exhibited high fluorescence emission, better photothermal heating, and singlet oxygen generation. The optical performance of nano-sized conjugated polymer at a single particle level has the potential to be an excellent theranostic agent by combining fluorescence imaging and phototherapy (Nagy-Simon et al. 2021).

Nano-sized polymeric particles are useful as carriers for therapeutic delivery. The advantages of therapeutic delivery using nanocarriers are the improvement of therapeutics' bioavailability and targeted delivery of therapeutics at disease sites (Figure 1.2). Although studies have shown outstanding performances of the nano-sized particles as therapeutic carriers, their clinical uses are still limited. Examples of common polymeric carriers for therapeutics delivery are [N-(2-hydroxypropyl) methacrylamide (HPMA), polyethylene glycol (PEG), poly N-vinyl pyrrolidone (PVP), cellulose derivatives, poly glutamic acid, polyethyleneimine (PEI), dextran, dextrin, chitosans, poly(l-lysine), poly(aspartamides), and poly(lactic-co-glycolic acid) (PLGA) (Alibolandi et al. 2016; Archana et al. 2014; Craparo et al. 2022; Ghaffarian et al. 2012; Gupta and Raghav 2019; Imperiale et al. 2019; Mekseriwattana et al. 2021; Parvez et al. 2020; Sajeesh et al. 2014; Shukla et al. 2021; Upadhyay et al. 2017; Yang et al. 2016).

Iles et al. synthesized a nanostructured system (~51nm) with modified biodegradable polymers (extracted from cashew gum) for the controlled release of alendronate sodium (a commonly used drug for the treatment of osteoporosis). The nanoparticles were characterized, and their gastric toxicity was investigated in rats. The controlled release of the drug (~420mins) reversed the gastric lesions and was found nontoxic in murine osteoblastic cells, even at high concentrations (Iles et al. 2021). Biswas et al. reviewed the application of nano-sized polymeric particles to prevent infectious diseases (e.g., tuberculosis). This review reports the current research on targeted and non-targeted delivery of anti-tubercular drugs with the help of nano-sized polymeric carriers to prevent multi-drug-resistant tuberculosis (Biswas et al. 2023). Several other biomedical applications of drug delivery using polymer nanocarriers are listed in Table 1.3.

TABLE 1.3
Biomedical Applications of Different Polymeric Nanomaterials

Nanoparticle/Carrier Used	Used Drug	Nanoparticle/Carrier Properties	*in vitro* Study/ *in vivo* Study	Ref.
[N-(2-hydroxypropyl) methacrylamide (HPMA)-Polylactic acid (PLA) Nano-Polymeric Micelles	Isoniazid, Rifampicin,	Mean size HPMA-PLA-INH conjugates 111.2 ± 2.34nm Zeta potential − 45.3 ± 2.30mV	*In vitro* drug release	(Upadhyay et al. 2017)
Dα-tocopheryl polyethylene glycol 1000 succinate	Amphotericin B, Paromomycin	199.4±18.9nm	J774A.1 Macrophage cells	(Parvez et al. 2020)
Chitosan, polyvinyl pyrrolidone (PVP), silver oxide nanoparticles		10–30nm	Cytotoxicity tested on mouse fibroblast cell line (L929 cells)	(Archana et al. 2014)
Nanocrystalline cellulose	Non-steroidal Anti-inflammatory Drugs (Etodolac, Ibuprofen, Diclofenac, Paracetamol)	1 to 10nm	*In vitro* drug release	(Gupta and Raghav 2019)
poly(L-glutamic acid) Nanorods	Doxorubicin hydrochloride	Length (300nm), Diameter (30nm)	*In vitro* studies on breast cancer cell line (MCF-7 cells), In vivo anti-tumor activity (Female BALB/c mice)	(Yang et al. 2016)
Cross-linked branched poly (ethylene imine)	RNA oligo-nucleotides	100nm	RNAi Activity, Transfection experiment [Human glioblastoma-astrocytoma (U87MG cells)], Cytotoxicity Assessment [Human cervix epithelial adenocarcinoma (HeLa cells)], Particle uptake [Mouse brain endothelial cell line (bEnd.3 cells)]	(Sajeesh et al. 2014)

Dextran-coated nano-graphene oxide	Aptamer	Single layered sheet, height (~ 1nm) width < 600nm	Uptake study: 4T1 amurine breast carcinoma, Breast cancer cell line (MCF-7)	(Alibolandi et al. 2016)
Brush polymer of hydrophobic polyurethane on hydrophilic dextrin	Dexamethasone	Micro clusters (2–5 μm)	Biocompatibility, *in vitro* cell-killing efficiency: Human cervix epithelial adenocarcinoma (HeLa cells), In vivo studies: mice	(Shukla et al. 2021)
Chitosan	Interferon alpha (IFNα)	47nm	Biocompatibility: Human epithelial cell line (Caco-2), Cellular uptake: Co-culture of Caco-2 & mucin-secreting HT29-MTX cell line	(Imperiale et al. 2019)
amphiphilic pegylated poly-ε-caprolactone/polyhydroxyethyl aspartamide graft copolymer	rapamycin	~2 μm	Cell viability assay: 16HBE cells	(Craparo et al. 2022)
Poly(lactic-co-glycolic acid)	Doxorubicin, Riboflavin analog	300nm	Breast cancer cell line (MCF-7)	(Mekseriwattana et al. 2021)
Alginate nanoparticles	Rifampicin, isoniazid, pyrazinamide, ethambutol.		*In vivo*	(Ahmad et al. 2006)
Nanoemulsion-based chitosan nanocapsules (CS-NC)	Bedaquiline	328 ± 35nm	*In vitro* drug release	(De Matteis et al. 2017)
Chitosan microparticles	Isoniazid, Rifabutin	5.9 ± 1.7 μm	*In vitro* studies: THP-1 cells (differentiated macrophages of rat).	(Cunha et al. 2019)
Chitosan microparticles	Isoniazid	(3.2 – 3.9) μm	*In vitro* drug release	(Oliveira et al. 2017)
alginate-cellulose nanocrystal hybrid nanoparticles (ALG-CNCNPs)	Rifampicin	(55 ± 20) nm	*In vitro* drug release studies	(Thomas et al. 2018)
Solid lipid nanoparticles (SLN)	Rifabutin	(99±4 – 186±12)nm	*In vitro* drug release studies	(Gaspar et al. 2015)

1.7 NUCLEIC ACID NANOSTRUCTURE FOR BIOLOGICAL APPLICATIONS

Nucleic acid (DNA and RNA) nanostructures are the innovations of modern nanotechnology with several biomedical applications. They are tools for detection, bio-sensing, probes, culture platforms, etc. Over the past few decades, robust nanostructures of nucleic acid have been created due to their vast composition and ability to interact with other bio-molecules (e.g., proteins and enzymes). The development of nucleic acid nanostructures with precise control has created a hope to mimic the functions of intracellular machinery and alter cellular activity *in vivo*. Kong et al. developed an octopus-like DNA structure that can capture eight HIV-1 genes efficiently through its sticky arms. This platform was shown as an efficient platform for detecting viral HIV-1 genes independent of temperature variations (Kong et al. 2020). Nam et al. created an X-shaped DNA nanostructure functionalized with photo-cross-linkable or cell adhesion moieties for 3D cell culture. The X-shaped DNA nanostructure was incorporated to form a hybrid hydrogel using polyethylene glycol diacrylate. Human ovarian cancer cell lines (SKOV-3) were encapsulated in various concentrations of X-shaped DNA nanostructure hydrogels and cultured for seven days. The study showed that hydrogels incorporated with X-shaped DNA nanostructure were able to adhere to cells, responsible for homogeneous cellular distribution, enhance cell proliferation rate, and improve cell viability (Nam et al. 2021). Another nucleic acid-based tetrahedral framework, DNA nanostructures (TFNAs), was explored for the treatment of fibrotic diseases. Four single-stranded DNA were self-assembled to form tetrahedral framework DNA nanostructures to inhibit the proliferation of the artificially induced fibroblast cells. *In vitro* experiments with rat lung epithelial-T-antigen negative cells (RLE-6TN) showed expression levels of collagen I, and fibronectin was suppressed and able to prevent the over-production of collagen I in fibrotic diseases condition (Zhang et al. 2020).

Gene expression regulation in the human breast cancer epithelial cell line (MDA-MB-231) has been shown using gold-conjugated structural ribonucleic acid (TectoRNA-AuNP) particles. Transfection of RNA samples using TectoRNA-AuNP particles (~10nm) showed their regulatory effect on the *CopGFP* (gene expressing green fluorescent protein) gene, excellent biocompatibility, and cytosolic delivery of the RNA molecules. These TectoRNA-AuNP particles hold promise to obtain an excellent delivery platform for the biologically active RNAs in the cells (Graczyk et al. 2021).

1.8 FUTURE PERSPECTIVES AND CONCLUSIONS

Applications of nanomaterials in biological and biomedical fields are fascinating. Engineered materials at the nanoscale have immense potential to provide solutions to early diagnostic challenges, deliver therapeutics precisely at the disease site, and treat the diseases. The global nanomedicine market has thus been growing steadily (Ray and Bandyopadhyay 2023). For instance, the market for cancer nanomedicine in 2023 increased almost three times from 2016. The market size of other nanomedicines, such as anti-microbial, anti-inflammatory, nanoneuromedicine, and cardiovascular, has also been growing steadily in this period. Applications of particulate matter in traditional medicine are well established, and current research enables us to think about the role of nanoparticulate matter in the activity of medicines. Carbon-based nanomaterials (e.g., single-walled CNTs, multi-walled CNTs, graphene oxide, and carbon nano onions) have the potential for various biomedical applications (e.g., therapeutic delivery, early diagnostics, biosensors, etc.). Ceramic-based bioactive nano-sized glass has shown the capability to form interfacial bonds with bones and tissues more efficiently than bulk materials and is recognized as useful in bone and tissue regeneration.

Applications of metal-based nanomaterials (Ag, TiO_2, ZnO, CuO, and MnO_2) in biomedical fields are versatile. Their anti-microbial properties make them suitable as filler in polymer matrices for use as food packaging materials. Semiconductor (quantum dots) nanocrystals (e.g., Fe_3O_4, Gd_2O_3, TiO_2, etc.) are commonly used to enhance the contrast of images, leading to early diagnosis of life-threatening diseases. Although several polymeric nanomaterials exhibited their superiority in targeted therapeutics delivery precisely, their clinical applications are still limited. Many scientific issues still need to be resolved before their clinical applications. A deeper understanding of the behavior of nanomaterial-based medicines or devices is required in an *in vivo* environment, which is quite complex and challenging. In the next 20 years, several nanomaterial-based medicines or devices currently under phase II clinical trial are expected to be commercialized for clinical applications.

ABBREVIATIONS

Ag_2O	Silver di-oxide
AIS	Silver indium sulfide
Al_2O_3	Alumina
BHT	Bai-Hu-Tang
CdS	Cadmium sulfide
CuS	Copper sulfide
CNO	Carbon Nano Onions
CNT	Carbon nano-tube
CO	Carbon quantum dots
CuO	Copper oxide
GO	Graphene oxide
Gd_2O_3	Gadolinium(III) oxide
IFNγ	Interferon γ
IHC	Immunohistochemistry
MnO_2	Manganese di-oxide
MnS	Manganese sulfide
MRI	Magnetic resonance imaging
MSCs	Mesenchymal stromal cells
NIR	Near-infrared
NiTi	Nickel-Titanium
OMCN	Oxidized mesoporous carbon nanospheres
PCL	Polycaprolactone
PEI	Polyethyleneimine
PBS	Phosphate buffer saline
PLA	Poly(L-lactide)
PLGA	Poly(lactic acid-co-glycolic acid)
PVP	Poly N-vinyl pyrrolidone
QDs	Quantum dots
SpaTa	Sparganii Rhizoma
TCM	Traditional Chinese Medicine
TiO_2	Titanium di-oxide
TNF-alpha	Tumor necrosis factor α
ZnO	Zinc oxide
ZrO_2	Zirconium di-oxide

REFERENCES

Ahmad, W., M. A. Khan, K. Ashraf, A. Ahmad, M. D. Ali, M. N. Ansari, Y. T. Kamal, S. Wahab, S. M. A. Zaidi, M. Mujeeb, and S. Ahmad. 2021. "Pharmacological evaluation of Safoof-e-Pathar Phori- A polyherbal unani formulation for urolithiasis." *Frontiers in Pharmacology* 12. doi: 10.3389/fphar.2021.597990.

Ahmad, Z., R. Pandey, S. Sharma, and G. K. Khuller. 2006. "Alginate nanoparticles as antituberculosis drug carriers: Formulation development, pharmacokinetics and therapeutic potential." *The Indian Journal of Chest Diseases and Allied Sciences* 48 (3): 171–176.

Akbarzadeh-Khiavi, M., M. Torabi, A. H. Olfati, L. Rahbarnia, and A. Safary. 2022. "Bio-nano scale modifications of melittin for improving therapeutic efficacy." *Expert Opinion on Biological Therapy* 22 (7): 895–909. doi: 10.1080/14712598.2022.2088277.

Akmal, M., A. Raza, M. M. Khan, M. I. Khan, and M. A. Hussain. 2016. "Effect of nano-hydroxyapatite reinforcement in mechanically alloyed NiTi composites for biomedical implant." *Materials Science and Engineering: C* 68: 30–36. doi: 10.1016/j.msec.2016.05.092.

Al-Ansari, M. M., A. J. A. Ranjit Singh, F. S. Al-Khattaf, and J. S. Michael. 2021. "Nano-formulation of herbo-mineral alternative medicine from linga chenduram and evaluation of antiviral efficacy." *Saudi Journal of Biological Sciences* 28 (3): 1596–1606. doi: 10.1016/j.sjbs.2020.12.005.

Alibolandi, M., M. Mohammadi, S. M. Taghdisi, M. Ramezani, and K. Abnous. 2016. "Fabrication of aptamer decorated dextran coated nano-graphene oxide for targeted drug delivery." *Carbohydrate Polymers* 155: 218–229. doi:10.1016/j.carbpol.2016.08.046.

Altwaim, S., M. Al-Kindi, N. AlMuraikhi, S. BinHamdan, and A. Al-Zahrani. 2021. "Assessment of the effect of silica calcium phosphate nanocomposite on mesenchymal stromal cell differentiation and bone regeneration in critical size defect." *The Saudi Dental Journal* 33 (8): 1119–1125. doi: 10.1016/j.sdentj.2021.03.008.

Antoniac, I. V., A. Antoniac, E. Vasile, C. Tecu, M. Fosca, V. G. Yankova, and J. V. Rau. 2021. "In vitro characterization of novel nanostructured collagen-hydroxyapatite composite scaffolds doped with magnesium with improved biodegradation rate for hard tissue regeneration." *Bioactive Materials* 6 (10): 3383–3395. doi: 10.1016/j.bioactmat.2021.02.030.

Archana, D., B. K. Singh, J. Dutta, and P. K. Dutta. 2014. "Chitosan-PVP-nano silver oxide wound dressing: in vitro and in vivo evaluation." *International Journal of Biological Macromolecules* 73: 49–57. doi: 10.1016/j.ijbiomac.2014.10.055.

Arshad, R., M. H. Kiani, A. Rahdar, S. Sargazi, M. Barani, S. Shojaei, M. Bilal, D. Kumar, and S. Pandey. 2022. "Nano-based theranostic platforms for breast cancer: A review of latest advancements." *Bioengineering (Basel)* 9 (7). doi: 10.3390/bioengineering9070320.

Babayevska, N., P. Florczak, M. WoÅºniak-Budych, M. Jarek, G. Nowaczyk, T. Zalewski, and S. Jurga. 2017. "Functionalized multimodal ZnO@Gd2O3 nanosystems to use as perspective contrast agent for MRI." *Applied Surface Science* 404: 129–137. doi: 10.1016/j.apsusc.2017.01.274.

Bal, Z., F. Korkusuz, H. Ishiguro, R. Okada, J. Kushioka, R. Chijimatsu, J. Kodama, D. Tateiwa, Y. Ukon, S. Nakagawa, E. C. Dede, M. Gizer, P. Korkusuz, H. Yoshikawa, and T. Kaito. 2021. "A novel nano-hydroxyapatite/synthetic polymer/bone morphogenetic protein-2 composite for efficient bone regeneration." *The Spine Journal* 21 (5): 865–873. doi: 10.1016/j.spinee.2021.01.019.

Balkrishna, A., S. K. Solleti, H. Singh, R. Singh, N. Sharma, and A. Varshney. 2021. "Biotite-Calx based traditional Indian medicine sahastraputi-abhrak-bhasma prophylactically mitigates allergic airway inflammation in a mouse model of asthma by amending cytokine responses." *Journal of Inflammation Research* 14: 4743–4760. doi: 10.2147/JIR.S313955.

Bhardwaj, S. K., M. Mujawar, Y. K. Mishra, N. Hickman, M. Chavali, and A. Kaushik. 2021. "Bio-inspired graphene-based nano-systems for biomedical applications." *Nanotechnology* 32 (50). doi: 10.1088/1361-6528/ac1bdb.

Bhowmick, T. K., A. K. Suresh, S. G. Kane, A. C. Joshi, and J. R. Bellare. 2009. "Physicochemical characterization of an Indian traditional medicine, Jasada Bhasma: detection of nanoparticles containing non-stoichiometric zinc oxide." *Journal of Nanoparticle Research* 11 (3): 655–664. doi: 10.1007/s11051-008-9414-z.

Bigham, A., A. H. Aghajanian, M. Movahedi, M. Sattary, M. Rafienia, and L. Tayebi. 2020. "A 3D nanostructured calcium-aluminum-silicate scaffold with hierarchical meso-macroporosity for bone tissue regeneration: Fabrication, sintering behavior, surface modification and in vitro studies." *Journal of the European Ceramic Society* 41 (1): 941–962. doi: 10.1016/j.jeurceramsoc.2020.07.073.

Biswas, B., T. K. Misra, D. Ray, T. Majumder, T. K. Bandyopadhyay, and T. K. Bhowmick. 2023. "Current therapeutic delivery approaches using nanocarriers for the treatment of tuberculosis disease." *International Journal of Pharmaceutics* 640: 123018. doi: 10.1016/j.ijpharm.2023.123018.

Bolibok, P., B. Szymczak, K. Roszek, A. P. Terzyk, and M. Wisniewski. 2021. "A new approach to obtaining nano-sized graphene oxide for biomedical applications." *Materials (Basel)* 14 (6). doi: 10.3390/ma14061327.

Chaudhary, A. 2011. "Ayurvedic bhasma: nanomedicine of ancient India--its global contemporary perspective." *Journal of Biomedical Nanotechnology* 7 (1): 68–69. doi: 10.1166/jbn.2011.1205.

Chen, H., Y. S. Rim, I. C. Wang, C. Li, B. Zhu, M. Sun, M. S. Goorsky, X. He, and Y. Yang. 2017. "Quasi-two-dimensional metal oxide semiconductors based ultrasensitive potentiometric biosensors." *ACS Nano* 11 (5): 4710–4718. doi: 10.1021/acsnano.7b00628.

Chen, W., X. Wang, B. Zhao, R. Zhang, Z. Xie, Y. He, A. Chen, X. Xie, K. Yao, M. Zhong, and M. Yuan. 2019. "CuS-MnS(2) nano-flowers for magnetic resonance imaging guided photothermal/photodynamic therapy of ovarian cancer through necroptosis." *Nanoscale* 11 (27): 12983–12989. doi: 10.1039/c9nr03114f.

Cheng, K., and W. S. Kisaalita. 2010. "Exploring cellular adhesion and differentiation in a micro-/nano-hybrid polymer scaffold." *Biotechnology Progress* 26 (3): 838–846. doi: 10.1002/btpr.391.

Craparo, E. F., S. E. Drago, F. Quaglia, F. Ungaro, and G. Cavallaro. 2022. "Development of a novel rapamycin loaded nano- into micro-formulation for treatment of lung inflammation." *Drug Delivery and Translational Research* 12 (8): 1859–1872. doi: 10.1007/s13346-021-01102-5.

Cunha, L., S. Rodrigues, A. M. Rosa da Costa, L. Faleiro, F. Buttini, and A. Grenha. 2019. "Inhalable chitosan microparticles for simultaneous delivery of isoniazid and rifabutin in lung tuberculosis treatment." *Drug Development and Industrial Pharmacy* 45 (8): 1313–1320. doi: 10.1080/03639045.2019.1608231.

Dash, M. K., N. Joshi, V. S. Dubey, K. N. Dwivedi, and D. N. S. Gautam. 2022. "Screening of anti-cancerous potential of classical Raudra rasa and modified Raudra rasa modified with hiraka bhasma (nanodiamond) through FTIR & LC-MS analysis." *Journal of Complementary and Integrative Medicine* 19 (3): 669–682. doi: 10.1515/jcim-2021-0410.

De Matteis, L., D. Jary, A. LucÃa, S. GarcÃa-Embid, I. Serrano-Sevilla, D. PÃ©rez, J. A. Ainsa, F. P. Navarro, and J. M. de la Fuente. 2017. "New active formulations against M. tuberculosis: Bedaquiline encapsulation in lipid nanoparticles and chitosan nanocapsules." *Chemical Engineering Journal* 340: 181–191. doi: 10.1016/j.cej.2017.12.110.

Deng, L., and B. Z. Wang. 2018. "A perspective on nanoparticle universal influenza vaccines." *ACS Infectious Diseases* 4 (12): 1656–1665. doi: 10.1021/acsinfecdis.8b00206.

Fathy Abo-Elmahasen, M. M., A. S. Abo Dena, M. Zhran, and S. A. H. Albohy. 2022. "Do silver/hydroxyapatite and zinc oxide nano-coatings improve inflammation around titanium orthodontic mini-screws? In vitro study." *International Orthodontics* 21 (1): 100711. doi: 10.1016/j.ortho.2022.100711.

Filip, D. G., V. A. Surdu, A. V. Paduraru, and E. Andronescu. 2022. "Current development in biomaterials-hydroxyapatite and bioglass for applications in biomedical field: A review." *Journal of Functional Biomaterials* 13 (4). doi: 10.3390/jfb13040248.

Frandsen, C. J., K. Noh, K. S. Brammer, G. Johnston, and S. Jin. 2013. "Hybrid micro/nano-topography of a TiO2 nanotube-coated commercial zirconia femoral knee implant promotes bone cell adhesion in vitro." *Materials Science and Engineering: C* 33 (5): 2752–2756. doi: 10.1016/j.msec.2013.02.045.

Fu Lin, C., C. Haur Kao, C. Yu Lin, Y. Wen Liu, and C. Hsiang Wang. 2020. "The electrical and physical characteristics of Mg-doped ZnO sensing membrane in EIS (electrolyte–insulator–semiconductor) for glucose sensing applications." *Results in Physics* 16: 102976. doi: 10.1016/j.rinp.2020.102976.

GarcÃa-CortÃ©s, M., M. T. FernÃ¡ndez-ArgÃ¼elles, J. M. Costa-FernÃ¡ndez, and A. Sanz-Medel. 2017. "Sensitive prostate specific antigen quantification using dihydrolipoic acid surface-functionalized phosphorescent quantum dots." *Analytica Chimica Acta* 987: 118–126. doi: 10.1016/j.aca.2017.08.003.

Gaspar, D. P., V. Faria, L. M. D. Goncalves, P. Taboada, C. Remunan lopez, and A. J. Almeida. 2015. "Rifabutin-loaded solid lipid nanoparticles for inhaled antitubercular therapy: Physicochemical and in vitro studies." *International Journal of Pharmaceutics* 497 (1): 199–209. doi: 10.1016/j.ijpharm.2015.11.050.

George Kerry, R., K. E. Ukhurebor, S. Kumari, G. K. Maurya, S. Patra, B. Panigrahi, S. Majhi, J. R. Rout, M. D. P. Rodriguez-Torres, G. Das, H. S. Shin, and J. K. Patra. 2021. "A comprehensive review on the applications of nano-biosensor-based approaches for non-communicable and communicable disease detection." *Biomaterials Science* 9 (10): 3576–3602. doi: 10.1039/d0bm02164d.

Ghaffarian, R., T. Bhowmick, and S. Muro. 2012. "Transport of nanocarriers across gastrointestinal epithelial cells by a new transcellular route induced by targeting ICAM-1." *J Control Release* 163 (1): 25–33. doi: 10.1016/j.jconrel.2012.06.007.

Giordo, R., Z. Wehbe, P. Paliogiannis, A. H. Eid, A. A. Mangoni, and G. Pintus. 2022. "Nano-targeting vascular remodeling in cancer: Recent developments and future directions." *Seminars in Cancer Biology* 86 (Pt 2): 784–804. doi: 10.1016/j.semcancer.2022.03.001.

Graczyk, A., R. Pawlowska, and A. Chworos. 2021. "Gold nanoparticles as carriers for functional RNA nanostructures." *Bioconjugate Chemistry* 32 (8): 1667–1674. doi: 10.1021/acs.bioconjchem.1c00211.

Gupta, R. D., and N. Raghav. 2019. "Nano-crystalline cellulose: Preparation, modification and usage as sustained release drug delivery excipient for some non-steroidal anti-inflammatory drugs." *International Journal of Biological Macromolecules* 147: 921–930. doi: 10.1016/j.ijbiomac.2019.10.057.

Hench, L. L. 2006. "The story of bioglass." *Journal of Materials Science. Materials in Medicine* 17 (11): 967–978. doi: 10.1007/s10856-006-0432-z.

Huang, Y., Y. Zhao, F. Liu, and S. Liu. 2015. "Nano traditional Chinese medicine: Current progresses and future challenges." *Current Drug Targets* 16 (13): 1548–1562. doi: 10.2174/13894501166661503 09122334.

Iles, B., I. R. de SÃ¡ GuimarÃ£es NolÃªto, F. F. Dourado, F. de O. S. Ribeiro, A. R. de AraÃºjo, T. M. de Oliveira, J. M. T. Souza, A. B. Barros, G. C. Sousa, A. C. de J. Oliveira, C. da S. Martins, M. de O. V. Veras, R. F. de C. LeitÃ£o, J. R. de S. de Almeida Leite, D. A. da Silva, and J. V. R. Medeiros. 2021. "Alendronate sodium-polymeric nanoparticles display low toxicity in gastric mucosal of rats and Ofcol II cells." *NanoImpact* 24: 100355. doi: 10.1016/j.impact.2021.100355.

Imperiale, J. C., I. Schlachet, M. Lewicki, A. Sosnik, and M. M. Biglione. 2019. "Oral pharmacokinetics of a chitosan-based nano-drug delivery system of interferon alpha." *Polymers (Basel)* 11 (11). doi: 10.3390/polym11111862.

Jiao, M., Y. Li, Y. Jia, C. Li, H. Bian, L. Gao, P. Cai, and X. Luo. 2020. "Strongly emitting and long-lived silver indium sulfide quantum dots for bioimaging: Insight into co-ligand effect on enhanced photoluminescence." *Journal of Colloid and Interface Science* 565: 35–42. doi: 10.1016/j.jcis.2020.01.006.

Kang, S.-K., G. Park, K. Kim, S.-W. Hwang, H. Cheng, J. Shin, S. Chung, M. Kim, L. Yin, J. C. Lee, K.-M. Lee, and J. A. Rogers. 2015. "Dissolution chemistry and biocompatibility of silicon- and germanium-based semiconductors for transient electronics." *ACS Applied Materials & Interfaces* 7 (17): 9297–9305. doi: 10.1021/acsami.5b02526.

Kantak, S., N. Rajurkar, and P. Adhyapak. 2019. "Synthesis and characterization of Abhraka (mica) bhasma by two different methods." *Journal of Ayurveda and Integrative Medicine* 11 (3): 236–242. doi: 10.1016/j.jaim.2018.11.003.

Khoobchandani, M., K. K. Katti, A. R. Karikachery, V. C. Thipe, D. Srisrimal, D. K. Dhurvas Mohandoss, R. D. Darshakumar, C. M. Joshi, and K. V. Katti. 2020. "New approaches in breast cancer therapy through green nanotechnology and nano-ayurvedic medicine - Pre-clinical and pilot human clinical investigations." *International Journal of Nanomedicine* 15: 181–197. doi: 10.2147/IJN.S219042.

Kong, J., Y. Wang, W. Qi, R. Su, and Z. He. 2020. "Enzyme-free visualization of nucleic acids during HIV infection by octopus-like DNA." *International Journal of Biological Macromolecules* 150: 122–128. doi: 10.1016/j.ijbiomac.2020.02.063.

Kotrannavar, V., R. Sarashetty, and V. Kanthi. 2013. "Physico-chemical analysis of Mayurapuccha Bhasma prepared by two methods." *Ancient Science of Life* 32 (1): 45–48. doi: 10.4103/0257-7941.113801.

Kumar, P., Y. E. Choonara, R. A. Khan, and V. Pillay. 2017. "The chemo-biological outreach of nano-biomaterials: Implications for tissue engineering and regenerative medicine." *Current Pharmaceutical Design* 23 (24): 3538–3549. doi: 10.2174/1381612823666170503144643.

Li, R., R. Gao, Y. Zhao, F. Zhang, X. Wang, B. Li, L. Wang, L. Ma, and J. Du. 2022. "pH-responsive graphene oxide loaded with targeted peptide and anticancer drug for OSCC therapy." *Frontiers in Oncology* 12: 930920. doi: 10.3389/fonc.2022.930920.

Liu, Y., D. Zhu, and J. L. Gilbert. 2021. "Sub-nano to nanometer wear and tribocorrosion of titanium oxide-metal surfaces by in situ atomic force microscopy." *Acta Biomaterialia* 126: 477–484. doi: 10.1016/j.actbio.2021.03.049.

Lu, J., K. Wang, W. Lei, Y. Mao, D. Di, Q. Zhao, and S. Wang. 2021. "Polydopamine-carbon dots functionalized hollow carbon nanoplatform for fluorescence-imaging and photothermal-enhanced thermochemotherapy." *Materials Science and Engineering: C* 122: 111908. doi: 10.1016/j.msec.2021.111908.

Maiti, D., X. Tong, X. Mou, and K. Yang. 2019. "Carbon-based nanomaterials for biomedical applications: A recent study." *Frontiers in Pharmacology* 9. doi: 10.3389/fphar.2018.01401.

Mamidi, N., R. M. V. Delgadillo, and A. Gonzalez-Ortiz. 2020. "Engineering of carbon nano-onion bioconjugates for biomedical applications." *Materials Science & Engineering. C, Materials for Biological Applications* 120: 111698. doi: 10.1016/j.msec.2020.111698.

Mekseriwattana, W., A. Phungsom, K. Sawasdee, P. Wongwienkham, C. Kuhakarn, P. Chaiyen, and K. P. Katewongsa. 2021. "Dual functions of riboflavin-functionalized poly(lactic-co-glycolic acid) nanoparticles for enhanced drug delivery efficiency and photodynamic therapy in triple-negative breast cancer cells." *Photochemistry and Photobiology* 97 (6): 1548–1557. doi: 10.1111/php.13464.

Mishra, A., A. K. Mishra, O. P. Tiwari, and S. Jha. 2014. "In-house preparation and characterization of an Ayurvedic bhasma: Praval bhasma." *Journal of Integrative Medicine* 12 (1): 52–58. doi: 10.1016/S2095-4964(14)60005-4.

Mouhat, F., F. X. Coudert, and M. L. Bocquet. 2020. "Structure and chemistry of graphene oxide in liquid water from first principles." *Nature Communications* 11 (1): 1566. doi: 10.1038/s41467-020-15381-y.

Mulik, S. B., and C. B. Jha. 2011. "Physicochemical characterization of an Iron based Indian traditional medicine: Mandura Bhasma." *Ancient Science of Life* 31 (2): 52–57.

Mushtaq, T., and A. Hannan. 2020. "Nano-materials: A scientific approach to validate an ancient dosage form the "Calx (Kushta) in Unani." 6: 32–37.

Nagy-Simon, T., O. Diaconu, M. Focsan, A. Vulpoi, I. Botiz, and A.-M. Craciun. 2021. "Pluronic stabilized conjugated polymer nanoparticles for NIR fluorescence imaging and dual phototherapy applications." *Journal of Molecular Structure* 1243: 130931. doi: 10.1016/j.molstruc.2021.130931.

Nair, R. V., P. J. Yi, P. Padmanabhan, B. Gulyas, and V. M. Murukeshan. 2020. "Au nano-urchins enabled localized surface plasmon resonance sensing of beta amyloid fibrillation." *Nanoscale Advances* 2 (7): 2693–2698. doi: 10.1039/d0na00164c.

Nam, K., B. I. Im, T. Kim, Y. M. Kim, and Y. H. Roh. 2021. "Anisotropically functionalized aptamer-DNA nanostructures for enhanced cell proliferation and target-specific adhesion in 3D cell cultures." *Biomacromolecules* 22 (7): 3138–3147. doi: 10.1021/acs.biomac.1c00619.

Oliveira, P. M., B. N. Matos, P. A. T. Pereira, T. Gratieri, L. H. Faccioli, M. S. S. Cunha-Filho, and G. M. Gelfuso. 2017. "Microparticles prepared with 50-190kDa chitosan as promising non-toxic carriers for pulmonary delivery of isoniazid." *Carbohydrate Polymers* 174: 421–431. doi: 10.1016/j.carbpol.2017.06.090.

Pal, D., and V. K. Gurjar. 2017. "Nanometals in Bhasma: Ayurvedic Medicine." In *Metal Nanoparticles in Pharma*, 389–415. Cham: Springer International Publishing.

Pandit, S., T. K. Biswas, P. K. Debnath, A. V. Saha, U. Chowdhury, B. P. Shaw, S. Sen, and B. Mukherjee. 1999. "Chemical and pharmacological evaluation of different ayurvedic preparations of iron." *Journal of Ethnopharmacology* 65 (2): 149–156. doi: 10.1016/s0378-8741(99)00003-3.

Paramita, P., M. Ramachandran, S. Narashiman, S. Nagarajan, D. K. Sukumar, T. W. Chung, and M. Ambigapathi. 2021. "Sol-gel based synthesis and biological properties of zinc integrated nano bioglass ceramics for bone tissue regeneration." *Journal of Materials Science: Materials in Medicine* 32 (1): 5. doi: 10.1007/s10856-020-06478-3.

Parvez, S., G. Yadagiri, A. Singh, A. Karole, O. P. Singh, S. Sundar, and S. L. Mudavath. 2020. "Improvising anti-leishmanial activity of amphotericin B and paromomycin using co-delivery in d-alpha-tocopheryl polyethylene glycol 1000 succinate (TPGS) tailored nano-lipid carrier system." *Chemistry and Physics of Lipids* 231: 104946. doi: 10.1016/j.chemphyslip.2020.104946.

Ping, Y., Y. Li, S. Lü, Y. Sun, W. Zhang, J. Wu, T. Liu, and Y. Li. 2020. "A study of nanometre aggregates formation mechanism and antipyretic effect in Bai-Hu-Tang, an ancient Chinese herbal decoction." *Biomedicine & Pharmacotherapy* 124: 109826. doi: 10.1016/j.biopha.2020.109826.

Polo, Y., J. Luzuriaga, J. Iturri, I. Irastorza, J. L. Toca-Herrera, G. Ibarretxe, F. Unda, J.-R. Sarasua, J. R. Pineda, and A. LarraÃ±aga. 2020. "Nanostructured scaffolds based on bioresorbable polymers and graphene oxide induce the aligned migration and accelerate the neuronal differentiation of neural stem cells." *Nanomedicine: Nanotechnology, Biology and Medicine* 31: 102314. doi: 10.1016/j.nano.2020.102314.

Raj, I., M. Mozetic, V. P. Jayachandran, J. Jose, S. Thomas, and N. Kalarikkal. 2018. "Fracture resistant, antibiofilm adherent, self-assembled PMMA/ZnO nanoformulations for biomedical applications: physico-chemical and biological perspectives of nano reinforcement." *Nanotechnology* 29 (30): 305704. doi: 10.1088/1361-6528/aac296.

Rajoria, K., S. K. Singh, and S. Dadhich. 2021. "Ayurvedic management in limb girdle muscular dystrophy - A case report." *Journal of Ayurveda and Integrative Medicine* 13 (1): 100486. doi: 10.1016/j.jaim.2021.07.002.

Rasheed, A., M. Naik, K. P. Mohammed-Haneefa, R. P. Arun-Kumar, and A. K. Azeem. 2014. "Formulation, characterization and comparative evaluation of Trivanga bhasma: a herbo-mineral Indian traditional medicine." *Pakistan Journal of Pharmaceutical Sciences* 27 (4): 793–800.

Raspantini, G. L., M. T. Luiz, J. P. Abriata, J. de Oliveira Eloy, M. M. Vaidergorn, F. da Silva Emery, and J. M. Marchetti. 2021. "PCL-TPGS polymeric nanoparticles for docetaxel delivery to prostate cancer: Development, physicochemical and biological characterization." *Colloids and Surfaces A: Physicochemical and Engineering Aspects* 627: 127144. doi: 10.1016/j.colsurfa.2021.127144.

Ray, S. S., and J. Bandyopadhyay. 2023. "Nanotechnology-enabled biomedical engineering: Current trends, future scopes, and perspectives." *Nanotechnology Reviews* 10 (1): 728–743. doi: 10.1515/ntrev-2021-0052.

Reddy, S., S. Wasnik, A. Guha, J. M. Kumar, A. Sinha, and S. Singh. 2012. "Evaluation of nano-biphasic calcium phosphate ceramics for bone tissue engineering applications: in vitro and preliminary in vivo studies." *Journal of Biomaterials Applications* 27 (5): 565–575. doi: 10.1177/088532 8211415132.

Rokaya, D., V. Srimaneepong, J. Qin, K. Siraleartmukul, and V. Siriwongrungson. 2019. "Graphene oxide/silver nanoparticle coating produced by electrophoretic deposition improved the mechanical and tribological properties of NiTi alloy for biomedical applications." *Journal of Nanoscience and Nanotechnology* 19 (7): 3804–3810. doi: 10.1166/jnn.2019.16327.

Sajeesh, S., T. Y. Lee, S. W. Hong, P. Dua, J. Y. Choe, A. Kang, W. S. Yun, C. Song, S. H. Park, S. Kim, C. Li, and D. K. Lee. 2014. "Long dsRNA-mediated RNA interference and immunostimulation: a targeted delivery approach using polyethyleneimine based nano-carriers." *Molecular Pharmaceutics* 11 (3): 872–884. doi: 10.1021/mp400541z.

Sarkar, P. K., and C. Das Mukhopadhyay. 2021. "Ayurvedic metal nanoparticles could be novel antiviral agents against SARS-CoV-2." *International Nano Letters* 11 (3): 197–203. doi: 10.1007/s40089-020-00323-9.

Sawah, D., M. Sahloul, and F. Ciftci. 2022. "Nano-material utilization in stem cells for regenerative medicine." *Biomed Tech (Berl)* 67 (6): 429–442. doi: 10.1515/bmt-2022-0123.

Shukla, A., A. P. Singh, and P. Maiti. 2021. "Injectable hydrogels of newly designed brush biopolymers as sustained drug-delivery vehicle for melanoma treatment." *Signal Transduction and Targeted Therapy* 6 (1): 63. doi: 10.1038/s41392-020-00431-0.

Srikanth, N., A. Singh, S. Ota, B. S. Galib, and K. S. Dhiman. 2018. "Chemical characterization of an Ayurvedic herbo-mineral preparation- Mahalaxmivilas Rasa." *Journal of Ayurveda and Integrative Medicine* 10 (4): 262–268. doi: 10.1016/j.jaim.2018.01.002.

Subhan, M. A., T. Ahmed, N. Uddin, A. K. Azad, and K. Begum. 2014. "Synthesis, characterization, PL properties, photocatalytic and antibacterial activities of nano multi-metal oxide NiOâ‹… CeO2â‹ … ZnO." *Spectrochimica Acta, Part A: Molecular and Biomolecular Spectroscopy* 136 Pt B: 824–831. doi: 10.1016/j.saa.2014.09.100.

Thakur, C. K., R. Neupane, C. Karthikeyan, C. R. Ashby, Jr. R. J. Babu, S. H. S. Boddu, A. K. Tiwari, and N. Moorthy. 2022. "Lysinated multiwalled carbon nanotubes with carbohydrate ligands as an effective nanocarrier for targeted doxorubicin delivery to breast cancer cells." *Molecules* 27 (21). doi: 10.3390/molecules27217461.

Thomas, D., M. S. Latha, and K. K. Thomas. 2018. "Synthesis and in vitro evaluation of alginate-cellulose nanocrystal hybrid nanoparticles for the controlled oral delivery of rifampicin." *Journal of Drug Delivery Science and Technology* 46: 392–399. doi: 10.1016/j.jddst.2018.06.004.

Umrani, R. D., and K. M. Paknikar. 2015. "Jasada bhasma, a zinc-based ayurvedic preparation: Contemporary evidence of antidiabetic activity inspires development of a nanomedicine." *Evidence-Based Complementary and Alternative Medicine* 2015: 193156. doi: 10.1155/2015/193156.

Upadhyay, S., I. Khan, A. Gothwal, P. K. Pachouri, N. Bhaskar, U. D. Gupta, D. S. Chauhan, and U. Gupta. 2017. "Conjugated and entrapped HPMA-PLA nano-polymeric micelles based dual delivery of first line anti TB drugs: Improved and safe drug delivery against sensitive and resistant mycobacterium tuberculosis." *Pharmaceutical Research* 34 (9): 1944–1955. doi: 10.1007/s11095-017-2206-3.

Wang, S., W. Wu, Y. Liu, C. Wang, Q. Xu, Q. Lv, R. Huang, and X. Li. 2022. "Targeted peptide-modified oxidized mesoporous carbon nanospheres for chemo-thermo combined therapy of ovarian cancer in vitro." *Drug Delivery* 29 (1): 1951–1958. doi: 10.1080/10717544.2022.2089298.

Wei, D., H. Yang, Y. Zhang, X. Zhang, J. Wang, X. Wu, and J. Chang. 2022. "Nano-traditional Chinese medicine: a promising strategy and its recent advances." *Journal of Materials Chemistry B* 10 (16): 2973–2994. doi: 10.1039/d2tb00225f.

Wei, T., J. Wang, X. Yu, Y. Wang, Q. Wu, and C. Chen. 2019. "Mechanical and thermal properties and cytotoxicity of Al(2)O(3) nano particle-reinforced poly(ether-ether-ketone) for bone implants." *RSC Advances* 9 (59): 34642–34651. doi: 10.1039/c9ra05258e.

Wu, Y., Y. Li, J. He, X. Fang, P. Hong, M. Nie, W. Yang, C. Xie, Z. Wu, K. Zhang, L. Kong, and J. Liu. 2019. "Nano-hybrids of needle-like MnO(2) on graphene oxide coupled with peroxymonosulfate for enhanced degradation of norfloxacin: A comparative study and probable degradation pathway." *Journal of Colloid and Interface Science* 562: 1–11. doi: 10.1016/j.jcis.2019.11.121.

Wu, Y.-Z., Y.-K. Shen, Y.-J. Chen, and J. Sun. 2020. "From ancient medicine to targeted nanocarrier: A sparganii rhizoma-derived nanoparticle for diagnostic imaging and endocrine therapy in cancer." *ACS Applied Bio Materials* 3 (4): 2028–2039. doi: 10.1021/acsabm.9b01158.

Xiang, Y., F. Song, L. Jiang, Z. Liu, and Y. Tu. 2022. "Novel fluorescent nano carbon quantum dots derived from lactarius hatsudake for high selective vitamin B12 detection." *Journal of AOAC International* 105 (5): 1350–1359. doi: 10.1093/jaoacint/qsac033.

Yang, S., Z. Wang, Y. Ping, Y. Miao, Y. Xiao, L. Qu, L. Zhang, Y. Hu, and J. Wang. 2020. "PEG/PEI-functionalized single-walled carbon nanotubes as delivery carriers for doxorubicin: synthesis, characterization, and in vitro evaluation." *Beilstein Journal of Nanotechnology* 11: 1728–1741. doi: 10.3762/bjnano.11.155.

Yang, S., F. Zhu, Q. Wang, F. Liang, X. Qu, Z. Gan, and Z. Yang. 2016. "Nano-rods of doxorubicin with poly (l-glutamic acid) as a carrier-free formulation for intratumoral cancer treatment." *Journal of Materials Chemistry B* 4 (45): 7283–7292. doi: 10.1039/c6tb02127a.

Yano, M. 1975a. "[Ayurveda. Caraka Samhita (1)]." *Nihon Rinsho* 33 (9): 2866–2870.

Yano, M. 1975b. "[Caraka Samhita]." *Nihon Rinsho* 33 (11): 3291–3295.

Zang, Y., L. Gong, L. Mei, Z. Gu, and Q. Wang. 2019. "Bi2WO6 semiconductor nanoplates for tumor radiosensitization through high-Z effects and radiocatalysis." *ACS Applied Materials & Interfaces* 11 (21): 18942–18952. doi: 10.1021/acsami.9b03636.

Zhang, M., X. Zhai, T. Ma, Y. Huang, C. Yan, and Y. Du. 2021. "Multifunctional cerium doped carbon dots nanoplatform and its applications for wound healing." *Chemical Engineering Journal* 423: 130301. doi: 10.1016/j.cej.2021.130301.

Zhang, R., L. Zhang, R. Yu, and C. Wang. 2023. "Rapid and sensitive detection of methyl parathion in rice based on carbon quantum dots nano-fluorescence probe and inner filter effect." *Food Chemistry* 413: 135679. doi: 10.1016/j.foodchem.2023.135679.

Zhang, T., Y. Gao, D. Xiao, J. Zhu, M. Zhou, S. Li, M. Zhang, Y. Lin, and X. Cai. 2020. "Nucleic acid based tetrahedral framework DNA nanostructures for fibrotic diseases therapy." *Applied Materials Today* 20: 100725. doi: 10.1016/j.apmt.2020.100725.

Zhang, Y., L. Wang, X. Xu, F. Li, and Q. Wu. 2018. "Combined systems of different antibiotics with nano-CuO against Escherichia coli and the mechanisms involved." *Nanomedicine (Lond)* 13 (3): 339–351. doi: 10.2217/nnm-2017-0290.

Zhang, Y., Y. Liu, T. Zhang, Q. Wang, L. Huang, Z. Zhong, J. Lin, K. Hu, H. Xin, and X. Wang. 2021. "Targeted thrombolytic therapy with metalâ€"organic-framework-derived carbon based platforms with multimodal capabilities." *ACS Applied Materials & Interfaces* 13 (21): 24453–24462. doi: 10.1021/acsami.1c03134.

Zhao, T., Y. Li, Y. Xia, S. S. Venkatraman, Y. Xiang, and X. Zhao. 2012. "Formation of a nano-pattering NiTi surface with Ni-depleted superficial layer to promote corrosion resistance and endothelial cell-material interaction." *Journal of Materials Science: Materials in Medicine* 24 (1): 105–114. doi: 10.1007/s10856-012-4777-1.

2 Micro and Nanostructures in Regenerative Medicine

Tejas Suryawanshi, Rohit Joshi, Shobha Shukla, and Abhijit Majumder

2.1 INTRODUCTION

2.1.1 Regenerative Medicine

Regenerative medicine, as the name suggests, is a treatment regime in which the regenerative potential of stem cells is exploited to treat various diseases, including the loss or damage of tissues due to dystrophy or trauma. The treatment strategy uses biomaterials as scaffolds, embryonic, adult, or induced stem cells as the cell source, and specific signals required for cell growth and lineage-specific differentiation. This combination of parameters is known as the triad of Tissue engineering and Regenerative Medicine (TERM). Advanced technologies of material sciences, cell and molecular biology, and fabrication techniques are employed in this emerging interdisciplinary field (Kolios and Moodley 2012). The term "Regenerative medicine" was coined by William Haseltine in 1999 to bring all the aforementioned areas under one roof (Hasetine 1999).

In the recent past, regenerative medicine has garnered a lot of attention for treating a number of critical health conditions worldwide such as cancer (Zhang et al. 2017; Mansouri et al. 2021), Parkinson's disease (Venkatesh and Sen 2017; Chi et al. 2019), Alzheimer's disease (Nakano et al. 2020; Hernández and García 2021), cardiovascular diseases (Guo et al. 2007; Guo et al. 2020), osteoporosis (Wang et al. 2006; Jiang et al. 2021), spine injuries (Caron et al. 2016; Tahmasebi and Barati 2022), and diabetes (El-Badri and Ghoneim 2013; Bhansali et al. 2017) by regenerating diseased or damaged tissues (Engel et al. 2008). Regenerative Medicine addresses both congenital and acquired defects which are otherwise difficult to treat with conventional methods (Mahla 2016). For example, the conventional mode of therapy in case of significant tissue/organ loss due to trauma or disease is transplantation; however, this is plagued with scarcity of donated organs and extreme immune complications. But these obstacles are most likely to be overcome by strategies such as generation and use of therapeutic stem cells, tissue engineering, and production of artificial organs. Many of these approaches have already completed the clinical trials and are currently in use (Colombo et al. 2017).

Similarly, regenerative medicine-based therapies for wound healing and orthopedics applications have been approved by the United States Food and Drug Administration (US FDA), leading to their commercial availability (Mao and Mooney 2015). One such example is Carticel, which is the first biological product approved by FDA in the orthopedic field. This therapy involves harvesting autologous chondrocytes from articular cartilage, expanding the same *in vitro*, and finally implanting the chondrocytes at the site of injury. After implantation, recovery of tissue was similar to that noticed by that of conventional ways such as microfracture and mosaicplasty techniques (Mao and Mooney 2015). Few other examples include Epicel, autologous keratinocytes for severe burn wounds; LaViv, autologous fibroblasts for improving nasolabial fold appearance (Colombo et al. 2017), and Celution, for diabetic foot ulcers (Mao and Mooney 2015). In the last few years, many cell-based products have been approved by different regulatory bodies for the treatment of various health conditions which are enlisted in Table 2.1.

DOI: 10.1201/9781003305583-2

TABLE 2.1
List of Products Approved by Regulatory Agencies and Their Applications in Regenerative Medicine

Sr. no.	Product Name	Health Condition	Year	Company	Source/Cell Type	Regulatory Approval	Reference
1	Carticel	Articular cartilage	1997	Genzyme	Autologous chondrocytes	US FDA	(Mao and Mooney 2015)
2	Apligraf	Diabetic foot ulcers	1998	Organogenesis	Allogenic keratinocyte with scaffold	US FDA	(Colombo et al. 2017)
3	ReliNethra®	Unilateral Limbal Stem Cell Deficiency	2008	Reliance Life Science	Limbal stem cells	CDSCO	(Chen 2016)
4	Osteocel Plus	Orthopedics	2009	NuVasive	Allogenic BM-MSCs	US FDA	(NCT00948532)
5	Chondro Celect	Cartilage defects	2009	TiGenix	Autologous Chondrocyte	EMEA	(Pereira Chilima et al. 2018)
6	Gintuit	Prostate cancer	2010	Organogenesis	Autologous Dendritic cells	US FDA	(Colombo et al. 2017)
7	Provenge	Prostate cancer	2010	Dendreon	Autologous Dendritic cells	US FDA	(Pereira Chilima et al. 2018)
8	Grafix	Wounds, soft tissue defects	2011	Osiris Therapeutics	Allogenic- MSC+ Placenta matrix	US FDA	(Lavery et al. 2014)
9	Cartistem	Traumatic Osteoarthritis	2012	Medipost	Allogenic-MSCs from UCB	KFDA	(Gottipamula et al. 2018)
10	Ducord	GVHD	2012	Duke University	Allogenic HPC cord blood	US FDA	(Gottipamula et al. 2018)
11	LaViv	Severe nasolabial fold wrinkles	2012	Fibrocell Science inc	Autologous fibroblast	US FDA	(Colombo et al. 2017)
12	Hemacord	GVHD	2013	New York Blood Center	Allogenic HPC cord blood	US FDA	(Gottipamula et al. 2018)
13	Epicel	Burn injuries	NA	Genzyme	Autologous keratinocytes	US FDA	(Mao and Mooney 2015)
14	AlloStem	Orthopedics	NA	AlloSource	Allogenic bone matrix with MSC	US FDA	(Pereira Chilima et al. 2018)
15	Holoclar	Eye disease	2015	Chiesi Farmaceutici S.p.A.	Limbal stem cells	EMEA	(Gottipamula et al. 2018)

(Continued)

TABLE 2.1 *(Continued)*

List of Products Approved by Regulatory Agencies and Their Applications in Regenerative Medicine

Sr. no.	Product Name	Health Condition	Year	Company	Source/Cell Type	Regulatory Approval	Reference
16	TEMCELL® HS Inj	Acute GVHD	2016	Allo- stem cell product	Mesoblast	Japanese Government regulatory approval	(Gottipamula et al. 2018)
17	Stempeucel CLI	Critical Limb Ischemia due to Buerger's disease	2017	Stempeutics Research Pvt Ltd	Allogenic BM-MSCs	DCGI	(Gottipamula et al. 2018)
18	ReliNethra® C	Conjunctival disorders	NA	Reliance Life Science	Limbal stem cells	CDSCO	(Pereira Chilima et al. 2018)
19	CardioRel®	Cardiac disorders	NA	Reliance Life Science	Autologous bone marrow-derived mesenchymal stem cells	CDSCO	(Chen 2016)
20	NeuroRel	Neurological disorders	NA	Reliance Life Science	Autologous bone marrow-derived mesenchymal stem cells	CDSCO	(Pereira Chilima et al. 2018)

(Abbreviations: BM-Bone marrow, MSCs-Mesenchymal Stem Cells, US FDA-United States Food and Drug Administration, KFDA-Korean federal drug administration EMEA-European medical agency, CDSCO- Central Drugs Standard Control Organization, DCGI- Drugs Controller General of India, GVHD- Graft-versus host disease, HPC- hematopoietic progenitor cell, UBC- Umbilical cord blood)

2.1.2 Stem Cells and Their Types

The delivery of stem cells for the structural and functional development of tissues is a fundamental model of regenerative medicine. Studies have reported the use of differentiated cells such as keratinocytes, hepatocytes, osteocytes, and chondrocytes for therapeutic purposes, as tabulated in Table 2.1. However, with the use of stem cells for repairing damaged tissues, a paradigm shift has taken place in the treatment strategy in the last few decades (Mahla 2016). Stem cells, due to their potential to differentiate into any type of cell and self-renewal capacity, have gained a lot of research interest globally.

Stem cells are uncommitted, undifferentiated cells that are capable of self-renewal and differentiation into specialized cells found in multicellular organisms. They are abundant in an embryonic stage which declines in number after birth. Based on their origin, stem cells are classified as embryonic stem cells (ESCs), adult stem cells (ASCs), and induced pluripotent stem cells (iPSCs) (Passier and Mummery 2003; Polak and Bishop 2006). ESCs are obtained from embryos (Rao 2017), ASCs are derived from specific tissue or organ such as the heart, adipose tissues, bone marrow, blood, liver, dental pulp, umbilical cord, and brain, whereas iPSCs are obtained from differentiated cells by molecular reprogramming using transcription factors namely Oct4 (Octamer binding transcription factor-4), klf4 (Kruppel Like Factor-4), Sox2 [(Sex determining region Y)-box 2] and c-myc (Takahashi and Yamanaka 2013). The usage of ESCs has several restrictions due to high immunological response and ethical concerns. However, this can be overcome by iPSCs. Human iPSCs were first established from dermal fibroblast but can also be obtained from cord blood and peripheral blood (Singh et al. 2016). Around 1608 clinical trials related to stem cells have been reported globally of which 599 studies have been completed to date (Trounson et al. 2011; Squillaro et al. 2016; Deinsberger et al. 2020; clinicaltrial.gov)

Other than the origin-based classification, stem cells are also grouped based on their differentiation potential; Totipotent, Pluripotent, Multipotent, and Unipotent (Ntege et al. 2020). Totipotency is the ability of a single cell to develop into a new organism. Totipotent cells can give rise to all the cells in body including extra-embryonic cells such as placental cells. Only zygote and early stage blastomeres are totipotent in nature. Stem cells isolated from inner cell mass of the blastocyst can give rise to all three germinal layers except the placenta. Hence, they are called pluripotent stem cells. ESCs and iPSCs are pluripotent in nature. Multipotent stem cells, as the name suggests, can differentiate into more than one cell types but not all. All adult stem cells (ASCs) are multipotent and have limited self-renewal and differentiation potential. For example, Hematopoietic stem cells are multipotent stem cells which give rise to specialized cells such as red blood cells (RBCs), white blood cells (WBCs), and platelets (Madonna 2016) but cannot differentiate into any non-hematopoietic type. There are no major ethical concerns and immunological response associated with ASCs, making them important tools in regenerative medicine (Gonzalez and Bernad 2012). Lastly, unipotent stem cells are the group of cells that can give rise to only one specific type of cell but have the property of self-renewal (Singh et al. 2016). For example, epidermal stem cells differentiating into skin (Blanpain and Fuchs 2006) and germ-line stem cell producing gametes (Spradling et al. 2011). This classification of different types of stem cells is shown in Figure 2.1. Despite their capacity for regeneration, stem cell-based therapies have not seen much growth in clinical use over the years because of a variety of challenges.

2.1.3 Challenges in Regenerative Medicines

In last few decades, stem cell-based therapies have progressed considerably. However, sporadic successes and research scale promises are yet to be converted into full-scale commercial standard-of-care applications due to some critical bottlenecks. Firstly, for tissue engineering

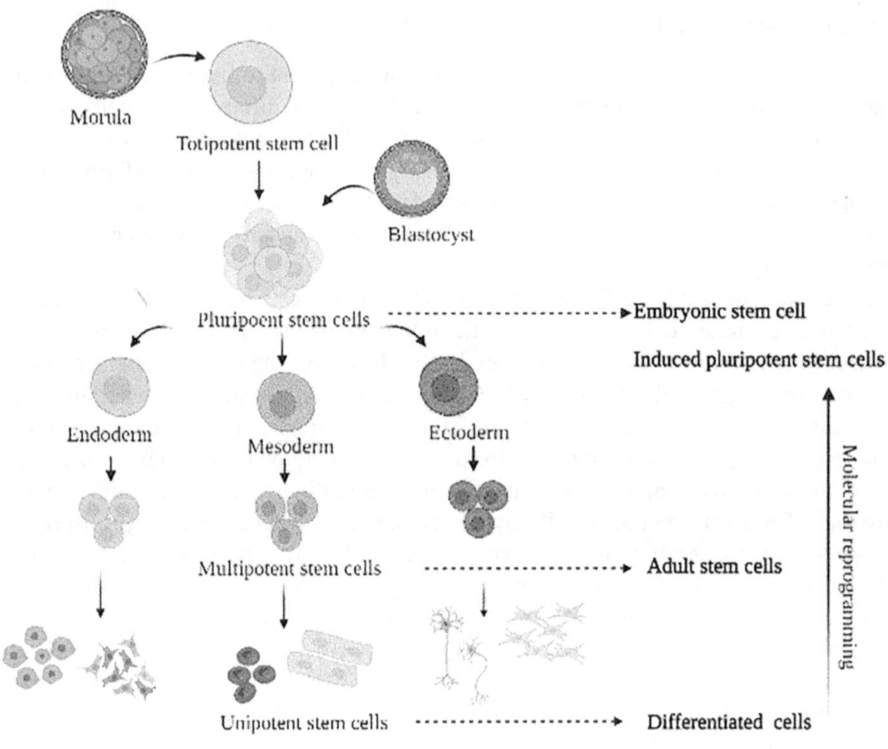

FIGURE 2.1 Types of stem cells based on origin and differentiation potential (Created in Biorender).

applications one needs a large number of stem cells which is not available in the native tissues. For example, for mesenchymal stem cells (MSCs) based therapies, a dose of ~100 million cells is required to treat a person of 70 kg (Ren et al. 2012). However, in bone marrow only 0.001–0.01% mononuclear cells are MSCs. Hence, to use bone marrow-derived hMSCs in clinics, *in vitro* expansion is the only available option. However, during long-term expansion, primary cells undergo senescence losing their proliferation and differentiation potential. Senescence also leads to unfavorable changes in cells such as change in cell shape and size, senescence associated secretion, nuclear abnormalities, and population heterogeneity (Chowdhury et al. 2010; Higuchi et al. 2013) leading to poor quality and quantity of expanded MSCs. When inside body, extracellular matrix provides the required microenvironment which decides the stem cell fate. However, mimicking such *in vivo* conditions *in vitro* is challenging leading to loss of cellular properties. Another challenge in cellular therapies is immunorejection during allogenic stem cell transplantation, as person's immune system may recognize the transplanted cell as a foreign material (Otsuka et al. 2020). These two factors together limit availability of stem cells for regenerative medicine. While iPSCs can offer a solution here, there are issues associated with the genetic instability of ESCs and iPSCs (Yoshihara et al. 2017) which are in turn strongly associated with cancers thus limiting their applications (Shen 2011). After the stem cells have been isolated, controlling and regulating the cellular process to successfully trigger differentiation into the desired cell type is another major challenge (Ikehara 2013). It was also observed that stem cells tend to undergo spontaneous differentiation *in vitro* which leads to heterogeneity of cultured cell, impairing their capacity for differentiation.

The poor differentiation potential thus restricts the efficiency of stem cell therapies (McKee and Chaudhary 2017).

Even if sufficient number of cells are obtained and used for cell therapies, after transplantation, maintaining the cell viability and targeting the transplanted cells to specific area and retaining them is important. Currently, fluorescent dyes and proteins have been used for cell tracking (Sah 2016). However, long-term tracking of stem cells and monitoring their differentiation *in vivo* has become a challenge which affects the success rate of the therapies. Moreover, it was also observed that the transplanted cells get entrapped in lung and vasculature after intravenous delivery that further restricts the success rate of therapies (Wang et al. 2015).

Additionally, though stem cell therapies are growing every year with improved efficacy rate, long-term effects are not well reported yet. Lastly, there are ethical challenges associated with use of stem cells, particularly hESCs in research (Lo and Parham 2009). These ethical questions and controversies have slowed down the hESC-based clinical therapies. Inspite of the significant number of successful studies on stem cell-based therapies, a major number of clinical trials have not yet acquired full regulatory approvals for validation as stem cell therapies due to scientific, legal, and ethical controversies (Aly 2020).

The potential application of stem cell-based therapeutics has been successfully demonstrated by recent advances in stem cell biology; however, the usage in regenerative medicine has been constrained by the problems outlined earlier. By offering an *in vivo*-like microenvironment, the majority of these issues could be resolved.

2.1.4 Importance of Mimicking in vivo Microenvironment

As stated previously, providing the *in* vivo-like microenvironment is crucial for maintaining the stemness of cells. Under physiological conditions, cells are always in contact with neighboring cells or with the ECM providing a holistic milieu for cell growth. As a result of this, several physical and chemical interaction such as cell-cell interaction and cell-ECM interaction occurs simultaneously between cells and their surroundings which controls cellular activities. It is now well established that stem cell behavior is regulated by genetic and molecular factors such as growth factors, low molecular weight proteins, and hormones which regulate the cellular response (Gangaraju et al. 2009; Li et al. 2021). These biochemical cues have attracted a lot of interest with the advancement of molecular and cellular biology, but the function of biophysical cues in stem cell research has been in oblivion for a long time. Julius Wolff was the first to introduce the importance of physical and mechanical signals in cell biology in 1892 when he described the alterations in bone formations in response to load bearing. However, it's only in the last few decades, after the emergence of mechanobiology, that researchers showed that the role of biophysical cues in cellular studies is inevitable. There are many such biophysical cues namely, material stiffness, topographical features, cell shape, and mechanical forces which can potentially influence stem cell behavior. Many studies have tried to stimulate the growth and differentiation of stem cells using physical cues in the absence of or in conjunction with inductive biochemical factors. For example, in lineage-specific differentiation, the stiffness of the substrates has been demonstrated to play a crucial role. If stem cells are cultured on a substrate mimicking the stiffness of particular tissue, they go into that tissue specific lineage differentiation. For instance, substrates with moduli in the range of 0.1–1kPa mimicking brain rigidity direct hMSCs into neuronal differentiation. Similarly substrate stiffness in the range of 8–17kPa and 25–40kPa mimics muscle and bone rigidity and favors myogenic and osteogenic differentiation respectively (Moghaddam et al. 2019). Other than substrate rigidity, 2D topology and 3D structures are also known to influence stem cell differentiation, albeit much less explored. Harrison was the first to demonstrate the influence of the underlying substratum on cell migration in 1911 when he grew cells on a spider web and found that the embryonic cells followed the fibers of the web. This phenomenon was called "stereotropism" or "thigmotaxis" (Harrison 1911). This emphasized the significance of topographical cues in regulating cellular processes and paved the foundation for a thorough investigation of topography-based control of cellular activities. The rest of this chapter will focus on understanding the effect of micro and nanoscale topographical features on stem cell behavior.

2.1.5 Topography

Extracellular Matrix (ECM) has a complex architecture of isotropic/anisotropic distribution of features such as fibers, pores, pits, and striations of different length scales (Yadav and Majumder 2021). For example, heart muscles are composed of sheets of aligned cardiac muscle cells and myocytes, bones have features such as submicron-scale lamellae to nanoscale collagen fibrils to molecular-sized minerals and proteins. Similar topological features are also observed in skin, neurons, blood vessels, and in almost all of the tissues (Li et al. 2014; Yadav and Majumder 2021). There is probably no tissue present in our body that does not have atypical topological features. Interactions of the cells with these micro and nanoscaled topographical cues are an integral part of many major processes such as cell adhesion, migration, orientation, proliferation, elongation, and differentiation, and in turn, tissue development and regeneration (Tudureanu et al. 2022). For instance, the inner wall of a blood vessel consists of endothelium cells organized in specific order on which muscle cells are aligned in order to maintain the functionality and microenvironment (Meer et al. 2009). Few other examples include corneal epithelium basement membrane (Abrams et al. 2000), heart myocardium (Kim et al. 2009), and microvilli of intestine which show presence of micro and nanoscale topographical pattern. The villi structure on the intestinal lining, which is highly complex, comprises pores and rough protruding structures in micro and nanoscale which helps in nutrients absorption (Totonelli et al. 2012). Figure 2.2 shows different topographical features from *in vivo* system. Thus, it is important to consider the effect of various tissue mimicking topological factors while culturing the cells *in vitro*.

One of the initial studies to explore the effect of topography on stem cell fate was first carried out by Curtis et al. in 1964. In this study, the response of cells to groove-like topographical features was reported where cell alignment in the direction of grooves was observed, known as contact guidance as reported by Paul Weiss in 1945 (Weiss 1945; Curtis and Varde 1964; McMurray et al. 2015). Size (Xie et al. 2021), height (Liu et al. 2016), width (Clark et al. 1990), depth (Clark et al. 1990; Matsuzaka et al. 1999), and spacing (Andersson et al. 2003) of topographical features appear to play a crucial role in regulating cell behavior. Initially, there were many controversies regarding the role of topography considering the cellular response as

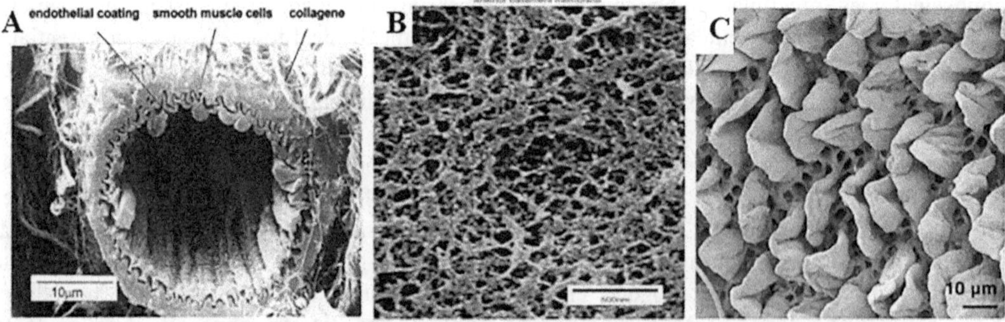

FIGURE 2.2 Micro and nanoscale topographical features in A) Blood vessel. Reprinted from Wu et al. Copyright (2018), with permission from Elsevier. B) Human corneal basement membrane. Reprinted from Last et al. Copyright (2009), with permission from Elsevier. C) Decellularized rat intestine. Reprinted from Totonelli et al. Copyright (2012), with permission from Elsevier.

the mere role of surface chemistry. This was disproved by experimental findings by Britland and group who used laminin as a chemical cue oriented at parallel and right angles to a topographic one that consists of grooves with varying depth. They showed that in the presence of chemical cue and topography, when the grooves were 500 nm deep or less, the rat dorsal root ganglia (DRG) nerve cells reacted mainly to the chemical cues. However, on deeper grooves, the topographic cue overrides the chemical cues resulting in orientation of about 80% of the cells along the topographical pattern confirming the role of topography (Britland et al. 1996). Later studies have shown that the cellular responses to topographical parameters are diverse and strongly dependent on the cell type and topological features. Initially, the effect of features with micrometer length scales was studied. However, it was observed that the ECMs with which the cells interact also have nanometer range features such as collagen fibers influencing cell behavior. With advancements in technology, now it is possible to fabricate structures with overlapping micro and nanoscale features, known as micro and nano topography (Cun and Hosta-Rigau 2020). It was observed that the microtopographical structures which include features larger than 10 μm control the overall morphology of cells whereas the nanotopographical structures influence the formation of filopodia and lamellipodia for subcellular sensing (Nguyen et al. 2016). Figure 2.3 presents some examples of micro/nanostructures that have been examined for different biological applications fabricated by microfabrication techniques.

Both micro and nanoscale topographical structures have been shown to influence cell adhesion and thus in turn affect various cell responses including morphology, proliferation, migration, endocytosis, gene expression, and differentiation (Salmasi 2015). Such customized micro and nanotopographical features can be employed to induce particular biological processes such as differentiation even in the absence of soluble biochemical cues. Therefore, these topographical features can be exploited as the crucial tools in cell biology experiments, tissue engineering, and

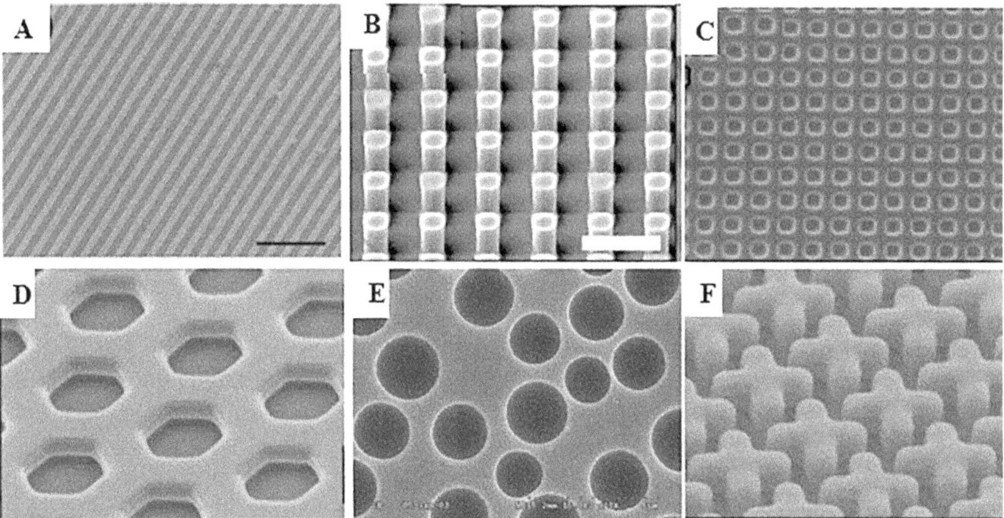

FIGURE 2.3 SEM micrographs of different types of micro and nanostructures A) Microgrooves (Scale bar: 50 μm). Reprinted from Altomare et al. Copyright (2010), with permission from Elsevier B) Square micropillar (Scale bar: 10 μm). Reprinted from Hasturk et al. Copyright (2019), with permission from Elsevier C)Micropit array. Reprinted from Seo et al. Copyright (2014), with permission from Elsevier D) Hexagonal pattern (Scale bar: 20 μm). Reprinted from Vasudevan et al. Copyright (2014), with permission from Elsevier. E) Pits (Scale bar: 5 μm). Reprinted from Wan et al. Copyright (2005), with permission from Elsevier F)Cross micropillars (Scale bar: 10 μm). Reprinted from Vasudevan et al. Copyright (2014), with permission from Elsevier.

cell therapy as they are cost effective, xeno-free, and mitigate the risk of unintended off target effects of externally added growth factors and soluble biochemicals (Cun and Hosta-Rigau 2020). It was also observed that when the bioengineered cells are transplanted *in vivo*, providing these soluble factors continuously for long duration becomes the major challenge as it affects the properties of transplanted cells and thus shows poor efficiency (Cun and Hosta-Rigau 2020). The topological conditioning of the cells may help the field in this aspect.

However, despite all the previously mentioned advantages, there exists one significant bottleneck. The working mechanism of topography-induced cellular response is still poorly understood and under exploration. Also, topography-induced response is mainly a surface phenomenon which needs to be extrapolated further using three-dimensional (3D) micro and nanostructures to get more holistic understanding. This is because a two-dimensional (2D) system has limited cell-cell interactions, lacks cellular organization, and is different from real *in vivo* system, emphasizing the need of 3D substrates known as scaffold. Scaffolds are materials that maintain the microenvironment for *in vitro* cell growth, proliferation, and differentiation, withstand external pressure, give structural support to the cells, and help in transfer of nutrients and metabolites (Bai et al. 2019).

Various materials and fabrication techniques are available to fabricate such 2D and 3D micro and nanostructures with different topographical features which are discussed in the next section of this chapter.

2.2 FABRICATION OF MICRO AND NANOSTRUCTURES

2.2.1 Materials Used for Fabrication of Micro and Nanostructures

Choice of materials is critical for the applications in regenerative medicine because materials are known to influence various cellular functions including adhesion, proliferation, migration, and differentiation of stem cells (Hubbell 1995; O'Brien 2011). Desired material and surface properties are achieved by suitably modifying surface functionality, wettability, biocompatibility, and biodegradability (Murphy et al. 2012).

The situation becomes further complex when such materials are used for fabrication of 2D/3D structures with micro/nanoscale topological cues. In such applications, materials should be suitable for the particular fabrication technique. For example, the materials should have enough mechanical strength to be used in micromachining or should be fast polymerizing to be used in photocuring techniques (more details can be found in Section 2.2.1.1). The material should not deform over time and should be able to give rise to structures with desirable resolution. As cells typically have low optical contrast, the material to be used should not interfere with the transmission optical microscopy in the phase contrast mode or should not autofluoresce for fluorescent microscopy (Curtis and Wilkinson 1997). Concentration of polymer, ligand density, structure, flexibility, stiffness, and porosity are few other physico-chemical factors that should be considered while designing/selecting the material (Bai et al. 2019). Different types of materials are currently used for topographical studies which include metals, inorganic compounds, and polymers (Bai et al. 2019). Metals such as gold, titanium (Postiglione et al. 2004), and stainless steel, as well as inorganic compounds such as silica, lithium niobate, and silicon nitride, are well reported in the literature for their application in the fabrication of topographical features (Curtis and Wilkinson 1997) (Table 2.2). However, polymers are a vast group of materials which are widely used to study cell adhesion, migration, proliferation, and differentiation which are discussed in detail in the next section.

2.2.1.1 Polymers

2.2.1.1.1 Natural Polymers

The majority of natural polymers utilized for cell culture involve proteins that make up the extracellular matrix (ECM). It includes materials such as collagen, fibronectin, laminin,

TABLE 2.2
Different Materials and Techniques Used for Fabrication of Micro and Nanostructures

Materials	Type of Topography	Fabrication Technique	Cell Type	References
Titanium	Grooves Taperd pits and inverted pyramids	Micromachining Microfabrication	Human gingival fibroblasts Rat calvarial osteoblasts (RCO), Porcine periodontal ligament epithelial cells (PLE)	(Oakley and Brunette 1993) (Hamilton et al. 2007)
Gold	Grating	Block copolymer micellar lithography	Mouse fibroblast	(Zhu et al. 2012)
Stainless steel	Periodic nanopatterns	Femtosecond laser ablation	Bm-hMSCs	(Martínez-Calderon et al. 2016)
Silicon	Micropillars	Photolithography	LRM55 astroglial cells	(Turner et al. 2000)
PMMA	Nanogrooves and pits Nanogratings	Photolithography Nanoimprint lithography	Osteoblasts PC12	(Biggs et al. 2009) (Ferrari et al. 2010)
PDMS	Periodic circle pattern Micropillars Microgrooves	Photolithography Photolithography Soft lithography	Fibroblasts, HeLa, and Primary hepatocytes 3T3 cells Human corneal stromal cells	(Kidambi et al. 2007) (Ghibaudo et al. 2009) (Bhattacharjee et al. 2020)
PCL	Nanopits	Electron beam lithography, Hot embossing	Human fibroblasts	(Dalby et al. 2004)
PLLA	Microgrooves	Soft lithography	PC12, Chick sympathetic neurites	(Li et al. 2008)
PLGA	Microgrooves Ridges/grooves	Laser ablation Photolithography	PC12 Chicken Dorsal root ganglion (DRG) cells	(Yao et al. 2009) (Li et al. 2018)
Gelatin	Microgrooves	Photo and soft lithography	Rat cardiomyocytes	(Navaei et al. 2017)
Alginate	Microgrooves and ridges	Micromolding	Human umbilical artery vascular smooth muscle cells, Cardiac myocytes	(Agarwal et al. 2015)
Silk	Microsphere array Microgrooves	Microsphere self-assembly, solvent casting Laser ablation	BM-MSCs C2C12 cells	(You et al. 2014) (Angelova et al. 2022)
Chitosan	Gratings, Isosceles Triangles, and Scalene triangles	Solvent casting on PDMS mold	RT4-SCs, Schwann cell line (nerve regeneration)	(Scaccini et al. 2021)
Hydroxyapatite	Micro/nanohybrid structures Honeycomb, Pillars, and Isolated islands	Template method, Hydrothermal treatment Microcasting	BM-hMSCs Adipose-derived stem cells	(Zhao et al. 2018) (Ramaswamy et al. 2021)

Abbreviations: PMMA- Polymethyl Methacrylate, PDMS-Polydimethyl Siloxane, PCL- Poly(caprolactone), PLLA- Poly-L-Lactic Acid, PLGA- Poly(lactic-co-glycolic acid) hydroxyapatite, and hyaluronic acid which are used as *in vitro* substrate for cell culture (Higuchi et al. 2012). Apart from this, biopolymers such as chitosan, cellulose, and silk fibroins obtained from natural sources like plants and insects are also reported to fabricate micro and nanostructures to study cellular response. These polymers show properties similar to native extracellular matrix and thus support cell adhesion, proliferation, and differentiation. Hydrogels such as alginate, gelatin, and agarose are biocompatible materials with high surface area and good adsorption capacity used for 3D printing of cells to generate tissues *in vitro*. Though natural polymers are preferred due to their inherent biocompatibility and bioactivity, there are certain issues associated with their use such as limited control over physiochemical properties, degradation rate, poor reproducibility, high variability, and challenges in purification, sterilization, and immunogenic interactions. However, nowadays few natural polymers such as chitosan and hyaluronic acid are well characterized and made commercially available which has helped to overcome the aforementioned issues to some extent (Croisier and Jérôme 2013; Silva et al. 2017). This emphasizes the need of materials with advanced properties to support cell growth and focuses on development of synthetic and semisynthetic polymers that provide high reproducibility with controlled properties (Dhandayuthapani et al. 2011; Silva et al. 2017).

2.2.1.1.2 Synthetic Polymers

Synthetic polymers are comprised of repeatable inert units which are synthesized from a non-biological source with a broad variety of structures and appropriate physical and chemical properties. It mainly includes polymers such as poly-lactic acid (PLA), poly- ε-caprolactone (PCL), poly (D, L-Lactic acid), poly glycolic acid (PGA), poly fumarates, poly vinyl alcohol (PVA), oligo(polyethylene)glycol, acrylate and epoxy-based polymers. Variety of composite materials such as Poly-Lactic-co-Glycolic acid (PLGA) has also been used in order to obtain the properties that are not shown by individual materials. Synthetic materials are cost effective, provide good reproducibility and mechanical properties, and thus have been widely used for fabrication of micro and nanostructures as enlisted in Table 2.2. Despite the good properties, induced immune response is a crucial issue associated with use of synthetic materials due to their hydrophobic nature. However, this can be solved by modifying the material surface properties (Ghasemi-Mobarakeh 2015).

Depending on the intended application, specific material is selected for fabrication of 2D and 3D micro/nanostructures which are used for studying cellular behavior. For example, ceramic materials are used for bone tissue engineering as it resembles properties similar to bone (Zhu et al. 2020) whereas antibacterial biopolymers such as chitosan are used for wound healing applications (Cui et al. 2021).

2.2.2 Techniques of Fabrication

Specific fabrication technique with suitable material is essential to fabricate desired micro and nanostructures to mimic *in vivo*-like microenvironment. The roughness of the underlying surface and the type of pattern are two key factors regulating the cell behavior which need to be considered during fabrication (Wieland et al. 2002; Nikkhah et al. 2012; Anselme et al. 2010). There are two main approaches for fabrication of ordered or randomly structured micro and nanotopographies namely, top-down approach and bottom-up approach (Madou 2002; Baek et al. 2021). These two approaches comprise of number of mechanical, chemical, and physical methods to fabricate nanostructures with different size and chemical composition. Typically, in bottom-up approach, nanoparticles and nanocrystals are used as building blocks and assembled into final micro/nanostructure. It includes techniques such as self-assembly, gas phase synthesis, sol-gel processing, sonochemical processing, hydrodynamic cavitation, and microemulsion processing (Sarvanan et al. 2008; Biswas et al. 2012; Liddle and Gallatin 2016). These techniques are well reviewed in these references (Eltom et al. 2019; Handrea-Dragan and Botiz 2021). In top-down approach, bulk material is processed using nanofabrication tools for synthesis of desired nanostructures. It includes methods such as lithography, nanodispensor, and reactive ion etching (Sarvanan et al. 2008; Ngô and Van de Voorde 2014; Kumar et al. 2018). In this chapter, we are mainly focusing on fabrication technologies for fabrication of 2D and 3D micro and nanostructures.

2.2.2.1 Techniques for Fabrication of 2D Micro and Nanostructures

2.2.2.1.1 Electrospinning

Electrospinning was introduced by Anton Formhals in the year 1930 for the production of micro and nanofibers of various materials using a high electric field and conducting solutions. Under high voltage, a solution of viscoelastic polymer is extruded through a conductive needle which leads to the formation of nanofibers. The micro and nanofibers produced by this method have varying properties such as size, porosity, alignment, and surface features depending on spinning condition. Figure 2.4 shows schematic of a typical electrospinning system and SEM images of aligned and random electrospun microfibers. Thus, this method is used for the production of a variety of scaffolds from natural and synthetic materials which can be used for the growth and differentiation of stem cells (Rim et al. 2012). The properties of these fibers can also be changed by adding bioactive molecules during the electrospinning process (Zhu et al. 2020). It is a simple technique that provides control in fabrication of fibers from micro to nanoscale. However, requirement of specific solvents, insufficient cell infiltration, and inhomogeneous cell distribution are few concerns with this technology (Hong et al. 2019). Recently biopolymers such as chitosan-based fabrication of pure and stable nanofibers have been reported without using the toxic organic solvents (Lemma et al. 2016). A new approach known as "Cell-electrospinning", which involves encapsulation of cells in polymeric materials for fabrication of 2D and 3D scaffolds is reported. It has gained good attention in regenerative medicine and tissue engineering due to its potential to fabricate a native extracellular matrix-mimicking environment which provides cell-friendly surroundings which promote cellular activitics (Hong et al. 2019). Many studies have reported electrospun micro/nanofibers for cartilage regeneration (Kadir et al. 2021), wound healing (Lanno et al. 2020), neural differentiation (Wang et al. 2009), and osteo differentiation (Yang et al. 2018).

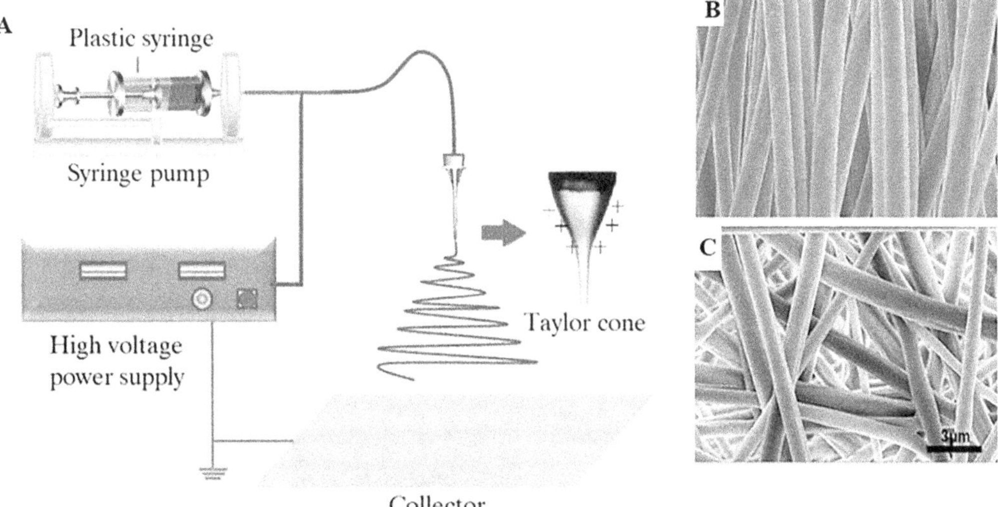

FIGURE 2.4 A) Schematic of electrospinning process (Chen et al. 2019). Reprinted with permission from MDPI. B, C) SEM micrographs of aligned and random PLLA electrospun microfibers (Scale: 3 μm). Reprinted from Xie et al. Copyright (2021), with permission from Elsevier.

2.2.2.1.2 *Photolithography*

Photolithography is one of the commonly used technologies for the generation of patterns and is applied in a variety of fields such as the manufacturing of integrated circuits (IC), microchips, and micro-electrical mechanical systems (MEMS) devices. It includes the use of light-sensitive polymer, known as photo-resist, and a photomask with a pattern which is exposed to ultraviolet (UV) light to get a specific pattern (Pimpin and Srituravanich 2012). Photomask allows the light to pass through specific areas to project an image on a surface. Photoresists are mainly classified into two types, namely positive photoresists (e.g., Azide quinone) and negative photoresists (e.g., SU-8). In the case of positive photoresists, after exposure to UV light, the material gets dissolved by the developer whereas, in negative photoresists, the unexposed material gets dissolved (Luo et al. 2020). Positive photoresists have an advantage over negative photoresists as the material gets dissolved after development, UV phototoxicity is not an issue. Typical photolithography includes the following major steps, positioning of the mask on spin-coated polymeric film, UV exposure, and development of resin (Ermis et al. 2018). Figure 2.5 illustrates a photolithography mechanism.

There are three main types of photolithography, namely, Contact, Proximity, and Projection lithography. The basic principle remains the same as explained earlier, however, the differences are based on the positioning of the photomask. In the case of contact lithography, as the name suggests, the mask is in direct contact with polymer film, and pressure is applied to get the pattern on film. Resolution up to 1 μm can be obtained by this method; however, getting uniform resolution is a challenge. In the case of proximity lithography, the mask is not in direct contact with polymer film but is present in close proximity because of which resolution is low. In projection lithography, there is a large gap in mask and polymer film however it uses an objective lens that collects the diffracted light and projects it on the substrate (Pimpin and Srituravanich 2012).

Due to the simple nature, variety of structures such as grooves (Biggs et al. 2008), pits (Biggs et al. 2009), and pillars (Wei et al. 2016) have been fabricated using photolithography and reported to study cellular behavior. The technique was also reported for fabrication of array of lines, triangle, circles, and squares for patterning of cardiomyocytes and osteocytes (Karp et al. 2006). Though one of the commonly used fabrication techniques, there are certain limitations of this technique such as the polymer must be photosensitive and fine structures such as curves cannot be fabricated.

2.2.2.1.3 *Soft Lithography*

Soft lithography mainly uses elastomeric materials such as Polydimethyl siloxane (PDMS) to fabricate micro and nanostructures (Ermis et al. 2018). It was first introduced by Bain and

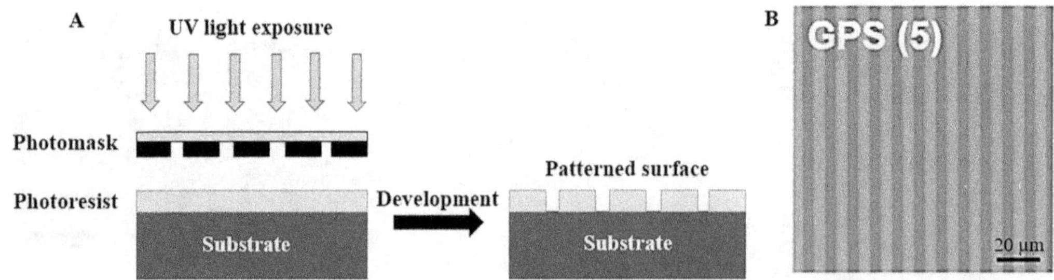

FIGURE 2.5 A) Schematic of photolithography B) SEM micrograph of GO-based patterned substrates (GPS) with microgrooves of 5 μm. Reprinted with permission from Yang et al. Copyright (2016). American Chemical Society.

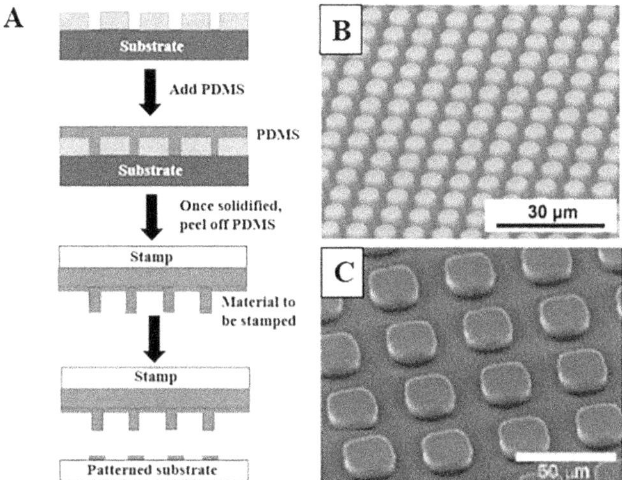

FIGURE 2.6 A) Schematic of soft lithography. SEM micrographs of B) micropillar array. Reprinted from Kameya et al. Copyright (2017), with permission from Elsevier and C)Round cornered square array on PDMS. Reprinted from Guan et al. Copyright (2006), with permission from Elsevier.

Whitesides in 1989 (Bain and Whitesides 1989; Whitesides et al. 2001). It mainly includes fabrication of patterned polymeric stamp which is used to transfer molecules in defined specific structures. Figure 2.6 represents a schematic of soft lithography. It provides structures in scale ranging from 30 nm to 100 µm. There are three main types of soft lithography namely micro stamping, stencil patterning, and microfluidic patterning (Betancourt and Brannon-Peppas 2006). This types of soft lithography are well explained by Nadine et al. (2022). This technique is simple, low cost, and has advantage to obtain structures with flexibility and curvatures which is not possible with photolithography. Due to this advantages, soft lithography is well reported for fabrication of structures for bone (Kiyama et al. 2018), cardiac (Annabi et al. 2013), and cartilage tissue engineering (Chou et al. 2013). However, distortion of elastomeric materials is one of the major challenge in soft lithography limiting its resolution (Pimpin and Srituravanich 2012; Krishna et al. 2016).

2.2.2.1.4 Electron Beam Lithography

Electron beam lithography (EBL) is one of the fundamental techniques, similar to photolithography, used for fabrication of patterns at nanoscale. In this technique, an electron beam is used which scans a material surface in order to obtain desired pattern. It consists of a chamber, an electron gun, and a column maintained in high vacuum. Commonly used electron sources are thermoionic emitters and thermal field emitters which have outputs in the range of 1 to 200 keV (Altissimo 2010). Typically electromagnetic lenses are used to focus the beam on an electron sensitive material and provide high resolution upto 5 nm which varies according to beam spot size (Hart et al. 2007). Depending on a solubility of material, area exposed to or not exposed to electron beam is removed during developing stage. Figure 2.7 shows schematic of EBL system and SEM images of structures fabricated by EBL. Due to its potential to provide nanoscale structure, this technique has a wide range of applications in development of ICs, photonic crystals, and nanofluidic devices (Altissimo 2010) and for fabrication of mold to study the topographical response of cells (Dalby et al. 2004). Main limitations of this technique are high cost of maintenance (Betancourt and Brannon-Peppas 2006) and requirement of clean room and electrical grounding in order to avoid charging effect (Altissimo 2010).

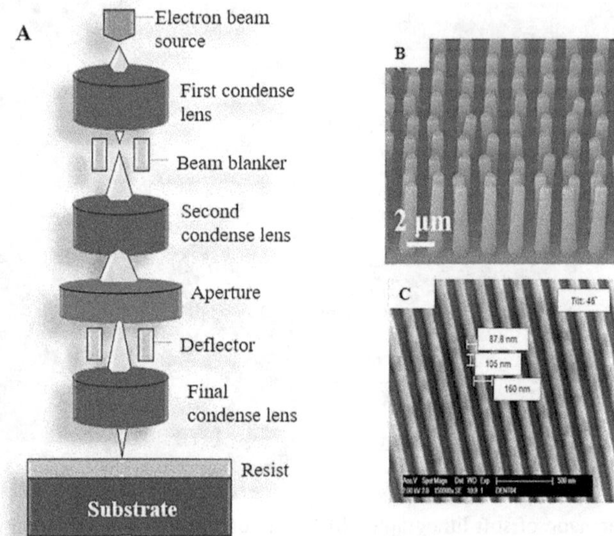

FIGURE 2.7 A) Schematic for E-beam lithography (Adapted from Pimpin and Srituravanich 2012). SEM micrographs of B) SU-8 micropillar. Reprinted from Ma et al. Copyright (2016), with permission from Elsevier and C) Nanogrooves on silicon wafer. Reprinted from Loesberg et al. Copyright (2007), with permission from Elsevier.

2.2.2.1.5 X-ray Lithography

X-ray lithography (XRL) was first introduced by H. Smith and Spears in 1972 and garnered the attention for the micro/nanofabrication due to its shorter wavelength and larger penetration depth than conventional photolithography (Bharti et al. 2022). In the XRL process, X-ray sensitive materials that change their dissolution rate in a specific solvent after irradiation are used to fabricate micro and nanostructures. It uses X-rays in a range of 0.5 to 4 nm to transfer pattern from mask to substrate. Commonly used X-ray sources are synchrotrons and laser induced plasma generators. Typically, the mask is few microns in thickness, and is usually composed of low-attenuation elements such as silicon carbide, silicon (Si), beryllium (Be), diamond, and silicon nitride (SiNx) in which patterns are created with heavy metals such as gold, tungsten, or tantalum which act as absorbing regions. Though resolution less than 30 nm can be achieved by this technique, deformation and bending of X-ray masks are one of the problems of XRL (Betancourt and Brannon-Peppas 2006; Chen and Pépin 2001; Bharti et al. 2022). Figure 2.8 shows schematic of X-ray lithography and a grating structure fabricated by XRL.

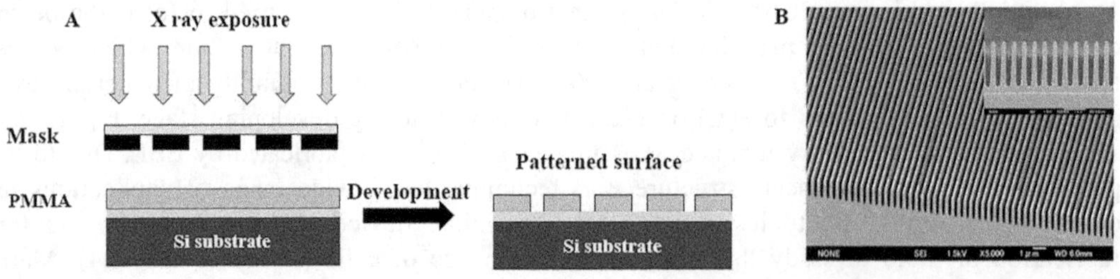

FIGURE 2.8 A) Schematic of X-ray lithography B) SEM micrograph showing 250 nm linewidth grating fabricated using X-ray lithography (Scale bar: 1 µm). Reprinted from Chenchen Luo et al., Copyright (2012), with permission from Elsevier (Luo et al. 2012).

2.2.2.1.6 Focused Ion Beam Lithography

Focused ion beam lithography (FIBL) is a maskless technique that provides high resolution, high density, high sensitivity, and high reliability. It uses an accelerated ion beam, mainly gallium ions that directly hit the sample surface. As the high-speed ions hit the substrate surface, energy is transmitted to atoms on the surface. This interaction between ions and substrate is regarded as a collection of elastic collision and inelastic collision events from the perspective of ion energy (Li et al. 2021). It leads to different reactions such as sputtering of neutral ionized and excited surface atoms, emission of electron, emission of photons, displacement of atoms in the solid, and chemical reactions. It is similar to EBL, but here electromagnetic lenses are replaced by electrostatic lenses due to high masses of ions. FIBL systems are also used for ion-beam-induced deposition which employs deposition of materials such as tungsten, platinum, and carbon (Hasan and Luo 2018). Due to short wavelength of ions, resolution in the range of 5–20 nm can be obtained by this technique which is of great interest in the field of tissue engineering and regenerative medicine to study the response of specific cells. However, it has limitation of low throughput and slow process (Pimpin and Srituravanich 2012). Figure 2.9 illustrates FIBL and a rectangular array structure fabricated by FIBL.

2.2.2.1.7 Nanoimprint Lithography

Nanoimprint lithography (NIL) is an advanced, but inexpensive nanofabrication technique that provides high resolution upto 5 nm (Pimpin and Srituravanich 2012). It was first introduced by S.Y. Chou as hot embossing technique (Chou et al. 1996). NIL uses hard mold to imprint onto a polymer film to provide nanoscale features. Molds are mainly made up of quartz and silicon. Use of hard material is preferred as they are stable at high temperature and results in minimal deformation of structure. However, lifetime of mold is limited as continuous heating/cooling, and high pressure tends to damage the mold. There are four types of NIL, namely, Thermal NIL, UV NIL, Laser assisted NIL, and Electrochemical NIL (Barcelo and Li 2016). Figure 2.10 illustrates different types of NIL.

In Thermal NIL, thermoplastic polymer is spin-coated on substrate on which mold is placed and pressure is applied. The film is then heated above its glass transition temperature which allows material flow and fills the mold structure. This is followed by lowering of temperature that allows solidification of material after which mold is removed. At last, the resist residual layer is removed by an etching process (Hasan and Luo 2018). Viscosity of the material is an issue in this technique which acts as limiting factor for minimizing pattern size and increasing feature density.

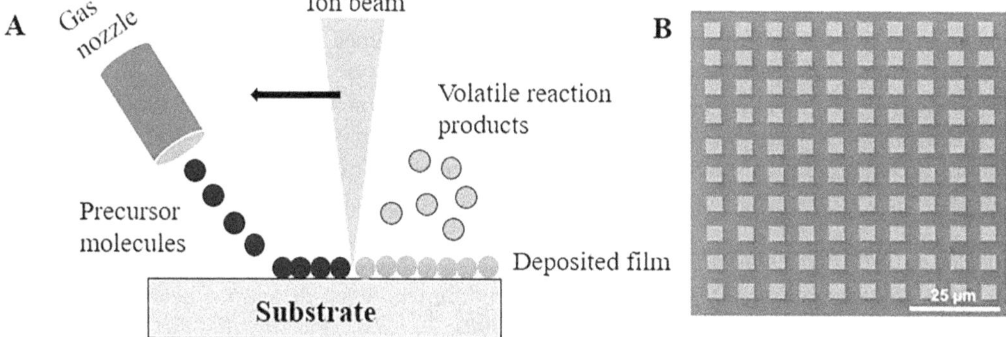

FIGURE 2.9 A) Schematic of focused ion beam lithography B) Pseudo-colored SEM micrograph of a W-C Cryo-deposit rectangular array, grown in a single Ga^+- irradiation exposure (Córdoba et al. 2019). Reprinted with permission.

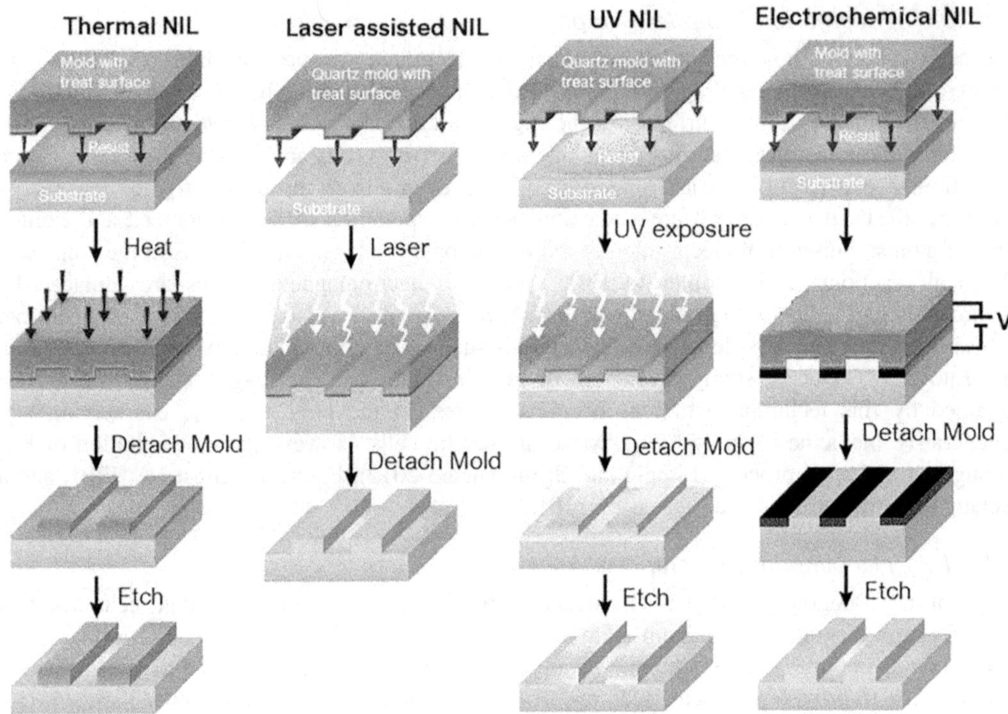

FIGURE 2.10 Schematic of nanoimprint lithography (Hasan and Luo 2018). Reprinted with permission.

In 1996, UV nanoimprint lithography was introduced in which, a low viscosity UV-curable monomer was used in order to enhance the fluidity of the material. On this monomer layer, transparent mold is placed which is directly exposed to the UV radiation. UV radiation results in crosslinking of monomer forming rigid polymer. It works at room temperature with reduced fabrication time and helps in reducing the imprint pressure and stress on mold (Pimpin and Srituravanich 2012).

Laser assisted nanoimprint lithography utilizes a single laser pulse to melt the polymer film on silicon or quartz substrate. A quartz mold then placed on it to pattern the nanostructures. The substrate is then allowed to cool down after which the mold is removed. It does not require etching process and allows submicrometer resolution down to 10 nm in less imprinting time such as 500 ns (Hasan and Luo 2018).

Electrochemical nanoimprinting is a technique that uses a mold fabricated from a superionic conductor. It is a resistless technique in which a voltage is applied between the mold and the target substrate through which current is allowed to flow when mold is in contact with the substrate. The strong electric flux from the protrusive parts of the mold to the substrate results in anodic oxidation of the substrate surface, corresponding to the protrusive parts of the mold, with the moisture present between the mold and the substrate. Finally, the substrate etching is done to release the nanostructures (Hasan and Luo 2018).

2.2.2.1.8 Dip Pen Nanolithography

Dip pen nanolithography (DPN) is a positive printing technique developed by Piner et al., for patterning nanoscale structures (Piner et al. 1999). It uses AFM tip as "nib", a solid-state substrate, and ink for fabrication. Ink is a material consisting of molecules that have chemical affinity to the substrate. The mechanism of DPN is based on transport of molecules from AFM tip to the substrate via capillary action which directly makes pattern in submicrometer dimension.

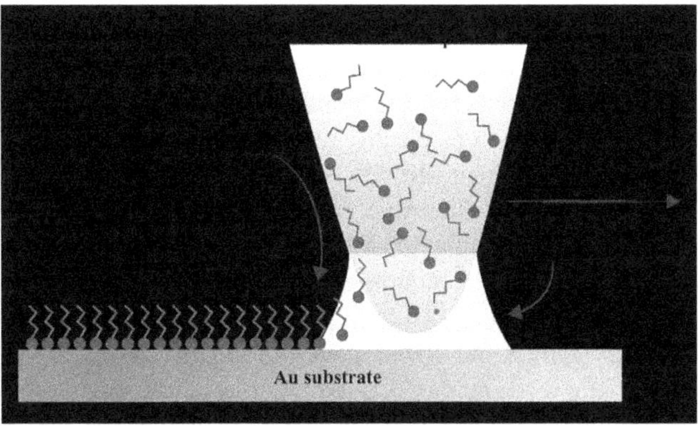

FIGURE 2.11 Schematic of dip pen nanolithography (Adapted from Piner et al. 1999).

Figure 2.11 represents schematic of DPN. It allows selective transport of different molecules at specific region in a particular structure. It provides good flexibility and control during fabrication and does not require a stamp and sophisticated instrumentation. Resolution down to 15 nm can be achieved by this technique (Piner et al. 1999). Number of materials such as alkanethiols and biomolecules such as proteins, lipids, and DNA are used as ink for patterning. Apart from conventional DPN, a multiplexed DPN is used nowadays which includes multiple tips that scan the substrate simultaneously in parallel manner which reduces the time of fabrication (Salaita et al. 2007). Curran et al. had reported arrays of nanodots of around 70 nm with chemical modified surfaces for controlled stem cell adhesion and differentiation without any exogenous supply of bioactive molecules (Curran et al. 2010).

2.2.2.2 Techniques for Fabrication of 3D Micro and Nanostructures

2.2.2.2.1 Stereolithography

As 2D substrates cannot mimic the 3D *in vivo* microenvironment, it is important to find methods for development of 3D micro and nanostructures. Stereolithography is a rapid prototyping 3D printing technology which uses high powered UV light for layer-by-layer fabrication. It uses computer-aided design (CAD) files to fabricate structures with precise internal architectures and external geometries, that resembles human tissue. The process includes focusing of UV light on a surface of photosensitive liquid which leads to photopolymerization of liquid forming a polymerized solid in that region. During fabrication, the first layer of photopolymerized polymer is attached to a base platform, which provides support for structures as they are fabricated. Once this layer is polymerized, the platform is moved to a defined step height for polymerization of the subsequent layer. The support platform is moved and each layer is separately cured repeatedly until the three-dimensional construction is completed (Skoog et al. 2014). Different materials such as trimethylene carbonate, polycaprolactone, poly(D,L-lactide), and poly(propylene fumarate) are reported for fabrication using stereolithography to study the response of mammalian cells (Skoog et al. 2014). This technique has advantages over other techniques as it provides freedom of design and is capable of fabricating complex structures with minimum feature sizes in micrometer scale with physiological relevance. One such example is composite scaffold of poly-D, L-lactic acid/polyethylene glycol/poly-D, L-lactic acid (PDLLA-PEG)]/hyaluronic acid (HA) which was fabricated using projection stereolithography for adipose stem cells to be used in cartilage tissue engineering (Sun et al. 2015). Although stereolithography has been used extensively for tissue engineering and regenerative medicine applications, its utilization has been restricted by its high cost and extended processing time.

2.2.2.2.2 Two Photon Lithography

Two Photon Lithography (TPL) is a powerful direct laser writing (DLW) technique used to fabricate 3D micro and nanostructures with sub-diffraction resolution. This technique has shown great potential in different fields such as electronics, communications, microoptics, biomedicine, metamaterials, microfluidic devices, and MEMS. It can generate scaffold at sub micrometric range with resolution of 100 nm and provides true free-standing architecture. It is considered as a powerful tool to study cell mechanobiology and helps to overcome problem with other fabrication techniques. In this technology, an IR femtosecond laser is passed through the photosensitive material which enables to write directly inside the material in confined regions by controlling the direction of the laser beam using a motorized stage (Chaudhary et al. 2017). Figure 2.12 shows schematic of TPL system. The working range of two photon-based polymerization is dependent on two factors: Two photon polymerization threshold which is an average power required to initiate the polymerization process and Burning threshold which is a power at which the material breaks down and the sample is destroyed (Ummethala et al. 2017; Nguyen and Narayan 2017; Lemma et al. 2019).

The mechanism of two photon polymerization is based on two photon absorption (TPA) where photoinitiator absorbs two photons simultaneously. It is a photochemical process initiated by the femtosecond laser beam. The laser beam is focused on the photosensitive resins by a high-numerical-aperture (NA) objective. It includes three steps namely the initiation process, the propagation process, and the termination process. In the initiation process, photoinitiators (PI) absorb the two photons and achieve the excited state (PI∗), which finally decomposes to free radicals (R·). It is preferred to use photoinitiators with long triplet lifetime for efficient polymerization process. The initiation process is followed by propagation process where the radicals combine with monomers (M) to produce monomer radicals (RMn·). In the termination process, two monomer radicals are combined and the photopolymerization process is terminated.

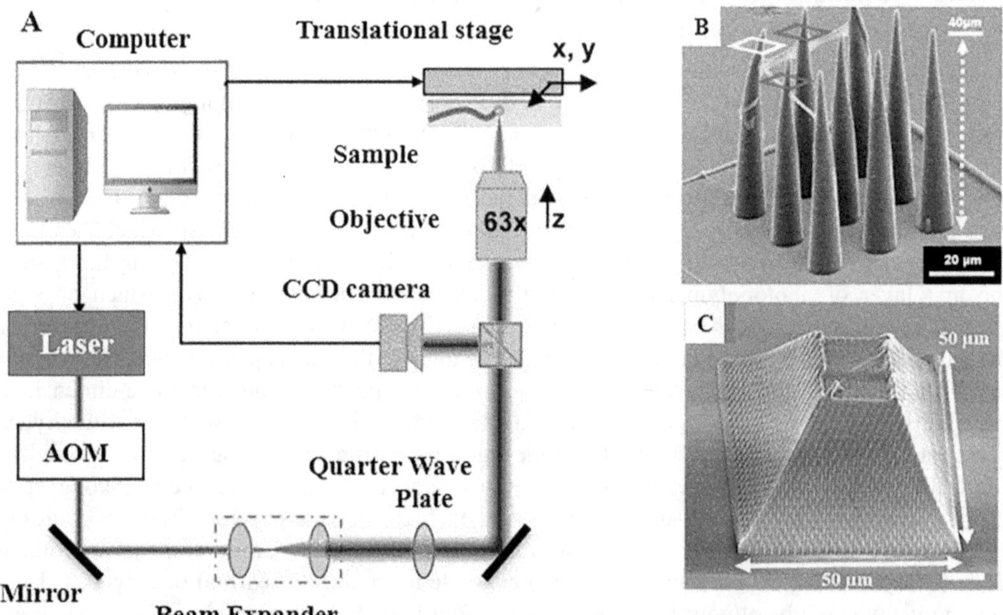

FIGURE 2.12 A) Schematic of two photon lithography (Adapted from Duc Nguyen et al. 2016 and Chaudhary et al. 2017) B) SEM micrographs of 3D microneedle array structures (Jaiswal et al. 2021) and C) 3D porous micropyramid (Jaiswal et al. 2021) fabricated by TPL. Reprinted with permission from Jaiswal et al. Copyright (2021) American Chemical Society.

$$\mathrm{PI} \xrightarrow{hv+hv} \mathrm{PI}^* \rightarrow \mathrm{R}\cdot + \mathrm{R}\cdot,$$
$$\mathrm{R}\cdot + \mathrm{M} \rightarrow \mathrm{RM}\cdot \xrightarrow{M} \mathrm{RMM}\cdot \cdots \rightarrow \mathrm{RM}_{n}\cdot,$$
$$\mathrm{RM}_{n}\cdot + \mathrm{RM}_{m}\cdot \rightarrow \mathrm{RM}_{n+m}\mathrm{R}.$$

FIGURE 2.13 Mechanism of two photon polymerization.

Figure 2.13 shows reactions involved in two photon polymerization. The polymerization process depends on radical concentration and starts only when the concentration is above the certain threshold (Zhou et al. 2015).

It can fabricate highly intricate structures with resolution down to 100 nm. Due to its high resolution and precisely controlled fabrication process, TPL is emerging as one of the important 3D fabrication techniques for tissue engineering and implant development. One of the examples is a poly (ethylene glycol) diacrylate (PEGDA) based 3D scaffold developed by Accardo et al. The scaffold was used for growth of Neuro2A cells which showed efficient colonization of cells on true free-standing architecture with formation of neuritic extension and interconnections extending its application in neural tissue engineering (Accardo et al. 2018). Despite the good properties, high maintenance cost, requirement of TPP compatible polymer and photoinitiator, and cytotoxicity due to unpolymerized materials have limited the application of this technique (Nguyen and Narayan 2017).

2.2.2.2.3 3D Bioprinting

With advancement in tissue engineering, additive manufacturing techniques such as bioprinting have gained more attention in last few years for *in vitro* tissue and organ development. Bioprinting is a technique to produce three-dimensional living organs using biomaterials, cells, and biomolecules. It allows homogenous distribution of cells and controls cell density and high-resolution cell deposition, which are considered major challenges with other tissue engineering techniques. Bioprinting uses bioink which primarily includes hydrogel with cells for tissue generation. Hydrogels are used for bioprinting due to their high-water content which supports cell entrapment and encapsulation. Thus, to be used as bioink, it is important to consider the choice of material, its chemical composition, and concentration. At present, there are three main bioprinting techniques namely, Inkjet bioprinting, Extrusion bioprinting, and Laser assisted bioprinting, illustrated in Figure 2.14 (Rider et al. 2018).

Inkjet bioprinting is a technique based on conventional inkjet printing (Li et al. 2020). It is a non-contact printing which uses bioink for ink cartridge dispensing by thermal, piezoelectric, and microvalve processes. Though this technique represents cost effective, high throughput way of

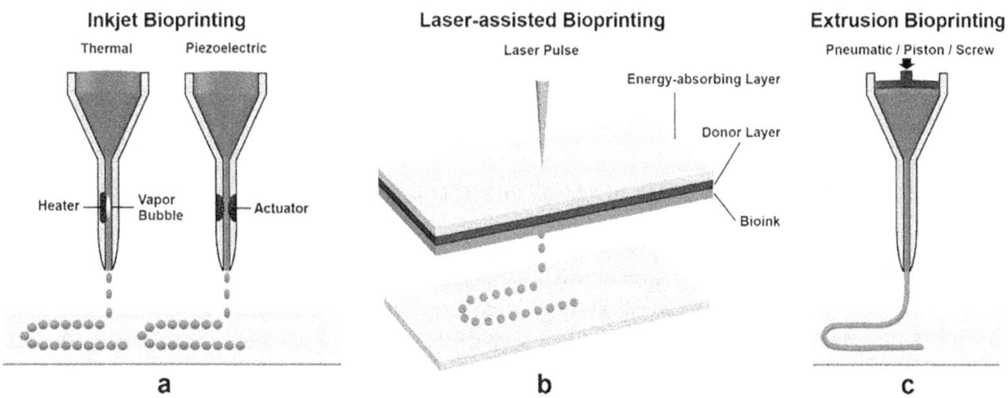

FIGURE 2.14 Schematic of a) Inkjet b) Laser assisted and c) Extrusion 3D bioprinting (Xie et al. 2020). Reprinted with permission.

bioprinting, its use is limited due to the ability to make only 2D structures. However, layer-by-layer method can be used to develop 3D structures from 2D structures (Kim et al. 2019). Despite its advantages, there are multiple challenges in inkjet bioprinting such as lack of structural integrity between droplets, nozzle clogging when used with high cell densities, non-uniform droplet size, and the possibility of cell damage due to high heating temperature or mechanical stress (Kim et al. 2019).

Extrusion bioprinting is most common type of 3D bioprinting. It was developed by Scott Crump in 1980 (Ramesh et al. 2021). It involves pneumatic and mechanical dispensing systems which extrude the bioink droplets sequentially under steady force by formation of cylindrical lines. It can be used to dispense high cell density bioinks with fast fabrication rate which becomes a major advantage of this technique over its counterparts. However, only viscous materials can be extruded by this method and it also had shown apoptotic effect on cells during printing due to pressure drop which limits its use. This technique has shown promising results in tissue engineering for development of bone tissue, muscular tissue, periodontal tissue, neural tissue, cartilage tissue, corneal tissue, cardiac tissue, and skin tissue (Rider et al. 2018; Kim et al. 2019).

Laser assisted bioprinting is a technique which uses laser as energy source to deposit biomaterials or cells onto a substrate (Ventura 2021). The lasers used are mainly nanosecond lasers with UV or near UV wavelengths. It mainly consists of three components: A pulsed laser source, donor layer or a ribbon coated with liquid biological material, and a receiving substrate. When a laser focuses on ribbon, it causes evaporation of liquid biological material which leads to droplet formation that reaches the receiving substrate. Once the cells are transferred to the receiving substrate, their growth is maintained by culture medium or polymer present in substrate. This method shows good resolution in pico to microscale. However, its efficiency is affected by different factors such as the energy of the laser source, thickness of the biological materials on the film, their rheological properties, the wettability of the substrate, the printing speed, and organization of the structure (Li et al. 2016).

2.2.3 Characterization of Micro and Nanostructures

The micro and nanostructures must be characterized after fabrication before assessing the cell response. Various methods are used for structural and surface characterization of fabricated micro and nanostructures in order to check their size, shape, design, charge, and chemical composition. Morphological features of structures are determined by optical microscopy, scanning electron microscopy (SEM), transmission electron microscopy (TEM), and atomic force microscopy (AFM). Chemical properties can be checked by Fourier transform infrared spectroscopy (FTIR), X-ray diffraction (XRD), Raman spectroscopy, X-ray photoelectron spectroscopy (XPS), and Brunauer-Emmett-Teller (BET) analysis (Hasan et al. 2018). Due to their advantages over bulk materials and enhanced properties with different shape, size, and design, these structures have gained considerable attention in biological and clinical field and have shown promising results in cell therapeutics especially stem cell technology.

2.3 EFFECT OF MICRO AND NANOSTRUCTURES ON CELLULAR BEHAVIOR

For proper regulation of cellular functions such as adhesion, proliferation, migration, differentiation, organ development, and tissue homeostasis integration of chemical and mechanical signals is essential (Jaalouk and Lammerding 2009). Cells sense their mechanical microenvironment and convert the mechanical cues into biochemical signals via the process called mechanotransduction. Dysregulation of mechanotransduction process has been linked to several diseases and pathological conditions such as cancer, lamilopathies, muscular dystrophy, osteoporosis, and other age-related degeneration (Ingber 2003; Jaalouk and Lammerding 2009). Thus, an in depth understanding of mechanotransduction process is needed not only for our fundamental knowledge but also for addressing these diseases. Out of the various mechanical cues that the cells receive from their microenvironment in various forms such as rigidity, shear stress, dynamic stretching, mechanical loading, surface topography, and 3D structures, here we will focus on the last two, i.e., surface topography and 3D structures.

2.3.1 Mechanism of Sensing 2D and 3D Micro/Nanostructures

How the cells sense the microstructure of their native tissue is an active area of research. Understanding cellular response to external cues such as topography has also been one of the key approaches in tissue engineering and regenerative medicine. This was well established after Paul Weiss gave experimental proof of contact guidance, an ability of a cell to orient and migrate bidirectionally in response to anisotropic topographical features, such as parallel grooves, and found to be physiologically relevant to many *in vivo* cellular processes (Thrivikraman et al. 2021). However, the mechanism behind such responses is not very clear yet. In last few years, many studies have reported possible mechanism by which cell respond to the topographical cues through direct or indirect mechanotransduction process (Paul et al. 2019; Naqvi and McNamara 2020). In the direct mechanotransduction process, cells sense the topography of the ECM and transmit this biophysical signal to the nucleus through series of proteins (Janson and Putnam 2015). For instance, integrins, the cell adhesion transmembrane proteins, bind to the ECM ligands and stimulate the formation of focal adhesions (FAs) which act as molecular clutches to the ECM (Doyle and Yamada 2016). These FAs are composed of proteins such as talin and vinculin which link integrin to cytoskeleton filaments. Finally, cytoskeleton filaments are physically connected to the nucleus through LINC (Linker of Nucleoskeleton and Cytoskeleton) complex (Méjat and Misteli 2010). As a result of this physical connection, cells directly control chromatin and thus gene expression, which subsequently decides cell fate. Besides direct mechanotransduction, topography also induces indirect mechanotransduction which activates biochemical cascades that result from cellular adhesion via activation of focal adhesion kinase (FAK), Rho/ROCK signaling, G protein-coupled receptors, calcium signaling, and mitogen-activated protein kinases (MAPK) (Anderson et al. 2016). Indirect mechanotransduction also alters cell fate by regulating gene expression. For example, extracellular signal regulated kinase (ERK)/MAPK signaling is one of the key regulators for MSCs differentiation (Naqvi and McNamara 2020). Thus, its important to gain in dept insight of how micro and nanostructures modulate vital cellular processes such as adhesion, proliferation, migration, and differentiation.

2.3.2 Examples of Micro and Nanostructures Regulating Cellular Behavior

As already mentioned, micro and nanostructure of tissue environment influences various critical cellular functions. Here, we provide a few of such examples. However, this list is indicative and not exhaustive. As this is an active area of research, new knowledge gets added almost every day.

2.3.2.1 Effect on Cellular Adhesion

For the survival and proper functioning of the majority of the cells in our body, adherence to their surrounding ECM, neighboring cells, or biological/non-biological surfaces is crucial (Schwarz and Bischofs 2005). Cellular adhesion essentially governs important cellular processes such as cell spreading, proliferation, migration, and differentiation (Anselme et al. 2010). Poor adhesion of cells may lead to cell quiescence or cell death, termed as Anoikis, which is a major challenge in cell transplantation therapies (Taddei et al. 2012). Surface topographies with features in the nano- to micro-range are one of the major factors in the strong regulation of cell adhesion, along with a variety of biochemical and mechanical cues. Such adhesion of cells to micro and nanostructures is mainly influence by specific surface charge, wettability, and protein adsorption ability (Ermis et al. 2018). From cellular aspect, many components are involved in the probing and sensing of the surrounding topographical microenvironment that triggers cell adhesion and morphogenesis, hence regulating cellular behavior. These include integrins in cell-ECM contact (McMurray et al. 2015), cadherins in cell-cell contact (Tzima et al. 2005), stretch-sensitive ion channels (Sachs 2010), receptor tyrosine kinases (Lemmon and Schlessinger 2011), and G protein-coupled receptors (Dufort et al. 2011). In the cell-ECM interplay, integrins are part of a larger supramolecular complex, known as focal adhesion (FA), that link the ECM to intracellular components (Nikkhah et al. 2012). At the nanoscale, integrin clustering mostly contributes to cell adhesion, but at the microscale, focal adhesion complex organization, and maturation are crucial for cell adherence (Nguyen et al. 2016).

For some applications such as implants, nonspecific cell adhesion may be undesirable. In such cases, specific topographical features might be helpful in preventing unwanted cell adhesion. For example, symmetric array of hexagonal pits has been reported to work as anti-adhesion surface by restricting focal adhesion formation and maturation (Robotti et al. 2018).

2.3.2.2 Effect on Cell Morphology and Nuclear Architecture

Cell morphology is a crucial element in cell-based therapies, that provides fundamental knowledge about the health and activity of cells. To perform cell specific functions, cells are often needed to have atypical shapes and arrangements. For example, neuronal cells can perform their duty of long-distance signal transduction due to their elongated structures. Cell alignment is another important step that is essential in various physiological processes such as embryogenesis, tissue maturation, and regeneration (Li et al. 2014). For example, muscle cells need to align and fuse together to form muscle fiber bundles so that they can apply force together. In this regard, micro and nanostructures have been demonstrated to be useful for controlling cell shape, structure, and size which in turn influences the cell fate. For example, it is well reported that cells growing on patterned anisotropic grooves and gratings show cellular alignment by orienting themselves in the direction of the pattern. Such alignment was observed in different cell lines such as human mesenchymal stem cells (hMSCs) (Yadav and Majumder 2022), neuroblastoma cells (SHSY5Y) (Yadav and Majumder 2022), skeletal muscle cells (C2C12) (Coyle et al. 2022), human corneal stromal cells (Bhattacharjee et al. 2020), and human vascular endothelial cells (hVECs) (Sun et al. 2017). Similar studies have been reported with micropillars (Murray et al. 2016), micropits (Seo et al. 2014), honeycomb pattern (Paun et al. 2018), etc.

A similar trend can be seen for nuclear size, shape, and orientation as well (Fei and Liu 2016; Liu et al. 2019; Bhattacharjee et al. 2020). This shows that topography not only influences the cellular morphology, but also influences nuclear morphology and architecture due to cytoskeletal reorganization which further leads to change in gene expressions (Ermis et al. 2018).

Although numerous studies have documented that the cells respond to substrate topography, an in depth molecular mechanisms is yet to be elicited. The problem becomes more critical as the impact of substrate topography on cell morphology is highly dependent on the cell type and varies with the geometry and dimension of the topographical features. Hence, more detailed study in this area is required with different geometries and cell types to reach any universal conclusion.

2.3.2.3 Effect on Cell Proliferation

Surface topography not just influences short timescale cellular behavior such as cell adhesion and spreading but also long timescale processes such as proliferation and differentiation (Anselme et al. 2010). It is reported by many studies that cells show enhanced proliferation on patterned surfaces such as grooves (Lu et al. 2008), nanopillars (Abdul Kafi et al. 2012), and nanofibers (Golizadeh et al. 2019) as compared to that of flat surface (Lu et al. 2008; Ermis et al. 2018). In one of such studies, mouse MSCs cultured on microgrooved surface showed 2–3 fold high proliferation with enhanced expression of cell surface markers CD29, CD44, and Sca-1 and pluripotency associated markers such as Oct3/4 and Nanog as compared to cells cultured on standard culture flask (Chaudhary and Rath 2017). In another study, growth kinetics of human corneal stromal cells was studied over 21 days, where patterned surface showed significantly more number of cells on day 7, 14, and 21 than control indicating the enhanced proliferation (Bhattacharjee et al. 2020). On the other hand, the topography also has been shown to reduce the cellular proliferation or have no effect on proliferation (Braber et al. 1996). For instance, no significant changes in proliferation rate were observed for endothelial cells such as HmVEC-d, HAEC, and HSaVEC whereas HUVEC showed 2-fold decrease in proliferation when cultured on groove and ridge nanotopographies (Liliensiek et al. 2010). This emphasized the key role of feature scale and its cell specific response in controlling the cellular behavior.

2.3.2.4 Effect on Cell Migration

Cell migration is a key process involved in embryonic development, wound healing, angiogenesis, and immune response (Trepat et al. 2012). Many physical factors such as surface stiffness, topography, electric field, and surface charges have been shown to impact the cellular migration (Cui et al. 2021). Among all, surface topography has been demonstrated to be an important factor as cells prefer specific topographical features for migration, the way ECM guides cell migration in *in vivo* condition (Ermis et al. 2018). Harrison was first to demonstrate the influence of the substratum on cell migration in 1911 when he cultured cells on a spider web and found that the embryonic cells followed the fibers of the web. This phenomenon was called stereotropism or physical guidance (Harrison 1911). Once adhered to the substrate, the cells move on the surface either to follow underlying microstructure, where it will first get polarized (i.e., wide front end and narrow rear) and extend its membrane in direction of the pattern leading to its migration (Anselme et al. 2010). Critical parameters for regulating the cellular migration include width, elevation, depth, bending angle, and curvature, which affect cell speed and directionality (Refaaq et al. 2020; Wang et al. 2020). Many studies have reported topographical features in micro- and nano-range, however grating and groove-like structures are predominantly reported for directed cell migration. For example, Polycaprolactone (PCL) microgrooves for MG63 and hMSCs migration (Zhang et al. 2015), grooves and elevations on Ti alloy for fibroblast migration (Kaiser et al. 2006), and PDMS microgrooves for vascular smooth muscle cells (VSMCs) and cervical cancer HeLa cells (Nagayama and Hanzawa 2022). Reports have also shown migration of different cell lines on wavelike pattern (Ge et al. 2020), quasi 3D pillars (Gorelashvili et al. 2014), micropits (Wang et al. 2020), micropillars (Krishnamoorthy et al. 2020), microposts (Hui and Pang 2019), nanoholes, and nanopillars (Cheng et al. 2021). The effect of local geometrical and mechanical anisotrophical cues in cell environment for directed cell migration is known as Ratchetaxis (Caballero et al. 2015). The overall effect of these micro and nanostructures majorly depends on shape and dimension of the structures, their arrangement, periodicity, or discontinuity which then controls the migration. Thus, understanding the effect of these micro/nanostructures on cell migration will be helpful in elucidating the mechanisms involved and its further implications in tissue engineering and regenerative medicine.

2.3.2.5 Effect on Cellular Differentiation

The self-renewal and differentiation potential of embryonic and adult stem cells have expanded their application in regenerative therapies and tissue repairs. Despite these properties, the inability to precisely control stem cell fate has been a major limiting factor in exploiting the therapeutic capacities of these cells. As discussed earlier in this chapter, this issue is mainly due to lack of *in vivo*-like microenvironment which encompasses of highly dynamic and intricate biochemical and biophysical cues to regulate stem cell differentiation (Nikkhah et al. 2012). Differentiation is a multistep process which can be triggered by chemical, architectural, and physical cues. Surface topography has been well reported as a rapid and inexpensive approach to better recapitulate the *in vivo* microenvironment and guide stem cell differentiation. It was observed that changes in topographical cues can push cells to lineage-specific differentiation. For example, in hMSCs topographical cues such as 15 nm pillars (Ermis et al. 2018) and 2 μm ridges (Abagnale et al. 2015) can upregulate osteogenic differentiation, 73 μm separation grooves (Ermis et al. 2018), and 15 μm ridges (Abagnale et al. 2015) gives adipogenic differentiation whereas nanogratings of 250 nm linewidth showed differentiation into neurogenic and myogenic lineage (Teo et al. 2013). The cellular processes that mainly direct topography-mediated cellular differentiation include integrin activated Focal Adhesion Kinase (FAK) pathway (Teo et al. 2013), RhoA/ROCK pathway (Seo et al. 2011), and Wnt/β-catenin signaling pathway (Rahman et al. 2018). Though micro/nanotopographical cues provide precise control of physical stimuli for cellular differentiation, it is still not clear which topographical parameters are modulating the cell response and thus need thorough investigation.

2.3.2.6 Effect on Cellular Secretome

Topographical features have been shown to influence immunomodulatory and anti-inflammatory properties of stem cells (e.g., hMSCs) which mainly attribute toward the secretome of the cells. Secretome is composed of all the bioactive factors secreted by cells such as cytokines, chemokines, growth factors, anti-inflammatory proteins, and extracellular vesicles (Makridakis et al. 2013). It is secreted in response to surrounding environment and has potential to stimulate the tissue repair without using the cells. The composition of secretome is complex and influenced by many chemical and physical factors. It varies based on type of stem cells, culture conditions, incubation time, and phase of cell growth. Though its therapeutic potential can be regulated by micro and nanostructures, there are other factors such as hypoxia conditioning, addition of biochemical stimuli, fluid shear, compression forces, and substrate stiffness which influence the secretome composition (Phelps et al. 2018).

Secretome of MSCs has garnered a lot of attention in last few years due to its advantages over cell-based therapies. It gives a cell-free approach that enhances the clinical efficiency which is limited in cell-based therapy due to: 1) High cost; 2) Failure of cell engraftment and differentiation after transplantation; 3) Retention of less cells at damaged site; 4) Maldifferentiation; and 5) Entrapment of MSCs in microvasculature (Eleuteri and Fierabracci 2019).

Secretome has large number of applications in treatment of skin wounds (e.g., burn injuries, chronic cutaneous ulcerations, and diabetic foot ulcers) (Kim et al. 2022), autoimmune diseases such as Systemic Lupus Erythematosus, Multiple Sclerosis, Graft-Versus-Host Disease, (Műzes and Sipos 2022), Covid-19 (Bari et al. 2020), and cancer (Phelps et al. 2018). A novel paradigm for cell-free therapeutics in regenerative medicine may thus be enabled by topography-induced secretome release. However, in depth study of how topography can enhance production of particular bioactive factor needs to be done.

2.4 FUTURE SCOPE

Majority of biological interfaces and tissues have structures ranging from nano to micrometer scale which play important roles in defining the structure and function of the biological systems. Understanding this, multiple studies have adopted the strategy of mimicking natural biointerfaces and tissues by fabricating more complex and hierarchical topographical features which can be used for wide range of applications such as cell-based therapies.

Stem cell-based therapies are considered as one of the promising disciplines with unprecedented clinical applications such as tissue engineering and regenerative medicine. To increase the effectiveness of stem cell-based therapies and make them applicable for a variety of health conditions, a significant number of stem cells must be made available. The development of topographical features in the micro and nanoscale, simulating the natural microenvironment, has been shown to be a potent tool for this aim providing specialized culture system that mimics the required microenvironment for long-term culture of cells. This approach can also be used to precisely control cell adhesion, morphology, proliferation, migration, and differentiation.

Another challenge restricting the cell-based therapies is the extensive use of costly media and bioactive factors for growth and differentiation of cells. The topographical cues have also been shown to induce cellular differentiation in the absence of the induction factor which will be helpful in reducing the use of expensive biochemical factors. Though a significant number of studies have been reported in the last few years for understanding the topography-induced cellular response and controlling the cell behavior through cell-substrate interactions, the molecular mechanism for the same is still not fully understood.

Previously numerous research has reported simple topological features such as grooves, grating, and pits to examine cellular behavior. However, intricate 2D and 3D structures should be investigated in the near future, as they will offer a more comprehensive technique to imitate the *in vivo* microenvironment with physiological relevance. It will have substantial effects on creating synthetic and implantable substrates with regulated properties for regenerative medicine and tissue engineering, in addition to advancing basic biological studies.

Furthermore, the use of micro and nanofabrication technologies can be extended beyond tissue engineering and regenerative medicine to develop *in vitro* tissue models for drug screening and toxicity testing. The topographical cues can also affect the unique immunomodulatory properties of stem cells, modulating their secretome composition and properties. This will open new doors for cell-free therapies for the treatment of diseases where poor cell homing and retention after transplantation present significant difficulties.

To summarize, micro and nanostructure-based system is a promising tool to control cell behavior and expand their applications in tissue engineering and regenerative medicine.

2.5 CONCLUSION

Micro and nanofabrication technologies have been used to address the current challenges in tissue engineering and regenerative medicine (TERM) such as limitation in mimicking the *in vivo* microenvironment, obtaining large number cells with high self-renewable and differentiation potential, and precisely controlling the cellular processes. These techniques have been widely used in the development of substrates and scaffolds which comprise precise topographical features in micro and nanoscale to meet the desired criteria and complexity of the native tissue architecture. Number of fabrication techniques such as photolithography, E-beam lithography, X-ray lithography, two photon lithography, and many others, are used to create such high resolution, reproducible structures with large surface area, and nano/microscale features that mimic the required microenvironment for growing mammalian cells, such as stem cells. In this chapter, we discussed how these topographical features influence vital cellular processes such as adhesion, migration, proliferation, and differentiation. Though these effects vary depending on different cell types, substrate material, and structure of topography, they have been found to have an impact on cellular behavior significantly. This information is helpful in designing suitable materials and platforms for regulating cellular response which can be expanded in TERM applications. In conclusion, micro and nanostructure-based system is a versatile approach for regulating cell behavior and extending the range of their uses in TERM.

REFERENCES

Abagnale, Giulio, Michael Steger, Vu Hoa Nguyen, Nils Hersch, Antonio Sechi, Sylvia Joussen, Bernd Denecke, et al. 2015. "Surface Topography Enhances Differentiation of Mesenchymal Stem Cells towards Osteogenic and Adipogenic Lineages." *Biomaterials* 61: 316–326. doi:10.1016/j.biomaterials.2015.05.030.

Abdul Kafi, Md, Waleed Ahmed El-Said, Tae Hyung Kim, and Jeong Woo Choi. 2012. "Cell Adhesion, Spreading, and Proliferation on Surface Functionalized with RGD Nanopillar Arrays." *Biomaterials* 33 (3): 731–739. doi:10.1016/j.biomaterials.2011.10.003.

Abrams, G. A., S. L. Goodman, P. F. Nealey, M. Franco, and C. J. Murphy. 2000. "Nanoscale Topography of the Basement Membrane Underlying the Corneal Epithelium of the Rhesus Macaque." *Cell and Tissue Research* 299 (1): 39–46. doi:10.1007/s004419900074.

Accardo, Angelo, Marie Charline Blatché, Rémi Courson, Isabelle Loubinoux, Christophe Vieu, and Laurent Malaquin. 2018. "Two-Photon Lithography and Microscopy of 3D Hydrogel Scaffolds for Neuronal Cell Growth." *Biomedical Physics and Engineering Express* 4 (2): 27009. doi:10.1088/2057-1976/aaab93.

Agarwal, Ashutosh, Yohan Farouz, Alexander Peyton Nesmith, Leila F. Deravi, Laura Mccain, and Kevin Kit Parker. 2015. "Muscle on a Chip **" 23 (30): 3738–3746. doi:10.1002/adfm.201203319. Micropatterning.

Altissimo, Matteo. 2010. "E-Beam Lithography for Micro-/Nanofabrication." *Biomicrofluidics* 4 (2). doi:10.1063/1.3437589.

Altomare, L., N. Gadegaard, L. Visai, M. C. Tanzi, and S. Farè. 2010. "Biodegradable Microgrooved Polymeric Surfaces Obtained by Photolithography for Skeletal Muscle Cell Orientation and Myotube Development." *Acta Biomaterialia* 6 (6): 1948–1957. doi:10.1016/j.actbio.2009.12.040.

Aly, Riham Mohamed. 2020. "Current State of Stem Cell-Based Therapies: An Overview." *Stem Cell Investigation* 7: 1–10. doi:10.21037/sci-2020-001.

Anderson, Hilary J., Jugal Kishore Sahoo, Rein V. Ulijn, and Matthew J. Dalby. 2016. "Mesenchymal Stem Cell Fate: Applying Biomaterials for Control of Stem Cell Behavior." *Frontiers in Bioengineering and Biotechnology* 4: 1–14. doi:10.3389/fbioe.2016.00038.

Andersson, Ann Sofie, Petra Olsson, Ulf Lidberg, and Duncan Sutherland. 2003. "The Effects of Continuous and Discontinuous Groove Edges on Cell Shape and Alignment." *Experimental Cell Research* 288 (1): 177–188. doi:10.1016/S0014-4827(03)00159-9.

Angelova, Liliya, Albena Daskalova, Emil Filipov, Xavier Monforte Vila, Janine Tomasch, Georgi Avdeev, Andreas H. Teuschl-Woller, and Ivan Buchvarov. 2022. "Optimizing the Surface Structural and Morphological Properties of Silk Thin Films via Ultra-Short Laser Texturing for Creation of Muscle Cell Matrix Model." *Polymers* 14 (13). doi:10.3390/polym14132584.

Annabi, Nasim, Kelly Tsang, Suzanne M. Mithieux, Mehdi Nikkhah, Afshin Ameri, Ali Khademhosseini, and Anthony S. Weiss. 2013. "Highly Elastic Micropatterned Hydrogel for Engineering Functional Cardiac Tissue." *Advanced Functional Materials* 23 (39): 4950–4959. doi:10.1002/adfm.201300570.

Anselme, Karine, Lydie Ploux, and Arnaud Ponche. 2010. "Cell/Material Interfaces: Influence of Surface Chemistry and Surface Topography on Cell Adhesion." *Journal of Adhesion Science and Technology* 24 (5): 831–852. doi:10.1163/016942409x12598231568186.

Baek, Dahee, Sang Hun Lee, Bong Hyun Jun, and Seung Hwan Lee. 2021. "Lithography Technology for Micro- and Nanofabrication." *Advances in Experimental Medicine and Biology* 1309: 217–233. doi:10.1007/978-981-33-6158-4_9.

Bai, Renu Geetha, Kasturi Muthoosamy, Sivakumar Manickam, and Ali Hilal-Alnaqbi. 2019. "Graphene-Based 3D Scaffolds in Tissue Engineering: Fabrication, Applications, and Future Scope in Liver Tissue Engineering." *International Journal of Nanomedicine* 14: 5753–5783. doi:10.2147/IJN.S192779.

Bain, Colin, and George M. Whitesides. 1989. "Modeling Organic Surfaces with Self Assembled Monolayers." *Advanced Materials* 28 (4): 506–512.

Barcelo, Steven, and Zhiyong Li. 2016. "Nanoimprint Lithography for Nanodevice Fabrication." *Nano Convergence* 3 (1). doi:10.1186/s40580-016-0081-y.

Bari, Elia, Ilaria Ferrarotti, Laura Saracino, Sara Perteghella, Maria Luisa Torre, and Angelo Guido Corsico. 2020. "Mesenchymal Stromal Cell Secretome for Severe COVID-19 Infections: Premises for the Therapeutic Use." *Cells* 9 (4): 5–9. doi:10.3390/cells9040924.

Betancourt, Tania, and Lisa Brannon-Peppas. 2006. "Micro-and Nanofabrication Methods in Nanotechnological Medical and Pharmaceutical Devices." *International Journal of Nanomedicine* 1 (4): 483–495. doi:10.2147/nano.2006.1.4.483.

Bhansali, Shobhit, Pinaki Dutta, Vinod Kumar, Mukesh Kumar Yadav, Ashish Jain, Sunder Mudaliar, Shipra Bhansali, et al. 2017. "Efficacy of Autologous Bone Marrow-Derived Mesenchymal Stem Cell and Mononuclear Cell Transplantation in Type 2 Diabetes Mellitus: A Randomized, Placebo-Controlled Comparative Study." *Stem Cells and Development* 26 (7): 471–481. doi:10.1089/scd.2016.0275.

Bharti, Amardeep, Alessio Turchet, and Benedetta Marmiroli. 2022. "X-Ray Lithography for Nanofabrication: Is There a Future?" *Frontiers in Nanotechnology* 4: 1–8. doi:10.3389/fnano.2022.835701.

Bhattacharjee, Promita, Brenton L. Cavanagh, and Mark Ahearne. 2020. "Effect of Substrate Topography on the Regulation of Human Corneal Stromal Cells." *Colloids and Surfaces B: Biointerfaces* 190: 110971. doi:10.1016/j.colsurfb.2020.110971.

Biggs, M. J. P., R. G. Richards, S. McFarlane, C. D. W. Wilkinson, R. O. C. Oreffo, and M. J. Dalby. 2008. "Adhesion Formation of Primary Human Osteoblasts and the Functional Response of Mesenchymal Stem Cells to 330 Nm Deep Microgrooves." *Journal of the Royal Society Interface* 5 (27): 1231–1242. doi:10.1098/rsif.2008.0035.

Biggs, Manus J. P., R. Geoff Richards, Nikolaj Gadegaard, Chris D. W. Wilkinson, Richard O. C. Oreffo, and Matthew J. Dalby. 2009. "The Use of Nanoscale Topography to Modulate the Dynamics of Adhesion Formation in Primary Osteoblasts and ERK/MAPK Signalling in STRO-1+ Enriched Skeletal Stem Cells." *Biomaterials* 30 (28): 5094–5103. doi:10.1016/j.biomaterials.2009.05.049.

Biswas, Abhijit, Ilker S. Bayer, Alexandru S. Biris, Tao Wang, Enkeleda Dervishi, and Franz Faupel. 2012. "Advances in Top-down and Bottom-up Surface Nanofabrication: Techniques, Applications & Future Prospects." *Advances in Colloid and Interface Science* 170 (1–2): 2–27. doi:10.1016/j.cis.2011.11.001.

Blanpain, Cédric, and Elaine Fuchs. 2006. "Epidermal Stem Cells of the Skin." *Annual Review of Cell and Developmental Biology* 22: 339–373. doi:10.1146/annurev.cellbio.22.010305.104357.

Braber, E. T. Den, J. E. De Ruijter, H. T. J. Smits, L. A. Ginsel, A. F. Von Recum, and J. A. Jansen. 1996. "Quantitative Analysis of Cell Proliferation and Orientation on Substrata with Uniform Parallel Surface Micro-Grooves." *Biomaterials* 17 (11): 1093–1099. doi:10.1016/0142-9612(96)85910-2.

Britland, Stephen, Cameron Perridge, Morgan Denyer, Hywell Morgan, and Chris Wilkinson. 1996. "Morphogenetic Guidance Cues Can Interact Synergistically and Hierarchically in Steering Nerve Cell Growth." *Experimental Biology Online - EBO* 1(2). doi:10.1101/pdb.caut639r.

Caballero, David, Jordi Comelles, Matthieu Piel, Raphaël Voituriez, and Daniel Riveline. 2015. "Ratchetaxis: Long-Range Directed Cell Migration by Local Cues." *Trends in Cell Biology* 25 (12): 815–827. doi:10.1016/j.tcb.2015.10.009.

Caron, Ilaria, Filippo Rossi, Simonetta Papa, Rossella Aloe, Marika Sculco, Emanuele Mauri, Alessandro Sacchetti, et al. 2016. "A New Three Dimensional Biomimetic Hydrogel to Deliver Factors Secreted by Human Mesenchymal Stem Cells in Spinal Cord Injury." *Biomaterials* 75: 135–147. doi:10.1016/j.biomaterials.2015.10.024.

Chaudhary, Jitendra Kumar, and Pramod C. Rath. 2017. "Microgrooved-Surface Topography Enhances Cellular Division and Proliferation of Mouse Bone Marrow-Derived Mesenchymal Stem Cells." *PLoS One* 12 (8): 1–22. doi:10.1371/journal.pone.0182128.

Chaudhary, Raghvendra P., Arun Jaiswal, Govind Ummethala, Suyog R. Hawal, Sumit Saxena, and Shobha Shukla. 2017. "Sub-Wavelength Lithography of Complex 2D and 3D Nanostructures without Two-Photon Dyes." *Additive Manufacturing* 16: 30–34. doi:10.1016/j.addma.2017.05.003.

Chen, Yen-Shun. 2016. "Mesenchymal Stem Cell: Considerations for Manufacturing and Clinical Trials on Cell Therapy Product." *International Journal of Stem Cell Research & Therapy* 3 (1): 1–12. doi:10.23937/2469-570x/1410029.

Chen, Yong, and Anne Pépin. 2001. "Nanofabrication: Conventional and Nonconventional Methods." *Electrophoresis* 22 (2): 187–207. doi:10.1002/1522-2683(200101)22:2<187::AID-ELPS187>3.0.CO;2-0.

Chen, Ke, Weimin Chou, Lichao Liu, Yonghui Cui, Ping Xue, and Mingyin Jia. 2019. "Electrochemical Sensors Fabricated by Electrospinning Technology: An Overview." *Sensors (Switzerland)* 19 (17). doi:10.3390/s19173676.

Cheng, Yijun, Shuyan Zhu, and Stella W. Pang. 2021. "Directing Osteoblastic Cell Migration on Arrays of Nanopillars and Nanoholes with Different Aspect Ratios." *Lab on a Chip* 21 (11): 2206–2216. doi:10.1039/d1lc00104c.

Chi, Heng, Yunqian Guan, Fengyan Li, and Zhiguo Chen. 2019. "The Effect of Human Umbilical Cord Mesenchymal Stromal Cells in Protection of Dopaminergic Neurons from Apoptosis by Reducing Oxidative Stress in the Early Stage of a 6-OHDA-Induced Parkinson's Disease Model." *Cell Transplantation* 28 (1_suppl): 87S–99S. doi:10.1177/0963689719891134.

Chou, S., P. R., Krauss, and Preston J. Renstrom. 1996. "Nanoimprint Lithography." *Journal of Vacuum Science & Technology B* 14 (6): 4129–4133. doi:10.1116/1.588605.

Chou, Chih Ling, Alexander L. Rivera, Takao Sakai, Arnold I. Caplan, Victor M. Goldberg, Jean F. Welter, and Harihara Baskaran. 2013. "Micrometer Scale Guidance of Mesenchymal Stem Cells to Form Structurally Oriented Cartilage Extracellular Matrix." *Tissue Engineering – Part A* 19 (9–10): 1081–1090. doi:10.1089/ten.tea.2012.0177.

Chowdhury, Farhan, Yanzhen Li, Yeh Chuin Poh, Tamaki Yokohama-Tamaki, Ning Wang, and Tetsuya S. Tanaka. 2010. "Soft Substrates Promote Homogeneous Self-Renewal of Embryonic Stem Cells via Downregulating Cell-Matrix Tractions." *PLoS One* 5 (12). doi:10.1371/journal.pone.0015655.

Clark, P., P. Connolly, A. S. G. Curtis, J. A. T. Dow, and C. D. W. Wilkinson. 1990. "Topographical Control of Cell Behaviour: II. Multiple Grooved Substrata." *Development* 108 (4): 635–644. papers3://publication/uuid/5B9FCDEA-F7EF-405C-8679-F2794EFBE4BF.

Colombo, Federica, Gianluca Sampogna, Giovanni Cocozza, SalmanYousuf Guraya, and Antonello Forgione. 2017. "Regenerative Medicine: Clinical Applications and Future Perspectives." *Journal of Microscopy and Ultrastructure* 5(1): 1. doi:10.1016/j.jmau.2016.05.002.

Córdoba, Rosa, Pablo Orús, Stefan Strohauer, Teobaldo E. Torres, and José María De Teresa. 2019. "Ultra-Fast Direct Growth of Metallic Micro- and Nano-Structures by Focused Ion Beam Irradiation." *Scientific Reports* 9 (1): 1–10. doi:10.1038/s41598-019-50411-w.

Coyle, Stephen, Bryant Doss, Yucheng Huo, Hemang Raj Singh, David Quinn, K. Jimmy Hsia, and Philip R. LeDuc. 2022. "Cell Alignment Modulated by Surface Nano-Topography – Roles of Cell-Matrix and Cell-Cell Interactions." *Acta Biomaterialia* 142: 149–159. doi:10.1016/j.actbio.2022.01.057.

Croisier, Florence, and Christine Jérôme. 2013. "Chitosan-Based Biomaterials for Tissue Engineering." *European Polymer Journal* 49 (4): 780–792. doi:10.1016/j.eurpolymj.2012.12.009.

Cui, Yang, Ying Yang, and Dong Qiu. 2021. "Design of Selective Cell Migration Biomaterials and Their Applications for Tissue Regeneration." *Journal of Materials Science* 56 (6): 4080–4096. doi:10.1007/s10853-020-05537-y.

Cui, Congjing, Shibin Sun, Shaohua Wu, Shaojuan Chen, Jianwei Ma, and Fang Zhou. 2021. "Electrospun Chitosan Nanofibers for Wound Healing Application." *Engineered Regeneration* 2: 82–90. doi:10.1016/j.engreg.2021.08.001.

Cun, Xingli, and Leticia Hosta-Rigau. 2020. "Topography: A Biophysical Approach to Direct the Fate of Mesenchymal Stem Cells in Tissue Engineering Applications." *Nanomaterials* 10 (10): 1–41. doi:10.3390/nano10102070.

Curran, Judith M., Robert Stokes, Eleanore Irvine, Duncan Graham, N. A. Amro, R. G. Sanedrin, H. Jamil, and John A. Hunt. 2010. "Introducing Dip Pen Nanolithography as a Tool for Controlling Stem Cell Behaviour: Unlocking the Potential of the next Generation of Smart Materials in Regenerative Medicine." *Lab on a Chip* 10 (13): 1662–1670. doi:10.1039/c004149a.

Curtis, A. S. G., and Malini Varde. 1964. "Control of Cell Behavior: Topological Factors." *Journal of the National Cancer Institute* 33 (1): 15–26. doi:10.1093/jnci/33.1.15.

Curtis, Adam, and Chris Wilkinson. 1997. "Topographical Control of Cells." *Biomaterials* 18 (24): 1573–1583. doi:10.1016/S0142-9612(97)00144-0.

Dalby, Matthew J., Nikolaj Gadegaard, Mathis O. Riehle, Chris D. W. Wilkinson, and Adam S. G. Curtis. 2004. "Investigating Filopodia Sensing Using Arrays of Defined Nano-Pits down to 35 Nm Diameter in Size." *International Journal of Biochemistry and Cell Biology* 36 (10): 2005–2015. doi:10.1016/j.biocel.2004.03.001.

Deinsberger, Julia, David Reisinger, and Benedikt Weber. 2020. "Global Trends in Clinical Trials Involving Pluripotent Stem Cells: A Systematic Multi-Database Analysis." *Npj Regenerative Medicine* 5 (1): 1–13. doi:10.1038/s41536-020-00100-4.

Dhandayuthapani, Brahatheeswaran, Yasuhiko Yoshida, Toru Maekawa, and D. Sakthi Kumar. 2011. "Polymeric Scaffolds in Tissue Engineering Application: A Review." *International Journal of Polymer Science* 2011 (ii). doi:10.1155/2011/290602.

Doyle, Andrew D., and Kenneth M. Yamada. 2016. "Mechanosensing via Cell-Matrix Adhesions in 3D Microenvironments." *Experimental Cell Research* 343(1): 60–66. doi:10.1016/j.yexcr.2015.10.033.

Duc Nguyen, Ho Hoai, Uwe Hollenbach, Ute Ostrzinski, Karl Pfeiffer, Stefan Hengsbach, and Juergen Mohr. 2016. "Freeform Three-Dimensional Embedded Polymer Waveguides Enabled by External-Diffusion Assisted Two-Photon Lithography." *Applied Optics* 55 (8): 1906. doi:10.1364/ao.55.001906.

Dufort, Christopher C., Matthew J. Paszek, and Valerie M. Weaver. 2011. "Balancing Forces: Architectural Control of Mechanotransduction." *Nature Reviews Molecular Cell Biology* 12 (5): 308–319. doi:10.1038/nrm3112.

El-Badri, Nagwa, and Mohamed A. Ghoneim. 2013. "Mesenchymal Stem Cell Therapy in Diabetes Mellitus: Progress and Challenges." *Journal of Nucleic Acids* 2013. doi:10.1155/2013/194858.

Eleuteri, Sharon, and Alessandra Fierabracci. 2019. "Insights into the Secretome of Mesenchymal Stem Cells and Its Potential Applications." *International Journal of Molecular Sciences* 20 (18). doi:10.3390/ijms20184597.

Eltom, Abdalla, Gaoyan Zhong, and Ameen Muhammad. 2019. "Scaffold Techniques and Designs in Tissue Engineering Functions and Purposes: A Review." *Advances in Materials Science and Engineering* 2019. doi:10.1155/2019/3429527.

Engel, Elisabeth, Alexandra Michiardi, Melba Navarro, Damien Lacroix, and Josep A. Planell. 2008. "Nanotechnology in Regenerative Medicine: The Materials Side." *Trends in Biotechnology* 26 (1): 39–47. doi:10.1016/j.tibtech.2007.10.005.

Ermis, Menekse, Ezgi Antmen, and Vasif Hasirci. 2018. "Micro and Nanofabrication Methods to Control Cell-Substrate Interactions and Cell Behavior: A Review from the Tissue Engineering Perspective." *Bioactive Materials* 3 (3): 355–369. doi:10.1016/j.bioactmat.2018.05.005.

Fei, Jie, and Ran Liu. 2016. "Drug-Laden 3D Biodegradable Label Using QR Code for Anti-Counterfeiting of Drugs." *Materials Science and Engineering C* 63: 657–662. doi:10.1016/j.msec.2016.03.004.

Ferrari, Aldo, Paolo Faraci, Marco Cecchini, and Fabio Beltram. 2010. "The Effect of Alternative Neuronal Differentiation Pathways on PC12 Cell Adhesion and Neurite Alignment to Nanogratings." *Biomaterials* 31 (9): 2565–2573. doi:10.1016/j.biomaterials.2009.12.010.

Gangaraju, Vamsi K., and Haifan Lin.2009. "MicroRNAs: key regulators of stem cells". *Nature Reviews Molecular Cell Biology*, 10 (2): 116–125. doi:10.1038/nrm2621.

Ge, Lu, Liangliang Yang, Reinier Bron, Janette K. Burgess, and Patrick Van Rijn. 2020. "Topography-Mediated Fibroblast Cell Migration Is Influenced by Direction, Wavelength, and Amplitude." *ACS Applied Bio Materials* 3 (4): 2104–2116. doi:10.1021/acsabm.0c00001.

Ghasemi-Mobarakeh, Laleh. 2015. "Structural Properties of Scaffolds: Crucial Parameters towards Stem Cells Differentiation." *World Journal of Stem Cells* 7 (4): 728. doi:10.4252/wjsc.v7.i4.728.

Ghibaudo, Marion, Léa Trichet, Jimmy Le Digabel, Alain Richert, Pascal Hersen, and Benoît Ladoux. 2009. "Substrate Topography Induces a Crossover from 2D to 3D Behavior in Fibroblast Migration." *Biophysical Journal* 97 (1): 357–368. doi:10.1016/j.bpj.2009.04.024.

Golizadeh, Mortaza, Afzal Karimi, Soheila Gandomi-Ravandi, Manouchehr Vossoughi, Mona Khafaji, Mohammad Taghi Joghataei, and Faezeh Faghihi. 2019. "Evaluation of Cellular Attachment and Proliferation on Different Surface Charged Functional Cellulose Electrospun Nanofibers." *Carbohydrate Polymers* 207: 796–805. doi:10.1016/j.carbpol.2018.12.028.

Gonzalez, Manuel A., and Antonio Bernad. 2012. "Characteristics of Adult Stem Cells." *Advances in Experimental Medicine and Biology* 741: 103–120. doi:10.1007/978-1-4614-2098-9_8.

Gorelashvili, Mari, Martin Emmert, Kai F. Hodeck, and Doris Heinrich. 2014. "Amoeboid Migration Mode Adaption in Quasi-3D Spatial Density Gradients of Varying Lattice Geometry." *New Journal of Physics* 16. doi:10.1088/1367-2630/16/7/075012.

Gottipamula, Sanjay, Samatha Bhat, Uday kumar, and Raviraja N. Seetharam. 2018. "Mesenchymal Stromal Cells: Basics, Classification, and Clinical Applications." *Journal of Stem Cells* 13 (1): 24–47.

Guan, Jingjiao, Nicholas Ferrell, L. James Lee, and Derek J. Hansford. 2006. "Fabrication of Polymeric Microparticles for Drug Delivery by Soft Lithography." *Biomaterials* 27 (21): 4034–4041. doi:10.1016/j.biomaterials.2006.03.011.

Guo, Jun, Guo Sheng Lin, Cui Yu Bao, Zhi Min Hu, and Ming Yan Hu. 2007. "Anti-Inflammation Role for Mesenchymal Stem Cells Transplantation in Myocardial Infarction." *Inflammation* 30 (3–4): 97–104. doi:10.1007/s10753-007-9025-3.

Guo, Yajun, Yunsheng Yu, Shijun Hu, Yueqiu Chen, and Zhenya Shen. 2020. "The Therapeutic Potential of Mesenchymal Stem Cells for Cardiovascular Diseases." *Cell Death and Disease* 11 (5). doi:10.1038/s41419-020-2542-9.

Hamilton, Douglas W., Babak Chehroudi, and Donald M. Brunette. 2007. "Comparative Response of Epithelial Cells and Osteoblasts to Microfabricated Tapered Pit Topographies in Vitro and in Vivo." *Biomaterials* 28 (14): 2281–2293. doi:10.1016/j.biomaterials.2007.01.026.

Handrea-Dragan, Madalina, and Ioan Botiz. 2021. "Multifunctional Structured Platforms: From Patterning of Polymer-Based Films to Their Subsequent Filling with Various Nanomaterials." *Polymers* 13 (3): 1–49. doi:10.3390/polym13030445.

Harrison, Ross G. 1911. "On the Stereotropism of Embryonic Cells". *Science*, 34 (870): 279–281. doi:10.1126/science.34.870.279.

Hart, Andrew, Nikolaj Gadegaard, Chris D. W. Wilkinson, Richard O. C. Oreffo, and Matthew J. Dalby. 2007. "Osteoprogenitor Response to Low-Adhesion Nanotopographies Originally Fabricated by Electron Beam Lithography." *Journal of Materials Science: Materials in Medicine* 18 (6): 1211–1218. doi:10.1007/s10856-007-0157-7.

Hasan, Rashed Md Murad, and Xichun Luo. 2018. "Promising Lithography Techniques for Next-Generation Logic Devices." *Nanomanufacturing and Metrology* 1 (2): 67–81. doi:10.1007/s41871-018-0016-9.

Hasan, Anwarul, Mahboob Morshed, Adnan Memic, Shabir Hassan, Thomas J. Webster, and Hany El Sayed Marei. 2018. "Nanoparticles in Tissue Engineering: Applications, Challenges and Prospects." *International Journal of Nanomedicine* 13: 5637–5655. doi:10.2147/IJN.S153758.

Hasetine, W.1999. "A brave new medicine. A conversation with William Haseltine. Interview by Joe Flower." *Health Forum J*, 42(4):28-30, 61–5A brave new medicine. A conversation with William Haseltine. Interview by Joe Flower., Health Forum J, 42, 28-30, 61-5

Hasturk, Onur, Menekse Ermis, Utkan Demirci, Nesrin Hasirci, and Vasif Hasirci. 2019. "Square Prism Micropillars on Poly(Methyl Methacrylate) Surfaces Modulate the Morphology and Differentiation of Human Dental Pulp Mesenchymal Stem Cells." *Colloids and Surfaces B: Biointerfaces* 178: 44–55. doi:10.1016/j.colsurfb.2019.02.037.

Hernández, A. E., and E. García. 2021. "Mesenchymal Stem Cell Therapy for Alzheimer's Disease." *Stem Cells International* 2021. doi:10.1155/2021/7834421.

Higuchi, Akon, Qing Dong Ling, Shih Tien Hsu, and Akihiro Umezawa. 2012. "Biomimetic Cell Culture Proteins as Extracellular Matrices for Stem Cell Differentiation." *Chemical Reviews* 112 (8): 4507–4540. doi:10.1021/cr3000169.

Higuchi, Akon, Qing Dong Ling, Yung Chang, Shih Tien Hsu, and Akihiro Umezawa. 2013. "Physical Cues of Biomaterials Guide Stem Cell Differentiation Fate." *Chemical Reviews* 113 (5): 3297–3328. doi:10.1021/cr300426x.

Hong, Jiyoung, Miji Yeo, Gi Hoon Yang, and Geunhyung Kim. 2019. "Cell-Electrospinning and Its Application for Tissue Engineering." *International Journal of Molecular Sciences* 20 (24). doi:10.3390/ijms20246208.

Hubbell, Jeffrey A. 1995. "Biomaterials in Tissue Engineering." *Bio/Technology* 13 (6): 565–576. doi:10.1038/nbt0695-565.

Hui, Jianan, and Stella W. Pang. 2019. "Cell Migration on Microposts with Surface Coating and Confinement." *Bioscience Reports* 39 (2): 1–18. doi:10.1042/BSR20181596.

Ikehara, Susumu. 2013. "Grand challenges in stem cell treatments". *Frontiers in Cell and Developmental Biology*, 1: 2. doi:10.3389/fcell.2013.00002.

Ingber, Donald E. 2003. "Mechanobiology and Diseases of Mechanotransduction." *Annals of Medicine* 35 (8): 564–577. doi:10.1080/07853890310016333.

Jaalouk, Diana E., and Jan Lammerding. 2009. "Mechanotransduction Gone Awry." *Nature Reviews Molecular Cell Biology* 10 (1): 63–73. doi:10.1038/nrm2597.

Jaiswal, Arun, Sweta Rani, Gaurav Pratap Singh, Mahbub Hassan, Aklima Nasrin, Vincent G. Gomes, Sumit Saxena, and Shobha Shukla. 2021. "Additive-Free All-Carbon Composite: A Two-Photon Material System for Nanopatterning of Fluorescent Sub-Wavelength Structures." *ACS Nano* 15 (9): 14193–14206. doi:10.1021/acsnano.1c01083.

Janson, Isaac A., and Andrew J. Putnam. 2015. "Extracellular Matrix Elasticity and Topography: Material-Based Cues That Affect Cell Function via Conserved Mechanisms." *Journal of Biomedical Materials Research - Part A* 103 (3): 1246–1258. doi:10.1002/jbm.a.35254.

Jiang, Yuhe, Ping Zhang, Xiao Zhang, Longwei Lv, and Yongsheng Zhou. 2021. "Advances in Mesenchymal Stem Cell Transplantation for the Treatment of Osteoporosis." *Cell Proliferation* 54 (1). doi:10.1111/cpr.12956.

Kadir, Nurul Dinah, Zheng Yang, Afizah Hassan, Vinitha Denslin, and Eng Hin Lee. 2021. "Electrospun Fibers Enhanced the Paracrine Signaling of Mesenchymal Stem Cells for Cartilage Regeneration." *Stem Cell Research and Therapy* 12 (1): 1–17. doi:10.1186/s13287-021-02137-8.

Kaiser, Jean Pierre, Andreas Reinmann, and Arie Bruinink. 2006. "The Effect of Topographic Characteristics on Cell Migration Velocity." *Biomaterials* 27 (30): 5230–5241. doi:10.1016/j.biomaterials.2006.06.002.

Kameya, Yuki. 2017. "Wettability Modification of Polydimethylsiloxane Surface by Fabricating Micropillar and Microhole Arrays." *Materials Letters* 196: 320–323. doi:10.1016/j.matlet.2017.03.103.

Karp, Jeffrey M., Yoon Yeo, Wenlinag Geng, Christopher Cannizarro, Kenny Yan, Daniel S. Kohane, Gordana Vunjak-Novakovic, Robert S. Langer, and Milica Radisic. 2006. "A Photolithographic Method to Create Cellular Micropatterns." *Biomaterials* 27 (27): 4755–4764. doi:10.1016/j.biomaterials.2006.04.028.

Kidambi, Srivatsan, Natasha Udpa, Stacey A. Schroeder, Robert Findlan, Ilsoon Lee, and Christina Chan. 2007. "Cell Adhesion on Polyelectrolyte Multilayer Coated Polydimethylsiloxane Surfaces with Varying Topographies." *Tissue Engineering* 13 (8): 2105–2117. doi:10.1089/ten.2006.0151.

Kim, Deok-ho, Elizabeth A. Lipke, Pilnam Kim, Raymond Cheong, Susan Thompson, and Michael Delannoy. 2009. "Nanoscale Cues Regulate the Structure and Function of Macroscopic Cardiac Tissue Constructs." *PNAS* 105 (2): 565–570. doi:10.1073/pnas.0906504107.

Kim, Hyerim, Chaewon Bae, Yun Min Kook, Won Gun Koh, Kangwon Lee, and Min Hee Park. 2019. "Mesenchymal Stem Cell 3D Encapsulation Technologies for Biomimetic Microenvironment in Tissue Regeneration." *Stem Cell Research and Therapy* 10 (1): 1–14. doi:10.1186/s13287-018-1130-8.

Kim, Ji Hyun, Denethia S. Green, Young Min Ju, Mollie Harrison, J. William Vaughan, Anthony Atala, Sang Jin Lee, John D. Jackson, Cory Nykiforuk, and James J. Yoo. 2022. "Identification and Characterization of Stem Cell Secretome-Based Recombinant Proteins for Wound Healing Applications." *Frontiers in Bioengineering and Biotechnology* 10: 1–17. doi:10.3389/fbioe.2022.954682.

Kiyama, Ryuji, Takayuki Nonoyama, Susumu Wada, Shingo Semba, Nobuto Kitamura, Tasuku Nakajima, Takayuki Kurokawa, Kazunori Yasuda, Shinya Tanaka, and Jian Ping Gong. 2018. "Micro Patterning of Hydroxyapatite by Soft Lithography on Hydrogels for Selective Osteoconduction." *Acta Biomaterialia* 81: 60–69. doi:10.1016/j.actbio.2018.10.002.

Kolios, George, and Yuben Moodley. 2012. "Introduction to Stem Cells and Regenerative Medicine." *Respiration* 85 (1): 3–10. doi:10.1159/000345615.

Krishna, Lekshmi, Kamesh Dhamodaran, Chaitra Jayadev, Kaushik Chatterjee, Rohit Shetty, S. S. Khora, and Debashish Das. 2016. "Nanostructured Scaffold as a Determinant of Stem Cell Fate." *Stem Cell Research and Therapy* 7 (1): 1–12. doi:10.1186/s13287-016-0440-y.

Krishnamoorthy, Srikumar, Zhengyi Zhang, and Changxue Xu. 2020. "Guided Cell Migration on a Graded Micropillar Substrate." *Bio-Design and Manufacturing* 3(1): 60–70. doi:10.1007/s42242-020-00059-7.

Kumar, Sanjay, Pulak Bhushan, and Shantanu Bhattacharya. 2018. *Fabrication of Nanostructures with Bottom-up Approach and Their Utility in Diagnostics, Therapeutics, and Others. Energy, Environment, and Sustainability*. doi:10.1007/978-981-10-7751-7_8.

Lanno, Georg Marten, Celia Ramos, Liis Preem, Marta Putrins, Ivo Laidmaë, Tanel Tenson, and Karin Kogermann. 2020. "Antibacterial Porous Electrospun Fibers as Skin Scaffolds for Wound Healing Applications." *ACS Omega* 5 (46): 30011–30022. doi:10.1021/acsomega.0c04402.

Last, Julie A., Sara J. Liliensiek, Paul F. Nealey, and Christopher J. Murphy. 2009. "Determining the Mechanical Properties of Human Corneal Basement Membranes with Atomic Force Microscopy." *Journal of Structural Biology* 167 (1): 19–24. doi:10.1016/j.jsb.2009.03.012.

Lavery, Lawrence A., James Fulmer, Karry Ann Shebetka, Matthew Regulski, Dean Vayser, David Fried, Howard Kashefsky, et al. 2014. "The Efficacy and Safety of Grafix® for the Treatment of Chronic Diabetic Foot Ulcers: Results of a Multi-Centre, Controlled, Randomised, Blinded, Clinical Trial." *International Wound Journal* 11 (5): 554–560. doi:10.1111/iwj.12329.

Lemma, Solomon Mengistu, Frédéric Bossard, and Marguerite Rinaudo. 2016. "Preparation of Pure and Stable Chitosan Nanofibers by Electrospinning in the Presence of Poly(Ethylene Oxide)." *International Journal of Molecular Sciences* 17 (11). doi:10.3390/ijms17111790.

Lemma, Enrico Domenico, Barbara Spagnolo, Massimo De Vittorio, and Ferruccio Pisanello. 2019. "Studying Cell Mechanobiology in 3D: The Two-Photon Lithography Approach." *Trends in Biotechnology* 37 (4): 358–372. doi:10.1016/j.tibtech.2018.09.008.

Lemmon, Mark A., and Joseph Schlessinger. 2011. "Cell Signaling by Receptor-Tyrosine Kinases" 141 (7): 1117–1134. doi:10.1016/j.cell.2010.06.011.Cell.

Li, Jianming, Helen McNally, and Riyi Shi. 2008. "Enhanced Neurite Alignment on Micro-Patterned Poly-L-Lactic Acid Films." *Journal of Biomedical Materials Research - Part A* 87 (2): 392–404. doi:10.1002/jbm.a.31814.

Li, Jipeng, Mingjiao Chen, Xianqun Fan, and Huifang Zhou. 2016. "Recent Advances in Bioprinting Techniques: Approaches, Applications and Future Prospects." *Journal of Translational Medicine* 14(1): 1–15. doi:10.1186/s12967-016-1028-0.

Li, Ching Wen, Brett Davis, Jill Shea, Himanshu Sant, Bruce Kent Gale, and Jayant Agarwal. 2018. "Optimization of Micropatterned Poly(Lactic-Coglycolic Acid) Films for Enhancing Dorsal Root Ganglion Cell Orientation and Extension." *Neural Regeneration Research* 13 (1): 105–111. doi:10.4103/1673-5374.224377.

Li, Yuhui, Guoyou Huang, Xiaohui Zhang, Lin Wang, Yanan Du, Tian Jian Lu, and Feng Xu. 2014. "Engineering Cell Alignment in Vitro." *Biotechnology Advances* 32 (2): 347–365. doi:10.1016/j.biotechadv.2013.11.007.

Li, Jianjun, Yufan Liu, Yijie Zhang, Bin Yao, Bin Enhejirigala, Zhao Li, Wei Song, et al. 2021. "Biophysical and Biochemical Cues of Biomaterials Guide Mesenchymal Stem Cell Behaviors." *Frontiers in Cell and Developmental Biology* 9: 1–13. doi:10.3389/fcell.2021.640388.

Li, Xinda, Boxun Liu, Ben Pei, Jianwei Chen, Dezhi Zhou, Jiayi Peng, Xinzhi Zhang, Wang Jia, and Tao Xu. 2020. "Inkjet Bioprinting of Biomaterials." *Chemical Reviews* 120 (19): 10793–10833. doi:10.1021/acs.chemrev.0c00008.

Li, Ping, Siyu Chen, Houfu Dai, Zhengmei Yang, Zhiquan Chen, Yasi Wang, Yiqin Chen, Wenqiang Peng, Wubin Shan, and Huigao Duan. 2021. "Recent Advances in Focused Ion Beam Nanofabrication for Nanostructures and Devices: Fundamentals and Applications." *Nanoscale* 13 (3): 1529–1565. doi:10.1039/d0nr07539f.

Liddle, J. Alexander, and Gregg M. Gallatin. 2016. "Nanomanufacturing: A Perspective." *ACS Nano* 10 (3): 2995–3014. doi:10.1021/acsnano.5b03299.

Liliensiek, Sara J., Joshua A. Wood, Jiang Yong,Robert Auerbach, Paul F. Nealey, and Christopher J. Murphy. 2010. "Modulation of Human Vascular Endothelial Cell Behaviors by Nanotopographic Cues." *Biomaterials* 31 (20): 5418–5426. doi:10.1016/j.biomaterials.2010.03.045.

Liu, Ruili, Qiong Liu, Zhen Pan, Xiangnan Liu, and Jiandong Ding. 2019. "Cell Type and Nuclear Size Dependence of the Nuclear Deformation of Cells on a Micropillar Array." *Langmuir* 35 (23): 7469–7477. doi:10.1021/acs.langmuir.8b02510.

Liu, Xiangnan, Ruili Liu, Bin Cao, Kai Ye, Shiyu Li, Yexin Gu, Zhen Pan, and Jiandong Ding. 2016. "Subcellular Cell Geometry on Micropillars Regulates Stem Cell Differentiation." *Biomaterials* 111: 27–39. doi:10.1016/j.biomaterials.2016.09.023.

Lo, Bernard, and Lindsay Parham. 2009. "Ethical Issues in Stem Cell Research." *Endocrine Reviews* 30 (3): 204–213. doi:10.1210/er.2008-0031.

Loesberg, W. A., J. te Riet, F. C. M. J. M. van Delft, P. Schön, C. G. Figdor, S. Speller, J. J. W. A. van Loon, X. F. Walboomers, and J. A. Jansen. 2007. "The Threshold at Which Substrate Nanogroove Dimensions May Influence Fibroblast Alignment and Adhesion." *Biomaterials* 28 (27): 3944–3951. doi:10.1016/j.biomaterials.2007.05.030.

Lu, Jing, Masaru P. Rao, Noel C. MacDonald, Dongwoo Khang, and Thomas J. Webster. 2008. "Improved Endothelial Cell Adhesion and Proliferation on Patterned Titanium Surfaces with Rationally Designed, Micrometer to Nanometer Features." *Acta Biomaterialia* 4 (1): 192–201. doi:10.1016/j.actbio.2007.07.008.

Luo, Chenchen, Yigui Li, and Sugiyama Susumu. 2012. "Fabrication of High Aspect Ratio Subwavelength Gratings Based on X-Ray Lithography and Electron Beam Lithography." *Optics and Laser Technology* 44 (6): 1649–1653. doi:10.1016/j.optlastec.2011.11.051.

Luo, Chaoyun, Chanchan Xu, Le Lv, Hai Li, Xiaoxi Huang, and Wei Liu. 2020. "Review of Recent Advances in Inorganic Photoresists." *RSC Advances* 10 (14): 8385–8395. doi:10.1039/c9ra08977b.

Ma, Yaqi, Yifan Xia, Jianpeng Liu, Sichao Zhang, Jinhai Shao, Bing Rui Lu, and Yifang Chen. 2016. "Processing Study of SU-8 Pillar Profiles with High Aspect Ratio by Electron-Beam Lithography." *Microelectronic Engineering* 149: 141–144. doi:10.1016/j.mee.2015.10.013.

Madonna, Rosalinda. 2016. "Biology and Function of Stem Cells," 3–7. doi:10.1007/978-3-319-25427-2_1.

Madou, Marc J. 2002. *Fundamentals of microfabrication: the science of miniaturization.* CRC Press.

Mahla, Ranjeet Singh. 2016. "Stem Cells Applications in Regenerative Medicine and Disease Therapeutics." *International Journal of Cell Biology* 2016. Hindawi Publishing Corporation. doi:10.1155/2016/6940283.

Makridakis, Manousos, Maria G. Roubelakis, and Antonia Vlahou. 2013. "Stem Cells: Insights into the Secretome." *Biochimica et Biophysica Acta - Proteins and Proteomics* 1834 (11): 2380–2384. doi:10.1016/j.bbapap.2013.01.032.

Mansouri, Vahid, Nima Beheshtizadeh, Maliheh Gharibshahian, Leila Sabouri, Mohammad Varzandeh, and Nima Rezaei. 2021. "Recent Advances in Regenerative Medicine Strategies for Cancer Treatment." *Biomedicine and Pharmacotherapy* 141: 111875. doi:10.1016/j.biopha.2021.111875.

Mao, Angelo S., and David J. Mooney. 2015. "Regenerative Medicine: Current Therapies and Future Directions." *Proceedings of the National Academy of Sciences of the United States of America* 112 (47): 14452–14459. doi:10.1073/pnas.1508520112.

Martínez-Calderon, M., M. Manso-Silván, A. Rodríguez, M. Gómez-Aranzadi, J. P. García-Ruiz, S. M. Olaizola, and R. J. Martín-Palma. 2016. "Surface Micro- and Nano-Texturing of Stainless Steel by Femtosecond Laser for the Control of Cell Migration." *Scientific Reports* 6: 1–10. doi:10.1038/srep36296.

Matsuzaka, K., X. F. Walboomers, J. E. De Ruijter, and J. A. Jansen. 1999. "The Effect of Poly-L-Lactic Acid with Parallel Surface Micro on Groove on Osteoblast-like Cells in Vitro." *Biomaterials* 20 (14): 1293–1301. doi:10.1016/S0142-9612(99)00029-0.

McKee, Christina, and G. Rasul Chaudhry. 2017. "Advances and challenges in stem cell culture". *Colloids and Surfaces B: Biointerfaces*, 159: 62–77. doi: 10.1016/j.colsurfb.2017.07.051.

McMurray, Rebecca J., Matthew J. Dalby, and P. Monica Tsimbouri. 2015. "Using Biomaterials to Study Stem Cell Mechanotransduction, Growth and Differentiation." *Journal of Tissue Engineering and Regenerative Medicine* 9 (5): 528–539. doi:10.1002/term.1957.

Meer, A. D., Van Der, A. A. Poot, M. H. G. Duits, J. Feijen, and I. Vermes. 2009. "Microfluidic Technology in Vascular Research." *Journal of Biomedicine and Biotechnology* 2009. doi:10.1155/2009/823148.

Méjat, Alexandre, and Tom Misteli. 2010. "LINC Complexes in Health and Disease." *Nucleus* 1 (1): 40–52. doi:10.4161/nucl.1.1.10530.

Moghaddam, Mehrdad M., Shahin Bonakdar, Mona R. Shariatpanahi, Mohammad A. Shokrgozar, and Shahab Faghihi. 2019. "The Effect of Physical Cues on the Stem Cell Differentiation." *Current Stem Cell Research & Therapy* 14 (3): 268–277. doi:10.2174/1574888x14666181227120706.

Murphy, Ciara M., Amos Matsiko, Matthew G. Haugh, John P. Gleeson, and Fergal J. O'Brien. 2012. "Mesenchymal Stem Cell Fate Is Regulated by the Composition and Mechanical Properties of Collagen-Glycosaminoglycan Scaffolds." *Journal of the Mechanical Behavior of Biomedical Materials* 11: 53–62. doi:10.1016/j.jmbbm.2011.11.009.

Murray, L. M., V. Nock, J. J. Evans, and M. M. Alkaisi. 2016. "The Use of Substrate Materials and Topography to Modify Growth Patterns and Rates of Differentiation of Muscle Cells." *Journal of Biomedical Materials Research - Part A* 104 (7): 1638–1645. doi:10.1002/jbm.a.35696.

Műzes, Györgyi, and Ferenc Sipos. 2022. "Mesenchymal Stem Cell-Derived Secretome: A Potential Therapeutic Option for Autoimmune and Immune-Mediated Inflammatory Diseases." *Cells* 11 (15). doi:10.3390/cells11152300.

Nadine, Sara, Ada Chung, Sibel Emir Diltemiz, Brooke Yasuda, Charles Lee, Vahid Hosseini, Solmaz Karamikamkar, et al. 2022. "Advances in Microfabrication Technologies in Tissue Engineering and Regenerative Medicine." *Artificial Organs* 46 (7): E211–E243. doi:10.1111/aor.14232.

Nagayama, Kazuaki, and Tatsuya Hanzawa. 2022. "Cell type-specific orientation and migration responses for a microgrooved surface with shallow grooves". *Bio-Medical Materials and Engineering*, 33 (5): 393–406. doi:10.3233/bme-211356.

Nakano, Masako, Kenta Kubota, Eiji Kobayashi, Takako S. Chikenji, Yuki Saito, Naoto Konari, and Mineko Fujimiya. 2020. "Bone Marrow-Derived Mesenchymal Stem Cells Improve Cognitive Impairment in an Alzheimer's Disease Model by Increasing the Expression of MicroRNA-146a in Hippocampus." *Scientific Reports* 10(1): 1–15. doi:10.1038/s41598-020-67460-1.

Naqvi, S. M., and L. M. McNamara. 2020. "Stem Cell Mechanobiology and the Role of Biomaterials in Governing Mechanotransduction and Matrix Production for Tissue Regeneration." *Frontiers in Bioengineering and Biotechnology* 8: 1–27. doi:10.3389/fbioe.2020.597661.

Navaei, Ali, Nathan Moore, Ryan T. Sullivan, Danh Truong, Raymond Q. Migrino, and Mehdi Nikkhah. 2017. "Electrically Conductive Hydrogel-Based Micro-Topographies for the Development of Organized Cardiac Tissues." *RSC Advances* 7 (6): 3302–3312. doi:10.1039/C6RA26279A.

Ngô, Christian, and Marcel H Van de Voorde. 2014. "Nanomaterials: Doing More with Less". *Nanotechnology in a Nutshell*: 55–70. doi: 10.2991/978-94-6239-012-6_4.

Nguyen, Alexander K., and Roger J. Narayan. 2017. "Two-Photon Polymerization for Biological Applications." *Materials Today* 20 (6): 314–322. doi:10.1016/j.mattod.2017.06.004.

Nguyen, Anh Tuan, Sharvari R. Sathe, and Evelyn K. F. Yim. 2016. "From Nano to Micro: Topographical Scale and Its Impact on Cell Adhesion, Morphology and Contact Guidance." *Journal of Physics Condensed Matter* 28 (18). doi:10.1088/0953-8984/28/18/183001.

Nikkhah, Mehdi, Faramarz Edalat, Sam Manoucheri, and Ali Khademhosseini. 2012. "Engineering Microscale Topographies to Control the Cell-Substrate Interface." *Biomaterials* 33 (21): 5230–5246. doi:10.1016/j.biomaterials.2012.03.079.

Ntege, Edward H., Hiroshi Sunami, and Yusuke Shimizu. 2020. "Advances in Regenerative Therapy: A Review of the Literature and Future Directions." *Regenerative Therapy* 14: 136–153. doi:10.1016/j.reth.2020.01.004.

O'Brien, Fergal J. 2011. "Biomaterials & Scaffolds for Tissue Engineering." *Materials Today* 14 (3): 88–95. doi:10.1016/S1369-7021(11)70058-X.

Oakley, C., and D. M. Brunette. 1993. "The Sequence of Alignment of Microtubules, Focal Contacts and Actin Filaments in Fibroblasts Spreading on Smooth and Grooved Titanium Substrata." *Journal of Cell Science* 106 (1): 343–354. doi:10.1242/jcs.106.1.343.

Otsuka, Ryo, Haruka Wada, Tomoki Murata, and Ken Ichiro Seino. 2020. "Immune Reaction and Regulation in Transplantation Based on Pluripotent Stem Cell Technology." *Inflammation and Regeneration* 40 (1). doi:10.1186/s41232-020-00125-8.

Passier, Robert, and Christine Mummery. 2003. "Origin and Use of Embryonic and Adult Stem Cells in Differentiation and Tissue Repair." *Cardiovascular Research* 58 (2): 324–335. doi:10.1016/S0008-6363(02)00770-8.

Paul, Colin D., Alex Hruska, Jack R. Staunton, Hannah A. Burr, Kathryn M. Daly, Jiyun Kim, Nancy Jiang, and Kandice Tanner. 2019. "Probing Cellular Response to Topography in Three Dimensions." *Biomaterials* 197: 101–118. doi:10.1016/j.biomaterials.2019.01.009.

Paun, Irina Alexandra, Roxana Cristina Popescu, Cosmin Catalin Mustaciosu, Marian Zamfirescu, Bogdan Stefanita Calin, Mona Mihailescu, Maria Dinescu, et al. 2018. "Laser-Direct Writing by Two-Photon Polymerization of 3D Honeycomb-like Structures for Bone Regeneration." *Biofabrication* 10 (2). doi:10.1088/1758-5090/aaa718.

Pereira Chilima, Tania D., Fabien Moncaubeig, and Suzanne S. Farid. 2018. "Impact of Allogeneic Stem Cell Manufacturing Decisions on Cost of Goods, Process Robustness and Reimbursement." *Biochemical Engineering Journal* 137: 132–151. doi:10.1016/j.bej.2018.04.017.

Phelps, Jolene, Amir Sanati-Nezhad, Mark Ungrin, Neil A. Duncan, and Arindom Sen. 2018. "Bioprocessing of Mesenchymal Stem Cells and Their Derivatives: Toward Cell-Free Therapeutics." *Stem Cells International* 2018 (iii). doi:10.1155/2018/9415367.

Pimpin, Alongkorn, and Werayut Srituravanich. 2012. "Reviews on Micro- and Nanolithography Techniques and Their Applications." *Engineering Journal* 16 (1): 37–55. doi:10.4186/ej.2012.16.1.37.

Piner, Richard D., Jin Zhu, Feng Xu, and Seunghun Hong. 1999. "'Dip-Pen' Nanolithography" 283 (January): 661–664.

Polak, Julia M., and Anne E. Bishop. 2006. "Stem Cells and Tissue Engineering: Past, Present, and Future." *Annals of the New York Academy of Sciences* 1068 (1): 352–366. doi:10.1196/annals.1346.001.

Postiglione, L., G. Di Domenico, L. Ramaglia, A. E. Di Lauro, F. Di Meglio, and S. Montagnani. 2004. "Different Titanium Surfaces Modulate the Bone Phenotype of SaOS-2 Osteoblast-like Cells." *European Journal of Histochemistry* 48 (3): 213–222.

Rahman, Saeed Ur, Joung Hwan Oh, Young Dan Cho, Shin Hye Chung, Gene Lee, Jeong Hwa Baek, Hyun Mo Ryoo, and Kyung Mi Woo. 2018. "Fibrous Topography-Potentiated Canonical Wnt Signaling Directs the Odontoblastic Differentiation of Dental Pulp-Derived Stem Cells." *ACS Applied Materials and Interfaces* 10 (21): 17526–17541. doi:10.1021/acsami.7b19782.

Ramaswamy, Yogambha, Iman Roohani, Young Jung No, Genevieve Madafiglio, Frank Chang, Furong Zhang, Zufu Lu, and Hala Zreiqat. 2021. "Nature-Inspired Topographies on Hydroxyapatite Surfaces Regulate Stem Cells Behaviour." *Bioactive Materials* 6 (4): 1107–1117. doi:10.1016/j.bioactmat.2020.10.001.

Ramesh, Srikanthan, Ola L. A. Harrysson, Prahalada K. Rao, Ali Tamayol, Denis R. Cormier, Yunbo Zhang, and Iris V. Rivero. 2021. "Extrusion Bioprinting: Recent Progress, Challenges, and Future Opportunities." *Bioprinting* 21: e00116. doi:10.1016/j.bprint.2020.e00116.

Rao, K. Suryanarayana. 2017. "Basics of Stem Cells and Preclinical Testing." *Biology, Engineering, Medicine and Science Reports* 3 (1): 17–20. doi:10.5530/bems.3.1.6.

Refaaq, F. M., X. Chen, and S. W. Pang. 2020. "Effects of Topographical Guidance Cues on Osteoblast Cell Migration." *Scientific Reports* 10 (1): 1–11. doi:10.1038/s41598-020-77103-0.

Ren, Guangwen, Xiaodong Chen, Fengping Dong, Wenzhao Li, Xiaohui Ren, Yanyun Zhang, and Yufang Shi. 2012. "Concise Review: Mesenchymal Stem Cells and Translational Medicine: Emerging Issues." *Stem Cells Translational Medicine* 1 (1): 51–58. doi:10.5966/sctm.2011-0019.

Rider, Patrick, Željka Perić Kačarević, Said Alkildani, Sujith Retnasingh, and Mike Barbeck 2018. "Bioprinting of Tissue Engineering Scaffolds." *Journal of Tissue Engineering* 9. doi:10.1177/2041731418802090.

Rim, Nae Gyune, Seok Joo Kim, Young Min Shin, Indong Jun, Dong Woo Lim, Jung Hwan Park, and Heungsoo Shin. 2012. "Mussel-Inspired Surface Modification of Poly(l-Lactide) Electrospun Fibers for Modulation of Osteogenic Differentiation of Human Mesenchymal Stem Cells." *Colloids and Surfaces B: Biointerfaces* 91 (1): 189–197. doi:10.1016/j.colsurfb.2011.10.057.

Robotti, Francesco, Simone Bottan, Federica Fraschetti, Anna Mallone, Giovanni Pellegrini, Nicole Lindenblatt, Christoph Starck, Volkmar Falk, Dimos Poulikakos, and Aldo Ferrari. 2018. "A Micron-Scale Surface Topography Design Reducing Cell Adhesion to Implanted Materials." *Scientific Reports* 8 (1): 1–13. doi:10.1038/s41598-018-29167-2.

Sachs, Frederick. 2010. "Stretch-Activated Ion Channels: What Are They?" *Physiology* 25 (1): 50–56. doi: 10.1152/physiol.00042.2009.

Sah, Jay Prakash. 2016. "Challenges of Stem Cell Therapy in Developing Country." *Journal of Stem Cell Research & Therapeutics* 1 (3): 96–98. doi:10.15406/jsrt.2016.01.00018.

Salaita, Khalid, Yuhuang Wang, and Chad A. Mirkin. 2007. n.d. "Nnano.2007.39," no. 1: 145–155.

Salmasi, Shima. 2015. "Role of Nanotopography in the Development of Tissue Engineered 3D Organs and Tissues Using Mesenchymal Stem Cells." *World Journal of Stem Cells* 7 (2): 266. doi:10.4252/wjsc.v7.i2.266.

Saravanan, P., R. Gopalan, and V. Chandrasekaran. (2008). "Synthesis and Characterisation of Nanomaterials." *Defence Science Journal* 58: 504–516. 10.14429/dsj.58.1671.

Scaccini, Luca, Roberta Mezzena, Alessia De Masi, Mariacristina Gagliardi, Giovanna Gambarotta, Marco Cecchini, and Ilaria Tonazzini. 2021. "Chitosan Micro-grooved Membranes with Increased Asymmetry for the Improvement of the Schwann Cell Response in Nerve Regeneration." *International Journal of Molecular Sciences* 22 (15). doi:10.3390/ijms22157901.

Schwarz, Ulrich S., and Ilka B. Bischofs. 2005. "Physical Determinants of Cell Organization in Soft Media." *Medical Engineering and Physics* 27 (9): 763–772. doi:10.1016/j.medengphy.2005.04.007.

Seo, Chang Ho, Katsuko Furukawa, Kevin Montagne, Heonuk Jeong, and Takashi Ushida. 2011. "The Effect of Substrate Microtopography on Focal Adhesion Maturation and Actin Organization via the RhoA/ROCK Pathway." *Biomaterials* 32 (36): 9568–9575. doi:10.1016/j.biomaterials.2011.08.077.

Seo, Chang Ho, Heonuk Jeong, Yue Feng, Kevin Montagne, Takashi Ushida, Yuji Suzuki, and Katsuko S. Furukawa. 2014. "Micropit Surfaces Designed for Accelerating Osteogenic Differentiation of Murine Mesenchymal Stem Cells via Enhancing Focal Adhesion and Actin Polymerization." *Biomaterials* 35 (7): 2245–2252. doi:10.1016/j.biomaterials.2013.11.089.

Shen, Zhiyuan. 2011. "Genomic Instability and Cancer: An Introduction." *Journal of Molecular Cell Biology* 3 (1): 1–3. doi:10.1093/jmcb/mjq057.

Silva, S. S., E. M. Fernandes, S. Pina, J. Silva-Correia, S. Vieira, J. M. Oliveira, and R. L. Reis. 2017. "2.11 Polymers of Biological Origin." *Comprehensive Biomaterials II* 2: 228–252. doi:10.1016/B978-0-12-803581-8.10134-1.

Singh, Vimal K., Abhishek Saini, Manisha Kalsan, Neeraj Kumar, and Ramesh Chandra. 2016. "Describing the Stem Cell Potency: The Various Methods of Functional Assessment and in Silico Diagnostics." *Frontiers in Cell and Developmental Biology* 4. doi:10.3389/fcell.2016.00134.

Skoog, Shelby A., Peter L. Goering, and Roger J. Narayan. 2014. "Stereolithography in Tissue Engineering." *Journal of Materials Science: Materials in Medicine* 25 (3): 845–856. doi:10.1007/s10856-013-5107-y.

Spradling, Allan, Margaret T. Fuller, Robert E. Braun, and Shosei Yoshida. 2011. "Germline Stem Cells." *Cold Spring Harbor Perspectives in Biology* 3 (11): 1–20. doi:10.1101/cshperspect.a002642.

Squillaro, Tiziana, Gianfranco Peluso, and Umberto Galderisi. 2016. "Clinical Trials with Mesenchymal Stem Cells: An Update." *Cell Transplantation* 25 (5): 829–848. doi:10.3727/096368915×689622.

Sun, Baoce, Kai Xie, Ting Hsuan Chen, and Raymond H. W. Lam. 2017. "Preferred Cell Alignment along Concave Microgrooves." *RSC Advances* 7 (11): 6788–6794. doi:10.1039/c6ra26545f.

Sun, Aaron X., Hang Lin, Angela M. Beck, Evan J. Kilroy, and Rocky S. Tuan. 2015. "Projection Stereolithographic Fabrication of Human Adipose Stem Cell-Incorporated Biodegradable Scaffolds for Cartilage Tissue Engineering". *Frontiers in Bioengineering and Biotechnology*, 3: 115. doi: 10.3389/fbioe.2015.00115.

Taddei, M. L., E. Giannoni, T. Fiaschi, and P. Chiarugi. 2012. "Anoikis: An Emerging Hallmark in Health and Diseases." *Journal of Pathology* 226: 380–393.

Tahmasebi, Fatemeh, and Shirin Barati. 2022. "Effects of Mesenchymal Stem Cell Transplantation on Spinal Cord Injury Patients." *Cell and Tissue Research* 389 (3): 373–384. doi:10.1007/s00441-022-03648-3.

Takahashi, Kazutoshi, and Shinya Yamanaka. 2013. "Induced Pluripotent Stem Cells in Medicine and Biology." *Development (Cambridge)* 140 (12): 2457–2461. doi:10.1242/dev.092551.

Thrivikraman, Greeshma, Alicja Jagie, Victor K. Lai, Sandra L. Johnson, Mark Keating, and Alexander Nelson. 2021. "Cell Contact Guidance via Sensing Anisotropy of Network Mechanical Resistance." *PNAS* 118 (29): 1–11. doi:10.1073/pnas.2024942118.

Teo, Benjamin Kim Kiat, Sum Thai Wong, Choon Kiat Lim, Terrence Y. S. Kung, Chong Hao Yap, Yamini Ramagopal, Lewis H. Romer, and Evelyn K. F. Yim. 2013. "Nanotopography Modulates Mechanotransduction of Stem Cells and Induces Differentiation through Focal Adhesion Kinase". *ACS Nano*, 7(6): 4785–4798. doi: 10.1021/nn304966z.

Totonelli, Giorgia, Panagiotis Maghsoudlou, Massimo Garriboli, Johannes Riegler, Giuseppe Orlando, Alan J. Burns, Neil J. Sebire, et al. 2012. "A Rat Decellularized Small Bowel Scaffold That Preserves Villus-Crypt Architecture for Intestinal Regeneration." *Biomaterials* 33 (12): 3401–3410. doi:10.1016/j.biomaterials.2012.01.012.

Trepat, Xavier, Zaozao Chen, and Ken Jacobson. 2012. "Cell Migration." *Comprehensive Physiology* 2 (4): 2369–2392. doi:10.1002/cphy.c110012.

Trounson, Alan, Rahul G. Thakar, Geoff Lomax, and Don Gibbons. (2011). "Clinical trials for stem cell therapies." *BMC Medicine*, 9. 10.1186/1741-7015-9-52.

Trounson, Alan, Thakar, Rahul G, Lomax, Geoff, & Gibbons, Don (2011). Clinical trials for stem cell therapies. *BMC Medicine*, 9 10.1186/1741-7015-9-52.

Tudureanu, Raluca, Iuliana M. Handrea-Dragan, Sanda Boca, and Ioan Botiz. 2022. "Insight and Recent Advances into the Role of Topography on the Cell Differentiation and Proliferation on Biopolymeric Surfaces." *International Journal of Molecular Sciences* 23 (14). doi:10.3390/ijms23147731.

Turner, A. M. P., N. Dowell, S. W. P. Turner, L. Kam, M. Isaacson, J. N. Turner, H. G. Craighead, and W. Shain. 2000. "Attachment of Astroglial Cells to Microfabricated Pillar Arrays of Different Geometries." *Journal of Biomedical Materials Research* 51 (3): 430–441. doi:10.1002/1097-4636(20000905)51:3<430::AID-JBM18>3.0.CO;2-C.

Tzima, Eleni, Mohamed Irani-Tehrani, William B. Kiosses, Elizabetta Dejana, David A. Schultz, Britta Engelhardt, Gaoyuan Cao, Horace DeLisser, and Martin Alexander Schwartz. 2005. "A Mechanosensory Complex That Mediates the Endothelial Cell Response to Fluid Shear Stress." *Nature* 437 (7057): 426–431. doi:10.1038/nature03952.

Ummethala, Govind, Arun Jaiswal, Raghvendra P. Chaudhary, Suyog Hawal, Sumit Saxena, and Shobha Shukla. 2017. "Localized Polymerization Using Single Photon Photoinitiators in Two-Photon Process for Fabricating Subwavelength Structures." *Polymer* 117: 364–369. doi:10.1016/j.polymer.2017.04.039.

Vasudevan, Ravikumar, Alan J. Kennedy, Megan Merritt, Fiona H. Crocker, and Ronald H. Baney. 2014. "Microscale Patterned Surfaces Reduce Bacterial Fouling-Microscopic and Theoretical Analysis." *Colloids and Surfaces B: Biointerfaces* 117: 225–232. doi:10.1016/j.colsurfb.2014.02.037.

Venkatesh, Katari, and Dwaipayan Sen. 2017. "Potential Cell Based Therapy for Parkinson's Disease." *Bentham Science Publishers, no. October* 2018: 326–347. doi:10.2174/1574888×12666161114122.

Ventura, Reiza Dolendo. 2021. "An Overview of Laser-Assisted Bioprinting (LAB) in Tissue Engineering Applications." *Medical Lasers* 10 (2): 76–81. doi:10.25289/ml.2021.10.2.76.

Wan, Yuqing, Yong Wang, Zhimin Liu, Xue Qu, Buxing Han, Jianzhong Bei, and Shenguo Wang. 2005. "Adhesion and Proliferation of OCT-1 Osteoblast-like Cells on Micro- and Nano-Scale Topography Structured Poly(L-Lactide)." *Biomaterials* 26 (21): 4453–4459. doi:10.1016/j.biomaterials.2004.11.016.

Wang, Xuan, Hua Wang, Fang He, and Jicong Zhang. 2020. "In Vitro Cell Migration through Three-Dimensional Interfaces of Varying Depths, Widths, and Curvatures on Micropatterned Polymer Surfaces." *ACS Applied Bio Materials* 3 (11): 7472–7482. doi:10.1021/acsabm.0c00697.

Wang, Han Bing, Michael E. Mullins, Jared M. Cregg, Andres Hurtado, Martin Oudega, Matthew T. Trombley, and Ryan J. Gilbert. 2009. "Creation of Highly Aligned Electrospun Poly-L-Lactic Acid Fibers for Nerve Regeneration Applications." *Journal of Neural Engineering* 6 (1). doi: 10.1088/1741-2560/6/1/016001.

Wang, Shan, Ling Guo, Jianfeng Ge, Lin Yu, Ting Cai, Ruiyun Tian, Yuyang Jiang, Robert CH Zhao, and Yaojiong Wu. 2015. "Excess Integrins Cause Lung Entrapment of Mesenchymal Stem Cells". *Stem Cells*, 33 (11): 3315–3326. doi: 10.1002/stem.2087.

Wang, Zhuo, James Goh, Shamal Das De, Zigang Ge, Hongwei Ouyang, Jeniffer Sue Wee Chong, Siew Leng Low, and Eng Hin Lee. 2006. "Efficacy of Bone Marrow-Derived Stem Cells in Strengthening Osteoporotic Bone in a Rabbit Model." *Tissue Engineering* 12 (7): 1753–1761. doi:10.1089/ten.2006.12.1753.

Wei, Jin, Jian Shi, Bin Wang, Yadong Tang, Xiaolong Tu, Emmanuel Roy, Benoit Ladoux, and Yong Chen. 2016. "Fabrication of Adjacent Micropillar Arrays with Different Heights for Cell Studies." *Microelectronic Engineering* 158: 22–25. doi:10.1016/j.mee.2016.03.008.

Wieland, Marco, Babak Chehroudi, Marcus Textor, and Donald M. Brunette. 2002. "Use of Ti-coated replicas to investigate the effects on fibroblast shape of surfaces with varying roughness and constant chemical composition". *Journal of Biomedical Materials Research* 60 (3): 434–444. doi:10.1002/jbm.10059.

Weiss, Paul. 1945. "Experiments on Cell and Axon Orientation in Vitro." *Journal of Experimental Zoology* 1 (1): 353–386.

Whitesides, George M., Emanuele Ostuni, Xingyu Jiang, and Donald E. Ingber. 2001. "Soft Lithography in Biology and Biochemistry." *Annual Review of Biomedical Engineering* 3: 335–373.

Wu, Jingxian, Changming Hu, Zengchao Tang, Qian Yu, Xiaoli Liu, and Hong Chen. 2018. "Tissue-Engineered Vascular Grafts: Balance of the Four Major Requirements." *Colloid and Interface Science Communications* 23: 34–44. doi:10.1016/j.colcom.2018.01.005.

Xie, Zelong, Ming Gao, Anderson O. Lobo, and Thomas J. Webster. 2020. "3D Bioprinting in Tissue Engineering for Medical Applications: The Classic and the Hybrid." *Polymers* 12 (8). doi:10.3390/POLYM12081717.

Xie, Jian, Hangqi Shen, Guangyin Yuan, Kaili Lin, and Jiansheng Su. 2021. "The Effects of Alignment and Diameter of Electrospun Fibers on the Cellular Behaviors and Osteogenesis of BMSCs." *Materials Science and Engineering C* 120: 111787. doi:10.1016/j.msec.2020.111787.

Yadav, Shital, and Abhijit Majumder. 2021. "Biomimicked Hierarchical 2D and 3D Structures from Natural Templates: Applications in Cell Biology." *Biomedical Materials (Bristol)* 16 (6). doi:10.1088/1748-605X/ac21a7.

Yadav, Shital, and Abhijit Majumder. 2022. "Biomimicked Large-Area Anisotropic Grooves from Dracaena Sanderiana Leaf Enhances Cellular Alignment and Subsequent Differentiation." *Bioinspiration and Biomimetics* 17 (5). doi:10.1088/1748-3190/ac7afe.

Yang, Xing, Yuanyuan Li, Wei He, Qianli Huang, Ranran Zhang, and Qingling Feng. 2018. "Hydroxyapatite/Collagen Coating on PLGA Electrospun Fibers for Osteogenic Differentiation of Bone Marrow Mesenchymal Stem Cells." *Journal of Biomedical Materials Research - Part A* 106 (11): 2863–2870. doi:10.1002/jbm.a.36475.

Yang, Kisuk, Jaehong Lee, Jong Seung Lee, Dayeong Kim, Gyeong Eon Chang, Jungmok Seo, Eunji Cheong, Taeyoon Lee, and Seung Woo Cho. 2016. "Graphene Oxide Hierarchical Patterns for the Derivation of Electrophysiologically Functional Neuron-like Cells from Human Neural Stem Cells." *ACS Applied Materials and Interfaces* 8 (28): 17763–17774. doi:10.1021/acsami.6b01804.

Yao, Li, Shenguo Wang, Wenjin Cui, Richard Sherlock, Claire O'Connell, Gopinath Damodaran, Adrienne Gorman, Anthony Windebank, and Abhay Pandit. 2009. "Effect of Functionalized Micropatterned PLGA on Guided Neurite Growth." *Acta Biomaterialia* 5 (2): 580–588. doi:10.1016/j.actbio.2008.09.002.

Yoshihara, Masahito, Yoshihide Hayashizaki, and Yasuhiro Murakawa. 2017. "Genomic Instability of IPSCs: Challenges Towards Their Clinical Applications." *Stem Cell Reviews and Reports* 13 (1): 7–16. doi:10.1007/s12015-016-9680-6.

You, Renchuan, Xiufang Li, Yu Liu, Guiyang Liu, Shenzhou Lu, and Mingzhong Li. 2014. "Response of Filopodia and Lamellipodia to Surface Topography on Micropatterned Silk Fibroin Films." *Journal of Biomedical Materials Research - Part A* 102 (12): 4206–4212. doi:10.1002/jbm.a.35097.

Zhang, Cheng Liang, Ting Huang, Bi Li Wu, Wen Xi He, and Dong Liu. 2017. "Stem Cells in Cancer Therapy: Opportunities and Challenges." *Oncotarget* 8 (43): 75756–75766. doi:10.18632/oncotarget.20798.

Zhang, Qing, Hua Dong, Yuli Li, Ye Zhu, Lei Zeng, Huichang Gao, Bo Yuan, Xiaofeng Chen, and Chuanbin Mao. 2015. "Microgrooved Polymer Substrates Promote Collective Cell Migration to Accelerate Fracture Healing in an in Vitro Model." *ACS Applied Materials and Interfaces* 7 (41): 23336–23345. doi:10.1021/acsami.5b07976.

Zhao, Cancan, Xiaoya Wang, Long Gao, Linguo Jing, Quan Zhou, and Jiang Chang. 2018. "The Role of the Micro-Pattern and Nano-Topography of Hydroxyapatite Bioceramics on Stimulating Osteogenic Differentiation of Mesenchymal Stem Cells." *Acta Biomaterialia* 73: 509–521. doi:10.1016/j.actbio.2018.04.030.

Zhou, Xiaoqin, Yihong Hou, and Jieqiong Lin. 2015. "A Review on the Processing Accuracy of Two-Photon Polymerization." *AIP Advances* 5 (3). doi:10.1063/1.4916886.

Zhu, Min, Guillaume Baffou, Nikolaus Meyerbröker, and Julien Polleux. 2012. "Micropatterning Thermoplasmonic Gold Nanoarrays To Manipulate Cell Adhesion". *ACS Nano* 6 (8): 7227–7233. doi:10.1021/nn302329c.

Zhu, Lisha, Dan Luo, and Yan Liu. 2020. "Effect of the Nano/Microscale Structure of Biomaterial Scaffolds on Bone Regeneration." *International Journal of Oral Science* 12 (1): 1–15. doi:10.1038/s41368-020-0073-y.

3 Nanocellulose for Biotechnology and Medicinal Applications

Prangan Duarah, Pranjal P. Das, and Mihir K. Purkait

3.1 INTRODUCTION

The world's need for energy, which is now fulfilled by fossil fuels, is growing due to the fast rise of the human population and industrial progress. Energy use is predicted to climb by almost 48% between 2012 and 2040 and six times over the course of this century (Debnath et al. 2021). As a result of excessive energy use and the resulting resource depletion, environmental sustainability and sustainable development are currently greater concern in the scientific community. In addition, the quick pace of development has resulted in significant pollution from industry that has degraded the environment (Das et al. 2021; Duarah et al. 2021; Das et al. 2021; Das et al. 2022). Utilizing naturally occurring renewable resources and eco-friendly biomass-based goods is a key approach for reducing environmental concerns (Haldar et al. 2022). The valorization of biomass is an approach that offers hope for achieving such goals. Waste materials, including agricultural by-products, forest residue, urban wastes, and industrial wastes, are sometimes referred to as biomass. Crop trimmings, vine pruning, roots, and fruit peels are some of the most common plant-based waste products in agricultural settings. Tree stumps, prunings, and roots are all examples of forest wastes. Agricultural wastes are typically produced quickly, with an average yearly growth rate of 5 to 10%, contributing to adverse environmental effects since the careless dumping or burning of such wastes pollutes the land, water, and air. Therefore, it is imperative to properly utilize waste biomass leftovers through conversion and valorization in order to reduce its negative environmental consequences (Debnath et al. 2022).

Agricultural waste and forest residue, in particular, offer a multitude of benefits to the environment as a result of their naturally abundant availability, renewability, cheap cost, and little energy required for conversion. In addition, the majority of plant wastes and plant-based materials include a sizeable quantity of cellulose, which serves as the primary constituent of their structures. The natural fibers of the residues are typically composed of cellulose, hemicellulose, and lignin; besides, cellulose may be recovered from them by means both physical and chemical (Duarah et al. 2020). In addition, the plentiful sources of cellulose derivatives typically utilized include agricultural by-products including remaining sections of food crops (straws of maize, rice, wheat, corn stover, and sunflower), and forest residue such as woody biomass. Utilizing these resources is viewed as an approach that has the potential to be considered sustainable because they do not pose a threat to the production of food. The valuable cellulose derivatives such as nanocellulose may be extracted from agricultural by-products and forest residues in an effective manner. Nanocellulose has attracted a rising amount of interest due to the remarkable attributes that it possesses, which include being the most abundant polymer on the planet, having an excellent aspect ratio, possessing superior mechanical capabilities, being renewable, biodegradable, and non-toxic. Nanocellulose is the

DOI: 10.1201/9781003305583-3

result of an extract from native cellulose, which may originate in plants, animals, and microorganisms. It is made of nanoscaled structural material, and its extraction relies on simple, scalable, and effective processes (Perumal et al. 2022; Debnath et al. 2022).

Nanocellulose's vast surface area, aspect ratio, mechanical properties, simplicity of surface modification, non-immunogenicity, and remarkable biocompatibility have drawn considerable attention for biomedical applications (Patil et al. 2022). The focus of this chapter is on the many approaches that may be used to synthesize nanocellulose, as well as the prospective uses of these methods in drug delivery and biosensing. Moreover, taking into consideration certain astounding discoveries brings to light the applicability of nanocellulose for tissue engineering and encapsulation. The chapter comes to a conclusion by discussing the potential applications of cellulose nanoparticles in the future as well as the obstacles that still need to be solved.

3.2 PROPERTIES OF NANOCELLULOSE AND PREPARATION STRATEGIES

3.2.1 Types of Nanocellulose

From the meter to the nanoscale scale, the hierarchy of cellulose found in plants is illustrated in Figure 3.1(a). Nanocellulose refers to cellulose-derived materials with a pore size of less than one nanometer. Nanocrystalline cellulose or cellulose nanocrystalline (CNC/NCC), cellulose nanofibers/nanofibrillated cellulose (CNF/NFC), and bacterial nanocellulose (BNC) are included in the nanofibers class. In contrast, microcrystalline cellulose (MCC) and microfibrils are included in the nanostructured materials class (Trache et al. 2020).

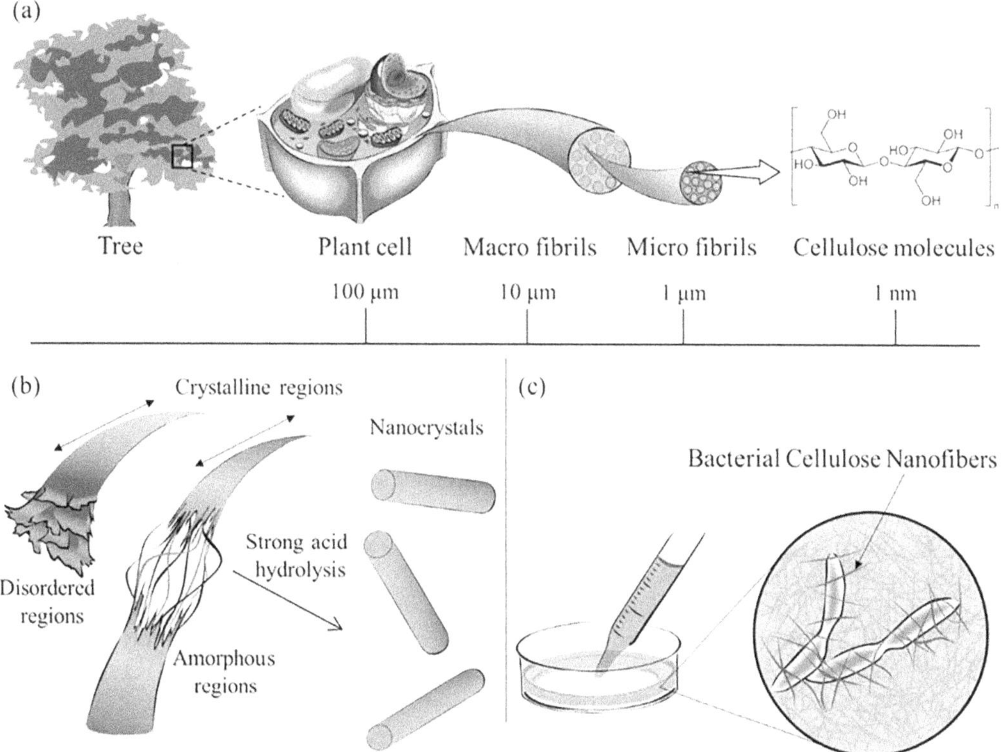

FIGURE 3.1 **(a)** From the meter to the nanoscale scale, the hierarchy of cellulose found in plants, **(b)** A graphic depicting the interaction between cellulose and strong acid to produce nanocellulose, **(c)** Bacterial nanocellulose (BNC) [Reproduced with permission from (Miyashiro et al. 2020)].

One of the forms of nanocellulose that has received the most attention is the NCC variety. Ranby,(1951) was the first to indicate that the degradation of cellulose fibers may be catalyzed by colloidal sulfuric acid. A number of different preparations of NCC were made from a wide variety of sources, including wood, olive tree, cotton, bacteria, microcrystalline cellulose, as well as lignocellulosic waste material (Mishra et al. 2019; Duarah et al. 2023). A variety of morphologies can be obtained in the NCC synthesized by acid hydrolysis of natural cellulose (Figure 3.1(b)). Hydrolysis of sulfuric acid under proper control produces a somewhat stable suspension, which is used to separate NCC from the cellulose (Elazzouzi-Hafraoui et al. 2008). Acid hydrolyzes the majority of the amorphous portion of native cellulose, resulting in a more crystallized product. At high temperatures, other acids than sulfuric acid were utilized to make water-soluble solutions: hydrobromic acids and hydrochloric and phosphoric acid (Zhou et al. 2017).

Cellulose nanofibres may be made in a similar way by pre-treating cellulose fiber breakdown and mechanical delamination. A group of researchers used high-pressure homogenization to first separate the CNF from bleached softwood fibers. Pear and apple, Helicteris isora, Oil Palm, Banana, Citrullus colocynthis seeds, and cassava peel have also been used to make cellulose nanofibers (Turbak et al. 1983). Because the procedure required a lot of energy, interest from industry and science declined. As a result, additional effective pre-treatments have been researched, including TEMPO catalytic oxidation, partial carboxymethylation, and enzymatic hydrolysis (Yousefi et al. 2011; Ul-Islam et al. 2015). Mechanical disintegration under high pressure, extrusion, friction grinding, cryopressure, micro fluidization, and high-intensity ultra-sonication are applied to the fibers after pre-treatment. NFC has also been extracted from sugarcane bagasse and curauá fibers, banana peels, and the lignocellulosic biomass of lemongrass. Numerous experimental techniques have also been devised to study the characteristics of nanocelluloses as well as how they might be modified. Kargarzadeh et al. (2018) gathered data on several pre-treatment techniques, including chemical and mechanical operations commonly utilized to produce CNFs and CNCs. It is possible to manufacture nanomaterials, including wood, from a variety of cellulosic sources. There are numerous new industrial uses for cellulose as a result of its nanoscale transformation, and this article provides an outline of the major markets that cellulose nanoparticles can touch.

Bacterial nanocellulose differs from other forms of nanocellulose in several ways. BNC, a potential natural polymer produced by bacteria, shares the same molecular structure as cellulose that is obtained from plants (Figure 3.1(c)). It has exceptional purity and water preservation qualities. Because of the variations in its physiological, morphological, and mechanical characteristics, the ideal approach for either of these two types of development depended on the application of BNC in the end. The cellulose that was produced by a culture that was agitated had a poor level of mechanical strength. A wider cultivation area and a prolonged cultivation period were necessary for a stable culture (Tyagi and Suresh 2016; Raghav et al. 2021). Some bacterial species can produce extracellular cellulose. For example, *Gluconacetobacter xylinus*, *Sarcina ventriculi*, and *Rhizobium* sp. have all been shown to produce cellulose. Gram-positive bacteria (e.g., *Sarcina ventriculi*) and Gram-negative bacteria (e.g., *Rhizobium and Salmonella)* have all been found to produce cellulose. *Agrobacterium, Sarcina*, and *Gluconacetobacter* are the cellulose-producing organisms that have received the most attention from researchers. To synthesize BNC in synthetic and non-synthetic mediums, oxidative fermentation at pH 3 to 7, temperature preservation between 25°C and 30°C, and the use of saccharides as a carbon source are all possible (Ul-Islam et al. 2015). A study found that bacterial cellulose (BC) was produced by *Gluconacetobacter xylinus* PTCC 1734 in sugar beet molasses, cheese whey, and conventional Hestrin–Schramm (HS) conditions. Sulfuric acid was used to further hydrolyze the synthesized BC in order to create BNC. Sugar beet molasses had the greatest BC content and output, followed by cheese whey, according to the study results (Salari et al. 2019).

3.2.2 Physical and Chemical Properties

Nanocellulose, as a natural nanoscaled material, has unique properties that distinguish it from traditional materials, such as special morphological features, crystalline nature, large surface area, rheological properties, liquid crystalline behavior, alignment and orientation, surface chemical reactivity, low cytotoxicity, mechanical reinforcement, barrier properties, and so on (Lin and Dufresne 2014a).

Both the ordered (crystalline) and disordered (amorphous) areas of the nanoparticle contribute to the nanocellulose's overall characteristics, and these qualities may be used to characterize the nanocellulose's mechanical properties. Disordered cellulose chains contribute to the bulk material's flexibility and plasticity, whereas ordered chains contribute to its stiffness and elasticity. When crystalline domains are blended with amorphous fraction, modulus of various kinds of nanocellulose is expected to be the end result of a blending rule. Since CNC has a greater proportion of crystalline regions, its stiffness and modulus ought to be superior to CNF and BC fibrils, which have crystalline and amorphous structures, respectively. Since the 1930s, researchers have been looking at crystalline cellulose's elastic modulus compared to other materials. CNC should have Young's modulus range of 100–200 GPa. This would make it comparable in strength to Kevlar (60–125 GPa) and may be even stronger than steel (200–220 GPa). Recent research has utilized atomistic simulations to evaluate the elastic modulus of NCC by employing the conventional uniform deformation methodology in addition to the nanoscale indentation method (similar to Kevlar). In a separate piece of research, Dri et al. employed the atomic structure model of cellulose in conjunction with quantum mechanics to arrive at an estimate for Young's modulus of NCC. They found that it was 206 GPa (similar to steel) (Dri et al. 2013). Based on several theoretical and experimental approaches, it was shown that the longitudinal modulus of cellulose microfibrils (including both CNF and BC) might vary widely. The typical modulus of cellulose microfibril is roughly 100 GPa. The elastic modulus of cellulose microfibrils was determined by a three-point bending experiment with atomic force microscopy tips. The longitudinal modulus of pulp CNF was reported to be 81 ± 12 GPa, based on measurements of the cellulose microfibrils' dimensions (Cheng et al. 2009). Using a Raman spectroscopic approach, researchers were able to determine the modulus of BC to be 114 GPa by looking at a shift in the middle location of the 1095 cm^{-1} Raman band. Nanocellulose may be employed as a load-bearing component in many host materials because of its outstanding mechanical capabilities. By enabling suitable stress transfer from the host material (matrix) to the reinforcing phase with homogenous dispersion as well as robust interfacial adhesion, the exposure to high nanocellulose can demonstrate potential nanoreinforcement (Hsieh et al. 2008).

Cellulose is a homopolysaccharide with a high molecular weight that is made up of β -1,4-anhydro-d-glucopyranose units. This makes cellulose an important structural component (Figure 3.2). These units adopt a chair conformation instead of lying perfectly in the plane of the structure, with consecutive glucose residues rotated through a 180° angle with respect to the molecular axis as well as hydroxyl groups in an equatorial position (Habibi et al. 2010). In Figure 3.2, intramolecular hydrogen bonds are largely formed between the OH group of the C3 carbon and the ring oxygen of the glucose unit (O5) that is nearby. Hydrogen bonds between nearby units' OH-6 primary hydroxyl and oxygen in position O3 as well as between OH-2 and oxygen in position O6 are seen in intermolecular cycles.

Nanocellulose's reactive surface has numerous active hydroxyl groups since cellulose has three per glucose unit. Each anhydroglucose unit has different hydroxyl group reactivities. Notably, hydroxyl groups in positions 2 and 3 functions as secondary alcohols, while hydroxyl groups in position 6 operate as primary alcohols. In fact, the carbon atom in the 6th position has just one alkyl group attached, but the carbons in the 2nd and 3rd places have two alkyl groups directly connected, resulting in steric effects due to the supramolecular structure of cellulose

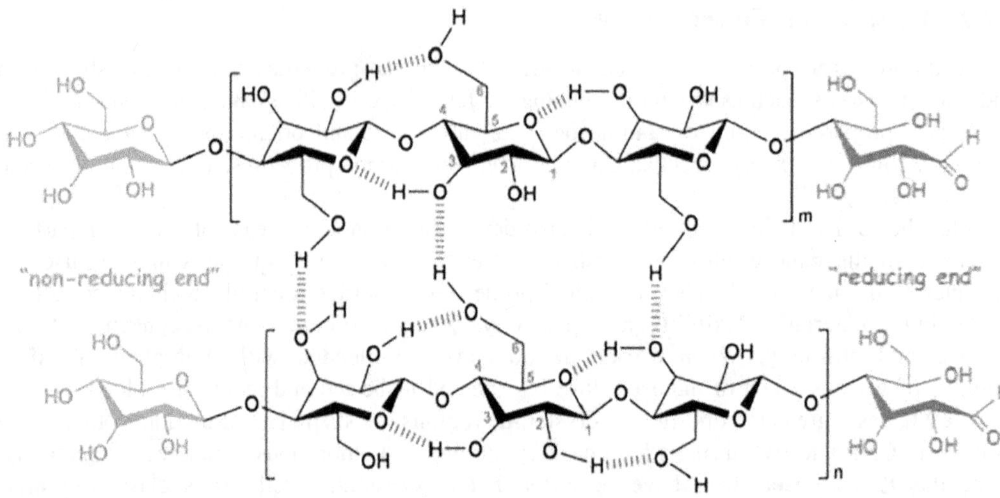

FIGURE 3.2 Pictorial representation of cellulose structure [Reproduced with permission from (Lin and Dufresne 2014a)].

and the reactant. On crystalline cellulose, the hydroxyl group at the 6th position reacts ten times faster than the others, whereas the 2nd position is twice as reactive as the 3rd. Reactants or solvents may affect the reactivity of nanocellulose's hydroxyl groups (including CNC). Recent research indicated that hydroxyl groups on CNC behaved as nucleophiles in the following order: $OH\text{-}C_6 = OH\text{-}C_2 > OH\text{-}C_3$ through etherification (de la Motte et al. 2011; Eyley and Thielemans 2014).

In addition to reactive groups, the surface chemistry of nanocellulose is also influenced by the presence of negative sulfate esters ($-OSO_3-$) on CNC, which have a significant impact. Condensation esterification (sulfation) between surface hydroxyls and an H_2SO_4 molecule is used to introduce surface sulfate esters during sulfuric acid hydrolysis. Because of the significant negative charge they contain, they result in the production of a colloidal solution that is aqueous and well-dispersed. Through the time and temperature of H_2SO_4 hydrolysis, the quantity of surface charge on CNC may be regulated. Biomedical applications, including the electrostatic adsorption of enzymes or proteins, are also made possible by the CNC's surface $-OSO_3-$ groups with negative charges (Lin and Dufresne 2014b).

3.2.3 Feedstock for Nanocellulose Preparation

The term "lignocellulosic biomass" refers to biomass that comes from plants. Biomass derived from plants has the greatest potential as a resource for the environmentally friendly production of cellulose. Wood and non-wood biomass are the two primary categories that may be utilized to describe the lignocellulosic material that makes up biomass. The predominant source of cellulose over the past few decades has been wood-based biomass, which typically accounts for 90–95% of the total generated cellulosic pulp (Pennells et al. 2020). Cellulose may also be derived from agricultural waste products which are considered as the secondary source. They offer a number of benefits over wood, including the fact that they can be renewed annually, they have a high biomass yield, they generate biomass quickly, and they have a high carbohydrate content. Other types of tertiary sources of cellulose include by-products from the beverages and food industry, bagasse, waste from municipal operations, and sludge from paper manufacturing. These wastes have the potential to be used as a feedstock in the production of nanocellulose, which would be beneficial from both an ecological and a financial point of view (García et al. 2016).

3.2.4 Conventional Preparation Method

The synthesis of nanocellulose consists of two stages: the first step is the pre-treatment of lignocellulosic feedstock in preparation for cellulose fiber separation, and the second stage is the breakdown of cellulose fiber that has been isolated to the nanoscale. The synthesis of nanocellulose has been attributed in the research to a variety of conventionally practiced technological modalities of various kinds. However, the type of nanocellulose that is being made, either CNF or CNC, will determine the technology that will be utilized in the preparation process.

Nanocellulose may be synthesized from cellulosic materials via acid hydrolysis, the simplest and oldest chemical technique. Cellulose crystals are preserved while the amorphous cellulose is broken down and removed by acid. While H_2SO_4, H_3PO_4, HCl, and HBr are commonly employed, H_2SO_4 is preferred because of its ability to both isolate CNC and disseminate it in a stable colloid system via esterification of the hydroxyl group by sulfates (Phanthong et al. 2018). The shape and size of nanocellulose are influenced by the reaction temperature and time, acid type and concentration, and the kind of acid. Numerous drawbacks include high water consumption and acidic wastewater production, prolonged processing times, high operational costs and maintenance needs, the potential for corrosion of the machineries, the creation of inhibitors, and a negative impact on the environment (Teo and Wahab 2020).

Nanocellulose may also be synthesized from lignocellulosic materials using the TEMPO oxidation technique. Nanocellulose's colloid solution stability is enhanced by TEMPO because it reduces surface charges (both negative and positive) on the fibers (Rana et al. 2021). To perform TEMPO-mediated oxidation therapy, bleaching agents like NaClO and catalysts like sodium bromide (NaBr) are commonly used (pH 9 to 11). The TEMPO oxidation method is preferable to conventional oxidation methods because it requires less energy, is more user-friendly, and produces milder reaction conditions. The limited oxidation states and potentially harmful compounds utilized are two of the method's drawbacks. (Zhang et al. 2020).

An additional chemical approach for synthesizing nanocellulose is the oxidation of ammonium persulfate (APS). In an acidic media and at relatively high temperatures, APS can form H_2O_2 and $SO4_2$ free radicals, which can efficiently solubilize the amorphous component of cellulose and lignin. (Ng et al. 2021). The yield and properties of nanocellulose are influenced by the APS concentration as well as other processing parameters, including temperature and treatment duration in this procedure (Pradhan et al. 2022). Nanocellulose synthesis on a large industrial scale is constrained by this method's need for lengthier processing times.

Ball milling is another common method that is utilized in the synthesis of nanocellulose. During this process, the cellulose suspension is being maintained in a hollow cylindrical container that is only partially occupied by balls (such as zirconia, ceramic, or metal). Because of the high-energy collisions between the balls as the container rotates, the cellulose fibers within are disintegrated (Nechyporchuk et al. 2016). Milling in a wet condition is preferable for nanocellulose synthesis in this procedure since it prevents defibrillation to an amorphous form. Milling with balls has several variables that affect nanocellulose formation. These include ball size, the weight ratio of the milled balls, time, and moisture content. Due of its high heat output and power consumption, the ball milling process has significant drawbacks that need to be addressed (Teo and Wahab 2020).

To produce CNF, cryocrushing is another mechanical method. During this method, fibers are immersed in water and cellulose absorbs water. Liquid nitrogen solidifies water-soaked cellulose, which is subsequently crushed (Nasir et al. 2017). When frozen cellulosic fibers are subjected to strong impact pressures, the ice crystals' pressure causes the cell wall to break, releasing nanofibers. The disadvantages of this procedure are its significant cost, excessive energy usage, sporadic nanocellulose dispersion, and limited recovery (Teo and Wahab 2020).

One other mechanical approach to CNF is known as high-shear grinding. The pulp is forced through two stones that are fixed in place while rotating in this way. (Nasir et al. 2017). The spacing between these stones may be altered, making it feasible to prevent congestion.

By using shear pressures and the individualization of pulp into nanoscale fibers, the grinder's fibrillation process breaks down the hydrogen bond and the structure of the cell wall. A high use of energy, overheating of the raw materials, poor nanocellulose recovery and homogeneity, and reduced CNF crystallinity are some of the downsides of the method.

3.2.5 Advance Preparation Method

Enzymatic processing, microwaves, deep eutectic solvent, cold plasma, pulsed-electric field, and electron beam irradiation are some of the most recent advanced processing techniques used in the production of nanocellulose.

In the range of 0.3–300 GHz, microwaves (MW) are non-ionizing waves with a wavelength of 1–1000 mm. For its efficiency and ease of use, MW has become a popular alternative to traditional heating in a number of regions. The liquefaction of lignocellulosic biomass may be used to produce CNFs and CNCs using MW energy. Compared to conventional liquefaction, MW-assisted liquefaction offers various advantages, including lower energy and chemical usage, faster cycle times, and lower treatment costs. Removing the lignin and hemicellulose from biomass using MW-aided liquefaction catalyzed by acid is a successful method for creating high-quality nanocellulose (Huang et al. 2017). Other chemical techniques of nanocellulose manufacturing can benefit from MW and ultrasonic irradiation applied in tandem as a process intensification tool to increase yield and speed up response time. In a hybrid technique, both ultrasound and microwaves have the potential to increase the effectiveness of mass transfer between cellulose fibrils. MW is also capable of accelerating the rate at which heat is transferred (Lu et al. 2019). Even though using MW requires a larger upfront expenditure, it has lower operating expenses than traditional heating, providing a quick return on investment. Other advantages of MW include quick processing times, minimal maintenance costs, energy efficiency, and environmental friendliness (Ekezie et al. 2017). MW is a potential new method for the manufacturing of nanocellulose because of its several advantages. A number of variables must be taken into account while developing a method for producing nanocellulose from biomass at an industrial scale utilizing MW. MW treatment causes complex thermal and chemical processes in lignocellulosic biomass. This necessitates a thorough analysis of the morphological and chemical alterations, the thermal degradation process, and the relationship between electromagnetic energy and MW processing parameters. Another challenge of MW treatment on lignocellulosic feedstock for nanocellulose synthesis is determining and understanding the relationship between MW process conditions, reaction medium conditions, and the dielectric properties of the lignocellulosic material, which includes its chemical composition, structure, and size. Emerging as a potential method for extracting nanocellulose from lignocellulosic biomass, enzymatic processing technology comprises many stages and works in a step-by-step manner. To begin, lignocellulosic biomass must be subjected to a pre-treatment in order to fractionate and retrieve the primary components, which may include cellulose, hemicellulose, and lignin. These primary components may be recovered by using either physical or chemical pre-treatment procedures. The enzymatic hydrolysis of fiber samples is subjected to regulated conditions in the second step. The material that has been prepared is dispersed throughout the buffer solution with the help of an enzymatic cocktail. The purpose of the enzymatic cocktail is to reduce the size of the cellulose polymer by breaking it down into smaller pieces. In the third phase, the fibers that have been treated with enzymes are homogenized (Ribeiro et al. 2019). This is done with a variety of different machines, including microfluidizers, ultrasonicators, and ultrafine grinders, amongst others. Cellulases are responsible for the great majority of instances of nanocellulose formation, despite the fact that a variety of enzymes have been used for this purpose. A group of enzymes known as cellulases is involved in accelerating the breakdown of cellulose into a more basic sugar. Cellulolytic microbes, such as those belonging to the genera *Aspergillus, Trichoderma,* and *Clostridium*, amongst others, are responsible for their products in their natural state (Michelin et al. 2020). Enzyme hydrolysis may be carried out under softer circumstances, such as moderate pressure, temperature, and pH. This is a benefit of enzyme hydrolysis. Aside from the obvious benefits of lessening the need for corrosion-resistant processing equipment and ensuring a safer operation, this can also save money in the long run. Environmental protection may be ensured by decreasing or eliminating chemical use during the process, which reduces or eliminates the amount of effluents that are generated. When chemicals are reduced or eliminated, the final stream of sugars may be used to make additional products (e.g. biofuels) that will aid in the implementation of a biorefinery model. Though there are several benefits over chemical techniques, the high price of enzymes is still an issue that has to be addressed. Thus, cost-cutting strategies, including enzyme immobilization and advancements in enzyme manufacturing or purification, can assist solve this challenge (Arantes et al. 2020).

When it comes to the creation of nanocellulose, electron beam irradiation (EBI) is a relatively new technology. Using this technology, lignocellulosic feedstock may be pre-treated to isolate cellulose fibers, and then fibrillation can be used to decrease cellulose fibers to nanometers in diameter. Using a linear accelerator, electron beam ionizing radiation is produced (Hassan et al. 2018). By extracting hydrogen from a glucose molecule and then degrading it, the EBI treatment generates free radicals, which may then be used in other processes. The oxidation or chain scission of cellulose is then induced by the free radicals. Chain scission may be caused by the breaking of glycosidic bonds at higher EBI dosages, but a cross-linking reaction may occur at lower EBI doses instead of chain scission (Lee et al. 2018).

Plasma technology, among the most recent green technologies, has just been used to create nanocellulose. Photons, metastable molecules, radicals, and charged particles like free electrons and ions make up the ionized gas, sometimes referred to as plasma or the fourth state of matter. Plasma is created by applying an electric field to electrodes with gas between them, either at atmospheric pressure or at a lower pressure. The properties of the plasma formed are influenced by the material and geometry of electrodes, electric power, and the composition of gases. CH_4, NH_3, N_2, O_2, Ar, and He are among the most often used gases (Liyanage et al. 2021).

A non-thermal developing technique known as the pulsed-electric field has been put to use for the processing of lignocellulosic materials. When doing PEF pre-treatment, a straightforward apparatus consisting of two electrodes is utilized. The non-thermal voltage pulses are applied to the lignocellulosic feedstock for extremely brief periods of time, ranging from nanoseconds to milliseconds, and the pulse amplitude can range anywhere from 0.1 to 100 kV/cm. Disruption as well as morphological changes in the biological membrane are driven on by PEF treatment, and this leads to a loss of semi-permeability (Kumar et al. 2020).

3.3 MEDICAL APPLICATIONS OF NANOCELLULOSE

3.3.1 Drug Delivery

It has been demonstrated that nanocellulose is an effective carrier for several routes of drug delivery, including the oral, transdermal, and local routes. According to the published research, the amount of time it takes for nanocellulose-based drugs to start releasing can range from few seconds to multiple hours (Kolakovic et al. 2011).

Extensive research has been done on the oral delivery of a variety of different forms of nanocellulose. NCC was employed as a support for anti-cancer medications including etoposide docetaxel, and paclitaxel by Jackson et al. (2011). The cationic surfactant cetyl trimethyl ammonium bromide (CTAB) was bound to the NCC surface, causing a concentration-dependent rise in the NCC zeta potential. According to the research findings, the medications were released gradually over a period of many days. NCC and chitosan, a cationic polymer, were used to create thin films by Mohanta et al. (2014). Electrostatic reactions took place between the chitosan amine groups and the NCC sulfonate groups in this instance. The anti-cancer medicines doxorubicin hydrochloride and curcumin, both water-soluble and insoluble, were put into films in this investigation. Hydrogen bonds and van der Waals contacts were shown to be the primary interactions between the two medications (doxorubicin and curcumin) and NCC. The researchers came to the conclusion that drugs that were both water-insoluble and soluble were continually released. NCC and MCC transporters were explored by Emaraa et al. (2016), who discovered that increasing the NCC load increased the solubility of a weakly water-soluble drug. NCC was shown to be a suitable carrier for weakly water-soluble medicines as a result of this investigation. Cellulose nanofibre, also known as CNF, is yet another fascinating material due to the fact that it possesses distinct physicochemical characteristics at various surfaces. CNF has a large specific surface area, which promotes a beneficial interaction between the medication and the CNF. Nanofibre-nanofibre interactions and high crystallinity increase the oxidative stability of oxygen-sensitive pharmaceuticals during storage; hence CNF films may be efficiently employed as pharmaceutical excipients in low humidity environments. CNF has also been utilized for the quick release of drugs in the form of tablets, capsules, and particles, in addition to being employed in the form of films for the control drug release (Gao et al. 2014).

Numerous studies have used CNF and inedible surfactants to encapsulate air bubbles using the Pickering technique, resulting in stable air bubbles. Pharmaceutical companies found these closed-cell, three-dimensional structures to be useful for delivering medications over time. An increase in the release of medicine owing to a larger surface area can be achieved by using aerogels that rapidly absorb liquid if the pockets are connected (Häbel et al. 2016).

Guo et al. (2017) synthesized MCC/alginate and CNF/alginate beads for the purpose of facilitating the release of metformin hydrochloride. Alginate acted as a carrier for the transport of drugs, but CNF offered enhanced mechanical and swelling characteristics. In comparison to MCC/alginate, the cumulative release via CNF/alginate beads was 10% larger and continued for 240 minutes. Patil et al. (2018) employed nanocomposites of maize starch, urea-formaldehyde, and CNF to inhibit dimethyl phthalate release. Although the primary release of dimethyl phthalate was severely impeded by CNF, the controlled release of the medication was successfully provided by this method. They came to the conclusion that the network present inside the starch matrix was responsible for the indirect pathway, which finally led to the release being delayed (~80–95% of the drug was released in a week). Nanocellulose alginate magnetic hydrogel beads (m-NCC) were produced by (Supramaniam et al. 2018) to provide ibuprofen for 30 to 330 minutes. It is possible to employ M-NCC to increase the mechanical strength and release behavior of a medicine while also utilizing MRI to target, identify, and perhaps treat malignant tissue. Researchers from Thomas et al. developed a hybrid polymer formulation of alginate and cellulose nanocrystals that have improved encapsulation efficiency (EE) and may be utilized to administer rifampicin orally under control. Mycobacterium tuberculosis is better treated with rifampicin-loaded polymer nanoparticles (Thomas et al. 2018). The regulated release of tetracycline hydrochloride (TCH) was investigated and published by Hivechi et al., who also created NCC-reinforced polycaprolactone nanofibres (PCL) for the study. The gradual release of a medication was the consequence of enhancing the quantity of NCC contained within the PCL nanofibers (Hivechi et al. 2019).

Nanocellulose aerogel scaffolds made from bacterial cellulose (BC) were developed by Valo et al. (2013) for oral administration and demonstrated sustained drug release. For a wide range of medicinal nanoparticle applications, these nanocomposite architectures have proven invaluable. Hydrogels developed by Ahmad et al. (2014) were found to be biocompatible and mucoadhesive BNC-g-polyacrylic acid hydrogels with a pH-responsive swelling behavior and regulated albumin administration at higher pH of 6.8.

3.3.2 Bio Sensing

Using a biosensor, biosensing is frequently used to detect biomolecules. These biosensors combine a physicochemical detector with biological elements like sweat, saliva, or other essential body fluids. These detectors, as was previously said, might be large pieces of machinery as well as compact, wearable, or portable gadgets. Due to its mobility, the convenience of testing, quick diagnoses, need for less sample preparation, etc., the latter has recently received increased attention.

Electrochemical biosensors take information from biological sources and convert it into analytical signals such as voltage or current. The quantifiable electrical response of these sensors is typically utilized in the process of detecting analyte concentrations. In other words, the higher the analyte concentration at the electrode, the stronger the response. Detecting adenine and guanine in cells helps with DNA sequencing, protein metabolism, and oxidative damage assessment. (Thangaraj et al. 2014). In addition, these bases are used to assess immunological insufficiency and contribute to the diagnosis of illnesses like as Alzheimer's, HIV infection, cancer, and others (Zhang and Zhang 2018). In order to determine if RNA or DNA contains adenine or guanine as a base, a thin film containing NC and SWCNHs was developed. Antifouling properties and strong catalytic activity were expected from the film generated as a result. Linear and cyclic sweep voltammetry were used to identify these bases. For adenine, the detection limit was 1.4×10^{-6} mol L^{-1} and 1.7×10^{-7} mol L^{-1} (Subhedar et al. 2021). Other applications included detecting the levels of these bases in synthetic human serum and fish sperm. Spectroscopy is the primary diagnostic tool for blood tests. The eccrine and apocrine glands of human skin are responsible for the production of sweat, which acts as a transport medium for many substances, including carbohydrates, electrolytes, proteins, metabolites, acids, and hormones (Gao et al. 2016). These are known as biomarkers, and they contain vital information that may be used to analyze genetic abnormalities or infections, track sports performance, and so on (Zhao et al. 2019). Due to its mobility, convenience of use, and minimal need for sample preparation, enzyme sensors have gained substantial interest in the field of electrochemical sensing. The fabrication of an electrochemical platform for biosensing applications may be done on wearables such as rings, patches, bracelets, and tattoos. Conducting polymers are used in biosensors to facilitate quicker electron transport and offer a solid surface on which to immobilize proteins. By covering the PANi/CNC nanocomposite with a thin layer of ionic liquid (IL), a recent study sought to use the synergistic effects of the CNCs to host enzymes and biomolecules in order to produce a sensitive enzymatic cholesterol biosensor (Abdi et al. 2019). For the most accurate measurement of cholesterol, the nanocomposite ensured that electrons could pass between the enzyme and the surface. In food processing and clinical diagnostics, the constructed biosensor showed high stability, reproducibility, as well as operational repeatability, indicating that it might be utilized for determining cholesterol levels. In relation to the electrocatalytic capabilities of nanocomposites with high porosity, it has been demonstrated that IL/conducting polymers and CNC structures have synergistic features that may be coupled such that this material is an excellent choice for containing enzymes and biomolecules. A smart wearable sensing gadget (Figure 3.3) has recently made a significant advance in the field of health monitoring (Gomes et al. 2020). For the purpose of detecting lactate in synthetic sweat, the wearable sensor made use of BNC as a substrate. Lactate oxidase (LOx) immobilization on BNC substrate enabled success. The electrochemical platform is made using Prussian blue nanocubes on carbon-based electrodes and LOx on BNC. Effectively fulfilling their role as electron mediators for H_2O_2 is the purpose for which Prussian blue nanocubes were constructed. The biosensor detected lactate in synthetic sweat at concentrations ranging from 1.0 to 24.0 mmol L^{-1} with good results.

The human body is exposed to a potentially lethal level of toxicity from the widespread presence of heavy metals and metal ions including Cu^{2+} ions, Ag^{+} ions, and others. The degree of the difficulties that might arise as a result of ingesting such ions can range from nausea and vomiting to neurological illnesses, damaged organs (including the liver and kidneys), and even death (Saleem and Lee 2014; Gao et al. 2016). Rh-2, a Rhodamine B derivative, was dissolved in BNC by Milindanuth and Pisitsak (2018) to develop test strips. When used as a template, BNC demonstrates outstanding biocompatibility with Rh-2. Spectroscopy may use these strips

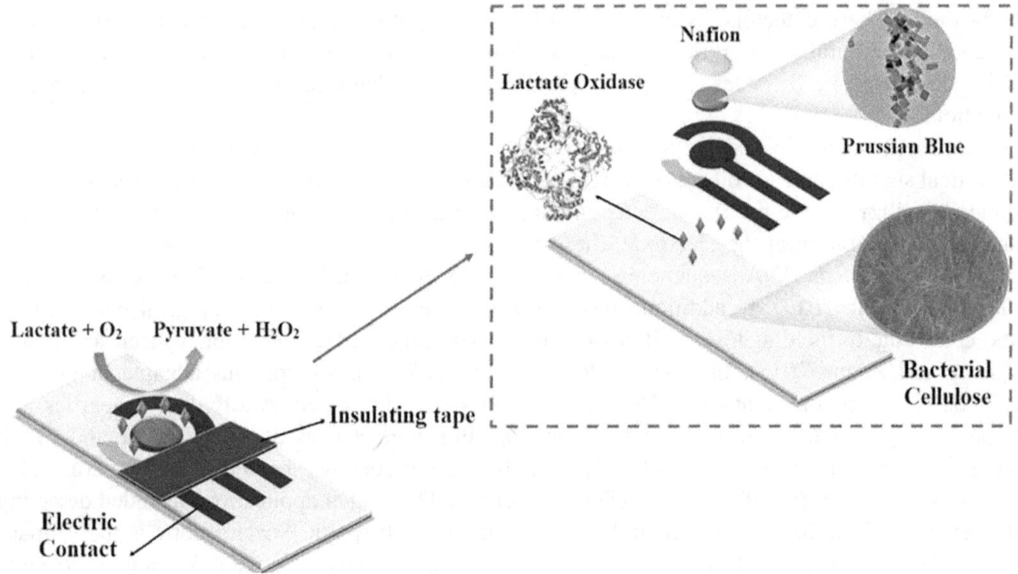

FIGURE 3.3 Smart wearable BNC-based biosensor [Reproduced with permission from (Gomes et al. 2020)].

because they allow color change to be detected with the naked eye. Cu^{2+} concentration and color intensity were shown to have a linear relationship. Copper ions caused the hue to shift from colorless to pink.

Through the utilization of NC, the sensor is transformed into a reasonably priced piece of analytical equipment that is not only disposable but also portable. Additionally, it serves as an excellent substrate for the creation of sensors that detect the presence of heavy metal ions. Colorimetric sensors based on NC grafted with 2,5-dithiourea (DTu) for the detection of Cu^{2+} and Ag^{+} ions are examples of such sensors (Guo et al. 2019). When the amount of Ag^{+} ions in the solution was increased, the color of the sensor changed from white to a yellow-red hue. When the amount of Cu^{2+} ions was increased, the color of the sensor changed from white to a light gray color. A 5 s reaction time was achieved after the ions were added. In addition to the test, the sensor showed potential in terms of speed and simplicity when identifying these ions in actual water. The use of fluorescent materials in composites and other sensor platform components allows the naked eye to detect color change. Spectroscopy might be made much simpler using this approach. It has proven possible to create NC-based colorimetric sensors using materials like carbon quantum dots (CQDs) and fluorescent elastase peptide. In order to create a fluorescent sensing platform for laccase enzyme detection, Ruiz-Palomero et al. (2017) fixed quantum dots of graphene that had been doped with nitrogen and sulfur onto TEMPO-oxidized NC hydrogels. It was stated that the composite has a detection limit of 0.048 U-mL1, which is quite low. It was discovered that these composite hydrogels are susceptible to the carcinogenic 2,4,5-trichlorophenol (TCP) that is present in herbicides and fungicides. During testing of the hydrogel against TCP in specified samples of water and red wine, high-intensity fluorescence was detected (Ruiz-Palomero et al. 2017). Fontenot et al. (2016) created an NC-elastase peptide biosensor to recognize the inflammatory illness marker Human Neutrophil Elastase (HNE). Fluorescent groups are produced when the peptide and biomarker react. NC has also been used to create many platforms for fluorescent bioimaging. In recent research, an elastase tripeptide-containing sensor was inserted into a layered wound dressing for the purpose of detecting HNE, with the top layer protecting the wound and the bottom layer supplying vital fluid flow (Fontenot et al. 2020). The dressing had to be taken off in order to be detected, and it had to be placed in front of a UV lamp source. High-intensity fluorescent light from the sensor would signal saturation in respect to the detection limit. This demonstrated efficient wound healing.

3.3.3 Tissue Engineering

The extracellular matrix required for specialized tissues may be effortlessly mimicked by a number of treatments that can produce the nanocellulosic in a 3D form. This idea is easily applicable to cell screening, cancer propagation, bone tissue engineering, etc. Recent research demonstrated that bone marrow's nanostructured collagen may be easily mimicked by nanocellulosic 3D structure hydrogels. A nanocellulosic fibril framework for skin tissue culture was utilized in order to produce a nanocellulosic bi-layered skin structure for the treatment of dermal and epidermal infections (Subhedar et al. 2021). Tissue engineering, which involves replacing damaged cells with brand-new cells, is related to wound healing. It is a sophisticated and well-orchestrated procedure carried out by a live body. The ability of porous and connected materials to deliver antibiotics or other medications to the desired location makes them suitable substrates for wound healing. Additionally, it guards against secondary infections brought on by external pathogens and serves as a barrier for microorganisms. A promising contender for wound healing is nanocellulose which incorporates glycosides, polysaccharides, proteins, and local anesthetics (Wiegand et al. 2015). Lysozyme nanofibers (LNFs)/NFC patch was developed by Silva et al. (2020) for use in wound healing applications. The white eggs served as the source of LNFs.

Nanocellulose is also being utilized in the field of dental tissue engineering in a significant way. For use in tissue engineering applications, Peng et al. (2021) created magnesium oxide (MgO) nanoparticles that included PCL/gelatin nanocellulose membranes using the coaxial electrospinning technique. Human periodontal ligament stem cells responded better to the new membrane's increased biocompatibility (hPDLSCs). Seven days of treatment with MgO nanoparticles resulted in an increase in ALP activity in PCL/gelatin nanocellulose membranes. Antibacterial properties of the newly created membrane were discovered. Furthermore, hPDLSCs membrane-treated groups showed greater overexpression of osteogenic-associated gene markers such as Col1 and Runx2 than the control group.

3.3.4 Encapsulation

When it comes to encapsulating bioactive compounds, nanocellulose is becoming increasingly popular because of its universally accepted safety and positive biological features, such as biocompatibility and nontoxicity. The antibacterial activities of nanocellulose in conjunction with other natural/active substances have been the subject of recent research.

Nanocellulose is an extremely important component of the delivery system. A team of researchers has only recently revealed that they were able to immobilize L. acidophilus 016 on a bacterial CNF using the adsorption-incubation approach and that the bacteria were able to live for up to 24 days. In addition, around 71% of the total population of lactobacillus was able to survive the preservation process (Jayani et al. 2020). Pectin, alginate, and nanocellulose were used in the preparation of a standardized probiotic tablet formulation by Huq et al. (2016). This formulation was successful in maintaining approximately 84% of the cell viability of Lactobacillus rhamnosus ATCC 9595 after being subjected to a number of different gastrointestinal media. After 42 days of storage, the only change that was observed was a decrease of between 0.4 and 0.2 log CFU per tablet. In order to distribute curcumin, (Gunathilake et al. 2018) developed a nonionic surfactant integrated CS/NC hydrogel using a chemical process. After being released in vitro, the medication retained its chemical property, as demonstrated by the UV-visible spectrum. Fascinatingly, Bacillus coagulans can thrive when bacterial nanocelluloses are used because they behave as a prebiotic. As a nanoscale prebiotic biopolymer, bacterial nanocellulose enhances the encapsulation of probiotics. Additional improvements in B. coagulans storage stability were achieved by employing a bionanocomposite made up of bacterial nanocelluloses, pectin, and Schizophyllum commune extract as a novel matrix (Khorasani and Shojaosadati 2016).

In a study that was published in 2020, Maleki et al. microencapsulated the probiotic Lactobacillus rhamnosus ATCC 7469 in a composite made of whey protein isolate, NC, and inulin (Maleki et al. 2020). The designed capsule significantly improved the probiotics' capacity to survive in settings that were simulated to be similar to those in the gastrointestinal tract. For the delivery of curcumin, (Wang et al. 2020) produced composite nanoparticles of soy protein isolate and nanocellulose. In the simulated gastrointestinal circumstances, the complex composite nanoparticles demonstrated improved encapsulation efficiency and controlled release. In the delivery method for the active chemicals, nanocellulose is crucial.

3.3.5 Other Applications

The healing of skin wounds was the first medicinal use of bacterial nanocellulose that was mentioned. Bacterial nanocellulose (BNC), which has hydrophilic qualities, has been found to be a superior hydrogel for the creation of dressings. However, it wasn't until 2006 that Polish researchers from Lodz University of Technology published the first results of their use of BNC membranes in the clinical treatment of second and third-degree burns. In the 1990s, numerous scientific groups and businesses conducted ground breaking research on its use as a never-dried wound dressing, primarily in Brazil and USA (Czaja et al. 2006). According to the authors, the never-dried cellulose membrane that is used as a wound dressing is represented by major properties that significantly contribute in the healing process of patients. Some of these qualities are as follows: 1) excellent conformability and adhesion to moving body parts; 2) cooling properties, in addition to the maintenance of an appropriate level of moisture in the surrounding area of the wound; 3) absorption of the wound's exudates; 4) transparency, which enables continuous clinical monitoring of the improvement in the patient's condition; and 5) O_2 permeability accelerating the process of the skin's natural regeneration. The investigations have shown that, compared to conventional treatment approaches, BNC membranes considerably accelerate the process of reepithelialization by facilitating the creation of scars by removing necrotic material and improving the growth of granulation tissue (Ludwicka et al. 2016). In addition, compared to control procedures, BNC dressings significantly reduced the amount of daily wound care required and significantly reduced patient discomfort. A Polish firm, BOWIL Biotech Ltd., has purchased the Lodz University of Technology-developed BNC product production technology and has since begun producing BNC wound dressings and BNC-assisted cosmetic goods under the brand name CELMAT®. Aside from this business, the medical market is already aware of a number other BNC-producing businesses. For pure, moist BNC membrane, the claimed outstanding wound healing results were clinically seen. However, there have been a lot of changes made to this dressing material recently as stated in the literature. The main objectives of this type of burns therapy are to close the wound as quickly and effectively as possible in order to speed up healing and relieve pain right away. Additionally, the correct wound care must keep it from drying up and contracting an infection. Therefore, the primary study areas are focused on enhancing the antibacterial characteristics of BNC, most frequently by creating nanocomposites with elements like zinc oxide or silver (adding nanosilver). Another approach is the use of various medications (antibiotics, analgesics) embedded in cellulose structures and released gradually throughout the patient's treatment with the intention of reducing pain or accelerating the healing process. The latter is also accomplished by combining BNC with substances that improve skin hydration, produce protective layers on the epidermis, or smooth and revitalize the skin, such as acrylic acid, chitosan, or hyaluronic acid. Souza et al. (2014) claim that the combination of BNC and AMSC membrane was successful and enabled for cell growth and distribution to the wounded tissue. Additionally, it was found that the delivered cells took part in the wound's regeneration technique and that, based on their structure, the membranes consisting of AMSC stimulated cellular growth and wound healing to varying degrees, suggesting that cellulose dressings can serve as an alternative to bioactive cures.

3.4 FUTURE PERSPECTIVES

Future study will largely require sustainable, environmentally friendly, and green extraction procedures, including enzymatic hydrolysis and mechanical extraction, in order to improve the approaches to adjust the features and qualities of NC. The majority of technologically sophisticated studies have been successfully completed on a lab size, but more research requires more exact optimization to be completed on an industrial scale. The life cycle assessment (LCA) of NC biocomposites should also be conducted while taking a number of environmental considerations into account. BNC is particularly intriguing; yet, this substitute is less practical due to the restricted scale-up of BNC goods to the commercial scale, which is connected to its existing high manufacturing costs for culture medium and slow processing. The development of cleaner methods for producing NC from wood/plant biomass, which offers non-toxic and morphologically distinct products for certain pharmaceutical and biomedical businesses, is very intriguing. In the near future, NC-based biomaterials may be developed as a potential alternative to address some issues with other biomedical materials because biomaterial advances in the biomedical area are also expanding quickly. Numerous medicinal plants' extracts have a variety of qualities in addition to being biocompatible and non-toxic. Therefore, it is likely that these natural extracts combined with NC will be studied in the future. Particularly intriguing is the fact that its biocompatibility may be altered by surface derivatization to induce hydrophobic, hydrophilic, and partly water-soluble properties.

3.5 CONCLUSIONS

This chapter offers a comprehensive analysis for the production, modifications, and uses of nanocellulose in the field of biomedicine, bringing the reader up to speed on the most recent findings. It possesses outstanding physicochemical qualities, including excellent mechanical properties, low density, biodegradability, and biocompatibility, nanocellulose. Due to these properties, this bio-polymer. has garnered the interest of a significant number of researchers. It is possible to derive nanocellulose from a diverse range of sources, including plants, microbes, and algae. There are three unique varieties of nanocellulose: CNC, CNF, and BNC. Each form has its own set of features. In contrast to CNFs, which have both amorphous and crystalline areas within their networks, CNCs take the form of needles and have a largely crystalline structure. The most pristine type of nanocellulose is called BNC. Chemical functionalization makes it possible to alter the characteristics of nanocellulose, which expands the material's range of potential uses in the biomedical field. Applications for functionalized nanocellulose include those in the fields of drug carriers and medicine, as well as those in the treatment of wastewater and the packaging of food, amongst other fields. The malleability of nanocellulose makes it an attractive candidate for use as a medication carrier. A variety of chemical and physical changes can be used to ensure the long-term and targeted administration of nanocellulose's active molecules. Nanocellulose contents and composites have an effect on the drug loading capacity. In addition to biosensing, encapsulation, and tissue engineering, nanocellulose has shown promising results. The tissue engineering potential of patches made of nanocellulose was found to be superior. Because of these observations, we came to the conclusion that nanocellulose is a promising material because it has the potential to be utilized to deliver a variety of compounds, and its antibacterial qualities are necessary for applications in the biomedical field. It is possible that in the not-too-distant future, more research will be carried out on the creation of novel hybrids that possess antibacterial capabilities without compromising their unique traits.

REFERENCES

Abdi, Mahnaz M., Rawaida Liyana Razalli, Paridah Md Tahir, Naz Chaibakhsh, Maryam Hassani, and Mahdi Mir. 2019. "Optimized fabrication of newly cholesterol biosensor based on nanocellulose." *International Journal of Biological Macromolecules* 126:1213–1222.

Ahmad, Naveed, Mohd Cairul Iqbal Mohd Amin, Shalela Mohd Mahali, Ismanizan Ismail, and Victor Tuan Giam Chuang. 2014. "Biocompatible and mucoadhesive bacterial cellulose-g-poly(acrylic acid) hydrogels for oral protein delivery." *Molecular Pharmaceutics* 11 (11):4130–4142. doi: 10.1021/mp5003015.

Arantes, Valdeir, Isabella K. R. Dias, Gabriela L. Berto, Barbara Pereira, Braz S. Marotti, and Carlaile F. O. Nogueira. 2020. "The current status of the enzyme-mediated isolation and functionalization of nanocelluloses: production, properties, techno-economics, and opportunities." *Cellulose* 27 (18): 10571–10630.

Cheng, Qingzheng, Siqun Wang, and David P. Harper. 2009. "Effects of process and source on elastic modulus of single cellulose fibrils evaluated by atomic force microscopy." *Composites Part A: Applied Science and Manufacturing* 40 (5):583–588.

Czaja, Wojciech, Alina Krystynowicz, Stanislaw Bielecki, and R Malcolm Brown Jr. 2006. "Microbial cellulose—the natural power to heal wounds." *Biomaterials* 27 (2):145–151.

Das, Pranjal P., Anweshan, and Mihir K. Purkait. 2021. "Treatment of cold rolling mill (CRM) effluent of steel industry." *Separation and Purification Technology* 274:119083. doi: 10.1016/j.seppur.2021.119083.

Das, Pranjal P., Piyal Mondal, Anweshan, A. Sinha, P. Biswas, S. Sarkar, and Mihir K. Purkait. 2021. "Treatment of steel plant generated biological oxidation treated (BOT) wastewater by hybrid process." *Separation and Purification Technology* 258:118013. doi: 10.1016/j.seppur.2020.118013.

Das, Pranjal P., Mukesh Sharma, and Mihir K. Purkait. 2022. "Recent progress on electrocoagulation process for wastewater treatment: A review." *Separation and Purification Technology* 292:121058. doi: 10.1016/j.seppur.2022.121058.

de la Motte, Hanna, Merima Hasani, Harald Brelid, and Gunnar Westman. 2011. "Molecular characterization of hydrolyzed cationized nanocrystalline cellulose, cotton cellulose and softwood kraft pulp using high resolution 1D and 2D NMR." *Carbohydrate Polymers* 85 (4):738–746.

Debnath, Banhisikha, Prangan Duarah, Dibyajyoti Haldar, and Mihir Kumar Purkait. 2022. "Improving the properties of corn starch films for application as packaging material via reinforcement with microcrystalline cellulose synthesized from elephant grass." *Food Packaging and Shelf Life* 34:100937. doi: 10.1016/j.fpsl.2022.100937.

Debnath, Banhisikha, Dibyajyoti Haldar, and Mihir Kumar Purkait. 2021. "A critical review on the techniques used for the synthesis and applications of crystalline cellulose derived from agricultural wastes and forest residues." *Carbohydrate Polymers* 273:118537. doi: 10.1016/j.carbpol.2021.118537.

Debnath, Banhisikha, Dibyajyoti Haldar, and Mihir Kumar Purkait. 2022. "Environmental remediation by tea waste and its derivative products: A review on present status and technological advancements." *Chemosphere* 300:134480. doi: 10.1016/j.chemosphere.2022.134480.

Dri, Fernando L., Louis G. Hector, Robert J. Moon, and Pablo D. Zavattieri. 2013. "Anisotropy of the elastic properties of crystalline cellulose Iβ from first principles density functional theory with Van der Waals interactions." *Cellulose* 20 (6):2703–2718.

Duarah, Prangan, Dibyajyoti Haldar, and Mihir Kumar Purkait. 2020. "Technological advancement in the synthesis and applications of lignin-based nanoparticles derived from agro-industrial waste residues: A review." *International Journal of Biological Macromolecules* 163:1828–1843. doi: 10.1016/j.ijbiomac.2020.09.076.

Duarah, Prangan, Dibyajyoti Haldar, Reeta Rani Singhania, Cheng-Di Dong, Anil Kumar Patel, and Mihir Kumar Purkait. 2023. "Sustainable management of tea wastes: resource recovery and conversion techniques." *Critical Reviews in Biotechnology*:1–20. doi: 10.1080/07388551.2022.2157701.

Duarah, Prangan, Dibyajyoti Haldar, V. S. K. Yadav, and Mihir Kumar Purkait. 2021. "Progress in the electrochemical reduction of CO2 to formic acid: A review on current trends and future prospects." *Journal of Environmental Chemical Engineering* 9 (6):106394. doi: 10.1016/j.jece.2021.106394.

Ekezie, Flora-Glad Chizoba, Da-Wen Sun, Zhang Han, and Jun-Hu Cheng. 2017. "Microwave-assisted food processing technologies for enhancing product quality and process efficiency: A review of recent developments." *Trends in Food Science & Technology* 67:58–69.

Elazzouzi-Hafraoui, Samira, Yoshiharu Nishiyama, Jean-Luc Putaux, Laurent Heux, Frédéric Dubreuil, and Cyrille Rochas. 2008. "The shape and size distribution of crystalline nanoparticles prepared by acid hydrolysis of native cellulose." *Biomacromolecules* 9 (1):57–65.

Emaraa, Laila H, Ahmed A. El-Ashmawya, Nesrin F. Tahaa, Khaled A. El-Shaffeib, El-Sayed M Mahdeyb, and Heba K. Elkhollyc. 2016. "Nano-crystalline cellulose as a novel tablet excipient for improving solubility and dissolution of meloxicam." *Journal of Applied Pharmaceutical Science* 6 (2):032–043.

Eyley, Samuel, and Wim Thielemans. 2014. "Surface modification of cellulose nanocrystals." *Nanoscale* 6 (14):7764–7779.

Fontenot, Krystal R., J Vincent Edwards, David Haldane, Elena Graves, Michael Santiago Citron, Nicolette T. Prevost, Alfred D. French, and Brian D. Condon. 2016. "Human neutrophil elastase detection with fluorescent peptide sensors conjugated to cellulosic and nanocellulosic materials: part II, structure/function analysis." *Cellulose* 23 (2):1297–1309.

Fontenot, Krystal R., J Vincent Edwards, David Haldane, Nicole Pircher, Falk Liebner, Sunghyun Nam, and Brian D. Condon. 2020. "Structure/function relations of chronic wound dressings and emerging concepts on the interface of nanocellulosic sensors." *Lignocellulosics* :249–278.

Gao, Jiali, Qing Li, Wenshuai Chen, Yixing Liu, and Haipeng Yu. 2014. "Self-assembly of nanocellulose and indomethacin into hierarchically ordered structures with high encapsulation efficiency for sustained release applications." *ChemPlusChem* 79 (5):725–731. doi: 10.1002/cplu.201300434.

Gao, Runjiao, Gang Xu, Lishuo Zheng, Yujia Xie, Minli Tao, and Wenqin Zhang. 2016. "A highly selective and sensitive reusable colorimetric sensor for Ag+ based on thiadiazole-functionalized polyacrylonitrile fiber." *Journal of Materials Chemistry C* 4 (25):5996–6006.

Gao, Wei, Sam Emaminejad, Hnin Yin Yin Nyein, Samyuktha Challa, Kevin Chen, Austin Peck, Hossain M. Fahad, Hiroki Ota, Hiroshi Shiraki, and Daisuke Kiriya. 2016. "Fully integrated wearable sensor arrays for multiplexed in situ perspiration analysis." *Nature* 529 (7587):509–514.

García, Araceli, Alessandro Gandini, Jalel Labidi, Naceur Belgacem, and Julien Bras. 2016. "Industrial and crop wastes: A new source for nanocellulose biorefinery." *Industrial Crops and Products* 93: 26–38.

Gomes, Nathalia Oezau, Emanuel Carrilho, Sergio Antonio Spinola Machado, and Livia Florio Sgobbi. 2020. "Bacterial cellulose-based electrochemical sensing platform: A smart material for miniaturized biosensors." *Electrochimica Acta* 349:136341.

Gunathilake, Thennakoon M Sampath Udeni, Yern Chee Ching, Cheng Hock Chuah, Hazlee Azil Illias, Kuan Yong Ching, Ramesh Singh, and Liou Nai-Shang. 2018. "Influence of a nonionic surfactant on curcumin delivery of nanocellulose reinforced chitosan hydrogel." *International Journal of Biological Macromolecules* 118:1055–1064.

Guo, Ting, Ying Pei, Keyong Tang, Xichan He, Jinbao Huang, and Fang Wang. 2017. "Mechanical and drug release properties of alginate beads reinforced with cellulose." *Journal of Applied Polymer Science* 134 (8). doi: 10.1002/app.44495.

Guo, Wei, Hui He, Hongxiang Zhu, Xudong Hou, Xingjuan Chen, Shile Zhou, Shuangfei Wang, Lingtao Huang, and Jiehan Lin. 2019. "Preparation and properties of a biomass cellulose-based colorimetric sensor for Ag+ and Cu2+." *Industrial Crops and Products* 137:410–418.

Häbel, Henrike, Helene Andersson, Anna Olsson, Eva Olsson, Anette Larsson, and Aila Särkkä. 2016. "Characterization of pore structure of polymer blended films used for controlled drug release." *Journal of Controlled Release* 222:151–158. doi: 10.1016/j.jconrel.2015.12.011.

Habibi, Youssef, Lucian A. Lucia, and Orlando J. Rojas. 2010. "Cellulose nanocrystals: chemistry, self-assembly, and applications." *Chemical Reviews* 110 (6):3479–3500.

Haldar, Dibyajyoti, Prangan Duarah, and Mihir Kumar Purkait. 2022. "Chapter 16 - Progress in the synthesis and applications of polymeric nanomaterials derived from waste lignocellulosic biomass." In *Advanced Materials for Sustainable Environmental Remediation*, edited by Dimitrios Giannakoudakis, Lucas Meili and Ioannis Anastopoulos, 419–433. Elsevier.

Hassan, Shady S., Gwilym A. Williams, and Amit K. Jaiswal. 2018. "Emerging technologies for the pretreatment of lignocellulosic biomass." *Bioresource Technology* 262:310–318.

Hivechi, Ahmad, S. Hajir Bahrami, and Ronald A. Siegel. 2019. "Drug release and biodegradability of electrospun cellulose nanocrystal reinforced polycaprolactone." *Materials Science and Engineering: C* 94:929–937. doi: 10.1016/j.msec.2018.10.037.

Hsieh, Y-C, H. Yano, M. Nogi, and S. J. Eichhorn. 2008. "An estimation of the Young's modulus of bacterial cellulose filaments." *Cellulose* 15 (4):507–513.

Huang, Xingyan, Cornelis F De Hoop, Feng Li, Jiulong Xie, Chung-Yun Hse, Jinqiu Qi, Yongze Jiang, and Yuzhu Chen. 2017. "Dilute alkali and hydrogen peroxide treatment of microwave liquefied rape straw residue for the extraction of cellulose nanocrystals." *Journal of Nanomaterials* 2017:1–9.

Huq, Tanzina, Khanh Dang Vu, Bernard Riedl, Jean Bouchard, Jaejoon Han, and Monique Lacroix. 2016. "Development of probiotic tablet using alginate, pectin, and cellulose nanocrystals as excipients." *Cellulose* 23 (3):1967–1978.

Jackson, J. K., K. Letchford, B. Z. Wasserman, L. Ye, W. Y. Hamad, and H. M. Burt. 2011. "The use of nanocrystalline cellulose for the binding and controlled release of drugs." *International Journal of Nanomedicine* 6:321–330. doi: 10.2147/ijn.s16749.

Jayani, T., B. Sanjeev, S. Marimuthu, and Sivakumar Uthandi. 2020. "Bacterial Cellulose Nano Fiber (BCNF) as carrier support for the immobilization of probiotic, Lactobacillus acidophilus 016." *Carbohydrate Polymers* 250:116965.

Kargarzadeh, Hanieh, Marcos Mariano, Deepu Gopakumar, Ishak Ahmad, Sabu Thomas, Alain Dufresne, Jin Huang, and Ning Lin. 2018. "Advances in cellulose nanomaterials." *Cellulose* 25 (4):2151–2189.

Khorasani, Alireza Chackoshian, and Seyed Abbas Shojaosadati. 2016. "Bacterial nanocellulose-pectin bionanocomposites as prebiotics against drying and gastrointestinal condition." *International Journal of Biological Macromolecules* 83:9–18.

Kolakovic, Ruzica, Leena Peltonen, Timo Laaksonen, Kaisa Putkisto, Antti Laukkanen, and Jouni Hirvonen. 2011. "Spray-dried cellulose nanofibers as novel tablet excipient." *Aaps Pharmscitech* 12 (4):1366–1373.

Kumar, Bikash, Nisha Bhardwaj, Komal Agrawal, Venkatesh Chaturvedi, and Pradeep Verma. 2020. "Current perspective on pretreatment technologies using lignocellulosic biomass: An emerging biorefinery concept." *Fuel Processing Technology* 199:106244.

Lee, Minwoo, Min Haeng Heo, Hyunho Lee, Hwi-Hui Lee, Haemin Jeong, Young-Wun Kim, and Jihoon Shin. 2018. "Facile and eco-friendly extraction of cellulose nanocrystals via electron beam irradiation followed by high-pressure homogenization." *Green Chemistry* 20 (11):2596–2610.

Lin, Ning, and Alain Dufresne. 2014a. "Nanocellulose in biomedicine: Current status and future prospect." *European Polymer Journal* 59:302–325. doi: 10.1016/j.eurpolymj.2014.07.025.

Lin, Ning, and Alain Dufresne. 2014b. "Surface chemistry, morphological analysis and properties of cellulose nanocrystals with gradiented sulfation degrees." *Nanoscale* 6 (10):5384–5393.

Liyanage, Sumedha, Sanjit Acharya, Prakash Parajuli, Julia L. Shamshina, and Noureddine Abidi. 2021. "Production and surface modification of cellulose bioproducts." *Polymers* 13 (19):3433.

Lu, Qilin, Linna Lu, Yonggui Li, Yuxin Yan, Zhaofeng Fang, Xin Chen, and Biao Huang. 2019. "High-yield synthesis of functionalized cellulose nanocrystals for nano-biocomposites." *ACS Applied Nano Materials* 2 (4):2036–2043.

Ludwicka, Karolina, Marzena Jedrzejczak-Krzepkowska, Katarzyna Kubiak, Marek Kolodziejczyk, Teresa Pankiewicz, and Stanislaw Bielecki. 2016. "Medical and cosmetic applications of bacterial nanocellulose." In *Bacterial Nanocellulose*, 145–165. Elsevier.

Maleki, Omid, Mohammad Alizadeh Khaledabad, Saber Amiri, Asghar Khosrowshahi Asl, and Sina Makouie. 2020. "Microencapsulation of Lactobacillus rhamnosus ATCC 7469 in whey protein isolate-crystalline nanocellulose-inulin composite enhanced gastrointestinal survivability." *LWT* 126: 109224.

Michelin, Michele, Daniel G. Gomes, Aloia Romaní, Maria de Lourdes TM Polizeli, and José A Teixeira. 2020. "Nanocellulose production: exploring the enzymatic route and residues of pulp and paper industry." *Molecules* 25 (15):3411.

Milindanuth, Panisa, and Penwisa Pisitsak. 2018. "A novel colorimetric sensor based on rhodamine-B derivative and bacterial cellulose for the detection of Cu (II) ions in water." *Materials Chemistry and Physics* 216:325–331.

Mishra, Shweta, Prashant S. Kharkar, and Anil M. Pethe. 2019. "Biomass and waste materials as potential sources of nanocrystalline cellulose: Comparative review of preparation methods (2016–Till date)." *Carbohydrate Polymers* 207:418–427.

Miyashiro, Daisuke, Ryo Hamano, and Kazuo Umemura. 2020. "A review of applications using mixed materials of cellulose, nanocellulose and carbon nanotubes." *Nanomaterials* 10 (2):186.

Mohanta, Vaishakhi, Giridhar Madras, and Satish Patil. 2014. "Layer-by-layer assembled thin films and microcapsules of nanocrystalline cellulose for hydrophobic drug delivery." *ACS Applied Materials & Interfaces* 6 (22):20093–20101.

Nasir, Mohammed, Rokiah Hashim, Othman Sulaiman, and Mohd Asim. 2017. "Nanocellulose: Preparation methods and applications." In *Cellulose-Reinforced Nanofibre Composites*, 261–276. Elsevier.

Nechyporchuk, Oleksandr, Mohamed Naceur Belgacem, and Julien Bras. 2016. "Production of cellulose nanofibrils: A review of recent advances." *Industrial Crops and Products* 93:2–25.

Ng, Law Yong, Ting Jun Wong, Ching Yin Ng, and Chiang Kar Mun Amelia. 2021. "A review on cellulose nanocrystals production and characterization methods from Elaeis guineensis empty fruit bunches." *Arabian Journal of Chemistry* 14 (9):103339.

Patil, Mayur D., Vishal D. Patil, Aditya A. Sapre, Tushar S. Ambone, A. T. Arun Torris, Parshuram G. Shukla, and Kadhiravan Shanmuganathan. 2018. "Tuning controlled release behavior of starch granules using nanofibrillated cellulose derived from waste sugarcane bagasse." *ACS Sustainable Chemistry & Engineering* 6 (7):9208–9217. doi: 10.1021/acssuschemeng.8b01545.

Patil, Tejal V., Dinesh K. Patel, Sayan Deb Dutta, Keya Ganguly, Tuhin Subhra Santra, and Ki-Taek Lim. 2022. "Nanocellulose, a versatile platform: From the delivery of active molecules to tissue engineering applications." *Bioactive Materials* 9:566–589. doi: 10.1016/j.bioactmat.2021.07.006.

Peng, Wenzao, Shuangshuang Ren, Yibo Zhang, Ruyi Fan, Yi Zhou, Lu Li, Xuanwen Xu, and Yan Xu. 2021. "MgO Nanoparticles-incorporated PCL/gelatin-derived coaxial electrospinning nanocellulose membranes for periodontal tissue regeneration." *Frontiers in Bioengineering and Biotechnology* 9:668428.

Pennells, Jordan, Ian D. Godwin, Nasim Amiralian, and Darren J. Martin. 2020. "Trends in the production of cellulose nanofibers from non-wood sources." *Cellulose* 27 (2):575–593.

Perumal, Anand Babu, Reshma B. Nambiar, J. A. Moses, and C. Anandharamakrishnan. 2022. "Nanocellulose: Recent trends and applications in the food industry." *Food Hydrocolloids* 127:107484. doi: 10.1016/j.foodhyd.2022.107484.

Phanthong, Patchiya, Prasert Reubroycharoen, Xiaogang Hao, Guangwen Xu, Abuliti Abudula, and Guoqing Guan. 2018. "Nanocellulose: Extraction and application." *Carbon Resources Conversion* 1 (1):32–43.

Pradhan, Dileswar, Amit K. Jaiswal, and Swarna Jaiswal. 2022. "Emerging technologies for the production of nanocellulose from lignocellulosic biomass." *Carbohydrate Polymers* 285:119258. doi: 10.1016/j.carbpol.2022.119258.

Raghav, N., Manishita R. Sharma, and John F. Kennedy. 2021. "Nanocellulose: A mini-review on types and use in drug delivery systems." *Carbohydrate Polymer Technologies and Applications* 2:100031. doi: /10.1016/j.carpta.2020.100031.

Rana, Ashvinder Kumar, Elisabete Frollini, and Vijay Kumar Thakur. 2021. "Cellulose nanocrystals: Pretreatments, preparation strategies, and surface functionalization." *International Journal of Biological Macromolecules* 182:1554–1581.

Ribeiro, Ruan S. A., Bruno C. Pohlmann, Veronica Calado, Ninoska Bojorge, and Nei Pereira Jr. 2019. "Production of nanocellulose by enzymatic hydrolysis: Trends and challenges." *Engineering in Life Sciences* 19 (4):279–291.

Ruiz-Palomero, Celia, Sandra Benítez-Martínez, M Laura Soriano, and Miguel Valcárcel. 2017. "Fluorescent nanocellulosic hydrogels based on graphene quantum dots for sensing laccase." *Analytica Chimica Acta* 974:93–99.

Ruiz-Palomero, Celia, M Laura Soriano, Sandra Benitez-Martinez, and Miguel Valcarcel. 2017. "Photoluminescent sensing hydrogel platform based on the combination of nanocellulose and S, N-codoped graphene quantum dots." *Sensors and Actuators B: Chemical* 245:946–953.

Salari, Mahdieh, Mahmood Sowti Khiabani, Reza Rezaei Mokarram, Babak Ghanbarzadeh, and Hossein Samadi Kafil. 2019. "Preparation and characterization of cellulose nanocrystals from bacterial cellulose produced in sugar beet molasses and cheese whey media." *International Journal of Biological Macromolecules* 122:280–288.

Saleem, Muhammad, and Ki-Hwan Lee. 2014. "Selective fluorescence detection of Cu2+ in aqueous solution and living cells." *Journal of Luminescence* 145:843–848.

Silva, Nuno H. C. S., Patrícia Garrido-Pascual, Catarina Moreirinha, Adelaide Almeida, Teodoro Palomares, Ana Alonso-Varona, Carla Vilela, and Carmen S. R. Freire. 2020. "Multifunctional nanofibrous patches composed of nanocellulose and lysozyme nanofibers for cutaneous wound healing." *International Journal of Biological Macromolecules* 165:1198–1210.

Souza, C. M. C. O., L. A. F. Mesquita, D. Souza, A. C. Irioda, J. C. Francisco, C. F. Souza, L. C. Guarita-Souza, M. R. Sierakowski, and K. A. T. Carvalho. 2014. "Regeneration of skin tissue promoted by mesenchymal stem cells seeded in nanostructured membrane." *Transplantation Proceedings*, 46, 1882–1886. 10.1016/j.transproceed.2014.05.066.

Subhedar, Aditya, Swarnim Bhadauria, Sandeep Ahankari, and Hanieh Kargarzadeh. 2021. "Nanocellulose in biomedical and biosensing applications: A review." *International Journal of Biological Macromolecules* 166:587–600.

Supramaniam, Jagadeesen, Rohana Adnan, Noor Haida Mohd Kaus, and Rani Bushra. 2018. "Magnetic nanocellulose alginate hydrogel beads as potential drug delivery system." *International Journal of Biological Macromolecules* 118:640–648. doi: 10.1016/j.ijbiomac.2018.06.043.

Teo, Hwee Li, and Roswanira Abdul Wahab. 2020. "Towards an eco-friendly deconstruction of agro-industrial biomass and preparation of renewable cellulose nanomaterials: A review." *International Journal of Biological Macromolecules* 161:1414–1430.

Thangaraj, Rajendiran, Subramanian Nellaiappan, Raja Sudhakaran, and Annamalai Senthil Kumar. 2014. "A flow injection analysis coupled dual electrochemical detector for selective and simultaneous detection of guanine and adenine." *Electrochimica Acta* 123:485–493.

Thomas, Deepa, M. S. Latha, and K. Kurien Thomas. 2018. "Synthesis and in vitro evaluation of alginate-cellulose nanocrystal hybrid nanoparticles for the controlled oral delivery of rifampicin." *Journal of Drug Delivery Science and Technology* 46:392–399. doi: 10.1016/j.jddst.2018.06.004.

Trache, Djalal, Vijay Kumar Thakur, and Rabah Boukherroub. 2020. "Cellulose nanocrystals/graphene hybrids—a promising new class of materials for advanced applications." *Nanomaterials* 10 (8):1523.

Turbak, Albin F., Fred W. Snyder, and Karen R. Sandberg. 1983. "Microfibrillated cellulose, a new cellulose product: properties, uses, and commercial potential." *Journal of Applied Polymer Science* 37:815–827.

Tyagi, Neha, and Sumathi Suresh. 2016. "Production of cellulose from sugarcane molasses using Gluconacetobacter intermedius SNT-1: optimization & characterization." *Journal of Cleaner Production* 112:71–80.

Ul-Islam, Mazhar, Shaukat Khan, Muhammad Wajid Ullah, and Joong Kon Park. 2015. "Bacterial cellulose composites: Synthetic strategies and multiple applications in bio-medical and electro-conductive fields." *Biotechnology Journal* 10 (12):1847–1861.

Valo, Hanna, Suvi Arola, Päivi Laaksonen, Mika Torkkeli, Leena Peltonen, Markus B. Linder, Ritva Serimaa, Shigenori Kuga, Jouni Hirvonen, and Timo Laaksonen. 2013. "Drug release from nanoparticles embedded in four different nanofibrillar cellulose aerogels." *European Journal of Pharmaceutical Sciences* 50 (1):69–77. doi: 10.1016/j.ejps.2013.02.023.

Wang, Songyan, Yuqing Lu, Xiao-kun Ouyang, and Junhong Ling. 2020. "Fabrication of soy protein isolate/cellulose nanocrystal composite nanoparticles for curcumin delivery." *International Journal of Biological Macromolecules* 165:1468–1474.

Wiegand, Cornelia, Sebastian Moritz, Nadine Hessler, Dana Kralisch, Falko Wesarg, Frank A. Müller, Dagmar Fischer, and Uta-Christina Hipler. 2015. "Antimicrobial functionalization of bacterial nanocellulose by loading with polihexanide and povidone-iodine." *Journal of Materials Science: Materials in Medicine* 26 (10):1–14.

Yousefi, Hossein, Mehdi Faezipour, Takashi Nishino, Alireza Shakeri, and Ghanbar Ebrahimi. 2011. "All-cellulose composite and nanocomposite made from partially dissolved micro-and nanofibers of canola straw." *Polymer Journal* 43 (6):559–564.

Zhang, Huan, Yuan Chen, Shanshan Wang, Liang Ma, Yong Yu, Hongjie Dai, and Yuhao Zhang. 2020. "Extraction and comparison of cellulose nanocrystals from lemon (Citrus limon) seeds using sulfuric acid hydrolysis and oxidation methods." *Carbohydrate Polymers* 238:116180.

Zhang, Lei, and Jing Zhang. 2018. "Multiporous molybdenum carbide nanosphere as a new charming electrode material for highly sensitive simultaneous detection of guanine and adenine." *Biosensors and Bioelectronics* 110:218–224.

Zhao, Jiangqi, Yuanjing Lin, Jingbo Wu, Hnin Yin Yin Nyein, Mallika Bariya, Li-Chia Tai, Minghan Chao, Wenbo Ji, George Zhang, and Zhiyong Fan. 2019. "A fully integrated and self-powered smartwatch for continuous sweat glucose monitoring." *ACS Sensors* 4 (7):1925–1933.

Zhou, Xufeng, Cong Chang, Yang Zhou, Lu Sun, Hua Xiang, Sijie Zhao, Liwei Ma, Guohua Zheng, Mingzhu Liu, and Hua Wei. 2017. "A comparison study to investigate the effect of the drug-loading site on its delivery efficacy using double hydrophilic block copolymer-based prodrugs." *Journal of Materials Chemistry B* 5 (23):4443–4454.

4 Synthesis of Novel Nanostructured Materials

Core-Shell and Hollow Metal

Suresh Mamidi and Mohammed Jamir Ahemad

4.1 INTRODUCTION

Nanomaterials have seen a surge in recent time due to their distinctive novel properties, making them desirable for many applications. One such class of nanostructured materials is the core-shell and hollow metal structures (CSHM) with a wide range of applications such as energy storage and conversion, catalysis, electronics, biomedicine, and sensors. Core-shell nanostructures have revolutionized these fields with their unique properties, making them crucial in the technological advancements. These nanoscale structures comprise of a central material (core) encased within a surrounding material (shell), creating a hybrid structure that combines the properties of both materials. This combination results in enhanced performance compared to traditional materials, making them suitable for various applications. In energy storage and conversion, core-shell nanostructures are deployed in Li-ion, Zn-air batteries, and supercapacitors, providing improved capacity and stability. In addition, they serve as contrast agents and drug delivery systems in medical imaging and therapy, improving diagnostic accuracy and therapeutic efficacy. Core-shell nanostructures are also utilized in catalysis, enhancing activity and selectivity for chemical reactions. Core-shell material-based electronics are used in LEDs and photovoltaics, improving efficiency and stability. In sensors, core-shell nanostructures are used to detect chemicals and biological molecules, providing highly sensitive and selective responses (Figure 4.1).

The fabrication of core-shell nanostructures involves growing a thin layer on the surface of a nanoscale inner core material, creating two-layer material with unique properties. These novel nanostructures are synthesized using various techniques such as electrodeposition, template-assisted methods, and solvothermal synthesis. The core-shell architecture comprises a central core material encompassed by a shell material, which provides stability and enhances the tunability of physical and chemical properties. This structure is widely used in catalytic reactions, where the core material offers the active site for catalytic reactions, and the shell material enhances stability and selectivity. Similarly, hollow metal nanostructures have a high surface area-to-volume ratio, making them ideal for battery electrode materials. The large surface area provides improved mass transport and efficient storage, enabling their potential use as supercapacitor electrodes and drug delivery vehicles.

However, synthesizing these nanostructures is a complex process that requires precise control over parameters such as temperature, reaction time, and precursor concentrations. This level of control is necessary to achieve the desired size, shape, and composition of the nanostructures. Therefore, synthesizing novel nanostructured materials such as CSHM holds great promise in various applications.

DOI: 10.1201/9781003305583-4

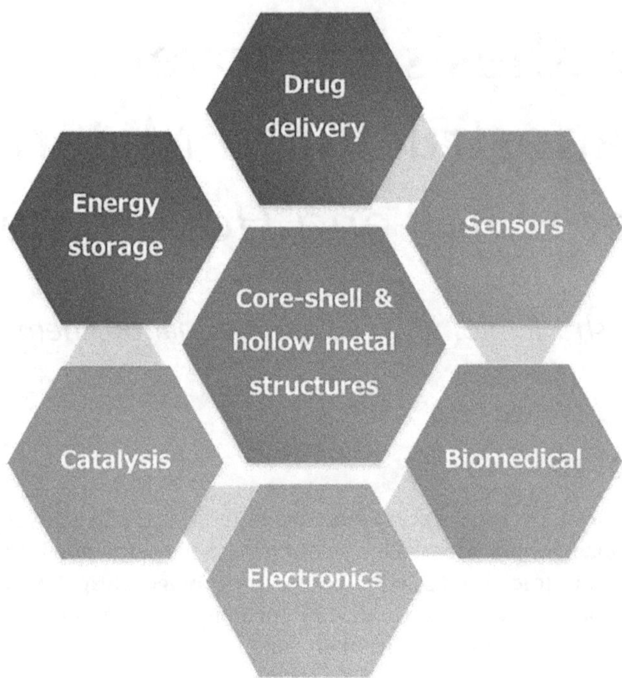

FIGURE 4.1 Various applications of CSHM.

4.2 SYNTHESIS METHODS FOR CSHM NANOPARTICLES

CSHM nanoparticles have unique physicochemical properties that make them desirable for a various applications. Synthesis methods for these structures include: 1) template-assisted methods: using template materials such as polymers, silica, or porous materials to shape the particles. 2) Electrochemical synthesis: involves the reduction of metal ions at electrodes. 3) Solvothermal methods: involves the reaction of metal precursors in organic solvents under high temperature and pressure. 4) Pulsed laser ablation: laser energy vaporizes a target material, and subsequent condensation leads to nanoparticles.

4.2.1 Fabrication of Hollow Metal Nanostructures through Template-Assisted Methods

Fabrication of hollow metal nanostructures through template-assisted methods is a widely explored technique in nanotechnology research. It involves the use of a template or sacrificial material as a mold to fabricate nanoscale metal structures with hollow interiors. This technique allows for precise control over the its dimensions, morphology, and outer layer thickness of the hollow structures, making it an attractive option for various applications.

Several template-assisted methods for fabricating hollow metal nanostructures (HMNS) include the galvanic replacement process and the hard and soft template methods. Each method utilizes a different type of template material, such as porous membranes, colloidal crystals, or block copolymers (Cao et al. 2016; Wu et al. 2013; Ahmad et al. 2019; Zhang et al. 2020; Abdelaal and Harbrecht 2016; Hossain et al. 2023).

In the galvanic replacement method, a metal precursor is electrochemically deposited into the pores of a sacrificial template, which is then dissolved to leave behind a hollow metal structure. This technique is beneficial for producing complex shapes and compositions. Ag@Au core-shell nanostructures were created by Ahmad et al. (2019) using a seed-mediated method.

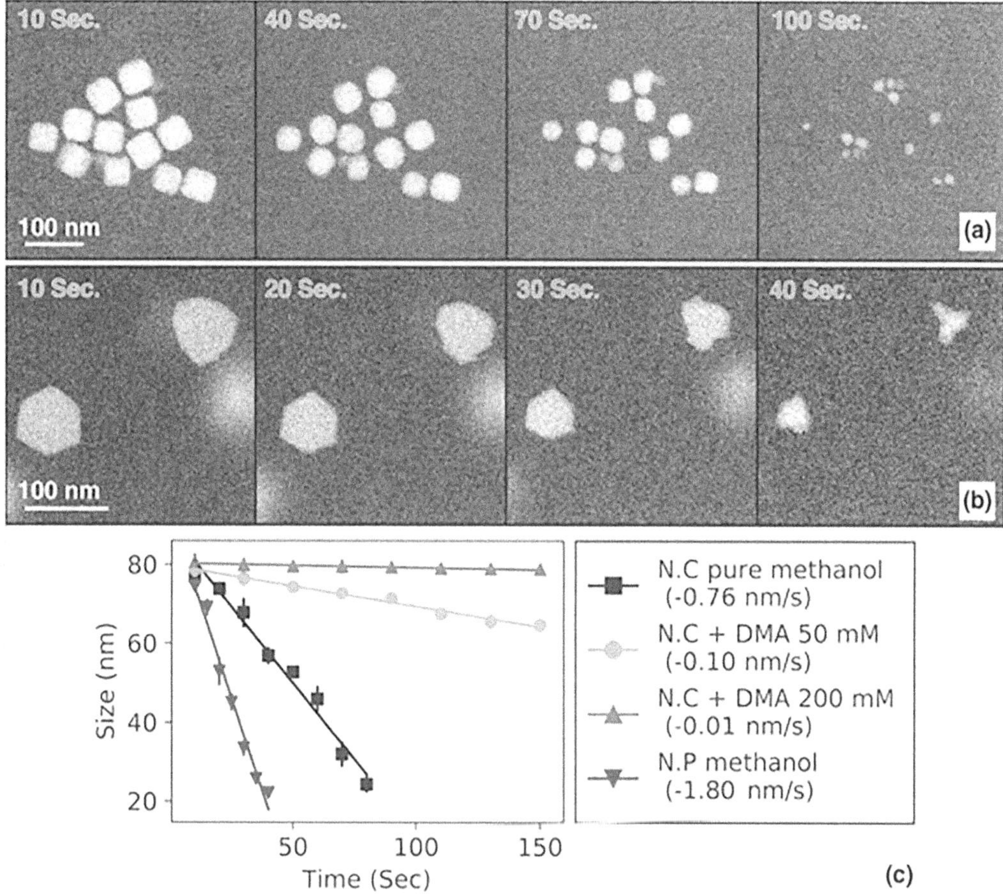

FIGURE 4.2 The dissolution of Ag template nanoparticles (a) Sequential STEM images captured in a methyl alcohol solvent, depicting the etching process induced by the electron beam over time. (b) Etching process on triangular and hexagonal Ag nanoplates using methyl alcohol at equivalent dosage levels as mentioned in (a). (c) Comparing the etching rates of silver nanoparticle templates in the presence and absence of DMA. (The image has been reproduced with permission from the source (Ahmad et al. 2019), Copyright (2019) American Chemical Society.).

Real-time monitoring of the nucleation and growth of nanostructures was conducted using in-situ scanning transmission electron microscopy (STEM), as shown in Figure 4.2. This method employed KI as a coordinating compound in methanol and dimethylamine (DMA). It also serves as a capping agent to decelerate the rate of chemical reactions and analyze the underlying mechanisms in the Ag@Au core-shell overgrowth process.

Next, the hard-template method involves depositing metal inside the pores of a hard-template material. After deposition, the template is removed by etching or calcination, leaving a hollow metal structure behind. To increase the efficiency of the oxygen reduction reaction (ORR) and Zn-air batteries, Qin et al. (2022) synthesized Fe-N with doped carbon hollow nanospheres. A rigid, sacrificial template made of silica (SiO_2) nanospheres was employed to create Fe-N-C via polymerization. After that, treatment with a 5M NaOH solution etched the SiO_2 away. Furthermore, in lithium-ion battery application, to increase the lithium storage capacity, Wu et al. (2013) synthesized rattle-type ball-in-ball V_2O_5 hollow microparticles using colloidal carbon particles as hard scaffolds. Core-shell composite microspheres made of carbon spheres@ vanadium-precursor (CS@V_2O_5) are first created using a one-step solvothermal technique.

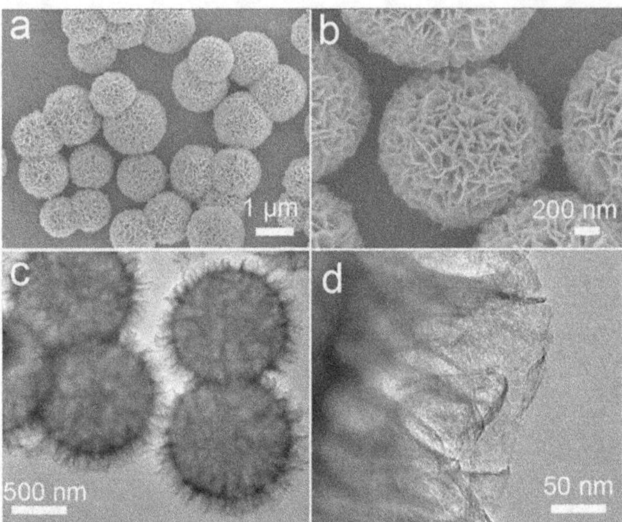

FIGURE 4.3 FESEM and TEM images of core-shell composite microspheres, comprising of CS@V-I (Image has been reproduced with permission from source (Wu et al. 2013), Copyright (2013) WILEY-VCH Verlag GmbH & Co. KGaA, Weinheim).

Solvothermally treated carbon spheres in isopropanol (IPA) and vanadium oxytriisopropoxide (VOT) were then annealed in air for 2 hours at 350 °C. Figure 4.3 illustrates the images of core-shell composite microspheres consisting of CS@V-I, as captured by FESEM and TEM. During the electrochemical characterization test, this CS@V_2O_5 core-shell exhibits improved cycling stability and rate capability.

The soft template method utilizes self-assembled structures, such as micelles or block copolymers, to template metal deposition into a hollow structure. This technique allows for the creation of hollow metal structures with controlled porosity and surface chemistry. The Fe-N-C nanosphere was developed using the soft template approach by Zhou et al. (2017). For the construction of the hollow-structured Fe/polyaniline (PANI)/polypyrrole (Ppy) composite, Triton X-100 was used as the soft template, while aniline and pyrrole were employed as the precursor materials for carbon and nitrogen, respectively. $FeCl_3$ salt served as the metal precursor.

4.2.2 Electrochemical Synthesis of CSHM

The electrochemical synthesis of core-shell and hollow metal nanoparticles is a highly versatile and promising approach to producing nanoscale materials with unique properties and applications. The growth of core-shell or hollow structures is achieved through the utilization of electrochemical reactions, which involve the deposition of metals onto a conductive substrate or template.

CSNS consist of a metallic core encased within a shell made of a distinct metal or non-metal material. The fabrication of CSNS can be achieved through a two-step electrochemical process, During this process, the substrate is first coated with the core metal, followed by the subsequent deposition of the shell material. By regulating the core and shell layers size, composition, and thickness, it is possible to finely adjust the properties of core-shell nanoparticles. Through electrochemical anodization with the aid of pulse sonication for the photoelectrochemical (PEC) water-splitting process, Khan and Qurashi (2018) created Ag/-Fe_2O_3/TiO_2 heterostructures following the schematic depicted in Figure 4.4. This method takes 2 hours and 20V to finish at room temperature. Additionally, the deposition rate rises with increased pulse sonication.

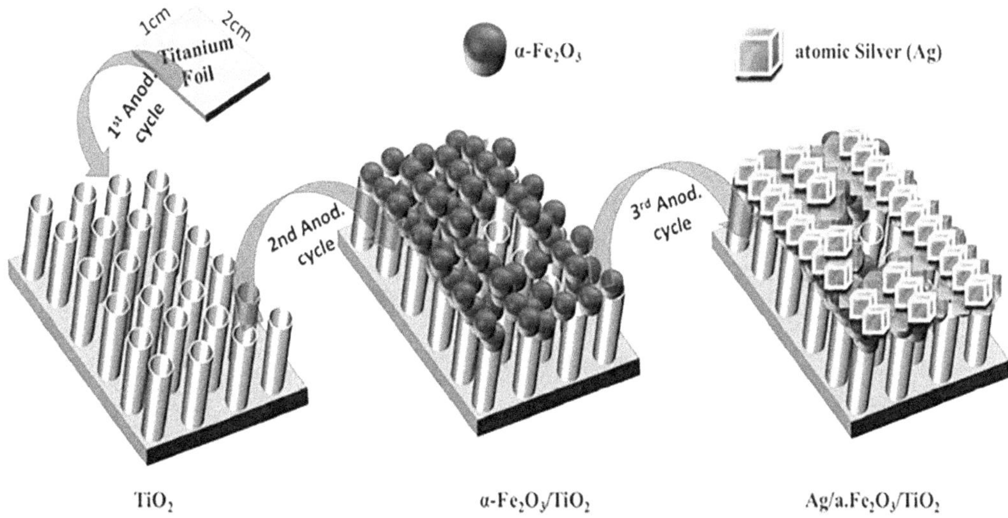

FIGURE 4.4 Schematic for preparation of Ag/-Fe_2O_3/TiO_2 heterostructures. (Image has been reproduced with permission from source (Khan and Qurashi 2018). Copyright (2018) American Chemical Society.).

Hollow metal nanoparticles, conversely, are characterized by their hollow interior, which offers a high surface area which attributes to superior catalytic activity and heightened sensing capabilities. The fabrication of hollow metal nanoparticles can be achieved through the electrochemical deposition of metal onto a sacrificial template, which is subsequently removed to leave behind a hollow structure. Ag@Ag_2S core@shell and hollow nanoparticles (NPs) were synthesized by Robinson and White (2019) using an electrochemical method. The structure of the nanoparticles varies depending on the applied potential, with higher overpotentials producing more hollow Ag/Ag_2S NPs and bigger void sizes. One of the key advantages of the electrochemical synthesis of CSHM nanoparticles is the ability to control their dimensions, morphology, and composition precisely. This method also offers high reproducibility and scalability, making it suitable for large-scale production.

4.2.3 Synthesis of Bimetallic CSHM Nanoparticles

Bimetallic core-shell and hollow metal nanoparticle synthesis have become an increasingly popular research topics in nanotechnology. Bimetallic nanoparticles are composed of two different metals, each with unique properties, which can be joined to produce materials with enhanced performance and new functionalities.

The fabrication of bimetallic CSNS involves the deposition of one metal on another metal to form a shell, creating a core-shell structure. This technique offers a high degree of control over the resulting nanoparticle dimensions, morphology, and composition. In addition, by varying the deposition conditions of the two metals, it is feasible to tailor the properties of the CSNS to meet specific application requirements. Wang et al. (2020) used an electrochemical deposition approach (localized surface plasmon resonance—LSPR) for the methanol oxidation process (MOR) to create plasmonic Au@Metal core-shell nanoparticles (Au@M, M = Rh, Pt, Pd, PtPd, and PdRh NPs). Platinum wires (300m) and saturated calomel electrodes (SCE) were deployed as counter and reference electrodes, and various molar concentrations of H_2PtCl_6, Na_2PdCl_4, and $RhCl_3$ were used as supporting electrolytes for deposition of Pt, Pd, and Rh on the Au surface, respectively, for the fabrication of Au@M NPs (Figure 4.5).

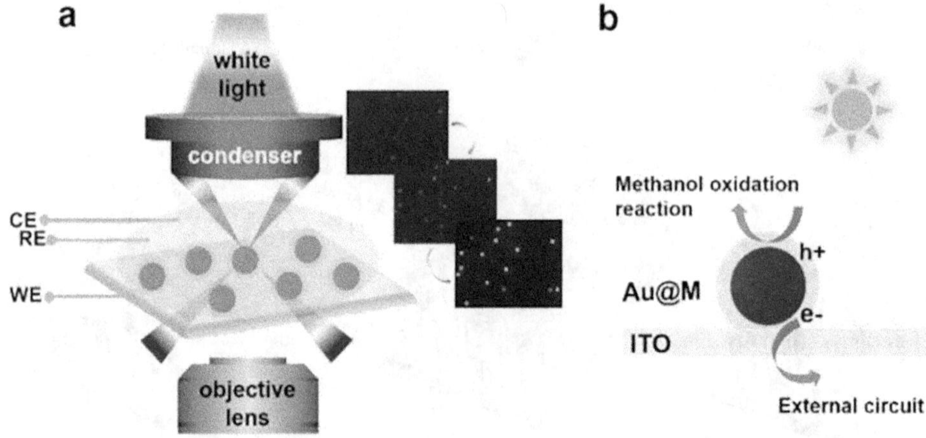

FIGURE 4.5 (a) Setup of dark field microscopy guided electrochemical synthesis of Au@M NPs and (b) the plasmonic accelerated MOR (Image has been reproduced with permission from source (Wang et al. 2020), Copyright (2020) American Chemical Society).

Hollow metal nanoparticles, on the other hand, are characterized by their unique morphology, which provides a high surface area and catalytic activity. The fabrication of hollow metal nanoparticles can be done through a variety of methods, including the use of sacrificial templates or selective etching of metal shells. Bimetallic hollow nanoparticles can also be produced by depositing two different metals onto a sacrificial template and subsequently removing the template to create a hollow structure. For the H_2 evolution reaction (HER), Lang et al. (2015) created Ni-Sn@C nanoparticles (NPs) with hollow core-shell structures using the CVD, etching, and sol-gel process. Triethylene glycol (TEG) was used to create Ni-Sn nanoparticles (NPs) in an N_2 environment at 170°C. Following the application of the SiO_2 coating, the component was calcined for 1 hour at 850 °C in an environment of $Ar/C_2H_2/H_2$ (6:2:2). Bimetallic hollow Ni-Sn@C NPs are obtained by chemically removing SiO_2 at 50°C for 3–6 hours in a solution of 2M NaOH.

The properties of bimetallic CSHM nanoparticles can be further improved by tuning the composition and dimension of the metal components. For example, bimetallic nanoparticles with a gold shell and a palladium core have been efficient catalysts for the oxidation of carbon monoxide. In addition, bimetallic hollow nanoparticles with a platinum-palladium composition have demonstrated enhanced electrocatalytic activity for fuel cell applications. Ruiz-Montoya et al. (2021) synthesized Au@Pdx/C catalysts (x = 0, 0.1, 0.8, and 1.6) for the oxidation of ethyl alcohol in an alkaline medium. $Au@Pd_x/C$ catalysts were made in three distinct atomic compositional ratios, each with a different quantity of palladium (x = 0.10, 0.80, and 1.60), while keeping the gold seed nucleus constant.

Overall, the synthesis of bimetallic core-shell and hollow metal nanoparticles offers a highly versatile and customizable approach to the design of nanomaterials with unique properties and applications. As research in this field continues to advance, bimetallic nanoparticles are expected to play an increasingly important role in areas such as catalysis, biomedicine, and renewable energy.

4.3 CONTROL OF SHELL THICKNESS AND COMPOSITION IN CORE-SHELL NANOPARTICLES

The properties of CSNS can be tuned by controlling its size, shape, and composition. One of the critical factors that affect the properties of CSNS is the thickness and composition of the shell

(García et al. 2013; Zou et al. 2019; Li et al. 2010; Liu et al. 2013; Guha et al. 2011). The shell can be tailored by tuning the deposition time, concentration, and reaction conditions during synthesis. The thickness and composition of the shell can also be tailored by using different methods of deposition, such as CVD, and electrochemical deposition

The thickness of the shell can influence the surface area and reactivity of the nanoparticles. A thicker shell can increase the nanoparticles' stability and reduce the core material's reactivity. On the other hand, a thinner shell can provide a larger surface area and increase the reactivity of the core material. Guha et al. (2011) synthesized Au-Ag core-shell nanoparticles in water at pH 6.94 utilizing the synthetic fluorescent dipeptide l-Ala-Trp, which exhibits ultra-sensitivity detection of HgII ions up to 9nM (1.8 ppb) concentration in water. For dye-sensitized solar cells (DSSCs), Liu et al. (2013) produced plasmonic Au@TiO_2 core-shell nanoparticles. They noticed that the voltage improved with a thicker TiO_2 shell while the current was enhanced with a thinner TiO_2 shell.

The composition of the shell can also alter the functionality of the CSNS. The choice of the shell material can influence the chemical and physical properties of the nanoparticle, such as catalytic activity, magnetic properties, and optical properties. For example, the use of gold as a shell material can improve the stability and biocompatibility of the CSNS. In contrast, the use of silver can enhance its antibacterial properties. Yang et al. (2017) produced customized Au@Ag NPs and investigated the antibacterial activity of these particles on human SH-SY5Y cells, Staphylococcus aureus, and E. coli. The findings demonstrate that the best antibacterial activity and outstanding biocompatibility are found in Au@Ag NPs with a thickness of 5nm or an Au:Ag ratio of 1:1.

The control of shell thickness and composition in core-shell nanoparticles (CSNS) is essential for the design of nanoparticles with specific properties and applications. The ability to tailor the properties of CSNS makes them a potential material for different applications such as drug delivery, biomedical imaging, and catalysis. As research in this field continues to advance, the introduction of new and innovative methods for controlling the shell thickness and composition of CSNS will further enhance their usefulness in a variety of fields.

4.4 SURFACE FUNCTIONALIZATION OF CSHM NANOPARTICLES

Surface functionalization of core-shell and hollow metal nanoparticles is an essential aspect of their synthesis that allows for controlling their physicochemical properties and subsequent applications. The nanoparticle surface plays a critical role in their reactivity, stability, and biocompatibility. Surface functionalization can be used to modify the surface of nanoparticles by attaching specific molecules or groups to improve their properties.

The functionalization of the surface of CSHM nanoparticles involves the attachment of organic molecules, polymers, or biomolecules on the surface of the nanoparticles. Several methods, such as chemical modification, electrostatic adsorption, or self-assembly, can achieve the objective of surface functionalization. The choice depends on the desired properties of the nanoparticles, the type of surface functionalization, and the intended application.

One of the most common types of surface functionalization is the addition of ligands on the nanoparticle surface. Ligands are molecules that can coordinate with the metal surface of nanoparticles and provide unique functionalities. For example, the use of thiol-containing ligands such as mercaptosuccinic acid can improve the stability and biocompatibility of the nanoparticles. In contrast, carboxylic acid-containing ligands can enhance their solubility and reactivity. Deming et al. (2015) synthesized different compositions of dodecyne-capped AuPd alloy nanoparticles showing excellent oxygen reduction reactions.

Another type of surface functionalization is the addition of polymers on the nanoparticle surface. Polymer coatings can provide a passive layer for the nanoparticles, improving their stability and biocompatibility. Polymer coatings can also provide functional groups that can be used for further functionalization. Liu et al. (2011) Fe_3O_4@PEGDMA, Fe_3O_4@P(EGDMA-co-MAA), and Fe_3O_4@P(MBAAm-co-HEMA) core-shell particles with various noble metal loadings (Au, Ag, and Pd) demonstrate an excellent catalytic properties.

Surface functionalization of hollow metal nanoparticles can also be achieved by selectively etching the metal shell to create functionalized surfaces. This can be done using a specific etching agent that selectively dissolves the metal shell, leaving behind a hollow core with a functionalized surface. The surface functionalization of CSHM nanoparticles is a crucial aspect of their synthesis, providing an opportunity to control their properties and tailor their functionalities for specific applications.

4.5 APPLICATIONS OF CSHM NANOPARTICLES IN BIOTECHNOLOGY, MEDICINE, AND HEALTHCARE

CSHM nanoparticles have gained significant interest in biotechnology, medicine, and healthcare due to their unique physicochemical properties and potential applications. These nanoparticles offer a range of benefits, such as high surface area-to-volume ratio, biocompatibility, and tuneable surface properties, which make them ideal for various applications.

One of the key applications of CSHM nanoparticles in biotechnology is in sensing and diagnostics. These nanoparticles can be functionalized with specific biomolecules such as antibodies or nucleic acids, allowing for highly sensitive and selective detection of biomolecules. For example, Au nanoparticles functionalized with antibodies can be used to detect biomarkers in blood samples, providing a non-invasive and sensitive diagnostic tool. Impedimetric immune sensors based on Au@Ag core-shell nanoparticles were produced by Rana et al. (2020) for the detection of hepcidin. These immune sensors exhibited satisfactory repeatability, stability, selectivity, and reusability.

In the field of medicine, CSHM nanoparticles have shown great potential for drug delivery. The hollow core of these nanoparticles can be used as a reservoir for drug molecules, while the surface can be functionalized to target specific cells or tissues. Additionally, the small size of these nanoparticles allows them to penetrate cells and tissues easily, improving drug efficacy and reducing side effects. For instance, silver nanoparticles have been shown to have antibacterial properties and can be used for the treatment of infections. Using near-IR irradiation of biological tissues, Kumal et al. (2018) produced Au-Ag-Au, core-shell-shell (CSS) nanoparticles with photothermal cleaving linkers. These nanoparticles can be deployed for specific drug delivery applications. In addition, the fluorescence quantification of the amount of siRNA release demonstrates that the photothermal cleaving effectiveness of the CSS nanoparticles is noticeably higher when compared to that of the gold nanoparticles created under conditions that are analogous to those used in this experiment.

Furthermore, core-shell and hollow metal nanoparticles can be used for imaging and therapy. These nanoparticles can be functionalized with contrast agents or therapeutic molecules, allowing for the targeted delivery of these agents to specific cells or tissues. For example, iron oxide nanoparticles functionalized with targeting molecules can be used for magnetic resonance imaging and for hyperthermia treatment of tumor cells. Zhang et al. (2009) Fe_3O_4 nanoparticles functionalized with folate-tetra(ethylene glycol)-poly(glycerol monoacrylate) (FA-TEG-PGA). It has good stability over a range of pH and salt concentrations, which relay their potential application in MRI applications and multi-modal bioimaging.

Core-shell and hollow metal nanoparticles can be used for biosensing and monitoring. The novel properties of these nanoparticles make them ideal for detecting biological events such as enzymatic reactions, pH changes, or protein-protein interactions. For instance, gold nanoparticles functionalized with enzymes can be used to detect glucose levels in diabetes patients. Yang et al. (2020) developed glucose-oxidase enzyme functionalized gold nanoparticles for glucose sensors in serum, showing excellent selectivity towards glucose with cholesterol, urea, L-cysteine, ascorbic acid, and galactose (Figure 4.6).

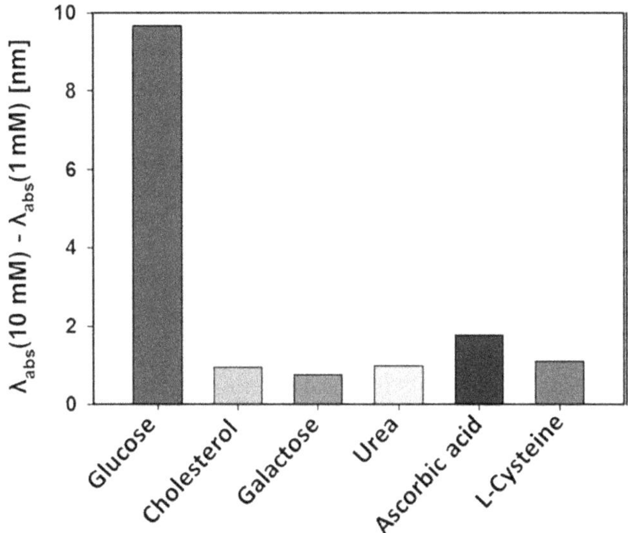

FIGURE 4.6 Selectivity of glucose-oxidase enzyme functionalized gold nanoparticles (Image is reproduced with permission from source (Yang et al. 2020). Copyright (2019) Springer Science+Business Media, LLC, part of Springer Nature.)

Overall, the applications of CSHM nanoparticles in biotechnology, medicine, and healthcare are numerous and diverse. As research in this field continues to progress, developing new and innovative applications of these nanoparticles will likely further advance the fields of biotechnology, medicine, and healthcare.

REFERENCES

Abdelaal, Haitham Mohammad, and Bernd Harbrecht. 2016. "Approachable Way to Synthesize 3D Silica Hollow Nanospheres with Mesoporous Shells via Simple Template-Assisted Technique." *ChemistrySelect* 1 (18): 5961–5966. 10.1002/slct.201601178.

Ahmad, Nabeel, Marta Bon, Daniele Passerone, and Rolf Erni. 2019. "Template-Assisted in Situ Synthesis of Ag@Au Bimetallic Nanostructures Employing Liquid-Phase Transmission Electron Microscopy." *ACS Nano* 13 (11): 13333–13342. 10.1021/acsnano.9b06614.

Cao, Zhenzhen, Hongyan Yang, Peng Dou, Chao Wang, Jiao Zheng, and Xinhua Xu. 2016. "Synthesis of Three-Dimensional Hollow SnO2@PPy Nanotube Arrays via Template-Assisted Method and Chemical Vapor-Phase Polymerization as High Performance Anodes for Lithium-Ion Batteries." *Electrochimica Acta* 209: 700–708. 10.1016/j.electacta.2016.01.158.

Deming, Christopher P., Albert Zhao, Yang Song, Ke Liu, Mohammad M. Khan, Veronica M. Yates, and Shaowei Chen. 2015. "Alkyne-Protected AuPd Alloy Nanoparticles for Electrocatalytic Reduction of Oxygen." *ChemElectroChem* 2 (11): 1719–1727. 10.1002/celc.201500252.

García, Stephany, Rachel M. Anderson, Hugo Celio, Naween Dahal, Andrei Dolocan, Jiping Zhou, and Simon M. Humphrey. 2013. "Microwave Synthesis of Au–Rh Core–Shell Nanoparticles and Implications of the Shell Thickness in Hydrogenation Catalysis." *Chemical Communications* 49 (39): 4241–4243. 10.1039/C3CC40387D.

Guha, Samit, Subhasish Roy, and Arindam Banerjee. 2011. "Fluorescent Au@Ag Core–Shell Nanoparticles with Controlled Shell Thickness and HgII Sensing." *Langmuir* 27 (21): 13198–13205. 10.1021/la203077z.

Hossain, Md Shahadat, Takeshi Furusawa, and Masahide Sato. 2023. "Sucrose-Derived Carbon Template-Assisted Synthesis of Zinc Oxide Hollow Microspheres: Investigating the Effect of Hollow Morphology on Photocatalytic Activity." *Inorganic Chemistry Communications* 148: 110376. 10.1016/j.inoche.2022.110376.

Khan, Ibrahim, and Ahsanulhaq Qurashi. 2018. "Sonochemical-Assisted In Situ Electrochemical Synthesis of Ag/α-Fe2O3/TiO2 Nanoarrays to Harness Energy from Photoelectrochemical Water Splitting." *ACS Sustainable Chemistry & Engineering* 6 (9): 11235–11245. 10.1021/acssuschemeng.7b02848.

Kumal, Raju R., Mohammad Abu-Laban, Prakash Hamal, Blake Kruger, Holden T. Smith, Daniel J. Hayes, and Louis H. Haber. 2018. "Near-Infrared Photothermal Release of SiRNA from the Surface of Colloidal Gold–Silver–Gold Core–Shell–Shell Nanoparticles Studied with Second-Harmonic Generation." *The Journal of Physical Chemistry C* 122 (34): 19699–19704. 10.1021/acs.jpcc.8b06117.

Lang, Leiming, Yi Shi, Jiong Wang, Feng-Bin Wang, and Xing-Hua Xia. 2015. "Hollow Core–Shell Structured Ni–Sn@C Nanoparticles: A Novel Electrocatalyst for the Hydrogen Evolution Reaction." *ACS Applied Materials & Interfaces* 7 (17): 9098–9102. 10.1021/acsami.5b00873.

Li, Zhong, Lisa A. Fredin, Pratyush Tewari, Sara A. DiBenedetto, Michael T. Lanagan, Mark A. Ratner, and Tobin J. Marks. 2010. "In Situ Catalytic Encapsulation of Core-Shell Nanoparticles Having Variable Shell Thickness: Dielectric and Energy Storage Properties of High-Permittivity Metal Oxide Nanocomposites." *Chemistry of Materials* 22 (18): 5154–5164. 10.1021/cm1009493.

Liu, Bin, Wei Zhang, Fengkai Yang, Hailiang Feng, and Xinlin Yang. 2011. "Facile Method for Synthesis of Fe3O4@Polymer Microspheres and Their Application As Magnetic Support for Loading Metal Nanoparticles." *The Journal of Physical Chemistry C* 115 (32): 15875–15884. 10.1021/jp204976y.

Liu, Wei-Liang, Fan-Cheng Lin, Yu-Chen Yang, Chen-Hsien Huang, Shangjr Gwo, Michael H. Huang, and Jer-Shing Huang. 2013. "The Influence of Shell Thickness of Au@TiO2 Core–Shell Nanoparticles on the Plasmonic Enhancement Effect in Dye-Sensitized Solar Cells." *Nanoscale* 5 (17): 7953–7962. 10.1039/C3NR02800C.

Qin, Ke, Zhenye Zhu, Fei-Xiang Ma, and Jiaheng Zhang. 2022. "Template-Assisted Synthesis of Iron–Nitrogen Co-Doped Carbon Hollow Nanospheres for Efficient Oxygen Reduction Reaction." *Journal of Electroanalytical Chemistry* 906: 116021. 10.1016/j.jelechem.2022.116021.

Rana, Shilpa, Anu Bharti, Suman Singh, Archana Bhatnagar, and Nirmal Prabhakar. 2020. "Gold-Silver Core-Shell Nanoparticle–Based Impedimetric Immunosensor for Detection of Iron Homeostasis Biomarker Hepcidin." *Microchimica Acta* 187 (11): 626. 10.1007/s00604-020-04599-8.

Robinson, Donald A., and Henry S. White. 2019. "Electrochemical Synthesis of Individual Core@Shell and Hollow Ag/Ag2S Nanoparticles." *Nano Letters* 19 (8): 5612–5619. 10.1021/acs.nanolett.9b02144.

Ruiz-Montoya, José G, Luiza M. S. Nunes, Angélica M Baena-Moncada, Germano Tremiliosi-Filho, and Juan Carlos Morales-Gomero. 2021. "Effect of Palladium on Gold in Core-Shell Catalyst for Electrooxidation of Ethanol in Alkaline Medium." *International Journal of Hydrogen Energy* 46 (46): 23670–23681. 10.1016/j.ijhydene.2021.04.159.

Wang, Hui, Wei Zhao, Yang Zhao, Cong-Hui Xu, Jing-Juan Xu, and Hong-Yuan Chen. 2020. "Real-Time Tracking the Electrochemical Synthesis of Au@Metal Core–Shell Nanoparticles toward Photo Enhanced Methanol Oxidation." *Analytical Chemistry* 92 (20): 14006–14011. 10.1021/acs.analchem.0c02913.

Wu, Hao Bin, Anqiang Pan, Huey Hoon Hng, and Xiong Wen (David) Lou. 2013. "Template-Assisted Formation of Rattle-Type V2O5 Hollow Microspheres with Enhanced Lithium Storage Properties." *Advanced Functional Materials* 23 (45): 5669–5674. 10.1002/adfm.201300976.

Yang, Longping, Wenjing Yan, Hongxia Wang, Hong Zhuang, and Jianhao Zhang. 2017. "Shell Thickness-Dependent Antibacterial Activity and Biocompatibility of Gold@silver Core–Shell Nanoparticles." *RSC Advances* 7 (19): 11355–11361. 10.1039/C7RA00485K.

Yang, Qingshan, Xia Zhang, Santosh Kumar, Ragini Singh, Bingyuan Zhang, Chenglin Bai, and Xipeng Pu. 2020. "Development of Glucose Sensor Using Gold Nanoparticles and Glucose-Oxidase Functionalized Tapered Fiber Structure." *Plasmonics* 15 (3): 841–848. 10.1007/s11468-019-01104-7.

Zhang, Furui, Xiaobing Hu, Eric W. Roth, Yonghwi Kim, and SonBinh T. Nguyen. 2020. "Template-Assisted, Seed-Mediated Synthesis of Hierarchically Mesoporous Core–Shell UiO-66: Enhancing Adsorption Capacity and Catalytic Activity through Iterative Growth." *Chemistry of Materials* 32 (10): 4292–4302. 10.1021/acs.chemmater.0c01065.

Zhang, Quan, Chenhong Wang, Lei Qiao, Husheng Yan, and Keliang Liu. 2009. "Superparamagnetic Iron Oxide Nanoparticles Coated with a Folate-Conjugated Polymer." *Journal of Materials Chemistry* 19 (44): 8393–8402. 10.1039/B910439A.

Zhou, Lihua, Chunli Yang, Jing Wen, Peng Fu, Yaping Zhang, Jian Sun, Huaqian Wang, and Yong Yuan. 2017. "Soft-Template Assisted Synthesis of Fe/N-Doped Hollow Carbon Nanospheres as Advanced Electrocatalysts for the Oxygen Reduction Reaction in Microbial Fuel Cells." *Journal of Materials Chemistry A* 5 (36): 19343–19350. 10.1039/C7TA05522F.

Zou, Jiasui, Min Wu, Shunlian Ning, Lin Huang, Xiongwu Kang, and Shaowei Chen. 2019. "Ru@Pt Core–Shell Nanoparticles: Impact of the Atomic Ordering of the Ru Metal Core on the Electrocatalytic Activity of the Pt Shell." *ACS Sustainable Chemistry & Engineering* 7 (9): 9007–9016. 10.1021/acssuschemeng.9b01270.

5 Carbon Nanomaterials

Novel Structures for Biological Applications

Anshul Rasyotra, Anupma Thakur, Bhagyashri Gaykwad, Vruddhi Jani, and Kabeer Jasuja

5.1 INTRODUCTION

Carbon is the 15th most abundant element in the earth's crest and the 4th most abundant element in the universe. It is the key component of deoxyribonucleic acid (DNA) and serves as the most common element of all known life on the planet. Carbon can exist in a variety of allotropes due to its electronic configuration ($1s^2$, $2s^2$, $2p^2$) and ability to bond together in diverse ways. The allotropes formed depend on the hybridization (sp^1, sp^2, sp^3), with common allotropes such as graphite and diamond having hybridization sp^2 and $sp^{3,}$ respectively. Carbon nanomaterials (C.N.s) have received tremendous interest in nanotechnology due to their unique properties and flexible dimensional structure. The study of carbon-based nanomaterials (C.N.s) for biomedical applications has attracted significant attention due to their unique physicochemical properties, including thermal, mechanical, electrical, optical, and structural diversity (Rauti et al. 2019; 'Carbon-Based Nanomaterials' 2008). C.N.s, including carbon nanotubes (CNTs), carbon nanofibers (CNFs), graphene oxide (G.O.), and graphene quantum dots (GQDs), mesoporous carbon, nanodiamonds, fullerenes have excellent electrical, thermal, and optical properties that make them promising materials for drug delivery, bioimaging, biosensing, biomedical scaffolds, medical implants, and tissue engineering applications (Mohajeri et al. 2019; Shannahan 2017; Pattnaik et al. 2016; Caruso et al. 2012; Master et al. 2013; Taratula et al. 2013; Wei et al. 2012). C.N.s, mainly CNTs and graphene, have been extensively studied because of their high surface-to-volume ratio, high adsorption ability, electrical and thermal conductivity, photothermal conversion capacity and supramolecular π-π stacking (Huang et al. 2011; Chao Wang et al. 2012; H. Wang et al. 2015; P. Zhao et al. 2012; S. Y. Wang et al. 2020). The surface functionalization of C.N.s enhances their drug loading/release capacity, their ability to target drug delivery to specific sites, and their dispersibility and suitability in biological systems (Y. C. Chen et al. 2013). However, all nanomaterials have several toxic effects that reduce their potential applications.

This chapter presents the research trends and state-of-the-art C.N.s being investigated for biomedical applications. This study also embraces the emerging biological applications of C.N.s, including targeted drug delivery, tissue engineering, biosensing, bioimaging, and photodynamic therapy.

5.2 RESEARCH TRENDS IN CARBON NANOMATERIALS AND THEIR COMPOSITES FOR BIOLOGICAL APPLICATIONS

Carbon-based nanomaterials are becoming attractive in the material science and engineering field (Cha et al. 2013; Z. Wang et al. 2016; Tiwari et al. 2016). Carbon-based materials have

DOI: 10.1201/9781003305583-5

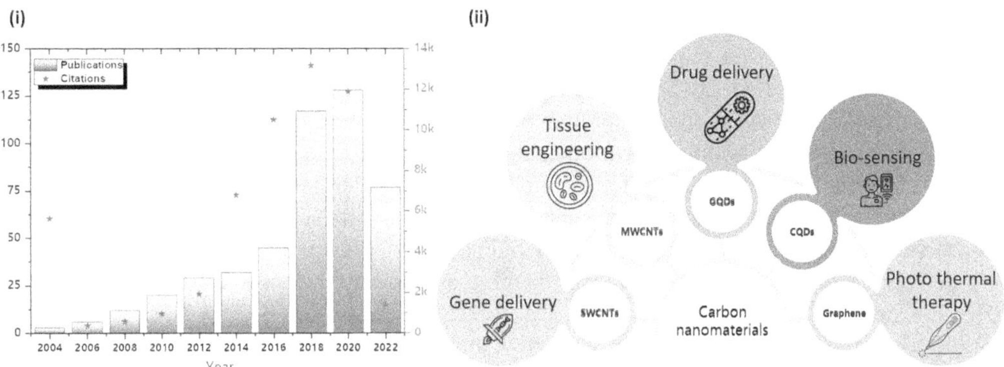

FIGURE 5.1 (i) Overview of research publications for the keyword "carbon-based nanomaterials for biological applications" in Web of science. (ii) Different types of carbon-based nanomaterials for biological applications.

recently become the largest family of emerging materials due to the existence of diverse allotropes of carbon, ranging from well-known allotropic phases such as amorphous carbon, graphite, and diamonds to newly discovered auspicious carbon nanotubes (CNTs), graphene oxide (G.O.), graphene quantum dots (GQDs), and fullerene (Hong et al. 2015). Each member of the carbon family has distinct properties that have been widely used in a variety of biological applications, such as biosensing, drug delivery, tissue engineering, imaging, diagnosis, and cancer therapy (Riley and Narayan 2021). This is supported by the number of publications and citations reported for the keyword "carbon-based nanomaterials for biological applications" in the Web of Science (Figure 5.1). CNTs are hollow cylinders made of graphitic sheets and are divided into single-walled carbon nanotubes (SWCNT) and multi-walled carbon nanotubes (MWCN) (MWCNT). The formation of multiwall CNTs from carbon arc discharge was studied in 1991 (Ebbesen and Ajayan 1992). After some years, the synthesis of single-wall CNTs was perceived (Kataura et al. 1999). SWCNTs, with a cylindrical nanostructure, are made by rolling up a single graphitic sheet with a high aspect ratio (Colomer et al. 2000). MWCNTs contain few graphitic layers in the rolling pattern, with an interlayer spacing of 3.4 Å (Ebbesen and Ajayan 1992). MWCNTs provide superior strength, flexibility, and electrical conductivity toward various biological entities due to their unique mechanical, electrical, and structural diversity, making them useful for sensing, medical diagnosis, and treatment of various diseases (Nimushakavi et al. 2022; Onyancha et al. 2021).

However, due to its unique intrinsic properties, graphene is considered the most interesting material among the various allotropes of carbon. Mouras and colleagues coined the term "graphene" in 1987 as "graphitic intercalation compounds (GIC)" (H. Sun et al. 2013). Over the last two decades, graphene research has exploded, and scientists have discovered a slew of novel properties. Graphene is defined as a graphitic planar sheet composed of a sp^2 hybridized carbon network with a carbon-carbon distance of 1.42 Å and an interlayer spacing of 3.4 Å (Chung et al. 2013). Graphene possesses a number of exceptional properties that contribute to its potential suitability for bio-applications. The prospect of simple functionalization leads to the enrichment of functional groups on its surface, facilitating the specific and selective detection of several biological segments. Furthermore, its extremely large surface area, chemical purity, and free electrons make it an ideal drug delivery candidate (J. Li et al. 2021; Pandit et al. 2021; Jahdaly et al. 2021). Furthermore, because of its adaptable behavior toward various fluorescent dyes, therapeutic agents, and other biomaterials, it is widely used for in-vivo imaging, cancer diagnosis, and treatment.

GQDs, a recently invented and appealing carbon family biomaterial, is defined as a zero-dimensional graphene sheet with a lateral dimension of less than 100 nm in one to a few layers (X. Li et al. 2015; Zhu et al. 2015). During the conversion of two-dimensional graphene sheets into GQDs, the GQDs endow excellent photoluminescence due to quantum confinement (Bacon et al. 2014). Surprisingly, GQDs outperform other fluorescent dyes or semiconductor quantum dots in terms of biocompatibility and photobleaching resistance. Furthermore, GQDs have amazing graphene properties, such as a large surface area and available electrons, making them an innovative nanomaterial for a wide range of biomedical applications such as imaging, targeted drug delivery, biomolecule sensing, cancer therapy, and so on (H. Sun et al. 2013; P. Tian et al. 2018; F. Chen et al. 2017).

With the growing research trends in carbon materials, the worldwide market is supposed to observe colossal development over the figure period attributable to its wide application in various end-use ventures, including biomedical applications (Prajapati et al. 2020). The carbon nanomaterials market is expected to be driven by its possible applications because of its excellent electrical conductivity. Becoming more robust and rigid primary materials combined with the rising application extent of carbon nanomaterials in the aeroplane, nanomedicines, shopper products, and water treatment industry is additionally expected to fuel market interest. MWCNTs addressed the most significant part of the carbon nanotubes market in 2021 (Nag et al. 2021).

The rising revenue for MWCNT in applications, such as clinical and clinical consideration, electrical gadgets, and optical devices, should fuel the advancement of the market during the figure time ('World Carbon Nanotubes Market Report 2022–2032: Demand For 2022). Clinical applications are projected to observe the quickest development of the carbon nanotubes market during the figure time frame regarding volume. Transdermal drug delivery is one of the most significant markets for CNTs in the clinical industry (Parveen et al. 2012; Jampilek and Kralova 2021; Ye et al. 2022). Multi-walled and twofold-walled CNTs have drug transporter properties through skin and infiltration improvement impacts (Ye et al. 2022).

Drug delivery is one of the most appealing application regions for nanotechnology. Carbon nanotubes can be utilized in drug transporter framework and proposition a high stacking and upgraded transdermal infiltration for hydrophobic medications (Saheeda et al. 2022; Hasnain and Nayak 2019; Bianco et al. 2005). Recently, utilizing the inherent properties of different carbon nanomaterials has been modified and extensively used in biomedicine, including applications for biosensing, drug delivery, and cancer therapy (S. D. Dutta et al. 2021).

5.3 BIOLOGICAL APPLICATIONS OF CARBON NANOMATERIALS

5.3.1 Biosensors

Electrochemical biosensors have received increased attention in recent years for their ability to detect metabolic disease biomarkers. A wide range of electrochemical CNT biosensors for detecting ions, metabolites, and protein biomarkers have been developed (Beitollahi et al. 2019). Enzymatic catalysis of an electroactive species-producing reaction can boost the electric signal generated by the electroactive surface area of CNTs. As a result, the enzymes immobilized on the working electrode recognize the target analyte, producing a current or contributing to voltage production, which is employed to design an electrochemical sensor. Several studies have been conducted to develop biomaterial-CNT hybrid systems, protein-linked CNTs, and nucleic acid-functionalized CNTs (Oliveira and Morais 2018). In one case, a novel portable electrochemical sensor was developed to detect both ascorbic acid and uric acid at the same time (Abellán-Llobregat et al. 2018). The sensitivity and reproducibility of the method were evaluated, and the results were promising in terms of low detection limits. This method has been successfully applied further to real-time sample analysis (Abellán-Llobregat et al. 2018). Another study used porous g-C_3N_4 (PCN) and MWCNTs to build an electrochemical sensor to detect uric acid.

The conductive properties of PCN were improved by using MWCNTs as conductive substrates because the PCN obtained by this method had the advantages of large specific surface area, excellent dispersion, favorable for electrocatalysis, and outstanding biocompatibility (Lv et al. 2019). A similar study used in-situ oxidation polymerization of N-(ferrocenyl formyl) pyrrole to create MWCNTs-based nanocomposites for the detection of ascorbic acid and uric acid (S. Zhang et al. 2019). The data obtained from actual physiological samples were quantified and plotted, and it agreed with the data obtained from conventional analytical methods (enzyme colourimetry kit).

Graphene has also been used to develop a wide range of biosensors for detecting biomarkers and biomolecules. In one study, a nanocomposite of iron oxide and polyvinyl alcohol was used to modify graphene for the detection of vitamin C (Das et al. 2018). The biomolecule was detected using the square wave voltammetry technique, which demonstrated that this material provided excellent results for vitamin C detection. An environmentally friendly and unlabeled immunosensor made of composite material for the detection of leptin was designed for the detection of nonalcoholic fatty liver disease (NAFLD) (Cai et al. 2019). By strongly coupling porous graphene surface plasma and anisotropic black phosphorus local surface plasma, this composite material was developed. The cross-linking of glutaraldehyde can effectively immobilize leptin antibodies (Cai et al. 2019). In another study, copper-nanoflower was used to decorate gold nanoparticles-graphene oxide nanofiber to detect glucose (Baek et al. 2020). According to their findings, the improved sensor can be used to monitor glucose levels in biological fluids for real-world clinical care site tests. Similarly, another highly sensitive glucose biosensor was developed by modifying a screen-printed carbon electrode with amine terminated MWCNTs/polyaniline/reduced graphene oxide/gold nanoparticles and then linking glucose oxidase on the electrode's surface (Maity et al. 2019). Furthermore, blood glucose detection in human serum samples was accomplished using this sensor.

Although carbon nanomaterials-based biosensors are promising, they have faced numerous practical challenges in biological applications. For example, the fabrication of biosensors typically involves a specific size and helicity of carbon nanomaterials (Joshi et al. 2021). Furthermore, their fabrication is not cost-effective in large-scale production of high-purity nanomaterials, which is the primary reason why current market prices of carbon nanomaterials are prohibitively expensive, limiting any realistic commercial applications. An enzyme must be immobilized on the surface of carbon nanomaterial-coated electrodes in carbon-based biosensors. As a result, biological receptors' biological activity, biocompatibility, and structural stability may be compromised. As a result, developing reversible, reusable, and long-lasting systems to replace current irreversible, disposable, and single-use devices continues to be difficult. More efforts are needed to develop carbon nanomaterials-based biosensors.

5.3.2 Drug Delivery

Several C.N.s have been synthesized and utilized to date for drug delivery applications. These are single-walled carbon nanotubes (SWCNTs), double-walled carbon nanotubes (DWCNTs), multi-walled carbon nanotubes (MWCNTs), graphene, graphene nanoribbons, nanodiamonds, fullerenes, porous carbon, and carbon dots. For efficient drug delivery, different physico-chemical properties – enzymatic degradation, biocompatibility, biological interaction, toxicity, and bio-corona, originating from C.N.s must be studied comprehensibly (Debnath and Srivastava 2021; Pattnaik et al. 2020).

Enzymatic degradation is important because it directly relates to the biological interactions of C.N.s with immunocompetent cells. SWCNTs are known to degrade *via* natural enzymatic catalysis; rate of which can be increased by horseradish peroxidase (HRP) and low amounts of hydrogen peroxide (H_2O_2) (Y. Zhao et al. 2011; Allen et al. 2008). For MWCNTs, layer-by-layer degradation has been observed with an overall rate of degradation depending on carboxylation. Alternatively, carbon dots (C.D.s), unlike other C.N.s, undergo peroxide-catalyzed degradation and exhibits unique degradation kinetics through enzyme oxidation for different charges (Srivastava et al. 2019; Farrera et al. 2014). To ensure that materials are susceptible to enzymatic degradation, assessment of different C.N.s to cells becomes a prerequisite for drug delivery applications. CNTs usually activate immune-related pathways and trigger the inflammation response of the body. In contrast, graphene interferes with the metabolic activity of macrophages by changing the potential of the mitochondrial membrane and increasing reactive oxygen species (ROS) in the vicinity (Medepalli et al. 2011; Bhattacharya et al. 2013). Hence, CNs drug delivery efficiency can be improved by reducing clearance and increasing retention time on or within the cells.

Similarly, to increase biocompatibility and enzymatic degradation, C.N.s can be coated with different biomolecules, lipids and proteins known as bio-corona (Docter et al. 2015). The additions of these new molecules to C.N.s can either increase or decrease the compatibility of the material with living systems. They can induce an immune response and inhibit enzymatic activity through cell surface receptors by conformational protein changes. For example, BSA coated on graphene and SWCNTs can decrease the inhibition efficiency of drug-metabolizing enzymes called CYP3A4 (Debnath and Srivastava 2021). Last, C.N.s are promising materials for drug delivery systems but are accompanied by unwanted toxic effects. The addition of metal impurities during synthesis, length, diameter, concentration, pristine surfaces, and charge of C.N.s are some common factors that can induce toxicity (Esfandiari et al. 2019; D. Dutta et al. 2007; X. Wang et al. 2012; Murphy et al. 2011; Kobayashi et al. 2017; Aldieri et al. 2013). However, these issues can be overcome by inducing different functional groups on the surface (Beg et al. 2018). C.N.s have been functionalized with various drugs such as ibuprofen, doxorubicin, hydroxyurea, chloroquine, and temozolomide to increase the time of localized drug action, improved cancer therapy, potential carriers for various drugs to target specific sites, act as a photosensitizer, improve bioavailability etc. (Debnath and Srivastava 2021).

5.3.3 Cancer Therapy

Despite the continuous advancement in medical science, cancer remains an uncertainly curable disease. Generally, cancer results as a consequence of gene mutations (Nigro et al. 1989). It is characterized by an abnormal growth of the cell cycle, resulting in the expansion of cancer cells in the body (Cheng et al. 2021). Common cancer treatments comprise surgical resection, chemotherapy, radiotherapy, and biological therapy. Chemotherapy is commonly considered for application due to its simplicity and its efficacy in curing a number of cancers. However, applying chemotherapy also brings some inevitable side effects, such as inhibiting the growth of tissues and cells of hair follicles, gastrointestinal tract cells, bone marrow, and inducing multi-drug resistance (MDR) (Q. Chen et al. 2015). In order to overcome these problems, nanomaterial-based chemotherapy has been studied for cancer treatment (Cheng et al. 2021).

Nanomaterials acquire unique properties such as high surface-to-volume ratio, electrical conductivity, mechanical strength, ability to interact with cells, optical properties, and fluorescence properties (Goenka et al. 2014). These properties indicate that nanomaterials can be utilized in cancer treatment. Carbon nanomaterials are widely used in the medical field because of their unique inherent properties and their biocompatibility compared to other nanomaterials. Due to the presence of $\pi - \pi$ stacking interactions, these materials offer strong binding force, self-assembly, controlled drug release, etc. (T. Chen et al. 2018). Several carbon nanomaterials such as graphene, fullerene, graphyne, carbon nanotubes, and carbon quantum dots are being studied for cancer treatment. The high planar surface makes graphene a useful candidate for drug delivery. Graphene is also considered to activate the human immune system. The presence of functional groups on the surface of graphene oxide (G.O.) makes them useful for cancer diagnosis and its therapy, tissue engineering, photodynamic therapy (PDT), and photothermal therapy (PTT) (Goenka et al. 2014).

Fullerene is also an allotrope of carbon, consisting of single or double-bonded carbon atoms. It has a closed mesh or a fused ring-type structure. Fullerenes are considered one of the non-toxic cancer therapeutics (Franskevych et al. 2019). Fullerenes can be used as an antioxidant, in drug delivery, PDT and PTT as well (Franskevych et al. 2019; J. Wang et al. 2006).

Carbon nanotubes are tubular forms of sp^2-hybradized carbon atoms. CNTs can induce immunity, resulting in the inhibition of tumor growth (Meng et al. 2010, 2008). Carbon quantum dots are commonly used in cancer imaging due to its excellent fluorescent and optical properties (Cheng et al. 2021). Despite possessing numerous useful properties, carbon nanomaterials are not much studied for clinical application due to its toxicity toward normal cells. However, with the advancement in cancer therapy research, these nanomaterials are believed to be researched with immense possibilities.

5.3.4 Tissue Engineering

Tissue engineering is an approach to produce or replace damaged tissues to repair body functions. Several people die every year while waiting for an organ transplant due to its unavailability. The field of tissue engineering aims to develop a biocompatible scaffold that can act as an artificial organ. This scaffold should also be compatible with other tissues and environments inside the body of individual (Figure 5.2) (Gomes et al. 2017). Bone is another important target for tissue engineering as its mechanical strength is equally important with its biocompatibility. In the last decade, the requirement for bone replacement has increased because of the skeletal damage in the bones that cannot be healed. Tissue engineering combined with nanotechnology is being looked at as a potential approach for tissue repair (Figure 5.3) (Eivazzadeh-Keihan et al. 2019).

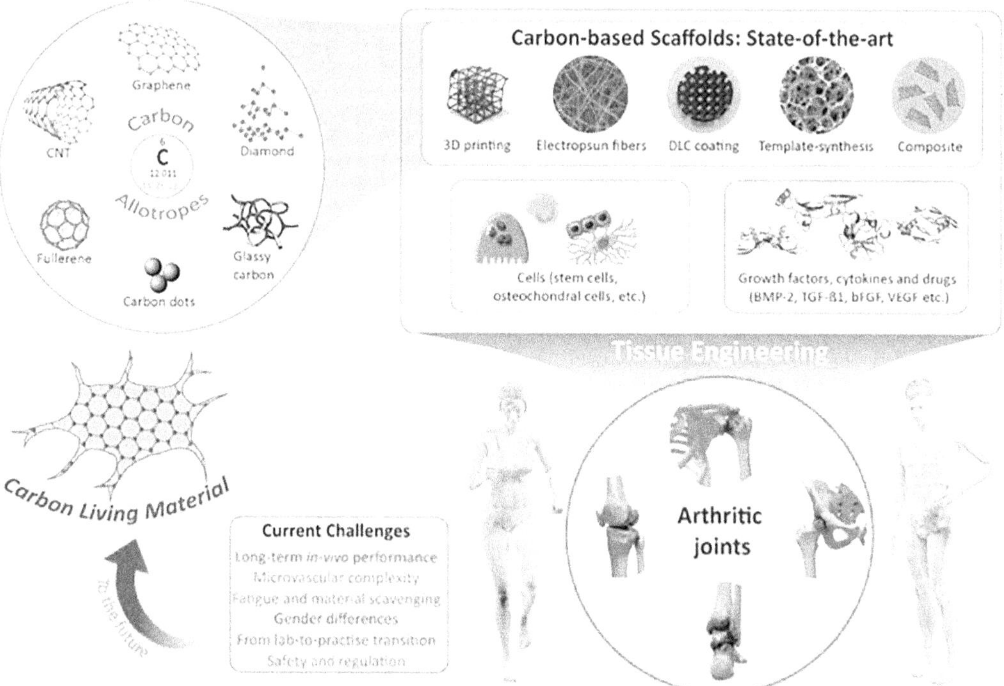

FIGURE 5.2 Carbon nanomaterials as a scaffold for tissue engineering. [Reprinted with permission from ref (Islam et al. 2022).].

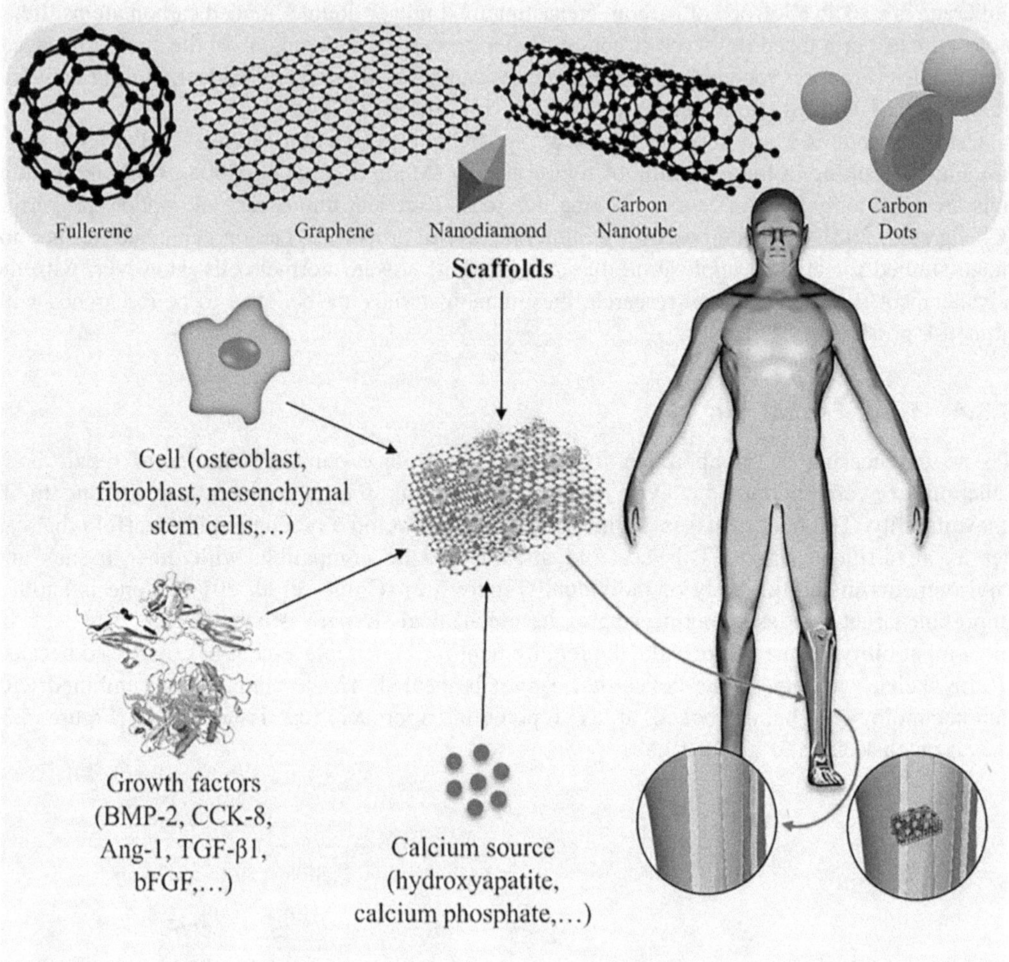

FIGURE 5.3 Carbon nanomaterials as a potential scaffold in bone tissue engineering. [Reprinted with permission from ref (Eivazzadeh-Keihan et al. 2019).].

Carbon nanomaterials are widely studied for biomedical applications due to their unique structures and excellent properties. Carbon nanotubes (CNTs) are also used to enhance the mechanical properties of biomaterials used for bone regeneration (Marrs et al. 2006). The single-walled carbon nanotubes (SWCNTs) with biodegradable polymer composite have been studied for bone tissue growth. In that study, it was found that suitable chemical composition plays an essential role in cell growth (Sitharaman et al. 2008). CNTs can also serve for protein transport either by adsorbing it or by covalently binding with it (Veetil and Ye 2009). They are also useful in targeted drug delivery in cancer treatment in order to prevent harm to other normal cells (Jin and Ye 2007). On the other hand, high surface area, high mechanical properties, and $\pi - \pi$ stacking interaction makes Graphene oxide (G.O.) one of the most suitable candidates for tissue engineering. The presence of G.O. in the polymer matrix makes it better for the mineralization of bone tissue (Eivazzadeh-Keihan et al. 2019). Stem cells are used in tissue growth, especially mesenchymal stem cells (MSCs), due to their ability to differentiate into osteoblasts, neuroblasts, fibroblasts, etc. Carbon dots are also used for the efficient osteogenic differentiation of rBMSCs (Shao et al. 2017). Biocompatibility, high surface-to-volume ratio, biodegradability, cell proliferation, and high mechanical properties make carbon nanomaterials a promising candidate for tissue engineering.

5.3.5 Antimicrobial Applications

Due to the excessive usage of antibiotics, microorganisms have started becoming resistant to several common antimicrobial agents. Therefore, developing new and more efficient antimicrobial agents has become essential. Several carbon nanomaterials such as carbon nanotubes (CNTs), graphene oxide (G.O.) nanoparticles, fullerenes prove their candidacy as antimicrobial agents. The surface area and the dimensions of carbon nanomaterials are important factors for their antimicrobial properties (Maleki Dizaj et al. 2015). With increasing surface area and reduction in size, their activity increases (Kang et al. 2008). Recently, it has been studied that the physical interaction between carbon nanomaterials and bacterial cells plays an important role in their antimicrobial properties (Maleki Dizaj et al. 2015; H. Chen et al. 2013). Single-walled carbon nanotubes (SWCNTs) were first studied for their antimicrobial activity (Kang et al. 2008, 2007). Later, both SWCNTs and MWCNTs were found to be capable of inhibiting different microorganisms, even after a short exposure time. CNT/ polymer composites show high antimicrobial properties against several microorganisms (Teixeira-Santos et al. 2021). SWCNTs and G.O. nanoparticles both show considerable microbicidal properties. Graphene and G.O. both prevent the growth of E. coli. Fullerenes also show antimicrobial activity against various bacteria such as E. coli, Salmonella, and Streptococcus spp (Teixeira-Santos et al. 2021). The membrane targeting, respiratory chain inhibition, and photosensitizing effects of the fullerene derivatives were suggested to be important factors for their antimicrobial properties (Deryabin et al. 2014). However, despite having excellent properties, carbon nanomaterials shall not be used for biomedical applications without addressing some important concerns related to their toxicity.

5.3.6 Bioimaging

Bioimaging is a non-invasive technique to visualize the biological process in real time. It is mainly used to monitor the physiological process in living cells/microorganisms, identify diseases, and understand human and animal anatomy. For that, different imaging sources, such as light, electrons, X-rays, fluorescence, ultrasound, etc., have been used. The carbon-based nanomaterial emerged as a suitable agent due to its high water solubility, excellent biocompatibility, and low cytotoxicity for bioimaging.

Carbon dots is the part of carbon-based nanomaterial which includes carbon dots (C.D.s), carbon quantum dots (CQDs), graphene quantum dots (GQDs), and carbonized polymer dots. They exhibit robust fluorescence, tuneable emission spectra, remarkable photostability, better aqueous solubility and no toxicity for in-vitro and in-vivo cell imaging. Over the years, C.D.s have been used for the imaging of various kinds of cells, such as cancer cells (Xue et al. 2018) neuron cells (Zhou et al. 2015), stem cells and cell organelles by using them as fluorescent probes in the microscopic analysis. In particular, the photoluminescent N-doped CDS have been used as a biomarker in confocal microscopy to visualize the hePG2 cells (Feng et al. 2015) and tumor cells (Figure 5.4(i,ii)) and fluorescent C.D.s were prepared by microwave digestion and utilized as a confocal imaging probe for capturing HeLa cells (Edison et al. 2016) surface functionalized CDS utilized as a bioimaging probe for in-vitro sensing of U87, M.G., 3T3, astrocyte and neuron cells (Kundu et al. 2018). GQDs with zero-dimensional derived from graphene was also investigated for bioimaging. GQDs have been synthesized by various synthesis routes; for e.g., the electrochemically exfoliated graphene used to synthesize highly fluorescence GQDs and used for stem cell imaging (L. Zhang et al. 2012). GQDs synthesized by the oxidation of graphite were investigated for the intracellular imaging of HeLa cells. Such

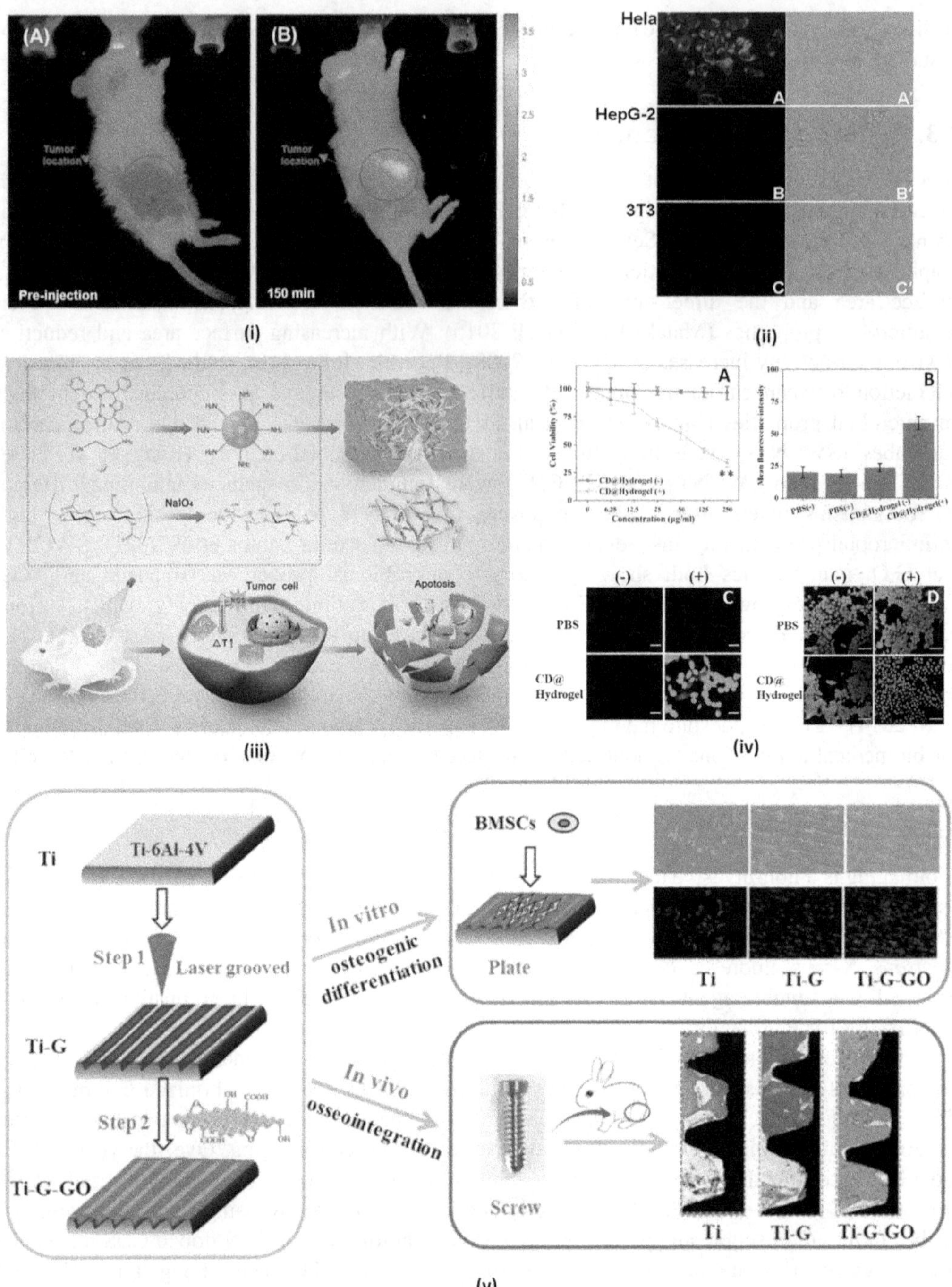

FIGURE 5.4 Bioimaging of tumor cells. (i) In-vivo fluorescent images of mice with tumor cells (WHIM4) before (A) and after (B) intravenous injection. The HA-nCQDs were detectable after 150 min at the tumor site. [Reprinted with permission from ref. (Karakoçak et al. 2021) Copyright 2022 American Chemical Society], (ii) Fluorescence images of cancer cells (human cells (Hella, and HePG2), and mouse cells (3T3) and corresponding bright field images. [Reprinted with permission from ref. (Y. Wang et al. 2013) Copyright 2013 American Chemical Society], (iii) Schematic diagram of the synthesis process and photodynamic therapy (PDT) of CD@Hydrogel. (iv) (A) cell viability of CD@Hydrogel with or without laser irradiation;

synthesized GQDs exhibit yellow-green fluorescence, a desirable characteristic of the imaging probe (L. Zhang et al. 2012).

Carbon nanotubes consist of single-walled carbon nanotubes (SWCNTs), and multi-walled carbon nanotubes (MWCNTs) with a 1-dimensional nanostructure. It exhibits outstanding properties for utilizing it as an imaging probe for fluorescence imaging due to its narrow band gap, Raman imaging owing to its strong Raman signal and exceptional photostability, and strong absorbance in NIR region utilized for photoacoustic imaging. Functionalized SWCNTs were used as NIR-II fluorescent imaging probes to visualize the mouse anatomy. It has been found that deep imaging could be possible with intact image contrast and feature size with incremental tissue thickness (Welsher et al. 2011). The multiplexed and multicolor raman imaging was carried out by SWCNTs with different isotopes. Here the cancer cells (human breast cancer cells and human glioblastoma cells) were detected and analyzed by raman imaging with different colors (Liu et al. 2008). Therefore, we can conclude that CNTs have a vast scope for utilization as an imaging probes, but there is a need to keep a close eye on the cytotoxic level. The 2-dimensional nanosheets of graphene, graphene oxide (G.O.) and graphene composites were also utilized for bioimaging owing to their π-π stacking. nano-graphene oxide (NGO) prepared by a modified Hummer method crafted with polyethylene glycol (PEG) was used as a fluorescence probe for Raji-B cell imaging (X. Sun et al. 2008). The reduced G.O. fabricated by protein-assisted route was employed for photoacoustic imaging of cancer cell in living mice (Sheng et al. 2013). Based on the several studies published so far, we can conclude that carbon-based materials have great potential for in-vitro and in-vivo cell imaging. Still, there is a need to take care of cell viability and cytotoxicity.

5.3.7 Photodynamic Therapy (PDT)

Photodynamic therapy (PDT) employs the concoction of light energy and photosensitized medication to destroy diseased cells such as cancer cells, acne and infection. The process of PDT involves irradiating suitable wavelength light over the photosensitizer. After that, it reacts with nearby oxygen molecules to create reactive oxygen species (ROS) with strong cytotoxicity. Such developed ROS used for the destruction of abnormal cells. The successful operation of PDT is mainly based on the type of photosensitizers because of controlling factors such as hydrophilicity, biocompatibility and stability. Extensive research has been done to minimize the issues related to that by employing carbon-based material. Carbon-based material includes carbon dots (C.D.s), graphene and its composites, graphene quantum dots (GQDs) and Carbon nanotubes. Several studies have been reported so far for utilizing previously mentioned carbon-based nanomaterial for PDT owing to their properties such as high stability, outstanding optical properties and good biocompatibility.

For the destruction of cancer cells, a photosensitizer molecule chlorin e6 (Ce6), was employed over polyethylene glycol (PEG)-functionalized graphene oxide (G.O.). Then, it was used to produce $_1O^2$ that was able to destroy human nasopharyngeal epidermal carcinoma K.B. cells (B. Tian et al. 2011). The single-walled carbon nanotubes (SWCNTs) were also used to generate $_1O^2$ and $O2^-$ ions (Murakami et al. 2012). Besides CNTs, the C3N4-modified carbon dots were used to restrict hypoxic tumor tissue (Zheng et al. 2016). Tuning the nitrogen content over the surface of graphene quantum dots was done to control the production of ROS (Ju et al. 2019).

(B) mean fluorescence intensities of ROS of different doses of PBS and CD@Hydrogel (C) intracellular ROS detection for different doses of PBS and CD@Hydrogel (scale bar = 50 μm); (D) staining assay of an animal cell(4T1 cell) treated with different doses of PBS and CD@Hydrogel, with or without laser irradiation, scale bar = 100 μm. [Reprinted with permission from ref. (Yue et al. 2022) Copyright 2022 American Chemical Society]. (v) Schematic representation of the synthesis of implant by grafting of graphene oxide(G.O.) coating over Ti–6Al–4 V plate and in-vitro and in-vivo testing of modified plates for osseointegration. [Reprinted with permission from ref. (Chenchen Wang et al. 2019) Copyright 2019 American Chemical Society].

Porphyrins were marked as a suitable photodynamic therapy agent, but low solubility and aggregation were constraining factors. Therefore, graphene quantum dots (GQDs) are used to enhance its properties. It demonstrated a significant phototoxic effect in the presence of white light in breast cancer (T-47D cell line) cells (Santos et al. 2021). Another work has been done recently by Yue et al to enhance porphyrin's properties. It has been done by synthesizing the injectable CD@HA hydrogel system, which effectively destroys cancer cells in light as shown in Figure 5.4(iii,iv). According to Figure 5.43), the designed hydrogel system (based on C.D.s and H.A.) used to produce singlet oxygen inhibiting tumor growth (Yue et al. 2022). Figure 5.44) depicts the ROS generation and in-vitro antitumor Effect in CD@HA Hydrogel. Along with the destruction of tumor cells by PDT, prevention of its relapse is also crucial. Therefore, work has been done to inhibit tumor growth in future. Functionalized nano-graphene has been loaded with tumor integrin $\alpha v \beta 6$ with HPPH photosensitizer. It stimulates the antitumor immunity inside the cells.

Despite the advantages associated with carbon-based material, there are still some limiting factors such as poor solubility and difficulty in controlling dose-dependent toxicity. The widely accepted mechanism of PDT stills suffers challenges due to its effect on the physiological activities of the organism due to uncontrolled cell penetration and damage caused due to cytotoxicity over the entire cell structure. Therefore, there is a need for extensive research that widely focuses on modifying photosensitizers with carbon-based nanomaterials.

5.3.8 Medical Implants

In general, a medical implant refers to the structure implanted inside the body to help or rejuvenate the damaged biological system or enhance the activity of the damaged biological structure. Traditionally, metals, alloys, polymers and ceramic materials have been used, but carbon-based material's biocompatibility has recently gained lots of attention. Medical implants comprise several kinds of implants, such as dental implants, hip implants, bone implants etc. Carbon fibers modified with polymers were tested for in-vitro and in-vivo implants in small animals. This study concludes that carbon-based material showed no toxicity or adverse effect around the implant area and also shows the potential to be used as a reservoir-based drug delivery system (Chua et al. 2021)

Metals such as zinc (Zn) and titanium (Ti) are used for medical implants. Still, they have associated challenges like corrosion, infection, coating dissolution, and delamination (Spear et al. 2022) carbon-based materials have been incorporated to resolve these problems. For addressing the bacterial infection, Lyu et al. demonstrated that comprising the ZnO nanotube/graphene oxide (G.O.) construct on the pure Zn metal enhanced the bactericidal phenomena (Lyu et al. 2019). Another study has been conducted where deposition of the graphene oxide(G.O.)- lysozyme(Lys) (G.O./Lys)n film was done by layer-by-layer deposition (LBL) technique (M. Li et al. 2019). Here Lys is used for improving the antibacterial properties and G.O. is used for enhancing the osteogenic profile. Osseointegration means the association of living bone with the surface of implant material, which is an important criterion for the success of the medical implant. For that, an intensive study has been done on the microstructural surface of the implant by culturing bone marrow stem cells (BMSCs) on the surface of the titanium alloy plate and Ti–G–GO modified plate (Chenchen Wang et al. 2019). Comparative study shows that G.O. modified plate shows better adhesion and osseointegration (Figure 5.4(v)). Carbon-based nanomaterial became a suitable candidate for medical implants owing to their mechanical, antimicrobial, and biological properties. Therefore, it requires much more attention to pave the way for further progression in this field.

5.4 CONCLUSION

In this chapter, we summarize the progress and recent advances in the application of carbon-based materials for biological applications. C.N.s having excellent optical, chemical and mechanical properties have been utilized in a plethora of applications. Depending on their size, shape and surface properties, different C.N.s interact differently with cells in the body resulting in different biological applications. Moreover, their application area can be broadened by introducing functionalization that suppresses the immune response and improves drug loading, targeting, cellular uptake etc. However, C.N.s faces the challenge of long-run toxicity, which has been improved somewhat using different functional groups on the surface. Systematic studies, standardization, and better formulation protocols are required to mitigate their adverse physiological response. More extensive research is still required to understand the long-term effects of these C.N.s on biological systems, including the human body, for the safe deployment of these materials as potential applicants.

REFERENCES

Abellán-Llobregat, A., C. González-Gaitán, L. Vidal, A. Canals, and E. Morall6n. 2018. "Portable Electrochemical Sensor Based on 4-Aminobenzoic Acid-Functionalized Herringbone Carbon Nanotubes for the Determination of Ascorbic Acid and Uric Acid in Human Fluids." *Biosensors and Bioelectronics* 109 (June). Elsevier: 123–131. doi:10.1016/J.BIOS.2018.02.047.

Aldieri, E., I. Fenoglio, F. Cesano, E. Gazzano, G. Gulino, D. Scarano, A. Attanasio, G. Mazzucco, D. Ghigo, and B. Fubini. 2013. "The Role of Iron Impurities in the Toxic Effects Exerted by Short Multiwalled Carbon Nanotubes (MWCNT) in Murine Alveolar Macrophages." Journal of Toxicology and Environmental Health, Part A 76 (18). Taylor & Francis: 1056–1071. doi:10.1080/15287394.2013.834855.

Allen, Brett L., Padmakar D. Kichambare, Pingping Gou, Irina I. Vlasova, Alexander A. Kapralov, Nagarjun Konduru, Valerian E. Kagan, and Alexander Star. 2008. "Biodegradation of Single-Walled Carbon Nanotubes through Enzymatic Catalysis." *Nano Letters* 8 (11). American Chemical Society: 3899–3903. doi:10.1021/NL802315H/SUPPL_FILE/NL802315H_SI_001.PDF.

Bacon, Mitchell, Siobhan J. Bradley, and Thomas Nann. 2014. "Graphene Quantum Dots." *Particle & Particle Systems Characterization* 31 (4). John Wiley & Sons, Ltd: 415–428. doi:10.1002/PPSC.201300252.

Baek, Seung Hoon, Jihyeok Roh, Chan Yeong Park, Min Woo Kim, Rongjia Shi, Suresh Kumar Kailasa, and Tae Jung Park. 2020. "Cu-Nanoflower Decorated Gold Nanoparticles-Graphene Oxide Nanofiber as Electrochemical Biosensor for Glucose Detection." *Materials Science and Engineering: C* 107 (February). Elsevier: 110273. doi:10.1016/J.MSEC.2019.110273.

Beg, Sarwar, Mahfoozur Rahman, Atul Jain, Sumant Saini, M. S. Hasnain, Suryakanta Swain, Sarim Imam, Imran Kazmi, and Sohail Akhter. 2018. "Emergence in the Functionalized Carbon Nanotubes as Smart Nanocarriers for Drug Delivery Applications." *Fullerenes, Graphenes and Nanotubes: A Pharmaceutical Approach*, January. William Andrew Publishing, 105–133. doi:10.1016/B978-0-12-813691-1.00004-X.

Beitollahi, Hadi, Fahimeh Movahedifar, Somayeh Tajik, and Shohreh Jahani. 2019. "A Review on the Effects of Introducing CNTs in the Modification Process of Electrochemical Sensors." *Electroanalysis* 31 (7). John Wiley & Sons, Ltd: 1195–1203. doi:10.1002/ELAN.201800370.

Bhattacharya, Kunal, Fernando Torres And6n, Ramy El-Sayed, and Bengt Fadeel. 2013. "Mechanisms of Carbon Nanotube-Induced Toxicity: Focus on Pulmonary Inflammation." *Advanced Drug Delivery Reviews* 65 (15). Elsevier: 2087–2097. doi:10.1016/J.ADDR.2013.05.012.

Bianco, Alberto, Kostas Kostarelos, and Maurizio Prato. 2005. "Applications of Carbon Nanotubes in Drug Delivery." *Current Opinion in Chemical Biology* 9 (6). Elsevier Current Trends: 674–679. doi:10.1016/J.CBPA.2005.10.005.

Cai, Jinying, Xiaodan Gou, Bolu Sun, Wuyan Li, Dai Li, Jinglong Liu, Fangdi Hu, and Yingdong Li. 2019. "Porous Graphene-Black Phosphorus Nanocomposite Modified Electrode for Detection of Leptin." *Biosensors and Bioelectronics* 137 (July). Elsevier: 88–95. doi:10.1016/J.BIOS.2019.04.045.

"Carbon-Based Nanomaterials." 2008. *Introduction to Nanoscience and Nanotechnology*, December. John Wiley & Sons, Ltd, 189–236. doi:10.1002/9781119096122.CH5.

Caruso, Frank, Taeghwan Hyeon, Vincent Rotello, Tennyson L. Doane, and Clemens Burda. 2012. "The Unique Role of Nanoparticles in Nanomedicine: Imaging, Drug Delivery and Therapy." *Chemical Society Reviews* 41 (7). The Royal Society of Chemistry: 2885–2911. doi:10.1039/C2CS15260F.

Cha, Chaenyung, Su Ryon Shin, Nasim Annabi, Mehmet R. Dokmeci, and Ali Khademhosseini. 2013. "Carbon-Based Nanomaterials: Multifunctional Materials for Biomedical Engineering." *ACS Nano*

7 (4). American Chemical Society: 2891–2897. doi:10.1021/NN401196A/ASSET/IMAGES/LARGE/NN-2013-01196A_0005.JPEG.

Chen, Fei, Weiyin Gao, Xiaopei Qiu, Hong Zhang, Lianhua Liu, Pu Liao, Weiling Fu, and Yang Luo. 2017. "Graphene Quantum Dots in Biomedical Applications: Recent Advances and Future Challenges." *Frontiers in Laboratory Medicine* 1 (4). No longer published by Elsevier: 192–199. doi:10.1016/J.FLM.2017.12.006.

Chen, Hanqing, Bing Wang, Di Gao, Ming Guan, Lingna Zheng, Hong Ouyang, Zhifang Chai, Yuliang Zhao, and Weiyue Feng. 2013. "Broad-Spectrum Antibacterial Activity of Carbon Nanotubes to Human Gut Bacteria." *Small* 9 (16). John Wiley & Sons, Ltd: 2735–2746. doi:10.1002/SMLL.201202792.

Chen, Qian, Hengte Ke, Zhifei Dai, and Zhuang Liu. 2015. "Nanoscale Theranostics for Physical Stimulus-Responsive Cancer Therapies." *Biomaterials* 73 (December). Elsevier: 214–230. doi:10.1016/J.BIOMATERIALS.2015.09.018.

Chen, Tao, Meixiu Li, and Jingquan Liu. 2018. "π-π Stacking Interaction: A Nondestructive and Facile Means in Material Engineering for Bioapplications." *Crystal Growth and Design* 18(5). American Chemical Society: 2765–2783. doi:10.1021/ACS.CGD.7B01503/ASSET/IMAGES/LARGE/CG-2017-01503A_0004.JPEG.

Chen, Yu Cheng, Xin Chun Huang, Yun Ling Luo, Yung Chen Chang, You Zung Hsieh, and Hsin Yun Hsu. 2013. "Non-Metallic Nanomaterials in Cancer Theranostics: A Review of Silica- and Carbon-Based Drug Delivery Systems." Science and Technology of Advanced Materials 14 (4). Taylor & Francis: 44407–44430. doi:10.1088/1468-6996/14/4/044407.

Cheng, Zhe, Maoyu Li, Raja Dey, and Yongheng Chen. 2021. "Nanomaterials for Cancer Therapy: Current Progress and Perspectives." *Journal of Hematology & Oncology 2021 14:1* 14(1). BioMed Central: 1–27. doi:10.1186/S13045-021-01096-0.

Chua, Corrine Ying Xuan, Hsuan Chen Liu, Nicola di Trani, Antonia Susnjar, Jeremy Ho, Giovanni Scorrano, Jessica Rhudy, et al. 2021. "Carbon Fiber Reinforced Polymers for Implantable Medical Devices." *Biomaterials* 271 (October 2020). Elsevier Ltd: 120719. doi:10.1016/j.biomaterials.2021.120719.

Chung, Chul, Young Kwan Kim, Dolly Shin, Soo Ryoon Ryoo, Byung Hee Hong, and Dal Hee Min. 2013. "Biomedical Applications of Graphene and Graphene Oxide." *Accounts of Chemical Research* 46 (10). American Chemical Society: 2211–2224. doi:10.1021/AR300159F/ASSET/IMAGES/LARGE/AR-2012-00159F_0016.JPEG.

Colomer, J. F., C. Stephan, S. Lefrant, G. van Tendeloo, I. Willems, Z. Kónya, A. Fonseca, Ch Laurent, and J. B. Nagy. 2000. "Large-Scale Synthesis of Single-Wall Carbon Nanotubes by Catalytic Chemical Vapor Deposition (CCVD) Method." *Chemical Physics Letters* 317 (1–2). North-Holland: 83–89. doi:10.1016/S0009-2614(99)01338-X.

Das, Trupti R., Suchit Kumar Jena, Rashmi Madhuri, and Prashant K. Sharma. 2018. "Polymeric Iron Oxide-Graphene Nanocomposite as a Trace Level Sensor of Vitamin C." *Applied Surface Science* 449 (August). North-Holland: 304–313. doi:10.1016/J.APSUSC.2018.01.173.

Debnath, Sujit Kumar, and Rohit Srivastava. 2021. "Drug Delivery With Carbon-Based Nanomaterials as Versatile Nanocarriers: Progress and Prospects." *Frontiers in Nanotechnology* 3 (April). Frontiers Media S.A.: 15. doi:10.3389/FNANO.2021.644564/BIBTEX.

Deryabin, Dmitry G., Olga K. Davydova, Zulfiya Zh Yankina, Alexey S. Vasilchenko, Sergei A. Miroshnikov, Alexey B. Kornev, Anastasiya v. Ivanchikhina, and Pavel A. Troshin. 2014. "The Activity of [60]Fullerene Derivatives Bearing Amine and Carboxylic Solubilizing Groups against Escherichia Coli: A Comparative Study." *Journal of Nanomaterials* 2014. doi:10.1155/2014/907435.

Docter, D., D. Westmeier, M. Markiewicz, S. Stolte, S. K. Knauer, and R. H. Stauber. 2015. "The Nanoparticle Biomolecule Corona: Lessons Learned – Challenge Accepted?" *Chemical Society Reviews* 44 (17). The Royal Society of Chemistry: 6094–6121. doi:10.1039/C5CS00217F.

Dutta, Debamitra, Shanmugavelayutham Kamakshi Sundaram, Justin Gary Teeguarden, Brian Joseph Riley, Leonard Sheldon Fifield, Jon Morrell Jacobs, Shane Raymond Addleman, George Alan Kaysen, Brij Mohan Moudgil, and Thomas Joseph Weber. 2007. "Adsorbed Proteins Influence the Biological Activity and Molecular Targeting of Nanomaterials." *Toxicological Sciences* 100 (1). Oxford Academic: 303–315. doi:10.1093/TOXSCI/KFM217.

Dutta, Sayan Deb, Keya Ganguly, Rajkumar Bandi, and Madhusudhan Alle. 2021. "A New Era of Cancer Treatment: Carbon Nanotubes as Drug Delivery Tools." *Nanotechnology in the Life Sciences*. Springer Science and Business Media B.V., 155–171. doi:10.1007/978-3-030-84262-8_6/COVER.

Ebbesen, T. W., and P. M. Ajayan. 1992. "Large-Scale Synthesis of Carbon Nanotubes." *Nature 1992 358:6383* 358 (6383). Nature Publishing Group: 220–222. doi:10.1038/358220a0.

Edison, Thomas Nesakumar Jebakumar Immanuel, Raji Atchudan, Mathur Gopalakrishnan Sethuraman, Jae Jin Shim, and Yong Rok Lee. 2016. "Microwave Assisted Green Synthesis of Fluorescent N-Doped Carbon Dots: Cytotoxicity and Bio-Imaging Applications." *Journal of Photochemistry and Photobiology B: Biology* 161. Elsevier B.V.: 154–161. doi:10.1016/j.jphotobiol.2016.05.017.

Eivazzadeh-Keihan, Reza, Ali Maleki, Miguel de la Guardia, Milad Salimi Bani, Karim Khanmohammadi Chenab, Paria Pashazadeh-Panahi, Behzad Baradaran, Ahad Mokhtarzadeh, and Michael R. Hamblin. 2019. "Carbon Based Nanomaterials for Tissue Engineering of Bone: Building New Bone on Small Black Scaffolds: A Review." *Journal of Advanced Research* 18 (July). Elsevier: 185–201. doi:10.1016/J.JARE.2019.03.011.

Esfandiari, Neda, Zeinab Bagheri, Hamide Ehtesabi, Zahra Fatahi, Hossein Tavana, and Hamid Latifi. 2019. "Effect of Carbonization Degree of Carbon Dots on Cytotoxicity and Photo-Induced Toxicity to Cells." *Heliyon* 5 (12). Elsevier: e02940. doi:10.1016/J.HELIYON.2019.E02940.

Farrera, Consol, Kunal Bhattacharya, Beatrice Lazzaretto, Fernando T. Andón, Kjell Hultenby, Gregg P. Kotchey, Alexander Star, and Bengt Fadeel. 2014. "Extracellular Entrapment and Degradation of Single-Walled Carbon Nanotubes." *Nanoscale* 6 (12). The Royal Society of Chemistry: 6974–6983. doi:10.1039/C3NR06047K.

Feng, Xin, Yaoquan Jiang, Jingpeng Zhao, Miao Miao, Shaomei Cao, Jianhui Fang, and Liyi Shi. 2015. "Easy Synthesis of Photoluminescent N-Doped Carbon Dots from Winter Melon for Bio-Imaging." *RSC Advances* 5 (40). Royal Society of Chemistry: 31250–31254. doi:10.1039/c5ra02271a.

Franskevych, Daria, Svitlana Prylutska, Iryna Grynyuk, Ganna Pasichnyk, Liudmyla Drobot, Olga Matyshevska, and Uwe Ritter. 2019. "Mode of Photoexcited C60 Fullerene Involvement in Potentiating Cisplatin Toxicity against Drug-Resistant L1210 Cells." *BioImpacts: BI* 9 (4). Tabriz University of Medical Sciences: 211–217. doi:10.15171/bi.2019.26.

Goenka, Sumit, Vinayak Sant, and Shilpa Sant. 2014. "Graphene-Based Nanomaterials for Drug Delivery and Tissue Engineering." *Journal of Controlled Release* 173 (1). Elsevier: 75–88. doi:10.1016/J.JCONREL.2013.10.017.

Gomes, Manuela E., Márcia T. Rodrigues, Rui M. A. Domingues, and Rui L. Reis. 2017. "Tissue Engineering and Regenerative Medicine: New Trends and Directions—A Year in Review." Tissue Engineering Part B: Reviews 23 (3). Mary Ann Liebert, Inc. 140 Huguenot Street, 3rd Floor New Rochelle, NY 10801 USA: 211–224. doi:10.1089/TEN.TEB.2017.0081.

Hasnain, Md Saquib, and Amit Kumar Nayak. 2019. "Carbon Nanotubes for Targeted Drug Delivery." *SpringerBriefs in Applied Sciences and Technology*. Singapore: Springer Singapore. doi:10.1007/978-981-15-0910-0.

Hong, Guosong, Shuo Diao, Alexander L. Antaris, and Hongjie Dai. 2015. "Carbon Nanomaterials for Biological Imaging and Nanomedicinal Therapy." *Chemical Reviews* 115(19). American Chemical Society: 10816–10906. doi:10.1021/ACS.CHEMREV.5B00008/ASSET/IMAGES/LARGE/CR-2015-00008F_0043.JPEG.

Huang, Huang Chiao, Sutapa Barua, Gaurav Sharma, Sandwip K. Dey, and Kaushal Rege. 2011. "Inorganic Nanoparticles for Cancer Imaging and Therapy." *Journal of Controlled Release* 155 (3). Elsevier: 344–357. doi:10.1016/J.JCONREL.2011.06.004.

Islam, Monsur, Andrés Díaz Lantada, Dario Mager, and Jan G. Korvink. 2022. Carbon-Based Materials for Articular Tissue Engineering: From Innovative Scaffolding Materials toward Engineered Living Carbon." *Advanced Healthcare Materials* 11 (1). John Wiley and Sons Inc: 2101834. doi:10.1002/adhm.202101834.

Jahdaly, Badreah Ali al, Mohamed Farouk Elsadek, Badreldin Mohamed Ahmed, Mohamed Fawzy Farahat, Mohamed M. Taher, and Ahmed M. Khalil. 2021. "Outstanding Graphene Quantum Dots from Carbon Source for Biomedical and Corrosion Inhibition Applications: A Review." *Sustainability 2021, Vol. 13, Page 2127* 13(4). Multidisciplinary Digital Publishing Institute: 2127. doi:10.3390/SU13042127.

Jampilek, Josef, and Katarina Kralova. 2021. "Advances in Drug Delivery Nanosystems Using Graphene-Based Materials and Carbon Nanotubes." *Materials 2021, Vol. 14, Page 1059* 14 (5). Multidisciplinary Digital Publishing Institute: 1059. doi:10.3390/MA14051059.

Jin, Sha, and Kaiming Ye. 2007. "Nanoparticle-Mediated Drug Delivery and Gene Therapy." *Biotechnology Progress* 23 (1). American Chemical Society (ACS): 32–41. doi:10.1021/BP060348J.

Joshi, Pratik, Rupesh Mishra, and Roger J. Narayan. 2021. "Biosensing Applications of Carbon-Based Materials." *Current Opinion in Biomedical Engineering* 18 (June). Elsevier: 100274. doi:10.1016/J.COBME.2021.100274.

Ju, Jian, Sagar Regmi, Afu Fu, Sierin Lim, and Quan Liu. 2019. "Graphene Quantum Dot Based Charge-Reversal Nanomaterial for Nucleus-Targeted Drug Delivery and Efficiency Controllable Photodynamic Therapy." *Journal of Biophotonics* 12 (6): e201800367. doi:10.1002/jbio.201800367.

Kang, Seoktae, Moshe Herzberg, Debora F. Rodrigues, and Menachem Elimelech. 2008. "Antibacterial Effects of Carbon Nanotubes: Size Does Matter!." *Langmuir* 24 (13). American Chemical Society: 6409–6413. doi:10.1021/LA800951V/SUPPL_FILE/LA800951V-FILE004.PDF.

Kang, Seoktae, Mathieu Pinault, Lisa D. Pfefferle, and Menachem Elimelech. 2007. "Single-Walled Carbon Nanotubes Exhibit Strong Antimicrobial Activity." *Langmuir* 23 (17). American Chemical Society: 8670–8673. doi:10.1021/LA701067R/SUPPL_FILE/LA701067R-FILE002.PDF.

Karakoçak, Bedia Begüm, Amine Laradji, Tina Primeau, Mikhail Y. Berezin, Shunqiang Li, and Nathan Ravi. 2021. "Hyaluronan-Conjugated Carbon Quantum Dots for Bioimaging Use." *ACS Applied Materials and Interfaces* 13 (1). American Chemical Society: 277–286. doi:10.1021/acsami.0c20088.

Kataura, H., Y. Kumazawa, Y. Maniwa, I. Umezu, S. Suzuki, Y. Ohtsuka, and Y. Achiba. 1999. "Optical Properties of Single-Wall Carbon Nanotubes." *Synthetic Metals* 103 (1–3). Elsevier: 2555–2558. doi:10.1016/S0379-6779(98)00278-1.

Kobayashi, Norihiro, Hiroto Izumi, and Yasuo Morimoto. 2017. "Review of Toxicity Studies of Carbon Nanotubes." *Journal of Occupational Health* 59 (5). John Wiley & Sons, Ltd: 394–407. doi:10.1539/JOH.17-0089-RA.

Kundu, Aniruddha, Jungpyo Lee, Byeongho Park, Chaiti Ray, K Vijaya Sankar, Wook Sung Kim, Soo Hyun Lee, Il Joo Cho, and Seong Chan Jun. 2018. "Facile Approach to Synthesize Highly Fluorescent Multicolor Emissive Carbon Dots via Surface Functionalization for Cellular Imaging." *Journal of Colloid and Interface Science* 513. Elsevier Inc.: 505–514. doi:10.1016/j.jcis.2017.10.095.

Li, Jie, Huamin Zeng, Zhaowu Zeng, Yiying Zeng, and Tian Xie. 2021. "Promising Graphene-Based Nanomaterials and Their Biomedical Applications and Potential Risks: A Comprehensive Review." *ACS Biomaterials Science and Engineering* 7 (12). American Chemical Society: 5363–5396. doi:10.1021/ACSBIOMATERIALS.1C00875/ASSET/IMAGES/LARGE/AB1C00875_0012.JPEG.

Li, Meng, Huaqiong Li, Qiongxi Pan, Chenyuan Gao, Yingying Wang, Shuoshuo Yang, Xingjie Zan, and Yifu Guan. 2019. "Graphene Oxide and Lysozyme Ultrathin Films with Strong Antibacterial and Enhanced Osteogenesis." *Langmuir*. American Chemical Society. doi:10.1021/acs.langmuir.9b00035.

Li, Xiaoming, Muchen Rui, Jizhong Song, Zihan Shen, and Haibo Zeng. 2015. "Carbon and Graphene Quantum Dots for Optoelectronic and Energy Devices: A Review." *Advanced Functional Materials* 25 (31). John Wiley & Sons, Ltd: 4929–4947. doi:10.1002/ADFM.201501250.

Liu, Zhuang, Xiaolin Li, Scott M. Tabakman, Kaili Jiang, Shoushan Fan, and Hongjie Dai. 2008. "Multiplexed Multicolor Raman Imaging of Live Cells with Isotopically Modified Single Walled Carbon Nanotubes." *Journal of the American Chemical Society* 130 (41): 13540–13541. doi:10.1021/ja806242t.

Lv, Jingjing, Caiwei Li, Shasha Feng, Shen Ming Chen, Yong Ding, Chenglong Chen, Qingli Hao, Ting Hai Yang, and Wu Lei. 2019. "A Novel Electrochemical Sensor for Uric Acid Detection Based on PCN/MWCNT." *Ionics* 25 (9). Institute for Ionics: 4437–4445. doi:10.1007/S11581-019-03010-8/TABLES/2.

Lyu, Hao, Zechao He, Yau Kei Chan, Xianhua He, Yue Yu, and Yi Deng. 2019. "Hierarchical ZnO Nanotube/Graphene Oxide Nanostructures Endow Pure Zn Implant with Synergistic Bactericidal Activity and Osteogenicity." *Industrial and Engineering Chemistry Research* 58 (42). American Chemical Society: 19377–19385. doi:10.1021/acs.iecr.9b02986.

Maity, Debasis, C. R. Minitha, and Rajendra Kumar Rajendra. 2019. "Glucose Oxidase Immobilized Amine Terminated Multiwall Carbon Nanotubes/Reduced Graphene Oxide/Polyaniline/Gold Nanoparticles Modified Screen-Printed Carbon Electrode for Highly Sensitive Amperometric Glucose Detection." *Materials Science and Engineering: C* 105 (December). Elsevier: 110075. doi:10.1016/J.MSEC.2019.110075.

Maleki Dizaj, Solmaz, Afsaneh Mennati, Samira Jafari, Khadejeh Khezri, and Khosro Adibkia. 2015. "Antimicrobial Activity of Carbon-Based Nanoparticles." *Advanced Pharmaceutical Bulletin* 5 (1). Tabriz University of Medical Sciences: 19–23. doi:10.5681/apb.2015.003.

Marrs, Brock, Rodney Andrews, Terry Rantell, and David Pienkowski. 2006. "Augmentation of Acrylic Bone Cement with Multiwall Carbon Nanotubes." *Journal of Biomedical Materials Research Part A* 77A (2). John Wiley & Sons, Ltd: 269–276. doi:10.1002/JBM.A.30651.

Master, Alyssa, Megan Livingston, and Anirban sen Gupta. 2013. "Photodynamic Nanomedicine in the Treatment of Solid Tumors: Perspectives and Challenges." *Journal of Controlled Release* 168 (1). Elsevier: 88–102. doi:10.1016/J.JCONREL.2013.02.020.

Medepalli, Krishnakiran, Bruce Alphenaar, Ashok Raj, and Palaniappan Sethu. 2011. "Evaluation of the Direct and Indirect Response of Blood Leukocytes to Carbon Nanotubes (CNTs)." *Nanomedicine: Nanotechnology, Biology and Medicine* 7 (6). Elsevier: 983–991. doi:10.1016/J.NANO.2011.04.002.

Meng, Jie, Jinhong Duan, Hua Kong, Li Li, Chen Wang, Sishen Xie, Shuchang Chen, Ning Gu, Haiyan Xu, and Xian da Yang. 2008. "Carbon Nanotubes Conjugated to Tumor Lysate Protein Enhance the Efficacy of an Antitumor Immunotherapy." *Small* 4 (9). John Wiley & Sons, Ltd: 1364–1370. doi:10.1002/SMLL.200701059.

Meng, Jie, Man Yang, Fumin Jia, Hua Kong, Weiqi Zhang, Chaoying Wang, Jianmin Xing, Sishen Xie, and Haiyan Xu. 2010. "Subcutaneous Injection of Water-Soluble Multi-Walled Carbon Nanotubes in Tumor-Bearing Boosts the Host Immune Activity." *Nanotechnology* 21 (14). IOP Publishing: 145104. doi:10.1088/0957-4484/21/14/145104.

Mohajeri, Mohammad, Behzad Behnam, and Amirhossein Sahebkar. 2019. "Biomedical Applications of Carbon Nanomaterials: Drug and Gene Delivery Potentials." *Journal of Cellular Physiology* 234 (1). John Wiley & Sons, Ltd: 298–319. doi:10.1002/JCP.26899.

Murakami, Tatsuya, Hirotaka Nakatsuji, Mami Inada, Yoshinori Matoba, Tomokazu Umeyama, Masahiko Tsujimoto, Seiji Isoda, Mitsuru Hashida, and Hiroshi Imahori. 2012. "Photodynamic and Photothermal Effects of Semiconducting and Metallic-Enriched Single-Walled Carbon Nanotubes." *Journal of the American Chemical Society* 134 (43): 17862–17865. doi:10.1021/ja3079972.

Murphy, Fiona A., Craig A. Poland, Rodger Duffin, Khuloud T. Al-Jamal, Hanene Ali-Boucetta, Antonio Nunes, Fiona Byrne, et al. 2011. "Length-Dependent Retention of Carbon Nanotubes in the Pleural Space of Mice Initiates Sustained Inflammation and Progressive Fibrosis on the Parietal Pleura." *The American Journal of Pathology* 178 (6). Elsevier: 2587–2600. doi:10.1016/J.AJPATH.2011.02.040.

Nag, Anindya, Md Eshrat E. Alahi, Subhas Chandra Mukhopadhyay, and Zhi Liu. 2021. "Multi-Walled Carbon Nanotubes-Based Sensors for Strain Sensing Applications." *Sensors 2021, Vol. 21, Page 1261* 21 (4). Multidisciplinary Digital Publishing Institute: 1261. doi:10.3390/S21041261.

Nigro, Janice M., Suzanne J. Baker, Antonette C. Preisinger, J. Milburn Jessup, Richard Hosteller, Karen Cleary, Sandra H. Bigner, et al. 1989. "Mutations in the P53 Gene Occur in Diverse Human Tumour Types." *Nature 1989 342:6250* 342 (6250). Nature Publishing Group: 705–708. doi:10.1038/342705a0.

Nimushakavi, Sahithi, Shagufta Haque, Rajesh Kotcherlakota, and Chitta Ranjan Patra. 2022. "Biomedical Applications of Carbon Nanotubes: Recent Development and Future Challenges." *Nanoengineering of Biomaterials*, February. John Wiley & Sons, Ltd, 353–388. doi:10.1002/9783527832095.CH29.

Oliveira, Thiago M. B. F., and Simone Morais. 2018. "New Generation of Electrochemical Sensors Based on Multi-Walled Carbon Nanotubes." *Applied Sciences 2018, Vol. 8, Page 1925* 8 (10). Multidisciplinary Digital Publishing Institute: 1925. doi:10.3390/APP8101925.

Onyancha, Robert Birundu, Uyiosa Osagie Aigbe, Kingsley Eghonghon Ukhurebor, and Perpetua Wanjiru Muchiri. 2021. "Facile Synthesis and Applications of Carbon Nanotubes in Heavy-Metal Remediation and Biomedical Fields: A Comprehensive Review." *Journal of Molecular Structure* 1238 (August). Elsevier: 130462. doi:10.1016/J.MOLSTRUC.2021.130462.

Pandit, Santosh, Karolina Gaska, Roland Kádár, and Ivan Mijakovic. 2021. "Graphene-Based Antimicrobial Biomedical Surfaces." *ChemPhysChem* 22 (3). John Wiley & Sons, Ltd: 250–263. doi:10.1002/CPHC.202000769.

Parveen, Suphiya, Ranjita Misra, and Sanjeeb K. Sahoo. 2012. "Nanoparticles: A Boon to Drug Delivery, Therapeutics, Diagnostics and Imaging." *Nanomedicine: Nanotechnology, Biology and Medicine* 8 (2). Elsevier: 147–166. doi:10.1016/J.NANO.2011.05.016.

Pattnaik, Satyanarayan, Y. Surendra, J. Venkateshwar Rao, and Kalpana Swain. 2020. "Carbon Family Nanomaterials for Drug Delivery Applications." *Nanoengineered Biomaterials for Advanced Drug Delivery*, January. Elsevier, 421–445. doi:10.1016/B978-0-08-102985-5.00018-8.

Pattnaik, Satyanarayan, Kalpana Swain, and Zhiqun Lin. 2016. "Graphene and Graphene-Based Nanocomposites: Biomedical Applications and Biosafety." *Journal of Materials Chemistry B* 4 (48). The Royal Society of Chemistry: 7813–7831. doi:10.1039/C6TB02086K.

Prajapati, Shiv Kumar, Akanksha Malaiya, Payal Kesharwani, Deeksha Soni, and Aakanchha Jain. 2020. "Biomedical Applications and Toxicities of Carbon Nanotubes." 10.1080/01480545.2019.1709492 45 (1). Taylor & Francis: 435–450. doi:10.1080/01480545.2019.1709492.

Rauti, Rossana, Mattia Musto, Susanna Bosi, Maurizio Prato, and Laura Ballerini. 2019. "Properties and Behavior of Carbon Nanomaterials When Interfacing Neuronal Cells: How Far Have We Come?" *Carbon* 143 (March). Pergamon: 430–446. doi:10.1016/J.CARBON.2018.11.026.

Riley, Parand R., and Roger J. Narayan. 2021. "Recent Advances in Carbon Nanomaterials for Biomedical Applications: A Review." *Current Opinion in Biomedical Engineering* 17 (March). Elsevier: 100262. doi:10.1016/J.COBME.2021.100262.

Saheeda, P., Y. M. Thasneem, K. Sabira, M. Dhaneesha, N. K. Sulfikkarali, and S. Jayaleksmi. 2022. "Multi-Walled Carbon Nanotubes/Polypyrrole Nanocomposite, Synthesized through an Eco-Friendly Route, as a Prospective Drug Delivery System." *Polymer Bulletin*, May. Springer Science and Business Media Deutschland GmbH, 1–21. doi:10.1007/S00289-022-04290-3/FIGURES/10.

Santos, Carla I. M., Laura Rodríguez-Pérez, Gil Gonçalves, Cristina J. Dias, Fátima Monteiro, Maria do Amparo F. Faustino, Sandra I. Vieira, et al. 2021. "Enhanced Photodynamic Therapy Effects of Graphene Quantum Dots Conjugated with Aminoporphyrins." *ACS Applied Nano Materials* 4 (12). American Chemical Society: 13079–13089. doi:10.1021/acsanm.1c02600.

Shannahan, Jonathan. 2017. "The Biocorona: A Challenge for the Biomedical Application of Nanoparticles." *Nanotechnology Reviews* 6 (4). Walter de Gruyter GmbH: 345–353. doi:10.1515/NTREV-2016-0098/ASSET/GRAPHIC/J_NTREV-2016-0098_CV_001.JPG.

Shao, Dan, Mengmeng Lu, Duo Xu, Xiao Zheng, Yue Pan, Yubin Song, Jinying Xu, et al. 2017. "Carbon Dots for Tracking and Promoting the Osteogenic Differentiation of Mesenchymal Stem Cells." *Biomaterials Science* 5 (9). The Royal Society of Chemistry: 1820–1827. doi:10.1039/C7BM00358G.

Sheng, Zonghai, Liang Song, Jiaxiang Zheng, Dehong Hu, Meng He, Mingbin Zheng, Guanhui Gao, et al. 2013. "Protein-Assisted Fabrication of Nano-Reduced Graphene Oxide for Combined Invivo Photoacoustic Imaging and Photothermal Therapy." *Biomaterials* 34 (21). Elsevier Ltd: 5236–5243. doi:10.1016/j.biomaterials.2013.03.090.

Sitharaman, Balaji, Xinfeng Shi, X. Frank Walboomers, Hongbing Liao, Vincent Cuijpers, Lon J. Wilson, Antonios G. Mikos, and John A. Jansen. 2008. "In Vivo Biocompatibility of Ultra-Short Single-Walled Carbon Nanotube/Biodegradable Polymer Nanocomposites for Bone Tissue Engineering." *Bone* 43 (2). Elsevier: 362–370. doi:10.1016/J.BONE.2008.04.013.

Spear, Rose L., Ruth E. Cameron, Richard Petersen, Riccardo Guazzo, Chiara Gardin, Gloria Bellin, Luca Sbricoli, et al. 2022. "Titanium Implant Osseointegration Problems with Alternate Solutions Using Epoxy/Carbon-Fiber-Reinforced Composite." *Metals* 11 (17). Springer Berlin Heidelberg: 127–133. doi:10.1007/s00256-022-04100-x.

Srivastava, Indrajit, Dinabandhu Sar, Prabuddha Mukherjee, Aaron S. Schwartz-Duval, Zhaolu Huang, Camilo Jaramillo, Ana Civantos, et al. 2019. "Enzyme-Catalysed Biodegradation of Carbon Dots Follows Sequential Oxidation in a Time Dependent Manner." *Nanoscale* 11 (17). The Royal Society of Chemistry: 8226–8236. doi:10.1039/C9NR00194H.

Sun, Hanjun, Li Wu, Weili Wei, and Xiaogang Qu. 2013. "Recent Advances in Graphene Quantum Dots for Sensing." *Materials Today* 16 (11). Elsevier: 433–442. doi:10.1016/J.MATTOD.2013.10.020.

Sun, Xiaoming, Zhuang Liu, Kevin Welsher, Joshua Tucker Robinson, Andrew Goodwin, Sasa Zaric, and Hongjie Dai. 2008. "Nano-Graphene Oxide for Cellular Imaging and Drug Delivery." *Nano Research* 1 (3): 203–212. doi:10.1007/s12274-008-8021-8.

Taratula, Oleh, Andriy Kuzmov, Milin Shah, Olga B. Garbuzenko, and Tamara Minko. 2013. "Nanostructured Lipid Carriers as Multifunctional Nanomedicine Platform for Pulmonary Co-Delivery of Anticancer Drugs and SiRNA." *Journal of Controlled Release* 171 (3). Elsevier: 349–357. doi:10.1016/J.JCONREL.2013.04.018.

Teixeira-Santos, Rita, Marisa Gomes, Luciana C. Gomes, and Filipe J. Mergulhão. 2021. "Antimicrobial and Anti-Adhesive Properties of Carbon Nanotube-Based Surfaces for Medical Applications: A Systematic Review." *IScience* 24 (1). Elsevier: 102001. doi:10.1016/J.ISCI.2020.102001.

Tian, Bo, Chao Wang, Shuai Zhang, Liangzhu Feng, and Zhuang Liu. 2011. "Photothermally Enhanced Photodynamic Therapy Delivered by Nano-Graphene Oxide." *ACS Nano* 5 (9): 7000–7009. doi:10.1021/nn201560b.

Tian, P., L. Tang, K. S. Teng, and S. P. Lau. 2018. "Graphene Quantum Dots from Chemistry to Applications." *Materials Today Chemistry* 10 (December). Elsevier: 221–258. doi:10.1016/J.MTCHEM.2018.09.007.

Tiwari, Jitendra N., Varun Vij, K Christian Kemp, and Kwang S. Kim. 2016. "Engineered Carbon-Nanomaterial-Based Electrochemical Sensors for Biomolecules." *ACS Nano* 10 (1). American Chemical Society: 46–80. doi:10.1021/acsnano.5b05690.

Veetil, Jithesh v., and Kaiming Ye. 2009. "Tailored Carbon Nanotubes for Tissue Engineering Applications." *Biotechnology Progress* 25 (3). American Chemical Society (ACS): 709–721. doi:10.1002/BTPR.165.

Wang, Chao, Xinxing Ma, Shuoqi Ye, Liang Cheng, Kai Yang, Liang Guo, Changhui Li, Yonggang Li, and Zhuang Liu. 2012. Protamine Functionalized Single-Walled Carbon Nanotubes for Stem Cell Labeling and In Vivo Raman/Magnetic Resonance/Photoacoustic Triple-Modal Imaging." *Advanced Functional Materials* 22 (11). John Wiley & Sons, Ltd: 2363–2375. doi:10.1002/ADFM.201200133.

Wang, Chenchen, Hongxing Hu, Zhipeng Li, Yifan Shen, Yong Xu, Gangqiang Zhang, Xiangqiong Zeng, et al. 2019. "Enhanced Osseointegration of Titanium Alloy Implants with Laser Microgrooved Surfaces and Graphene Oxide Coating." *ACS Applied Materials and Interfaces* 11 (43). American Chemical Society: 39470–39483. doi:10.1021/acsami.9b12733.

Wang, Hui, Yubing Sun, Jinhui Yi, Jianping Fu, Jing Di, Alejandra del Carmen Alonso, and Shuiqin Zhou. 2015. "Fluorescent Porous Carbon Nanocapsules for Two-Photon Imaging, NIR/PH Dual-Responsive Drug Carrier, and Photothermal Therapy." *Biomaterials* 53 (June). Elsevier: 117–126. doi:10.1016/J.BIOMATERIALS.2015.02.087.

Wang, Jiangxue, Chunying Chen, Bai Li, Hongwei Yu, Yuliang Zhao, Jin Sun, Yufeng Li, et al. 2006. "Antioxidative Function and Biodistribution of [Gd@C82(OH)22]n Nanoparticles in Tumor-Bearing Mice." *Biochemical Pharmacology* 71 (6). Elsevier: 872–881. doi:10.1016/J.BCP.2005.12.001.

Wang, Shang Yu, Hong Zhi Hu, Xiang Cheng Qing, Zhi Cai Zhang, and Zeng Wu Shao. 2020. "Recent Advances of Drug Delivery Nanocarriers in Osteosarcoma Treatment." *Journal of Cancer* 11 (1). Ivyspring International Publisher: 69–82. doi:10.7150/JCA.36588.

Wang, Xiang, Jian Guo, Tian Chen, Haiyu Nie, Haifang Wang, Jiajie Zang, Xiaoxing Cui, and Guang Jia. 2012. "Multi-Walled Carbon Nanotubes Induce Apoptosis via Mitochondrial Pathway and Scavenger Receptor." *Toxicology in Vitro* 26 (6). Pergamon: 799–806. doi:10.1016/J.TIV.2012.05.010.

Wang, Yong, Jia Tong Chen, and Xiu Ping Yan. 2013. "Fabrication of Transferrin Functionalized Gold Nanoclusters/Graphene Oxide Nanocomposite for Turn-on Near-Infrared Fluorescent Bioimaging of Cancer Cells and Small Animals." *Analytical Chemistry* 85 (4). American Chemical Society: 2529–2535. doi:10.1021/ac303747t.

Wang, Zonghua, Jianbo Yu, Rijun Gui, Hui Jin, and Yanzhi Xia. 2016. "Carbon Nanomaterials-Based Electrochemical Aptasensors." *Biosensors and Bioelectronics* 79 (May). Elsevier: 136–149. doi:10.1016/J.BIOS.2015.11.093.

Wei, Alexander, Jonathan G. Mehtala, and Anil K. Patri. 2012. "Challenges and Opportunities in the Advancement of Nanomedicines." *Journal of Controlled Release* 164 (2). Elsevier: 236–246. doi:10.1016/J.JCONREL.2012.10.007.

Welsher, Kevin, Sarah P. Sherlock, and Hongjie Dai. 2011. "Deep-Tissue Anatomical Imaging of Mice Using Carbon Nanotube Fluorophores in the Second Near-Infrared Window." *Proceedings of the National Academy of Sciences of the United States of America* 108 (22): 8943–8948. doi:10.1073/pnas.1014501108.

'World Carbon Nanotubes Market Report 2022-2032: Demand For'. 2022. Accessed December 6. https://www.globenewswire.com/en/news-release/2022/06/21/2465789/28124/en/World-Carbon-Nanotubes-Market-Report-2022-2032-Demand-for-MWCNTs-Increasing-SWCNTs-Gaining-Market-Traction.html.

Xue, Mingyue, Jingjin Zhao, Zhihua Zhan, Shulin Zhao, Chuanqing Lan, Fanggui Ye, and Hong Liang. 2018. "Dual Functionalized Natural Biomass Carbon Dots from Lychee Exocarp for Cancer Cell Targetable Near-Infrared Fluorescence Imaging and Photodynamic Therapy." *Nanoscale* 10 (38). Royal Society of Chemistry: 18124–18130. doi:10.1039/c8nr05017a.

Ye, Lijuan, Weibin Chen, Yuan Chen, Yuqin Qiu, Jun Yi, Xiaofang Li, Qiuxiao Lin, and Bohong Guo. 2022. "Functionalized Multiwalled Carbon Nanotube-Ethosomes for Transdermal Delivery of Ketoprofen: Ex Vivo and in Vivo Evaluation." *Journal of Drug Delivery Science and Technology* 69 (March). Elsevier: 103098. doi:10.1016/J.JDDST.2022.103098.

Yue, Juan, Peng Miao, Li Li, Ruhong Yan, Wen Fei Dong, and Qian Mei. 2022. "Injectable Carbon Dots-Based Hydrogel for Combined Photothermal Therapy and Photodynamic Therapy of Cancer." *ACS Applied Materials and Interfaces*. doi:10.1021/acsami.2c15428.

Zhang, Liming, Yadong Xing, Nongyue He, Yi Zhang, Zhuoxuan Lu, Jianping Zhang, and Zhijun Zhang. 2012. "Preparation of Graphene Quantum Dots for Bioimaging Application." *Journal of Nanoscience and Nanotechnology* 12 (3). American Scientific Publishers: 2924–2928. doi:10.1166/JNN.2012.5698.

Zhang, Sen, Feng Xu, Zhan Qing Liu, Ya Shao Chen, and Yan Ling Luo. 2019. "Novel Electrochemical Sensors from Poly[N-(Ferrocenyl Formacyl) Pyrrole]@multi-Walled Carbon Nanotubes Nanocomposites for Simultaneous Determination of Ascorbic Acid, Dopamine and Uric Acid." *Nanotechnology* 31 (8). IOP Publishing: 085503. doi:10.1088/1361-6528/AB53BB.

Zhao, Peng, Lihong Wang, Changshan Sun, Tongying Jiang, Jinghai Zhang, Qiang Zhang, Jin Sun, Yihui Deng, and Siling Wang. 2012. "Uniform Mesoporous Carbon as a Carrier for Poorly Water Soluble Drug and Its Cytotoxicity Study." *European Journal of Pharmaceutics and Biopharmaceutics* 80 (3). Elsevier: 535–543. doi:10.1016/J.EJPB.2011.12.002.

Zhao, Yong, Brett L. Allen, and Alexander Star. 2011. "Enzymatic Degradation of Multi-walled Carbon Nanotubes." *Journal of Physical Chemistry A* 115 (34). American Chemical Society: 9536–9544. doi:10.1021/JP112324D/SUPPL_FILE/JP112324D_SI_001.PDF.

Zheng, Di Wei, Bin Li, Chu Xin Li, Jin Xuan Fan, Qi Lei, Cao Li, Zushun Xu, and Xian Zheng Zhang. 2016. "Carbon-Dot-Decorated Carbon Nitride Nanoparticles for Enhanced Photodynamic Therapy against Hypoxic Tumor via Water Splitting." *ACS Nano* 10 (9): 8715–8722. doi:10.1021/acsnano.6b04156.

Zhou, Nan, Zeyu Hao, Xiaohuan Zhao, Suraj Maharjan, Shoujun Zhu, Yubin Song, Bai Yang, and Laijin Lu. 2015. "A Novel Fluorescent Retrograde Neural Tracer: Cholera Toxin B Conjugated Carbon Dots." *Nanoscale* 7 (38). Royal Society of Chemistry: 15635–15642. doi:10.1039/c5nr04361a.

Zhu, Shoujun, Yubin Song, Xiaohuan Zhao, Jieren Shao, Junhu Zhang, and Bai Yang. 2015. "The Photoluminescence Mechanism in Carbon Dots (Graphene Quantum Dots, Carbon Nanodots, and Polymer Dots): Current State and Future Perspective." *Nano Research 2014 8:2* 8 (2). Springer: 355–381. doi:10.1007/S12274-014-0644-3.

6 Inorganic Nanomaterials

Synthesis and Functionalization for Medical and Biotechnological Applications

Tushar Das, Shubham Raj, Subhadip Roy, and Subrata Das

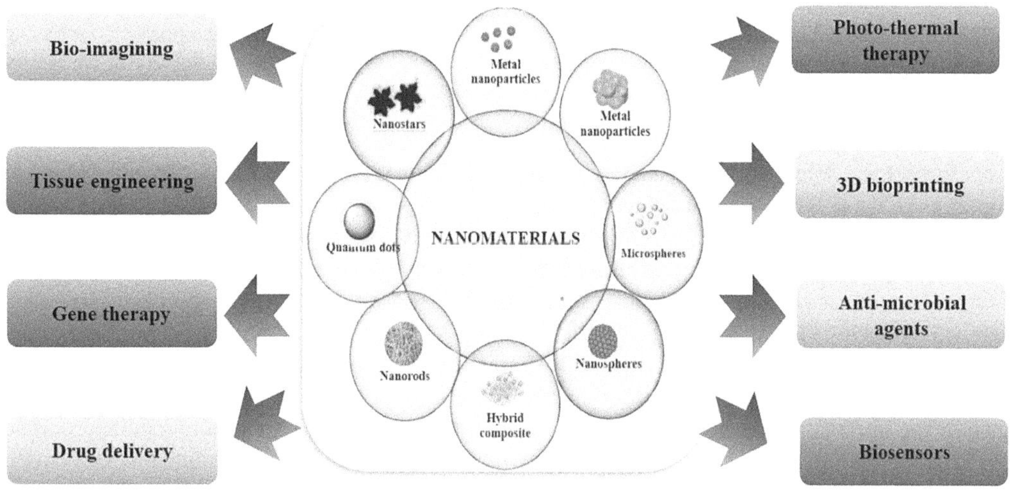

6.1 INTRODUCTION

Nanoparticles (NPs) represent a plethora of extremely small particles with a size of approximately one billionth of a meter. These NPs are either made from metallic, non-metallic, or bi-metallic sources. In addition, NPs can also be formed from organic precursors like carbon-based sources, etc. Depending upon the type of precursor and methodology opted for, the nanostructure arrangement in NPs displays different microscopic shapes like the formation of nanorods, cones, spheres, clusters, tubes, wires, spindles, core-shell structures, etc. A brief overview of the different shapes of NPs is summarized in Figure 6.1. Besides the difference in the physical states, NPs give rise to various properties like the quantum hall effects, high electron mobility, and surface-oriented reactivities, making them suitable for use in different areas of materials-driven based research. Engineering these NPs in biomedical sectors like biotechnology, pharmaceuticals, and medicinal chemistry shows a prospective future in developing contemporary nanomedicines, photosensitive therapies, drug carriers, stabilizers, and other theragnostic and diagnostic agents. Furthermore, the non-toxic, hydrophilic, and biocompatible nature of NPs offers justified roles in patient care. To date, many

DOI: 10.1201/9781003305583-6

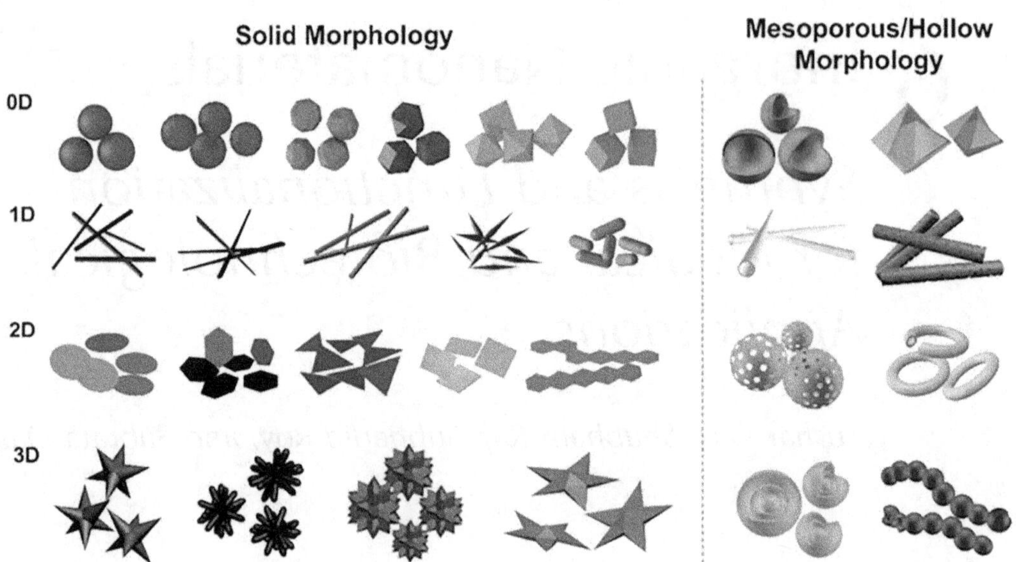

FIGURE 6.1 Illustration showing the various surface morphologies of solid and mesoporous NPs having 0D, 1D, 2D, and 3D dimensions. (Wu et al.; Nanoscale, 2016, 8, 1237 © The Royal Society of Chemistry (Wu et al. 2016)).

discussions have been made in understanding and controlling the intrinsic physicochemical properties of these NPs. In this chapter, we summarized its applicability in different areas of the healthcare sector and will develop interest among different reader groups belonging to chemistry and biology domains about the recent progress of NPs as biomedical aid.

One of the most popular techniques for creating 0D NPs from inorganic sources is based on chemical synthesis (Thomas and O'Brien 2007; Rao et al. 2007). The ability to synthesize NPs chemically has advanced significantly across different research groups since late 2007. The chapter classifies and briefly explains the different methodologies available for synthesizing NPs since much of the intrinsic dimensionalities and extrinsic properties depend on selecting an appropriate approach for their synthesis. These concepts of nanomaterial structure and their property-based applications will significantly influence future design and fabrication prospects. The chapter critically examines some of the critical challenges during NPs synthesis and emphasizes more in summarizing their contributing roles in biomedical-based research. All colloidal and nanostructured materials are synthesized from the same basic principles. Still, the primary difference is in its properties, determined by their physical states (Morphology and the number of contributing atoms). The principles of colloidal chemistry can be applied to comprehend the emergence and expansion of these nanoscale ensembles. In a nutshell, under a typical supersaturated condition, atoms undergoing the crystallization process bear unfavorable surface energy offset by gradually generated lattice energies of the molecules, leading to nucleation, but under the nanoscale. Since there is a limitation of only a few surface-residing atoms, it is observed that most of the adopted procedures involved in NPs use ligands. These molecules inhibit their growth to the microscale. Except for the core-shell methods for synthesizing bi-metallic ones, which form NPs inside micelles or vesicles, most NPs synthesis is carried out either by utilizing controlled limits of constituents or by using highly diluted media, Figure 6.2(A). The system's extrinsic behavior, which allows for the formation of NPs when the temperatures suddenly rise and fall, is yet another concept that helps explain the nucleation process. It is seen that adding a cold solvent to the mixes made at high temperatures allows for better nucleation, and this concept can be achieved in Figure 6.2(B).

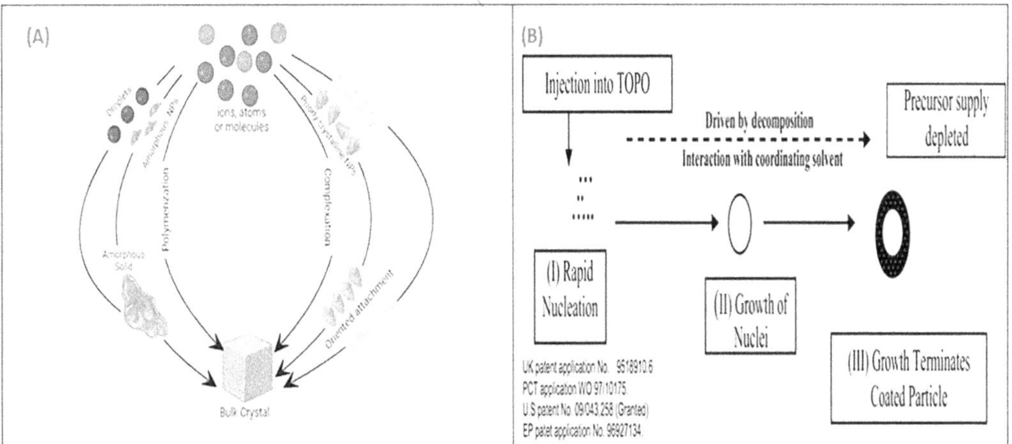

FIGURE 6.2 **(A)** Diagram showing the classical growth mechanism pattern of crystals by monomeric additions of ions, atoms, and molecules (gray curve), while the non-classical model of crystal growth occurs via the addition of nanocrystals, nanoparticles (crystalline, amorphous), colloidal droplets, oligomers (black curve), (Yoreo et al., Science, 2015, 349, 6247 (De Yoreo et al. 2015)). **(B)** Nucleation and growth of nanoparticles due to perturbation of the system.

Despite these complicated processes of NPs synthesis, it is necessary to control and modify the NP structures to have consistent geometries, chemical compositions, and surface properties that lead to their use in various scientific domains. The different techniques that are employed in creating stable NPs with high purity, surface modification, and nanometer (nm) size distributions are summarized in this context.

6.2 FABRICATION OF INORGANIC NANOPARTICLES

The production of NPs can be categorized broadly into two techniques (Tulinski and Jurczyk 2017). The conventional method relies on attrition and other forms of comminution to produce fine particles using a 'top-down' approach. While 'bottom-up' synthesis methods have been extensively used in recent years (Scheu et al. 2006). But both methods play pivotal roles in today's material synthesis and, most likely, in developing the future of nanotechnology (Zweck et al. 2008). Moreover, both these methods have their benefits and drawbacks. The materials made using either of these methods have remarkably miscellaneous characteristics. In addition to direct atomic manipulation, several other physical, chemical, and mechanical methods also exist for fabricating nanomaterials (Wang et al. 2003; Yeadon et al. 1997).

6.2.1 Physical Methods

6.2.1.1 Vapor Deposition

The vapor phase coating technique involves transferring materials at an atomic level. Unlike chemical vapor deposition (CVD), the major difference lies in the state of the raw material used, i.e., solid form. At the same time, in the case of CVD, the compounds are directed into the reaction chamber in gaseous form. Sputtering/pulsed laser deposition (PLD) is another commonly used technique alongside physical vapor deposition (PVD). PVD is usually carried out in inert assembly and involves multiple steps like (i) evaporation, (ii) carriage, (iii) reaction, and (iv) deposition (Tsuzuki and McCormick 2004; Joseph et al. 2005). The material that must be deposited concisely is bombarded using a high-energy source like an electron or ion beam such that the atoms vaporize and travel in straight lines from the target to the coating substrate.

The thin films PVD methods produce are used in optoelectronic, magnetic, and microelectronic device fabrication.

6.2.1.2 Pulsed Laser-Assisted Deposition

In the 18th century, Schulze first showed that silver salts, when exposed to light, become dark; this marked the beginning of the photochemical synthesis of NPs. Much attention nowadays is given to this method due to its versatility and convenience, along with some distinctive benefits like space-selective fabrication. The photophysical (top-down) and photochemical (bottom-up) approaches can be used to categorize photo-induced synthetic procedures. PLD is another straightforward and adaptable technique employed to date that can be used for depositing thin films for various nanomaterials, including metals, semiconducting materials, nitrides, and oxides, along with organic compounds, among many more (Chrisey and Hybler 1994). In this process, high-power short-pulsed laser radiation is exposed to solid targets placed in a chamber. The high-power laser causes atom-by-atom removal from the substrate through vaporization under non-equilibrium. The method allows depositions of both multilayer structures and supports film growth. Nanoscale metal oxides like thin films, multi-layered materials for superconductivity, and so on can be formed using the PLD method. The primary mechanism of PLD methods is based on density-based deposition, i.e., materials close to the laser sources possess high density with particle size < 5 nm. As a result, these particles acquire stable lower surface energy states. To limit specific agglomeration among the NPs, inert gases are used to drive away the particles from the surrounding region of the source. Moreover, the rate at which the evaporation occurs and the pressure at which the gas enters the chamber determines the particle size formed (Kulkarni 2014).

The main advantages of photochemical synthesis are 1) high spatial resolution; 2) reaction selectivity; 3) reaction controllability; and 4) shape selectivity. Example: light-induced modulation of size and shape of silver to nano prism shows an effective way to synthesize NPs using direct photoreduction (Millstone et al. 2009; Pastoriza-Santos and Liz-Marzán 2008; Jin et al. 2001). The method offers no reducing agents to be used in various mediums for growth induction like polymer films, glasses, cells, etc. Additionally, the photoactive reagents create photoirradiation intermediates that decrease the existing metal substrate to form an M^0 state. In contrast to direct photoreduction, this approach enables flexible excitation wavelength selection coupled with quick and efficient creation of metal NPs.

Numerous research advocate laser ablation of solids in liquids, primarily because it simplifies experimentation. Since only NPs composed of the target material and the fluid are present without reducing agents or counterions, numerous studies suggest it allows laser ablation of solids in liquids. Laser ablation uses laser energy to ablate solid target materials. This method evaporates light-absorbing material by concentrating intense energy on a solid surface. A brief overview of the different NPs syntheses using laser ablation is shown in Table 6.1.

6.2.1.3 High-Energy Ball Milling Method

A mechanical ball mill is one of the simplest ways to produce powdered NPs from metals and alloys. Other types of mills also include planetary, vibratory, rod, and tumbler mixers (Gorrasi and Sorrentino 2015). The method operates simultaneously via one or more containers to obtain large quantities of NPs. Based on the amount required, the number of containers is decided. Hardened steel or tungsten carbide balls are added within each container, combined with powder or metal oxide flakes. Initially, materials have large sizes and shapes. Generally, a 1:2 mass ratio of substance to the ball is used, i.e., when the container is more than half full, milling efficiency is reduced. The use of heavy balls increases the collisional impact energy. It is during the collisions; the temperature rises to 1100 °C. Lowering the temperatures favors the development of amorphous particles. Care should be taken since the constantly active surfaces formed gases like O_2, N_2, etc., that might act as a contaminant source. Sometimes cryo-cooling is employed to disperse

TABLE 6.1
Representing the Synthesis of Different NPs Using Different Laser Sources

Type	Laser source	Pulse width	Wavelength (nm)	Frequency (Hz)	Target material	Size of NPs (nm)	Ref
ZnO NPs	Nd-YAG	100 nm	355	10	Zn foil	5–19	(Amendola and Meneghetti 2009)
ZnO NPs		<10 ns	532	6	Zn foil	80.76–102.54	(Gondal et al. 2009)
ZnO NPs		10 ns	532	10	Zn	35	(Al-Dahash et al. 2018)
ZnO NPs		1.0 ms	1064	5	Zn plate	40–119	(Mintcheva et al. 2018)
CuO NPs		7 ns	1064	5	Cu	8–10	(Abdulateef et al. 2016)
CuO NPs		5 ns	532	10	Cu foil	9–26	(Gondal et al. 2013)
CuO NPs		7 ns	1064	5	Cu	8–10	(Abdulateef et al. 2016)
TiO_2 NPs	Ytterbium Doped Fiber Laser	–	–	–	Ti plate	5–25	(Boutinguiza et al. 2013)
TiO_2 NPs	Nd-YAG	4.5 ns	1062	500	Ti	4–35	(Singh et al. 2016)
AuNPs		8 ns	1062	10	–	~30	(Maciulevičius et al. 2013)
AuNPs		8 ns	1064	1	Au plate	15.12–9.5	(Al-Azawi and Bidin 2015)
AuNPs		7 ns	1064	15	Au sheet	~23.5	(Khumaeni et al. 2017)
AuNPs		8.5 nm	1064	5	Au plate	50	(Wender et al. 2011)
AgNPs		20 ns	1064	5	Ag	8.5	(Hajiesmaeilbaigi et al. 2005)
AgNPs		10 ns	532	10	Ag	2–5	(Pyatenko et al. 2004)
AgNPs		10 ns	532	30	Ag	22	(Tajdidzadeh et al. 2014)

the heat created. During the milling process, liquids can also be used. A few hundred revolutions per minute rotate the containers around their axis. In addition, they can rotate about a central axis and are therefore referred to as 'planetary ball mills.'

6.2.1.4 Electrospinning and Lithography

Electrospinning is another technique to fabricate nanofibers for various materials, including polymers. Under electrospinning, coaxial electrospinning is currently the most frequently used technique, wherein the spinneret consists of two coaxial capillaries, within which two viscous liquids or a single viscous liquid, in the shell, and a non-viscous liquid, as the core, can be used to create core-shell nanoarchitectures in the presence of aqueous solution. Coaxial electrospinning is an efficient and straightforward 'top-down' method for producing large quantities of core-shell ultrathin fibers. These ultrathin nanomaterials can be stretched to several centimeters (cm) lengths. The technique offers the creation of core-shell and hollow polymer, inorganic, organic, and hybrid materials (Kumar et al. 2014).

Nanoarchitectures can also be produced using lithography, focused beam of light, or electrons for 3D fabrication of nanoarchitecture. The two main lithography techniques are mask lithography and maskless lithography. Unlike photolithography, nanoimprint, and soft lithography, masked nanolithography transfers nanopatterns over a significant surface area using a specified mask or template (Pimpin and Srituravanich 2012; Szabó et al. 2013; Kuo et al. 2003). Focused ion beam, scanning probe, and electron beam lithography are all other types of mask lithography (Matsumoto et al. 2020). In maskless lithography, any nanopattern can be written without a mask.

6.2.1.5 Arc Discharge Method

Most carbon-based nanomaterials like carbon nanotubes (CNT), fullerenes, few-layer graphene (FLG), etc (Zhang et al. 2019) are produced using this method. The method allows control over both structural morphologies, especially for fullerene-like materials. In this, two graphitic rods are usually inserted closed under helium pressure. The presence of helium irradicates oxygen allowing the formation of fullerene. Vaporization of the graphitic rods is driven viz., arc discharges between the ends (Lieber and Chen 1994). Additionally, the environment where the arc discharge takes place significantly impacts how the nanomaterial's shape and structural morphology are controlled. Different carbonaceous materials like multiwall CNT and other related materials can be collected from the different regions during the arc discharge process due to the difference in the growth characteristics of nanomaterials (Liang et al. 2012). Apart from the graphitic electrodes used, carbonaceous nanomaterials formation is also influenced under different atmospheric conditions like 'flower-like' and 'bud-like' single-walled carbon nanoribbons (SWCNHs) can be generated under CO and CO_2 environments. Besides this method can also be used to get graphene nanostructures. It is observed that graphene sheets produced viz. hydrogen arc discharge method possess superior electrical conductivity and thermal stability compared to graphene obtained using other methods like argon arc discharge. A brief overview of the principle of arc discharge is shown in Figure 6.3.

6.2.2 Chemical Methods

6.2.2.1 Hydrothermal Method and Solvothermal Method

This section will discuss another important fabrication method for designing and preparing nanostructures using hydrothermal approaches. The term is derived from two Greek words, 'hydros' meaning 'water' and 'thermos' meaning 'heat'. It can be summed up as any heterogeneous reaction that occurs under high pressure and temperature conditions with aqueous solvents or mineralizers to dissolve and recrystallize substances generally intractable under regular circumstances. Both monodispersed and extremely homogenous NPs and nanohybrid and

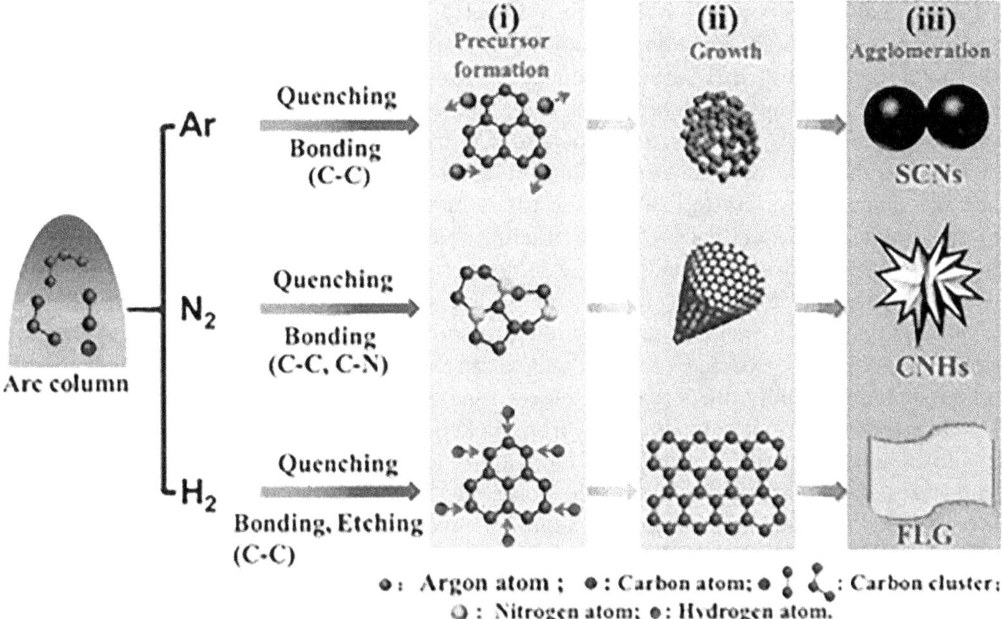

FIGURE 6.3 Illustration of arc discharge method for creating carbon-based nanomaterials on the interior of a chamber utilizing various gases. (Zhang et al. Carbon 2019, 142, 278–284 (Zhang et al. 2019a)).

nanocomposite materials can be synthesized using this method. For the processing of sophisticated materials, this can be a crucial method. Midst the chemical routes employed for synthesizing inorganic NPs, hydrothermal and solvothermal plans provide the most reliable and cheap way. Generally, it is observed that inorganic NPs like metal oxides, chalcogenides, and organic NPs are produced using this method. Moreover, it is necessary to maintain reaction conditions like temperature, pH, and concentration of the inorganic NPs to enhance the rate of nucleation and particle size distribution. Likewise, Chang et al. reported a hydrothermal synthesis route for obtaining a monodisperse octahedral complex of gold NPs (AuNPs) from aqueous dispersions of tetra chloroauric (III) acid trihydrate ($HAuCl_4$) using trisodium citrate (TSC) and cetyltrimethylammonium bromide (CTAB) at temperature ~110 °C for 6, 12, 24, 48, and 72 hr. The obtained gold nanocrystal had an average size of 30, 60, 90, 120, and 150 nm, respectively. The cationic reagent, CTAB, controls the surface morphologies, while TSC enhances the (111) faces of the nanocrystal (Chang et al. 2008). Li et al. obtained highly crystalline TiO_2 nano anatase solvothermal using diethyl ether at temperatures as low as 100 °C, wherein titanium butoxide [$Ti(OBu)_4$] was used as a source of titania and acetic acid as a growth initiator (Li et al. 2009). In this, the use of diethyl ether not only promotes rapid hydrolyzation but also allows the conversion of amorphous titania to its crystalline form under low temperatures. Liu et al. reported a scalable synthesis for monodispersed, highly crystalline Wurtzite-type ZnO nanorods via a hydrothermal route. The average particle size is below 50 nm and has a high aspect ratio of 20–30 nm. Herein, zinc nitrate [$Zn(NO_3)_2.6H_2O$] was used as a precursor for Zn, which, when mixed with sodium hydroxide (NaOH), produced an alkaline solution, which was further placed in a mobile phase containing water: ethanol (1:6) and ethylenediamine as a reducing agent. The desired product was obtained by heating the components for 20 hrs at 180 °C in a Teflon-lined autoclave (Liu and Zeng 2003). Hydrothermal methods potential lies in using low-cost, large-volume equipment, energy-saving nucleation control, and high dispersity. Advanced material processing is also minimized, making it environment-friendly.

In the majority of cases, it is seen that used solvents can behave differently under hydrothermal conditions depending upon their chemical structure at critical, supercritical, and subcritical temperatures, pH variation, dielectric constant, coefficient of expansion, viscosity, density, etc. Commonly used metal sulfides, oxides, nitrates, and chlorides with metallic centers like Ag, Mn, Ni, Cu, Zn, Co, and Cd having different oxidation states can be converted to their monovalent forms using this method. The stability of Cd and Zn salts in a wide pH range allows crystallization in various mediums. While in the case of Phosphate-buffered saline (PBS), crystal growth in alkaline solutions is only possible over a narrow pH range. For optimal working conditions, a hydrothermal autoclave should have the following qualities: 1) resistance to acids, bases, and oxidizers; 2) ease of assembly and disassembly; 3) sufficient length to achieve the desired pressure and temperature range; and 4) tough enough to withstand extreme pressure and temperature experiments for prolonged periods without damage. Usually, thick glass cylinders, quartz cylinders, and high-strength alloys such as 300 series (austenitic) stainless steel, iron, nickel, cobalt-based super alloys, titanium, and its alloys are preferred in autoclave fabrication (Higuita and Vargas 2017). Figure 6.4 summarizes the hydrothermal approach; a sealed stainless-steel autoclave with a pressure range of (50–500 psi) is used to heat aqueous mixes of metal substrates over the boiling point of water or other solvents. In reaction, the autoclave's internal pressure rises sharply above air pressure. High temperature and pressure work together synergistically to create highly crystalline materials in a single process, negating the need for post-annealing procedures. The type and concentration of the precursors, the stabilizing agents, the solvent, the reaction temperature, and the reaction duration greatly influence the products. According to conventional wisdom, the growth mechanism is as follows: Early in the process, NPs arise

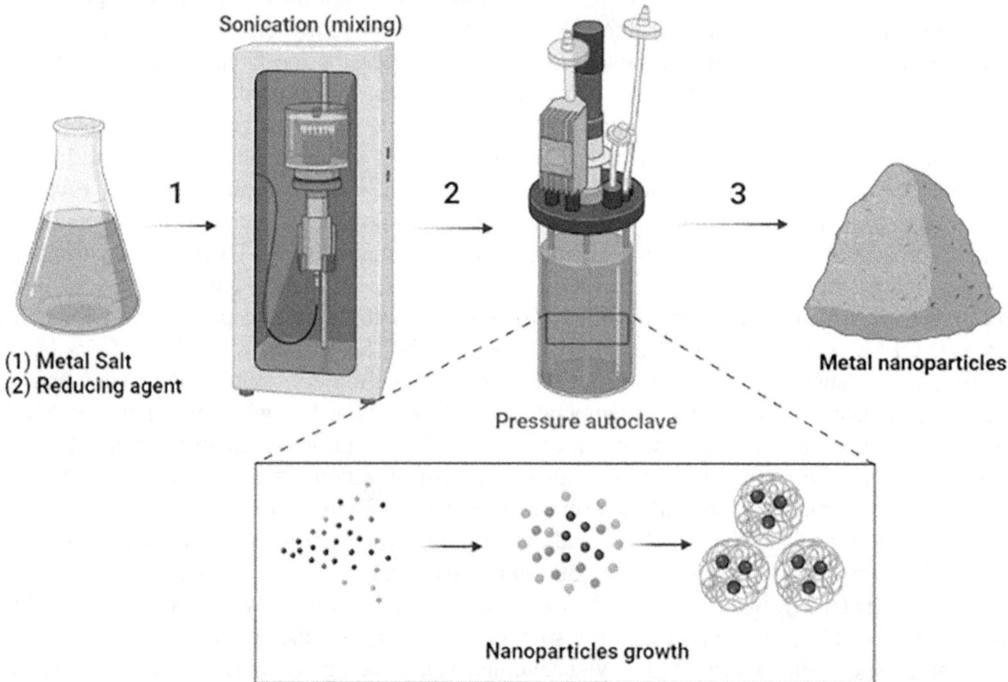

FIGURE 6.4 Picture representing the overview of the hydrothermal setup in different steps after the addition of varying mono, di, tri, etc. valent metal salts along with suitable agent 1) sonication using probe or bath to achieve suitable mixing; 2) hydrothermal autoclave setup transform into for synthesis inset representing the growth of metal NPs inside the autoclave under elevated temperature; 3) isolation of metal NPs.

as seeds, which combine and develop into microspheres. The amount of reducing agent present determines the ultimate morphology, and the spheres transform into instantaneous shapes that all other microcrystals.

6.2.2.2 Sol-Gel Method for Fabrication of NPs

The sol-gel process is based on a wet chemical approach for producing nanostructures, especially for metal oxide NP synthesis. In this, the precursors are usually dissolved in water/alcohol and then heated and stirred to form a gel via a process known as hydrolysis/alcoholysis. This method results in a wet gel; therefore, it must be dried before use. Due to low reaction temperature, the sol-gel process provides the advantage of a low-cost technology with good uniformity over the chemical composition of the end products. The sol-gel method is frequently used for molding materials, such as when fabricating ceramics, and as a bridge between thin films of metal oxides. This method is used in various technological applications like electronics, optical, surfaces, biosensors, pharmaceuticals, and separation technologies. The essential idea behind the sol-gel technique lies in creating a homogenous sol from the precursors and its final conversion into the gel. It is seen that the drying method has a significant role in the properties of the gel. In precise, 'removal-of-solvent' is chosen based on the application for which the gel is intended. Dried gels are more used in industries for designing surface coatings, building insulations, and manufacturing clothes. A brief overview of the sol-gel method is illustrated in Figure 6.5. The sol-gel technique can be used for the preparation of 1) the manufacturing of

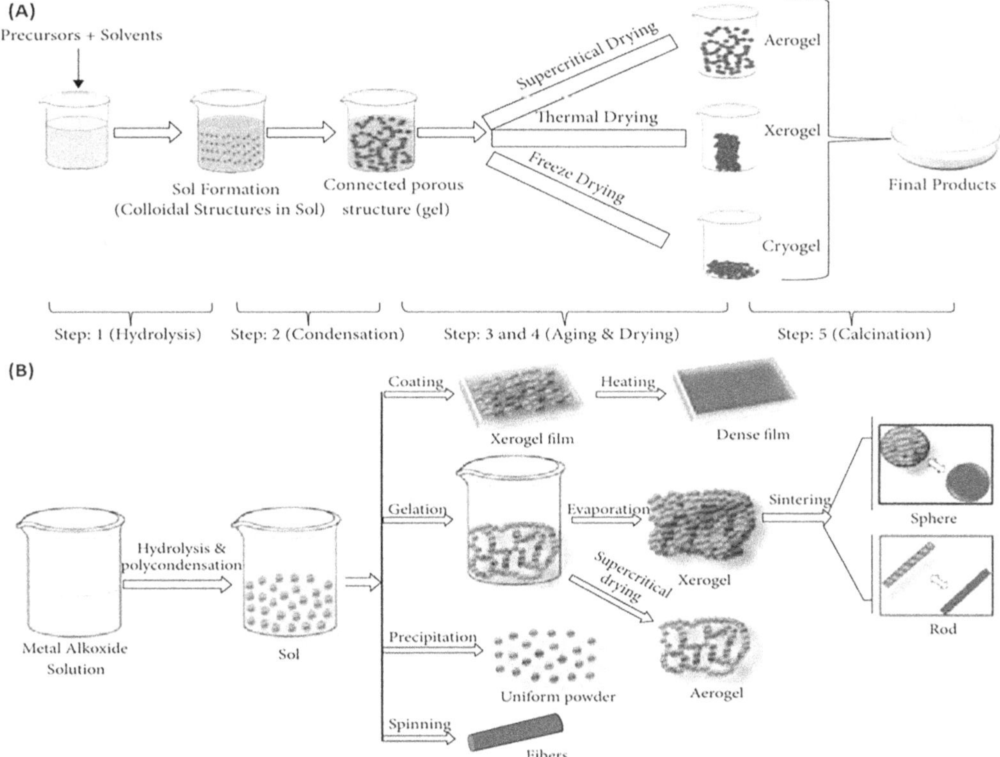

FIGURE 6.5 Brief illustration of the sol-gel technique used to fabricate metal NPs. **(A)** Scheme representing the different steps of the sol-gel process: from precursor to final product. **(B)** Picture depicting the detailed overview of different techniques that can be done using the sol-gel method and the products obtained from each process (Bokov et al. Adv. Mater. Sci. Eng. 2021, 1–21 (Bokov et al. 2021)).

optical components with complicated shapes, 2) excellent product purity, 3) simplicity of the process, 4) high production efficiency, 5) the ability to synthesize uniform compounds in the form of composite oxides, 6) the ability to plan and regulate chemical composition to achieve homogeneity, 7) the ability to use the product with unique shapes like fibers and aerogels, 8) high surface coverage, 9) synthesis of amorphous materials with thin layers, 10) Materials with modified physical properties, such as low thermal expansion coefficient, low UV absorption, and high optical transparency, 11) atom profitability, 12) precursors with high chemical reactivity, 13) precision control of materials morphology, 14) production of porous and rich materials containing organic or, polymeric compounds.

Several parameters, like the activity of metals influence the hydrolysis and condensation reactions (sol-gel process), the water/metal ratio, pH, temperature, solvent type, and additives used. Another aspect to consider is the employment of catalysts to regulate the rate and volume of condensation and hydrolysis reactions. These processing parameters can be changed to produce materials with different microstructures and surface chemistry. The 'sol' can be further processed to create ceramic materials in various shapes. A substrate can be coated with spin coating or dipped in coating to produce thin films, usually, a wet 'gel' forms when the 'sol' is put into a mold. After additional drying and heating, the 'gel' transforms into dense ceramic or glass particles. Xie et al. created chromium-doped TiO_2 NPs with different Cr ion concentrations (1–10%) using the sol-gel technique (Xie et al. 2019). Here, nitric acid was used as a complexing agent to combine tetra butyl titanate (TBOT), a source of titanium, with chromium acetate (CAs), a source of chromium, to create Cr-doped TiO_2 NPs. Acetic acid speeds up the synthesis process; Figure 6.6(A) summarizes these processes. Dörner et al. reported a low-cost sol-gel approach for producing loosely clustered CuO NPs, in which a precursor phase containing copper-carbonate species was selected to pay clustered CuO NPs (Dörner et al. 2019). This method did not require adding additives; the mechanism summary is shown in Figure 6.6(B). Beig et al. also reported a low-cost sol-gel approach for the synthesis of ZnO NPs using zince acetate and almunium nitrate as precursor (Beig et al. 2022). Similarly, ZnO NPs were created by Soleimani et al. using an auto-combustion method based on sol-gel. Here, organic complexing agents (reduction agents) were heated to form a gel, which was then calcined to make metal nanocomposites, Figure 6.6(D) (Soleimani et al. 2018). Metal nitrates were used as oxidizers due to their role as precursors. Figure 6.6(E)–(F) shows published reports by Phromma et al. and Basnet et al. that offer practical ways to manufacture metal NPs (Phromma et al. 2020; Basnet and Chatterjee 2020). In these investigations, TiO_2 particles were created by the wet ball milling sol-gel (WBMS) approach. The following is a description of the chemical reaction that occurs during the formulation of metal NPs using the sol-gel technique is illustrated below:

$$M(OR)_x + mH_2O \rightarrow M(OR)_{x-m} + mROH(OH)$$

[If x=m, two-stage reaction, total hydrolysis during water/alcohol condensation]

Step 1

$$2M(OR)_{x-m}(OH)_m \rightarrow (OH)_{m-1}(OR)_{x-m} - M - O - M(OR)_{x-m}(OH)_{m-1} + H_2O(OH)$$

Step 2

$$2M(OR)_{x-m}(OH)_m \rightarrow (OH)_{m-1} - 1(OR)_{x-m} - M - O - M(OR)_{x-m} + ROH(OR)$$

Overall reaction

$$M(OR)_x + x/2H_2O \rightarrow M(OR)_{x/2} + xROH(OR)$$

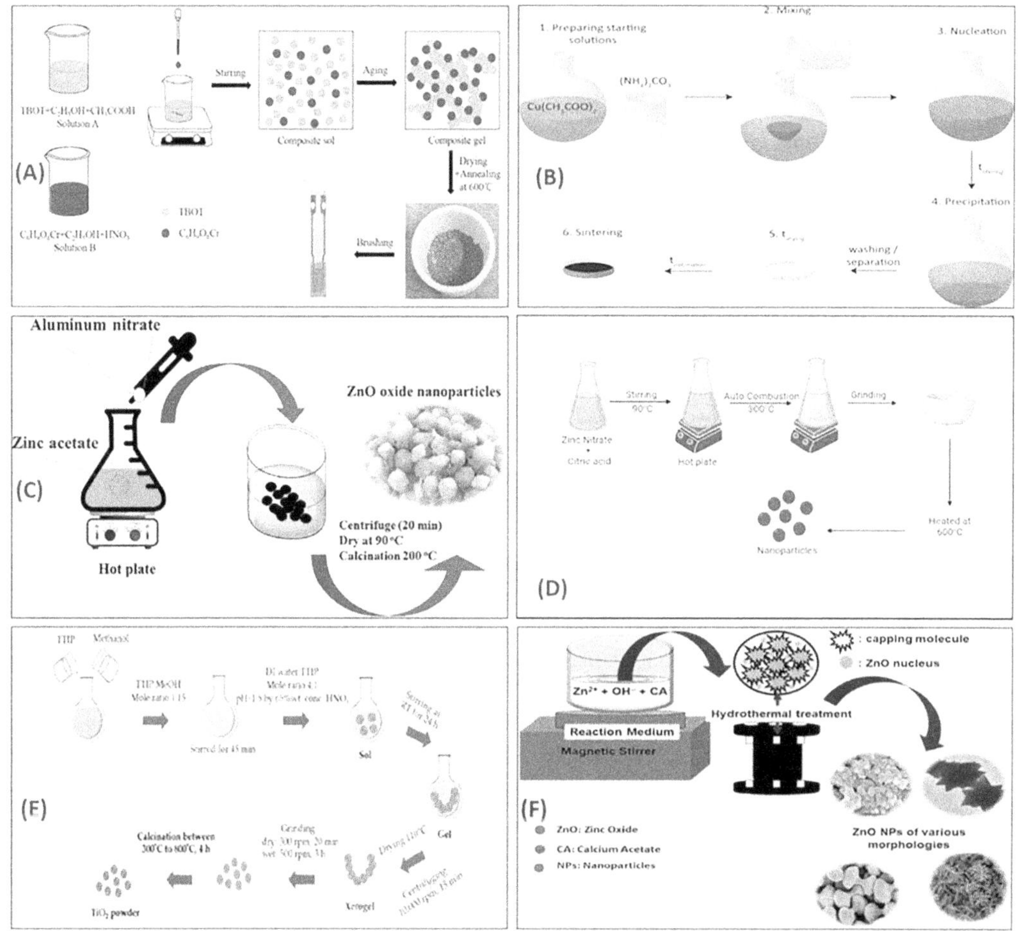

FIGURE 6.6 Examples of synthesis of NPs using sol-gel method **(A)** Picture representing the synthesis of TiO_2 nanoparticles with a p-type response. (Li et al. Front. Mater. 2019, 6, 96 Xie, Li, Sun, Dong, Fatima, Wang, Yao and Haidry (Xie et al. 2019)). **(B)** Schematic representation of sol-gel synthesis and subsequent calcination process. (Dörner et al. Sci Rep. 2019, 9, 11758 (Dörner et al. 2019)). **(C)** Synthesis of ZnO nanoparticles by sol-gel method employing zinc acetate and aluminum nitrate as precursors. (Beig et al. Environ Chem Lett.2022, 20, 2709–2726 (Beig et al. 2022)). **(D)** Methodology representing the synthesis of sol-gel-based auto-combustion synthesis of ZnO nanoparticles (Soleimani et al. Appl. Phys. A. 2018, 124, 128 (Soleimani et al. 2018)). **(E)** TiO_2 particles were synthesized by the wet ball milling sol-gel (WBMS) method. (Phromma et al. Appl. Sci. 2020, 10, 993 (Phromma et al. 2020)). **(F)** Growth mechanism ZnO NPs induced by capping agents (Basnet et al. Nano-Structures & Nano-Objects, 22, p.100426 (Basnet and Chatterjee 2020)).

Metal-oxygen-metal bonds are created in non-aqueous sol-gel synthesis either due to alkyl through aldol-like condensation, halide elimination, ether elimination, ester elimination, or another method.

6.2.2.3 Phase Transfer Methods

Phase-transfer mediated synthesis involves transporting reactants from a polar to a non-polar media (and vice versa), followed by post-modifying them in a non-polar medium. The production of metallic NPs of different sizes and shapes in the organic phase has frequently been done using this method, either directly or via transporting NPs from the aqueous phase to the organic phase. The latter approach provides the advantage of making metallic NPs in an aqueous phase without

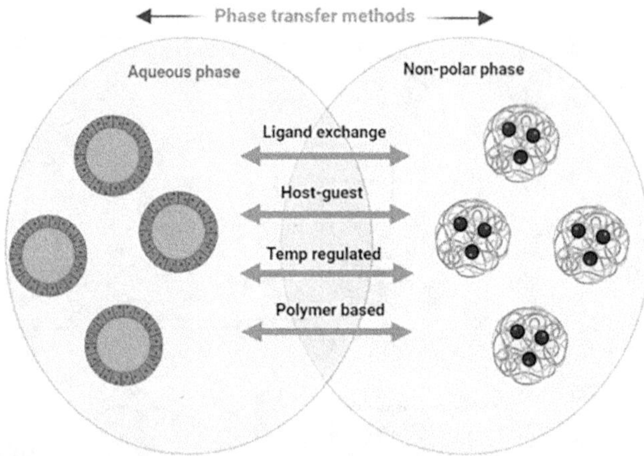

FIGURE 6.7 Overview of the phase transfer methods used for the synthesis of NPs.

using expensive organometallics. On the other hand, the low interfacial energy seen in non-polar organic solvents may provide more precise control throughout the solution and surface processing. Phase transfer has always been a traditional method for preparing organic-sols of metal NPs in non-polar organic media like hexane, benzene, chloroform, and toluene were constrained by the poor solubility of corresponding metal ions. Phase transfer is also useful for fabricating metallic NPs with the relevant ligands to produce ordered, tightly packed arrays and nanoparticulate thin films. Nanoparticle solubility is determined by how the solvent and stabilizing agent interact. The capping agents hydrophilicity or hydrophobicity can be managed by altering them. Controlled ligand attachment enables the NPs to keep their original characteristics while altering their solubility behavior. Figure 6.7 provides a summary of phase transfer in general. Post-synthetic phase transfer of NPs between immiscible liquids is necessary in many applications. Substitution or ligand exchange has been the most popular approach in these circumstances.

6.2.3 Sustainable Methods

The use of harsh reducing agents, which leads to high-energy consumption, low yield, environmental damage, and increased expense, is believed to be the foundation of the chemical and physical approaches. Furthermore, the microorganisms (fungi, yeast, bacteria, algae, etc.) used to create the NPs via biological methods have already been covered elsewhere (Ghosh et al. 2021). Often it is observed that the use of microorganisms has proven to be dangerous due to their inherent pathogenicity concerns; and maintenance. An alternative to these methods is using the 'green synthesis' approach. An overview of various 'green synthesis' techniques is provided in the following paragraph. This technique, which may be used to make NPs from plants or their components, is more affordable, less complicated, and environmentally friendly. The bioreduction of metallic ions is accomplished by the phytocompounds in plant extracts, like terpenoids, polyphenols, and polyols. As a result of their nanoscale size, the resultant NPs exhibit outstanding antibacterial, catalytic, and antioxidant characteristics (Kharissova et al. 2013; Savithramma et al. 2011; Abdel-Aziz et al. 2014; Edison and Sethuraman 2012). Along with this, several other elements that affect the environmentally friendly production of NPs are considered and described. Metal ions are transformed into their elemental form, which has a size of ~1–100 nm when plants or plant components are used in bioreduction (Hussain et al. 2015). Several NPs, including zinc oxide, silver, gold, iron oxide, and palladium can be produced using green synthesis (Khalil et al. 2014; Zeiri et al. 2014; Yang et al. 2009; Huang et al. 2014; Elumalai et al. 2015; Mittal et al. 2013).

6.3 BIO-MEDICAL APPLICATION OF NANOMATERIALS

Ever since the conceptualization of physics and biology, sharing an everyday standpoint, the beginning of the 20th century marks the conjuncture between chemistry and biology, i.e., a shift from plants-based derivatives to synthetic or semisynthetic alternatives was observed that accounted for global industrialization of the medicinal research for improved patient and life care. Recent statistical analysis shows a new addition to this group: the incorporation of nanotechnology. One can already imagine how this concept will modernize the overall patient care system and will be able to maintain exponential progress in different platforms of biomedical application. Usually, the variability in the physical and chemical properties of NPs allows for their surface engineering that is used in developing novel tools for biomedical-oriented applications. In contrast, the high biocompatibility with NPs also safeguards its use in in-vivo models. So far numerous attempts have shown prospective expansion in using these NPs ultra-small materials as therapeutic carriers for poorly soluble and environmentally susceptible drugs. Still, such therapies' success is relatively less than conventional therapeutic agents. This can be reasoned from the fact that complex heteromeric or homomeric structure shows variability in their pharmacokinetic and pharmacodynamic distribution of NPs. The present section emphasizes the role of different NPs in different biomedical applications. As previously observed, NPs results indicate that NPs are solid colloidal particles with a size range of up to 1000 nm. Besides, NPs also offer numerous advantages over their pristine parts, including a higher surface/volume (S/V) ratio and improved magnetic properties. The interest in using these features of NPs in biomedical applications such as targeted medication administration, bioimaging, and biosensors has progressively developed in recent years. Furthermore, NPs have prevailed in a variety of applications such as drug delivery, hyperthermia, bioimaging, cell labeling, and gene transport due to their superior qualities such as chemical inertness, non-toxicity, high biocompatibility, high saturation, magnetization, and high magnetic susceptibility. Finally, the future challenges and potential applications are highlighted in this section to give a clear overview of the various bio-oriented applications of nanomaterials, Figure 6.8.

6.3.1 EXAMPLES OF NANOMATERIALS FOR BIOMEDICAL APPLICATION

Over the last few decades, rapid advancements in nanotechnology alongside widespread applications of NPs in interdisciplinary fields, including materials science, energy, and medicine, have taken prospective developments. Inorganic NPs, like Fe_2O_3 NPs, AuNPs, quantum dots (QDs), and other rare earth NPs, possess intrinsic magnetic, optical, and electrical properties that have paved the way for developing newer biomedical technologies. The various physical properties of inorganic NPs, like the superparamagnetic nature, surface plasmon resonance (SPR) ability of various metal NPs, photoluminescence (eq. QDs), and up-converted luminescence in NPs can be modified by engineering its size, shape, composition, and structure of the inorganic core to create effective therapies. Besides, the large S/V ratio of NPs allows appropriate decoration of NP's surface to regulate their in-vivo behaviors. It provides multiple surface binding sites that enable NPs to combine with active targeting agents to facilitate imaging and therapy. Considering the above salient features of NPs, it can be presumed that these materials can be an ideal platform for developing multifunctional theragnostic nano-vehicles and eventually realizing 'precision medicine' and 'personalized medicine' (Fakruddin et al. 2012). In fact, according to some clinically approved reports, inorganic NPs like iron oxide NP-based ferumoxytol (Feraheme®) is used to treat iron deficiency anemias in people suffering from chronic kidney failures (Vadhan-Raj et al. 2013). Ferucarbotran (Resovist®) is a magnetic resonance imaging (MRI) contrast generation agent used to detect and characterize small liver lesions (Reimer and Balzer 2003). Other inorganic NPs have moved rapidly into clinical trials in recent years, like in the case of Aurimmune CYT-6091 (phase II) agent for the treatment of solid tumors, which consists of colloidal AuNPs bound to an

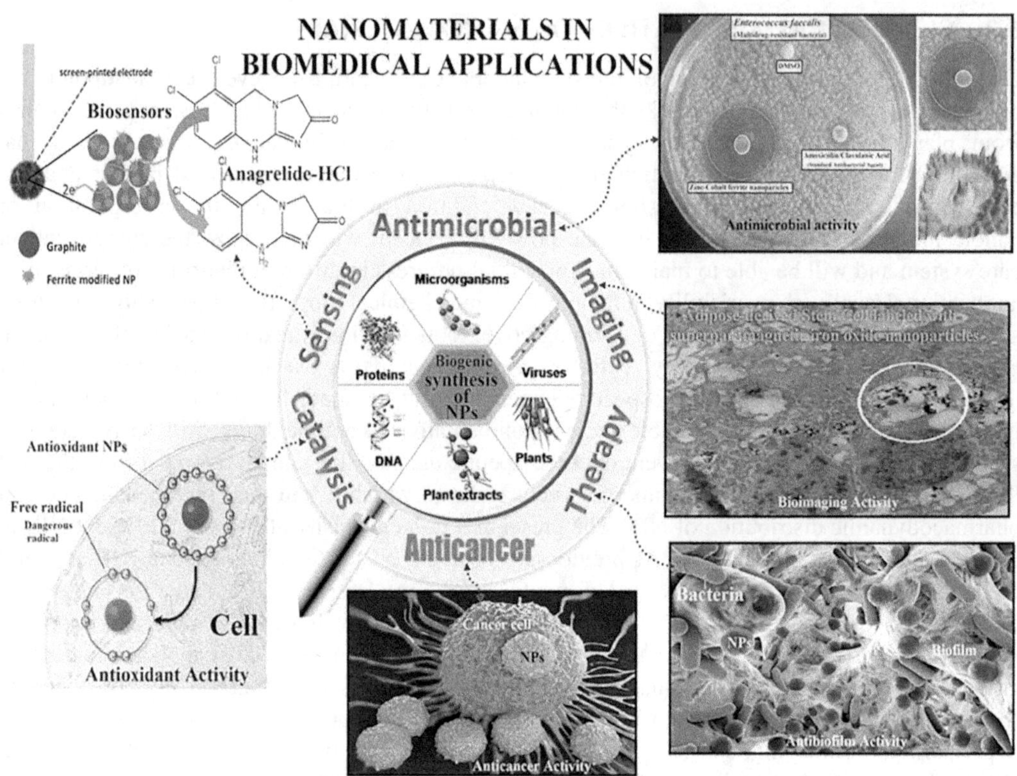

FIGURE 6.8 Picture representing the application of nanomaterials in various biomedical fields. (Elkodous et al. Colloids and Surfaces B: Biointerfaces, 2019, 180, 411–428 (Abd Elkodous et al. 2019)).

immune-evading component (poly (ethylene glycol)) and tumor necrosis factor (TNF) alpha. Nonetheless, the clinical transformation of NPs agents still lags behind the rapid growth of applications. Like in the case of in-vivo applications, it is unavoidable to be concerned about the toxicity of NPs (NPs), but the primary reason for this is that the in-vivo behaviors of NPs are quite complex and are influenced by many factors and parameters. In general, nanoscience and nanotechnology work harmoniously, having enormous potential to advance many fields apart from those mentioned here. The present section only addresses some of the most rapid onset developments in NPs biomedical care, like gene biosensing, bioimaging, and drug delivery. To gain better interest among our readers regarding the insights into the structure-activity relationship, the section broadly cites examples of different NPs systems like metallic NPs, bi-metallic or alloy NPs, metal oxide NPs, and magnetic NPs used in nanobiotechnology. Midst the different types of NPs, metal NPs (MNPs) have multiple roles in diagnosis and therapy due to their inherent physicochemical properties. The particle size tunability and large surface area in MNPs generate characteristic features entirely distinct from those of their macro-sized structures. There are several methods for synthesizing the MNPs that are discussed earlier. The top-down method creates bare metallic NPs, which are susceptible to further aggregation and therefore, unsuitable for biomedical applications. In contrast, the bottom-up strategy is based on solid-state, liquid-state, gas-phase, biological, microfluidic-technology-based, and other techniques for synthesizing MNPs that provide controlled growth of NPs. Chemical reduction in a 'bottom-up' fashion is the most widely preferred method for synthesizing MNPs due to its reproducibility, cost-effectiveness, and produces particles in a uniform form. The biological nanoparticle synthesis method has recently gained popularity due to its lack of toxicity, low cost, sustainability, and eco-friendliness. Furthermore, due to the metallic origin, MNPs display significant optical absorption due to localized surface phenomena known as plasmon resonance. These NPs are primarily investigated for potential new biomedical uses. Many utilities are made possible by the large nature of absorption bands in certain MNPs like gold, silver, and copper that show their bands in the visible region of the electromagnetic spectrum. They are mostly used for light-induced drug delivery, therapy, and detection as well (Giner-Casares et al. 2016). Likewise, iron oxide nanoparticles (FeNPs) are widely used in bioimaging, medication delivery, gene transfer, hyperthermia treatment, and MRI contrast-generating agents (Vallabani and Singh 2018). Long et al. reported the role of silver nanoparticles (AgNPs) in delivering medications, wound dressing, antibacterial agents, and cancer therapy (Long et al. 2015). AuNPs are utilized mostly for bioimaging, photothermal therapy, and drug delivery (Elahi et al. 2018). Moreover, the intrinsic ability of certain NPs to produce reactive oxygen species, like zinc oxide nanoparticles (ZnNPs), is mostly used in the cosmetics industry. Such MNPs also have found their use in antimicrobial and anticancer applications. Cadmium and Copper-based NPs have also been used in several biomedical applications (Rubilar et al. 2013).

6.3.2 Biosensing Application

Nanomaterials have proven their worth in biosensing applications, as they have in many other fields. A drastic performance improvement, like a multi-order reduction in detection limits and a significant improvement in sensitivity, can be achieved by strategically employing nano-objects. The high-specific surface of nanomaterials allows for the immobilization of more bioreceptor units than is possible with conventional materials. Immobilization strategies for conjugating the bio-specific entity intimately onto such nanomaterials remain a persistent obstacle. Therefore, one of the critical factors in creating a trustworthy biosensor is the method used for the immobilization of bio-elements. A brief overview of the effective strategies for bio-functionalizing nanomaterials is earlier reported by Putzbach et al. electrostatic interaction, stacking, entrapment in polymers, and Van der Waals forces are some common examples of non-covalent approaches between nanomaterials and living things (Putzbach and Ronkainen 2013). These guidelines protect all the unique characteristics of nanomaterials and biomolecules. Attaching biomolecules to nanomaterials, viz., covalent bonds offers advantages over other methods like increased stability, reproducibility, and decreased unspecific physisorption. For example, cross-linking, click chemistry, and traditional amide coupling reactions can all be used to create covalent bonds. The biomolecule may also become unintentionally anchored; such influences may cause a false recognition event. Binding biological species to surfaces through supramolecular or coordinated interactions has gained widespread acceptance in recent years. Likewise, Biotin and Avidin (or streptavidin) are the most well-known examples used in biosensor engineering (Wilchek and Bayer 1988). Furthermore, reversing the immobilization process and regenerating the transducer element is a major benefit of such nanosystems compared to other alternative detection methods. The biosensor's reproducibility is ensured by its constituent parts, such as the functionalized surface. As far the chemical composition is concerned, almost all nanomaterials can be functionalized directly (in some cases, during synthesis itself) or coated with functional polymers without altering their specific properties. However, in certain cases, the recognition event is only sometimes directly detectable by the transduction technique, i.e., it presents a unique challenge for, e.g., affinity biosensors such as the immunoreaction between an antigen and its antibody or the hybridization of complementary DNA strands. Thus, additional bispecific components (e.g., secondary antibodies/DNA strands, aptamers, etc.) can be labeled for optical or electrochemical transduction to overcome such anomalies. Perhaps, the unique properties of specific nanomaterials have also contributed to the development of 'label-free' transduction and clear signals amplification when used as labels (Biju et al. 2010).

The biosensor consists of an analytical platform employed for examining biological samples when present at the trace level. The biosensing concept is based on transforming biological, chemical, or biochemical stimuli into a detectable output. The three main fundamental units of a biosensor include 1) the bio-element, usually composed of nucleic acids, enzymes, cells, aptamers, or antibodies; 2) the transducer element (e.g., optical, electronic, pyroelectric, piezoelectric, gravimetric materials, etc.); and 3) the electronic unit comprising of an amplifier, a processor, and a display assembly. The global incidence rate of diseases and disorders has increased tremendously; hence, to limit such uncontrolled spread, detecting such anomalies in its initial stage is highly desirable. In addition, the possibility of monitoring diseases or disorder therapy with biosensors offers hope for personalized treatment. For these reasons, developing a simplified, sensitive, and cost-effective detection technique that provides additional information about a specific disease is still desirable. As detection labels, inorganic NPs are extensively utilized in developing biosensors for point-of-care (POC) diagnostics. The physicochemical stabilities offered by these NPs, like the ease of synthesis, rapid surface functionalization, and high S/V ratio, not only allow them to be used for signal transduction but also can be used to improve the sensor's detection capabilities for target analytes. Furthermore, their discrete physical and optical properties implement them to produce detectable signals in the form of color change, magnetic induction, fluorescence, and Raman signals, which are well-suited for the detection of different analytes types. The paragraph elaborates on using inorganic nanomaterials, like metal (Au and Ag), fluorescence (QDs), and magnetic NPs, in the development sectors of POC diagnostics. Depending upon their chemical composition, almost all nanomaterials can be outfitted with appropriate functions through direct functionalization or coating with functional polymers without compromising unique properties (De Crozals et al. 2016). This functionalization not only makes it possible to immobilize bioreceptor units in a reproducible fashion, but it also has the potential to improve the material's biocompatibility. The transduction technique typically employed in biosensing devices does not detect the recognition event. The immunoreactions between an antigen and its antibody and the hybridization of complementary DNA strands are well-established examples of affinity biosensors. Furthermore, secondary antibodies or DNA strands labeled for optical or electrochemical transduction are required as additional bio-specific components. Some nanomaterial's unique properties contributed to the development of 'label-free' transduction techniques, and their use as labels can significantly boost the intensity of a signal.

Among the different NPs, AuNPs have been the subject of the most research due to their inherent optical properties for sensing, detection, and imaging applications. The techniques based on the synthesis of AuNPs continue to evolve, allowing for greater control over their size and shape. AuNPs are smart candidates for drug delivery, diagnostic, and photothermal therapeutic applications apart from their role in sensing. For instance, with basic spectroscopy, AuNPs can amplify Rayleigh and Raman signals for obtaining chemical information on numerous previously inaccessible organisms. In addition, the modulation of the size /shape of AuNPs is an attractive way to develop new biosensors. It will continue to be utilized as a tool for future fabrication of novel devices. The surface and core properties of AuNPs are modified for their applications in molecular recognition, chemical sensing, and imaging. Even before applying the fabricated colloidal AuNPs to clinical trial research, numerous obstacles like its reproducibility, dependability, scalability, and durability during assays must be surmounted. It is anticipated that the output of multidisciplinary research will yield new fundamental scopes for AuNPs-based nanotechnology in the mere future. Herein, we have summoned AuNPs as a nano platform for generating various *in-vitro* and *in-vivo* biocompatible sensors. The bioimaging and therapeutic implications of these novel nanomaterials are also discussed here. The optical properties of the AuNPs, which are highly dependent on their size, shape, composition, and surface coating, make them particularly intriguing. These optical properties of AuNPs have been used in sensing biological molecules and cells due to their sensitivity and selectivity in response to the biological environment. Moreover, for targeting biological targets, such as DNA, RNA, cells, metal ions, small organic compounds, and proteins, a variety of

AuNP formulations have been fabricated. We have used multiple examples to analyze and discuss the most recent developments in the respective field. Detecting DNA, aptamers, and oligonucleotides has received significant attention over the past few years due to its importance in medical research, diagnosis, and the food and drug industries. Many assays identify specific sequences by hybridization of immobilized probes onto the surface of the target analyte that is pre-modified with a covalently linked optical probe. Numerous research groups have created DNA, aptamer, and oligonucleotide detection strategies based on chemically functionalized AuNPs. These simple methods employ facile NPs surface functionalization chemistry and typically do not require expensive equipment. Numerous DNA sensing systems are formulated using AuNPs, to improve the detection limit and sensitivity. Li et al. showed that single and double-stranded oligonucleotides have characteristic adsorption propensities on AuNPs. Moreover, the adsorption of single-stranded DNA stabilizes and prevents aggregation of AuNPs. Since the color of AuNPs is determined by its surface plasmon resonance (SPR) and the NPs aggregation state, one can design simple colorimetric hybridization assays by exploiting the difference in electrostatic properties between single-stranded and double-stranded DNA. The assay can be used for the detection of sequence-specifically untagged oligonucleotides. It is intended for visual detection at concentrations as low as 100 femtomolar (fM), and the authors demonstrated that it can be readily adapted to detect single-base discrepancies between probe and target (Li and Rothberg 2004). They further report the fluorescent-based assay for DNA hybridization by taking advantage of the ability to adsorb only single-stranded DNA onto the negatively charged AuNPs. This method makes detecting target sequences in complex mixtures of DNA and single-base mismatches in DNA sequences simple and robust (Li and Rothberg 2004b). In another report, colorimetric adenosine biosensors developed on the antizyme-directed assemblage of AuNPs can also be used as a detection platform. Aptazymes are derived from DNAzymes, containing an adenosine aptamer motif that modulates the activity of DNAzyme through allosteric interactions in the presence of adenosine. On the other hand, the presence of adenosine activates the aptazyme that cleaves the substrate strand and disrupts the formation of NPs aggregate. Since aptamers can target a wide range of important analyte classes, they can be converted into aptazyme based biosensors (Liu and Lu 2006). Elghanian et al. showed the construction of a colorimetric detection platform using AuNPs to detect polynucleotide molecules (Elghanian et al. 1997). The high-specific colorimetric polynucleotide detection approach was based on mercaptoalkyloligonucleotide-modified AuNP probes, i.e., when a single-stranded target oligonucleotide (30 bases) was introduced into a solution containing the necessary probes, it led in the development of a polymeric network of NPs and a red-to-pinkish/purple color change. The freezing and thawing of the solutions aided hybridization, and denaturation of these hybrid materials revealed transition temperatures across a small range that permitted distinction of a variety of defective targets. When the hybridization mixture was transferred to a reverse-phase silica plate, it dried with a visible blue tint. The unoptimized system can identify oligonucleotides as small as 10 fM (Elghanian et al. 1997). Mariotti et al. showed the preparation of a DNA-based SPR affinity biosensor (Mariotti et al. 2002). The biosensor was used to identify genetically modified organisms (GMOs). Single-stranded DNA (ssDNA) probes were immobilized on the SPR devices sensor chip, and the hybridization between the immobilized probe and the complementary sequence target was measured. The probe sequences were internal to the 35 S promoter and NOS terminator sequences, which are inserted elements in the GMO genome that regulate transgene expression. The technique was optimized using synthetic oligonucleotides before being applied to real-world sample analysis. Polymerase chain reaction (PCR) was used to amplify samples carrying transgenic target sequences, which were subsequently identified using the SPR biosensor. The procedure presented in this example is based on high pH dissociation and physical separation of the two strands using magnetic particles coated with immobilized streptavidin (Mariotti et al. 2002). AuNPs have also been employed to improve the responsiveness of SPR biosensors. Numerous research groups have used functionalized AuNPs in immunoassays to demonstrate their biosensing amplification. The AuNPs-enhanced SPR approach has been used at industrial levels to build cholera cell detection tests. Several groups have also developed an Au-amplified SPR sandwich immunoassay to detect human immunoglobulin at the picomolar level. According to these results, the detection limit of the AuNPs-amplified SPR biosensor system may be increased further by tuning the setup settings. Introducing functionalized AuNPs to improve biosensor responsiveness will open new possibilities for single-molecule interaction detection biosensors and DNA diagnostics. We briefly explain different SPR biosensor setups and then present a generic way for employing AuNPs to improve SPR biosensor responsiveness. In this regard, Matusi et al. fabricated an SPR sensor on an Au substrate that was created to detect low molecular weight analyte (Matsui et al. 2005). The sensing mechanism was based on the swelling of the imprinted polymer gel due to the analyte binding event inside the polymer gel, causing an increase in the distance between the AuNPs and the substrate, causing a dip in an SPR curve to move to a higher SPR angle. On a sensor chip covered with allyl mercaptan, a combination of acrylic acid, N-isopropyl acrylamide, N, N'-methylene bisacrylamide, and AuNPs was radically polymerized in the presence of dopamine as model template species. The modified sensor chip demonstrated an increased SPR angle in the presence of dopamine concentrations. Additionally, AuNPs were proven effective for increasing signal intensity (change in SPR angle) compared to a sensor chip without AuNPs. The analyte binding process and the resulting swelling appeared reversible, allowing the disclosed sensor chip to be used again and again (Matsui et al. 2005). Wang et al. designed a novel label-free amplified SPR biosensor for miRNA detection using AuNPs labeled with DNA. AuNPs were used for SPR amplification and terminal hybridization of DNA. The DNA-hybridized AuNPs further triggered the hybridization of other probes resulting in the formation of a DNA super sandwich, with a detection limit reaching up to 8 fM (Wang et al. 2016). Apart from their outstanding optical properties AuNPs are also used as redox couples due to their intrinsic ability to exchange electrons between wide ranges of bioactive species, making them suitable for the fabrication of electrochemical sensors. In conventional electrochemical sensors, the formed species are either oxidized or reduced by the electrode to produce electrochemical signals. But its major drawback is that the analyte molecules must first diffuse to the electrode, wherein a significant loss occurs in the solution. The inherent ability of AuNPs to relocate in the redox center of enzymes and their ability to regenerate biocatalysts by transferring electrons to the electrode increases the overall electrochemical signal response.

QDs, luminescent semiconducting NPs, are yet some other type of nanomaterial to be used in making bio-analytical tools. The most researched colloidal QDs are based on cadmium-based chalcogenides (S, Se, Te) (Murray et al.; Park et al.; Reiss et al.) that offer a very broad absorption wavelength due to their difference in the band gaps of these semiconducting materials, i.e., the larger the particle, the narrower is the band gap (Murray et al. 1993; Park et al. 2007; Reiss et al. 2009). These cause distinct emission wavelength patterns arising from the electron-hole exciton combination. Furthermore, the availability of a wide range of emission wavelengths for QDs of varying diameters enables efficient multiplexed analysis using conventional optical transduction. However, Murphy et al. discovered that the structural defects in the crystal lattice can capture exiting electrons or holes, resulting in nonradiative relaxation (Murphy 2002). To circumvent this problem, Jaiswal et al. developed core/shell semiconductors having a larger band gap range to passivate surface defects and improve quantum yields and photostability (Jaiswal et al. 2002). QDs are now available with inert or biocompatible coatings that provide functional groups for immobilization. QDs, also known as semiconductor NPs, are particularly relevant for multimodal, multifunctional, and multiplexed imaging of biomolecules, cells, tissues, and animals. QDs have unique optical properties, including size-dependent tunable absorption and emission in the visible and near-infrared region (NIR) regions, narrow emission and broad absorption bands, high photoluminescence quantum yields, large one- and multi-photon absorption cross-sections, remarkable photostability. Multimodal imaging probes are created by interfacing the unique optical properties of QDs with magnetic or radioactive materials. In addition, the crystalline structure of QDs expands the purview of X-ray and TEM imaging with high contrast. As long as the photophysical recombination event remains unaffected, virtually any biomolecule can be affixed to these nanocrystals. Förster resonance energy transfer (FRET) is a well-known photophysical technique that involves the nonradiative transfer of excitation energy from excited fluorophores to a neighboring ground-state acceptor. FRET is used in numerous domains, including analytical chemistry, protein conformation research, and biological experiments. QDs may increase FRET because of their broad excitation spectra and symmetric photoemission. Recent FRET investigations utilizing luminescent QDs have addressed fundamental concerns and developed targeted applications with potential biological applications, such as sensor design and protein conformation studies. Compared to conventional dyes, QDs offer several distinct advantages as energy donors. QD-based FRET nanosensors will be especially appealing for intracellular sensing due to their high photobleaching thresholds and a significant reduction in direct excitation of dye and fluorescent protein acceptors, which could allow for long-term monitoring of intracellular processes. FRET-based QDs may find uses in monitoring a wide range of intracellular and extracellular markers; however, success will be determined by the ability to devise easy and reproducible ways for delivering compact conjugates to subcellular areas and cell membranes. Sadly, limited publications describing FRET-based QDs are available due to a lack of necessary control experiments to prove that energy transfer was the only mode of interaction and no other uncontrolled processes. As more QD materials are prepared and more QD bioconjugates are designed with reasonable control over conjugate architecture and functionality, a better understanding of energy transfer applied to luminescent QDs and the refinement of specific FRET-based QD sensing assemblies will be achieved. This approach is notably utilized for optical DNA and oligonucleotide sensors (Freeman et al. 2013). Bioluminescence Resonance Energy Transfer (BRET) is another type of nonradiative energy transfer that induces QD fluorescence. In this, light-emitting protein label transfers energy to QDs, in the presence of an external excitation light source. FRET and BRET, as well as charge transfer quenching and chemiluminescence resonance energy transfer (CRET), are the most used techniques in biosensing applications using QDs as optical transducers (Algar et al. 2010). QDs have shown their suitability in bioanalysis as transducer units or optical identifiers. It is therefore, possible that QD-based biosensors will undergo a rapid evolution. The described principle and examples regarding using QDs in bioanalysis suggest that combining different nano-objects and their specific characteristics is a promising avenue for developing biosensors with novel and highly effectual transduction techniques.

6.3.2.1 Metal oxide NPs (MONPs)

MONPs with various morphologies are used as electroanalytical devices due to their size, stability, and large surface area, these MONPs, exhibit a variety of electrical and photochemical characteristics. The primary function of MONPs in electroanalysis is improving the conductive sensing interface (Katz et al. 2004a). The rapid electron transfers between the transducer and analyte molecule in the presence of MONPs led to their classification as 'electronic conductors' and 'electrocatalysts.' Highly Biocompatible MONPs are typically used for immobilization and are further used to fabricate immunosensors, enzymes, and DNA sensors. In contrast, the semiconducting NPs are predominantly used as markers and tracers in electrochemical studies. Different techniques like physical adsorption, electrodeposition, chemical covalent bonding, and electropolymerization are used to accomplish high affinity of MONPs to the surface of the working electrode (Wang and Hu 2009).

6.3.2.2 Titanium dioxide (TiO_2) NPs

TiO_2 NPs are ceramic nanomaterials that play an indispensable role in electrochemical sensing due to their dependability, biocompatibility, high conductivity, and low cost. TiO_2 is chemically stable under both acidic and alkaline conditions and has an efficient catalytic activity that can be used to reduce small organic molecules (He and Hu 2011). Other forms of TiO_2, like NPs, nanotubes, nanofibers, and nanoneedles, have also shown promising electrode materials for developing electrochemical sensing devices (Kafi et al. 2011). TiO_2 nanotube arrays (TNTAs) are used for biomolecule detection due to their superb electron transfer behavior, large surface area, and availability of active reaction sites (Liao et al. 2015). While the low solubility of TiO_2 NPs and its instability on the electrode surface results in low sensitivity of the electrochemical sensor. Thus, increasing the efficacy of electrode systems containing catalytic materials like enzymes or NPs of noble metals on TiO_2 nanostructures is an area to be looked at. Likewise, Palladium nanoparticles (PdNPs) deposited on the surface of a titanium dioxide-silicon carbide nanohybrid (TiO_2-SiC) by chemical reduction (Pd@TiO_2-SiC) modified glassy carbon electrode (GCE) show a promising electrode material for quantification of hydroquinone and bisphenol, with detection limits achieving up to 5.50 and 4.30 nM, respectively (Yang et al. 2014). TiO_2 nanocomposites with shape-complementary cavities and a vast surface area can house highly selective molecules within their recognition cavities. A spin-coated molecularly imprinted polymers (MIP) – TiO_2 NPs/multiwalled CNTs nanocomposite was produced on the PGE (pencil graphite electrode) surface for ultra-trace measurement of enantiomeric sensing of aspartic acid isomers. Another MIP employing TiO_2 and AuNPs is a more accurate technique for determining 4-nonylphenol (Huang et al. 2011). Hu et al. showed light-enhanced electrochemical sensing for bisphenol A (BPA) using carbon-doped TiO_2 nanotube arrays decorated with AuNPs (TiO_2/Au NTAs) that allowed the detection of BPA under UV irradiation (Hu et al. 2016). The electrode continuously provided a new surface and increased photocurrent caused by BPA's utilization of holes.

6.3.2.3 Cerium dioxide (CeO_2) NPs

As a biosensor and anticancer agent, CeO_2 NPs or nanoceria have been utilized in biomedical applications (Renu et al. 2012; Nesakumar et al. 2013). Like Sol-gel-derived nanoceria coatings deposited on Au electrodes, platinum-coated glass plates, and even filter paper, followed by immobilization with glucose oxidase (GOx), are used as effective glucose sensors that do not require a mediator (Patil et al. 2012). Similarly, Cholesterol oxidase (ChOx) immobilized sol-gel derived nanoceria film can measure cholesterol concentration in serum samples. Rabbit-immunoglobulin antibodies (r-IgGs) and bovine serum albumin (BSA) immobilized nanoceria film on an indium–tin oxide-coated glass substrate can detect neurotoxin (ocher-toxin-A) (Kaushik et al. 2009). The ceria associated with praseodymium demonstrates efficient oxygen sensor activity. The nanoceria and chitosan composite matrix designed to immobilize single-stranded DNA probes can be used as a DNA biosensor for the colorectal cancer gene.

6.3.2.4 Silicon dioxide (SiO_2) NPs

The use of silica (SiO_2) NPs in the biomedical field can be accredited to their unique properties, like large specific surface area, pore volume ratio, controllable particle size, and high biocompatibility making them suitable for potential use in drug delivery and biosensors (Wang et al. 2015). Specific molecular receptors, such as proteins, have been employed in silica-based biosensors to achieve analyte selectivity (for example, glucose oxidase and horseradish peroxidase-loaded mesoporous silica are used for glucose detection) (Wei et al. 2002). Myoglobin and hemoglobin were immobilized on mesoporous silica-modified electrodes for H_2O_2 and NO_2-sensors in an independent investigation (Dai et al. 2004). Because of their excellent substrate specificity, enzymes, and proteins are likely chosen as the recognition unit. However, these

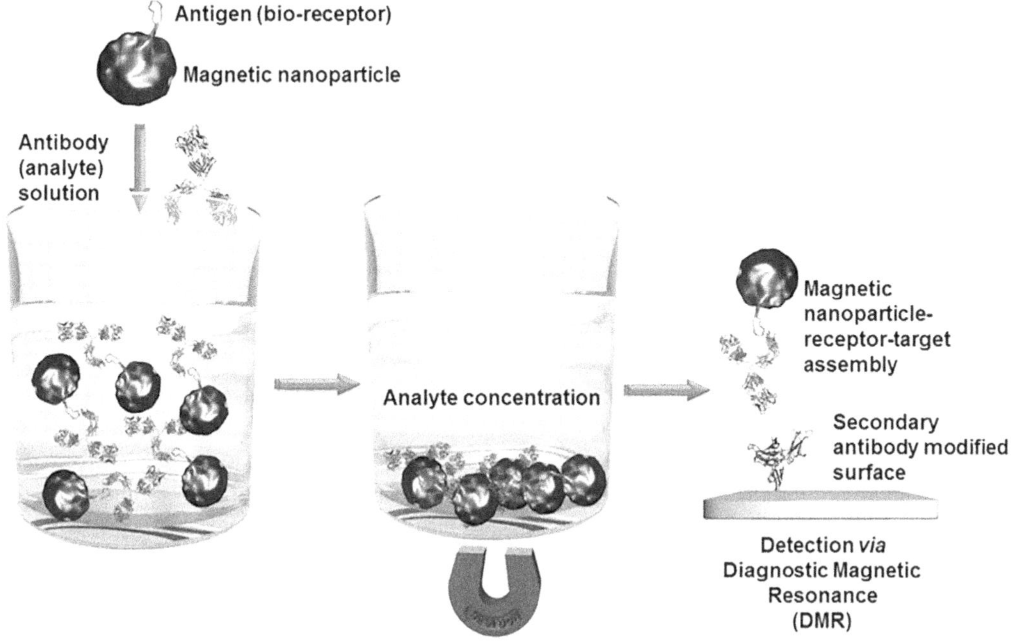

FIGURE 6.9 Image depicting analyte concentration in complicated liquids using bioreceptor-modified magnetic nanoparticles. Following the recognition event, the collected analytes are reverted through the precipitation of magnetic NPs under the influence of an external magnetic field (Holzinger et al., Frontiers in Chemistry 2014, 2 (Holzinger et al. 2014)).

biomolecules need long-term stability and are relatively pricey. In contrast to the molecular imprinting method, selectivity is obtained by modulating the diffusional penetration of analytes into surface-functionalized mesopores rather than by synthesizing shapes or selective recognition receptors. The materials ability to respond differently to ATP, ADP, and AMP (adenosine triphosphate, diphosphate, and monophosphate, respectively) is owing to a combination of the bulkiness of the grafted group and the steric limitations imposed by the materials pore size. ATP can suppress the materials fluorescence to a greater extent than the other two species, and minor anions such as phosphate, bromide, and chloride did not affect the sensor. In comparing the ATP sensitivity of aminomethyl anthracene MSNs and aminomethyl anthracene grafted on fumed silica, the MSN-based matrix demonstrated a 100-fold increase in ATP sensitivity (Descalzo et al. 2005).

6.3.2.5 Magnetic NPs

Another type of sensors and biosensors are based on magnetic NPs that have gained considerable attention over recent years. Tiny magnetic NPs can be built into transducer components or disseminated throughout the sample before being drawn to the biosensors active detecting surface by the influence of an external magnetic field. Recent trends in MNPs as sensors and biosensors are described and discussed, emphasizing their analytical figures of merit. Future directions and potential applications for magnetic NP-based sensors and biosensors are also shown here. To date, considerable inputs have been made towards developing magnetic NPs due to their excellent properties like small size, difference in the physiochemical properties, high biocompatibility, and cheaper costs. Generally, magnetic NPs with a size range of 10–20 nm show the highest super magnetic behavior, making them a quick response system under the influence of an external magnetic field. Moreover, such materials also offer high sensitivity and stability in sensors, allowing them to detect trace levels of analytes from clinical samples, environmental samples, etc.

Among the various magnetic NPs, Fe_3O_4 NPs are mostly used in developing biosensors. However, issues of using Fe_3O_4 are its dipolar attraction nature and large S/V ratio may lead to the aggregation of these NPs. These limitations can be well overcome by functionalization.

Recent progress utilizes more magnetic NPs as biosensors in cancer diagnosis. Magnetic NPs are mostly used in these areas because of their inherent ability to conjugate with recognition elements like antibodies that can act as a detection system for determining overexpressed biomarkers. Additionally, these NPs also offer exceptional sensitivity toward detecting cancer cells. Magnetic NPs also generate strong electromagnetic fields on their surfaces, enabling radiative features such as scattering and absorption, which are used in optical imaging. So far, many types of sensors, such as SPR, colorimetric, fluorescent, electrical, electrochemical, and BioBarcode assay sensors, have been developed based on these magnetic NP features (Figure 6.9).

Tian et al. created red-fluorescent AuNPs using diluted egg whites and microwave irradiation. It was discovered that live HepG2 cancer cells could activate fluorescence signals of Au nanoclusters, whereas normal cells were unaffected. The fluorescence capacity increased significantly with time. The MTT results also showed that the nanocluster was biocompatible with cells, as more than 85% of cells in the concentration range of 0–1.2 mg/mL were viable after 24 hours. The findings suggest that similar research could be utilized to detect cancer cells early on (Tian et al. 2017).

6.3.3 Nanomedicine and drug delivery

Development and maneuvering of NPs for better patient care have offered early diagnosis of diseases and specific treatment. As an obvious outcome of nanotechnology, nanomedicine has provided numerous medical developments wherein NPs are used as delivery vehicles. This is usually done to improve the pharmacokinetic and pharmacodynamic properties of the drug molecule. Moreover, the slow drug release from the NPs allows for high sustainability and supports prolonged drug release to specific sites. The main advantages of nanocarriers are shown in Figure 6.10.

Additionally, the solubility of the drugs can be improved by encapsulating them inside the core of the NPs. At present liposome-based drug delivery systems have made it to the market, whereas polymer-based micellar drug delivery systems are yet to be marketed but have entered the final stages of clinical trials. The progress in using inorganic NPs as drug delivery systems is relatively less. However, the blooming strength of nanofabrication has made progressive developments in the preparation and fabrication of NPs that can be used for drug loading and unloading. Inorganic NPs made from Au, Ni, Pt, and Fe offer several advantages, like small size, contrast generation, and heat production, making them more suitable than conventional drug carriers that only act as reservoirs for drugs. Nonetheless, NPs, due to their small size, can readily pass through vessels and capillaries to distribute drugs to the site of action, i.e., increasing the bioavailability of the drugs. Some of the widely used applications of NP-based drug delivery include their use in drug and gene delivery, diabetes treatment, protein detection, tissue engineering, tumor eradication, DNA structure analysis, and cancer therapy are all aspects of nanomedicine (Li et al. 2019). Current advances in NPs offer a broad range of clinical applications. NPs, as observed, surpass the limitations of free therapeutics and provides the movement of the therapeutically active molecules through various biological barriers like micro-environment, systemic and cellular layers without altering their efficacy. Another benefit of using NPs is that it surpasses patient heterogeneity, i.e., offers personalized therapies. Much of the modern-day research is based on optimizing the delivery platform, i.e., a single uniform-size nanoparticle can deliver a wide range of drugs and pass through different biological layers. The use of inorganic NPs for precision-based drug delivery has considerably increased over the years, resulting in the formation of personalized care. The present paragraph focuses on

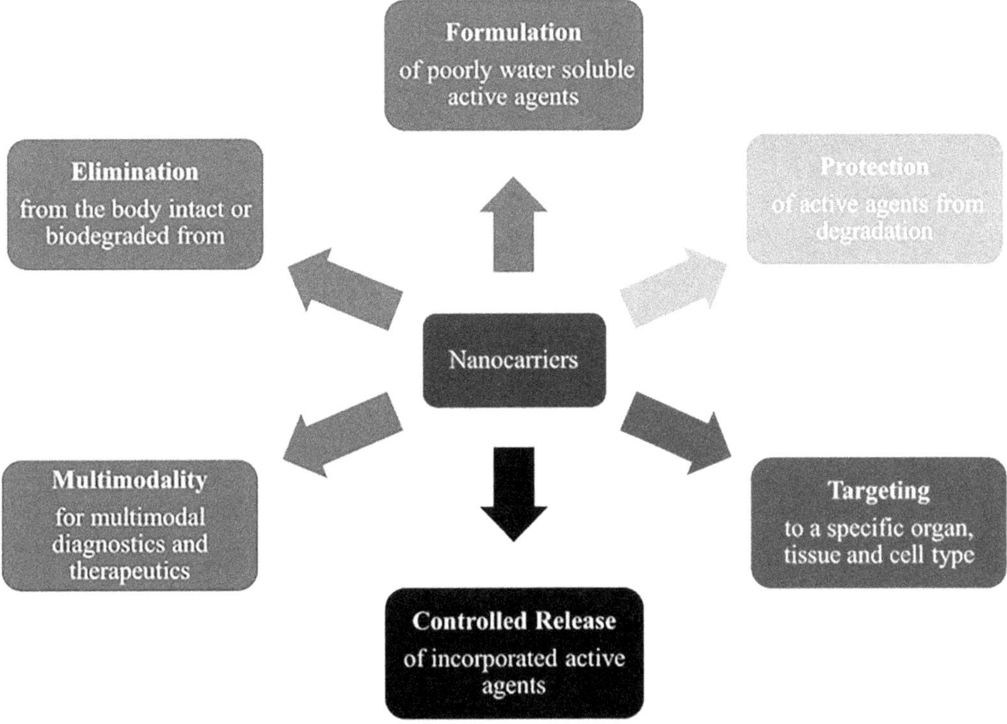

FIGURE 6.10 Advantages of nanocarrier formulations in medicine. (Chen et al., Chemical Reviews, 2016, 240, 114–124 (Chen et al. 2016)).

summarizing the role of NPs in drug delivery. A brief overview of the uses of NPs in drug delivery is highlighted below in Figure 6.11.

Diabetes is characterized by the inability to create enough insulin for glucose management or the body cells inability to utilize this insulin produced, which has been a severe threat to people's health. Diabetes can be classified into two major types: 1) Diabetes mellites and 2) Diabetes insipidus. Among the two, diabetes mellitus is the largest cause of diabetes that impacts healthcare. It is caused by a decrease in insulin production by β-cells, resulting in a larger accumulation of glucose in the body. At the same time, the other diabetes insipidus is a sporadic disorder characterized by the frequent loss of large amounts of body fluids by urination. Pathophysiology of diabetes suggests improper functioning of vasopressin, an antidiuretic hormone, from the pituitary's posterior lobe, resulting in uncontrolled bodily fluid release. The role of NPs in treating diabetes mellitus is through effective insulin delivery. Some of the different types of NPs intended to be used for the delivery of insulin are 1) Polymeric biodegradable NPs, 2) Ceramic NPs, 3) Dendrimers, 4) Polymeric micelles, 5) Liposomes, etc. To treat diabetes mellitus, insulin, a polypeptide of 51 amino acids, is typically given parenterally. Regrettably, most patients refuse to receive injections because they hurt so much. The oral administration of insulin has drawbacks because it is a protein that needs to be processed before being absorbed into the bloodstream. However, the intestinal epithelium is the major barrier to medication absorption since medicines cannot cross epithelial cells and enter the circulatory system. Owing to this worry, it was claimed to enhance medication paracellular transport (Ghorbani et al. 2019). Several compounds that promote intestinal permeability, including chitosan, are utilized to support the absorption of macromolecules. Chitosan NPs have enhanced protein molecule absorption in-vivo to a greater level. Because chitosan is less soluble at pH levels higher than 6.5, chitosan NPs are employed to administer insulin. Cancer is one of

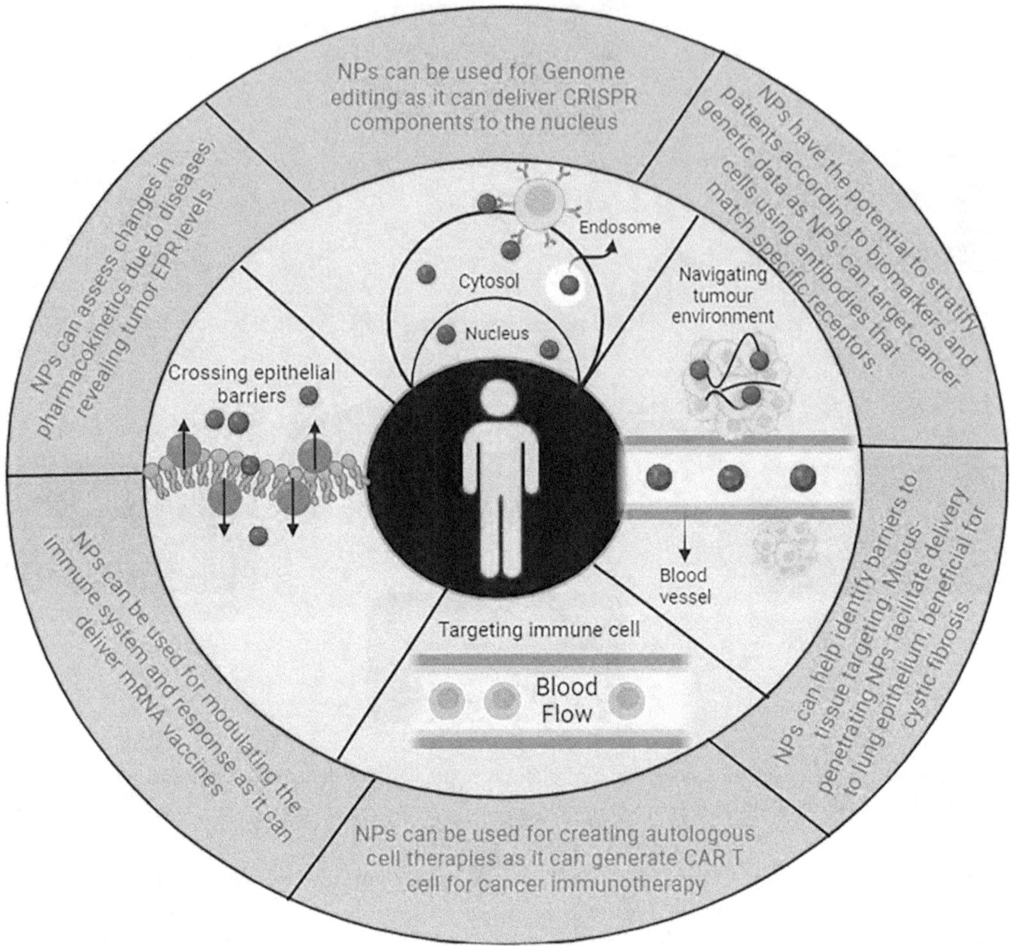

FIGURE 6.11 Overview of some biological hurdles NPs can overcome and precision medicine applications that may benefit from NPs (outer ring). As discussed in this Review, intelligent NP designs that improve delivery can potentially increase precision medicine performance, thereby hastening clinical translation. (Mitchell et al., Nature Reviews Drug Discovery, 2020, 20, 101–124 (Mitchell et al. 2020)).

the leading causes of death, and it is becoming more and more common. Chemotherapy, radiation, and surgery are the only cancer care options. Recent cancer therapies encounter several significant obstacles, including a generic systemic dispersion of anti-tumor medications, an insufficient concentration of pharmaceuticals reaching the tumor, and an inadequate capacity to track therapy response. Many problems result from ineffective drug distribution, including multidrug resistance. Based on specific properties, NPs are applied to the treatment of cancer. Developing beneficial systems may benefit from manipulating the surface characteristics and other physicochemical aspects of NPs (Kunjachan and Jose 2010). Targeting cancer cells occurs using two distinct strategies: inactive targeting and active targeting. Nanoparticle-induced tumor cell inactivation depends on the IPR effect. The half-life of the NPs affects this delivery (Jo et al. 2017). As a result, the goal choice is based on its enormous quantity. Although it is claimed to be active, polymeric NPs may use targeting as its likely delivery method to reach cancer cells with chemotherapy medications (Xie et al. 2018). Nanotechnology has a critical role to play in the treatment of neurodegenerative diseases. Many nanocarriers, such as dendrimers and liposomes, are the subject of intensive investigation for therapeutic medication delivery to

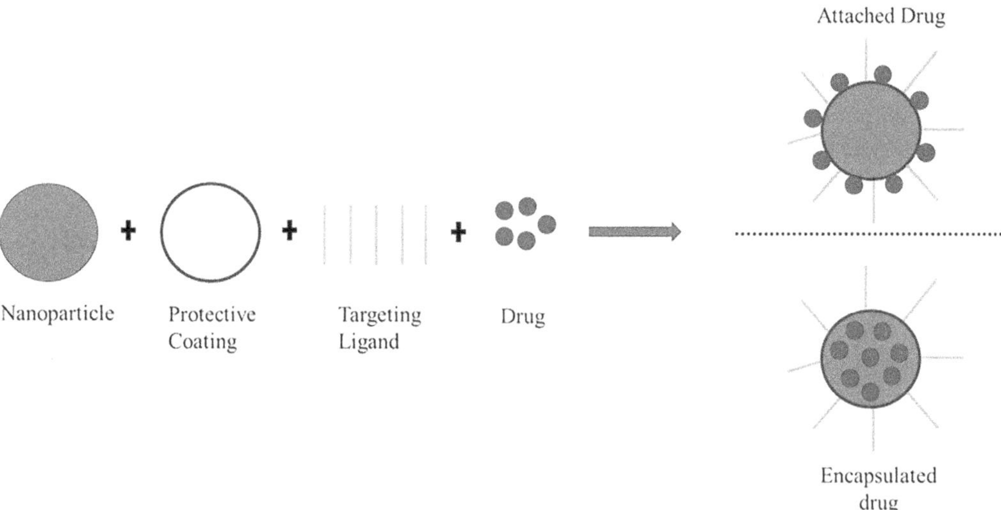

FIGURE 6.12 Schematic of drug loading options in targeted drug delivery.

the central nervous system. Targeted drug delivery aims to deliver the treatment to the correct spot, increasing the amount of medication supplied at the needed site while minimizing unwanted effects. Magnetic NPs are used in targeted medicine delivery to deliver the medication to a specific location. Figure 6.12 shows how to functionalize magnetic NPs so medicine can be conjugated to the surface or encapsulated within the nanoparticle. Magnetic NPs, such as Au or polymers, are frequently coated with a biocompatible coating. After administration, an external magnetic field drives the drug/nanoparticle complex to the specific spot. The medication is released by changes in pH, temperature, or osmolality, as well as by enzyme activity (Berry and Curtis 2003; Pankhurst et al. 2003; Tartaj et al. 2003).

The application of AgNPs in several biomedical domains, including drug transport, diagnostics, and anticancer agents, has emerged, according to research. Franco-Molina et al. explored using colloidal silver as an anticancer agent on human breast cancer cells (Franco-Molina et al. 2010). To ascertain whether a lethal impact had occurred, MCF-7 human breast cancer cells were cultured with various colloidal AgNP concentrations. The findings demonstrated that colloidal silver has a dose-dependent cytotoxic effect on a human breast cancer cell type. Apoptosis was found to be the mechanism of cell death, with no impact on healthy cells. Colloidal silver is effective against cancer. In this regard, the AgNPs-chitosan nanocarrier was created by Sanpui et al., and its cytotoxic impact on human colon cancer cell lines was investigated (Sanpui et al. 2011). The present study used HT 29 colon cancer cell lines. The findings indicated the possibility of using low-concentration AgNPs-chitosan nanocarriers in cancer therapy by inducing cell death in HT 29 colon cancer cells. Since the cancer cells can withstand chemotherapy, multidrug resistance (MDR) cancer poses a serious problem for disease treatment. Studies on the anti-tumor effects of AgNPs and modified AgNPs were conducted both in-vitro and in-vivo. The cell-penetrating peptide TAT was used to modify the AgNPs. The outcomes demonstrated that the modified AgNPs were the cause of the improved cellular uptake. Also, it was demonstrated that nanosilver anti-tumor effects were concentration-dependent. Both AgNPs and modified AgNPs could stop cell line proliferation, according to the in-vitro investigation. The in-vivo tests revealed that AgNPs with TAT modifications could stop tumor growth. According to this study, TAT-nanosilver may be used to treat MDR cancer (Sanpui et al. 2011). Benyettou et al. looked at using AgNPs as a drug delivery technology to efficiently transport both doxorubicin and alendronate to cervical cancer cells (HeLa) (Benyettou et al. 2015). A bacterial antibiotic called doxorubicin is frequently used to treat cancer. Alendronate is a potent bio-phosphonate medication intended to slow down or stop the breakdown of bone tissue. Bio-phosphonates are thought to lessen pain and fracture risk in patients with particular cancer types. Alendronate molecules attached to the AgNPs by chelating bisphosphonate moieties are coated on them. Drug attachment is possible at the ammonium location. Both doxorubicin and Rhodamine B were employed to test the toxicity and cell permeability of the AgNP drug-based system. The cellular uptake studies showed that Rhodamine B conjugated NPs enter cells via macropinocytosis and endocytosis. Furthermore, doxorubicin-modified NPs were proven to have better anticancer activity than doxorubicin alone or alendronate and were successfully delivered to HeLa cells. Alendronate-coated AgNPs can be modified to accept other medications, providing a different means of delivering medication to cancer cells (Benyettou et al. 2015). Researchers are becoming increasingly interested in Fe-Pt NPs for potential biological applications due to their unique chemical and magnetic properties, which include superparamagnetism, high coercivity, chemical stability, oxidation resistance, and biocompatibility (Shi et al. 2015; Chiang et al. 2007). These characteristics have led to a rise in research on Fe-Pt NPs potential applications in biomedical fields such as hyperthermia (Seehra et al. 2010), contrast agents for MRI (Wu et al. 2015; Lai et al. 2012), drug delivery/cancer therapy (Seemann et al. 2015; Zheng et al. 2015), and biosensors (Moghimi and Leung 2013). Fuchigami et al. investigated the effects of magnetic porous Fe-Pt NPs loaded with an anticancer medicine for targeted drug delivery on gastric and lung cancer cell lines (Fuchigami et al. 2012). The Fe-Pt-Dox capsules destroyed cancer cell lines by over 70% under the influence of an external magnetic field, resulting in an anti-proliferative effect. Dox realized enabling the release of Dox to cancer sites of the cancer cell lines was destroyed by the Fe-Pt-Dox capsules, which were led by a magnetic field to stop the proliferation of both cell lines cancer cells. Moreover, the impact of various Fe-Pt nanoparticle surface coatings on brain tumor cells was examined by Wu et al. (2012). The Fe-Pt NPs were coated with oleic acid/oleylamine (OA/OA) and L-cysteine. Compared to Fe-Pt-Cys NPs, which showed a minimal or nonexistent cytotoxic effect, Fe-Pt-OA/OA-coated NPs suppressed glioma cell proliferation. Moreover, the surface coatings caused glioma cell development to be suppressed. TiO_2, or titanium dioxide, has also been extensively explored. Like it has been utilized in many industries, including toothpaste, cosmetics, and the pharmaceutical and cosmetics industries. TiO_2 is an interesting material for biomedical applications because of its multiple distinguishing characteristics, which include high biocompatibility, low toxicity, chemical stability, and photocatalytic properties. Because of their unique properties, TiO_2 NPs have been studied extensively in various biological applications, including medication administration, PDT, cell imaging, biosensors, and genetic editing (Fei Yin et al. 2013; Du et al. 2015; Idris et al. 2014; Tu et al. 2010). Qin et al. investigated whether adding the medicine to the NPs would influence their effectiveness (Qin et al. 2011). TiO_2 NPs were loaded with doxorubicin (DOX) using two different techniques: the first was a non-covalent complexation (TiO_2/DOX), and the second was a covalent conjugation (TiO2-DOX). Using a rat glioma (C6) cell line as a model, these two approaches cytotoxicity, cellular uptake, and intracellular dispersion were evaluated. The findings revealed that cytotoxicity was higher when DOX was loaded using the non-covalent complexation (TiO_2/DOX) approach than when DOX was free or covalently conjugated (TiO2-DOX). The confocal laser scanning microscopy results revealed that the nucleus contained DOX produced by the complexation process (TiO_2/DOX). The cytoplasm was where most of the TiO_2-DOX produced during the conjugation process was found. This study highlights the significance of the drug-loading strategy for TiO_2 NPs. In parallel, TiO_2 NPs were also studied by Chen et al. to create a new pH-responsive daunorubicin (DNR) drug delivery system. DNR is a cancer-fighting medication, but because of its severe side effects, its use is restricted (Chen and Zhang 2012). The release of DNR was found to be enhanced upon lowering the pH from 7.4 to 5.0. DNR-TiO_2 nanocomposites demonstrated better cellular absorption of DNR when compared to DNR alone. However, the DNR-TiO_2

nanocomposites were discovered to be more hazardous. The mechanism of DNR release can be described by its pH; at pH 7.4, DNR remains conjugated to TiO_2 NPs, but as the pH of the surrounding solution decreases, protonation of the medicine occurs. The medication is chemisorbed due to protonation, which changes the surface charge of TiO_2. As a result, the cells are exposed to a high concentration of DNR, causing them to perish. Venkatasubbu et al. also looked at the possibility of using TiO_2 NPs as nanocarriers for targeted medicine delivery (Venkatasubbu et al. 2013). TiO_2 NPs were surface modified with polyethylene glycol (PEG) to improve biocompatibility, and they were then functionalized to folic acid ligands for active targeting of folate receptors found in cancer cells. The TiO_2-PEG-FA composite surface was treated with the drug paclitaxel (PTX). In-vitro studies on cytotoxicity and drug release were conducted. In cytotoxicity experiments, HepG2 cell lines from human liver cancer were employed. According to the drug release study, PTX is initially released in a burst but gradually becomes a continuous release. According to the results of the cytotoxicity assays, both the TiO_2-PEG-FA and TiO_2-PEG-FA-PTX complexes were poisonous to the cells. The TiO_2-PEG-FA-PTX combination has a greater harmful effect on cancer cells. The cancer cells were more toxically affected by the TiO_2-PEG-FA-PTX combination. These researches demonstrated how TiO_2 NPs could be used as drug delivery nanocarriers.

6.4 CONCLUSION

Nanomaterials represent a unique class of small-sized particles with a size range1–100 nm. The present chapter primarily explains the different mechanisms involved in nanoparticle growth from its pristine components. Moreover, the most widely accepted methods for synthesizing various nanoparticles, like physical, mechanical, and chemical methods, are briefly discussed to develop an interest in people from different material and biomedical science areas. It is seen that based on the size, uniformity, and other physicochemical properties of NPs, their applications arc intended in different areas of the biomedical field. Among the diverse range of applications of nanoparticles, the present work sheds light is correlating the different types of nanoparticles used in biosensing and drug delivery due to their vast industrial applications. This chapter will develop interest among the researchers to work with nanobiotechnology, wherein formulation and applications can be well correlated.

ABBREVIATIONS

1 Nanoparticles NPs
2 Chemical vapor deposition CVD
3 Pulsed laser deposition PLD
4 Physical vapor deposition PVD
5 Nanometer nm
6 Centimeter Cm
7 Carbon nanotubes CNT
8 Few-layer graphene FLG
9 Single-walled carbon nanoribbons SWCNHs
10 Gold NPs AuNPs
11 Trisodium citrate TSC
12 Cetyltrimethylammonium bromide CTAB
13 Phosphate-buffered saline PBS
14 Pounds per square inch psi
15 Quantum dots QDs
16 Surface/volume S/V
17 Metal NPs MNPs

18 Iron oxide NPs FeNPs
19 Magnetic resonance imaging MRI
20 Silver NPs AgNPs
21 Zinc oxide NPs ZnNPs
22 Point-of-care POC
23 Surface plasmon resonance SPR
24 femtomolar fM
25 Genetically modified organisms GMOs
26 single-stranded DNA ssDNA
27 Polymerase chain reaction PCR
28 Förster Resonance Energy Transfer FRET
29 Bioluminescence Resonance Energy Transfer BRET
30 Chemiluminescence Resonance Energy Transfer CRET
31 Metal oxide NPs MONPs
32 Palladium NPs PdNPs
33 Molecularly imprinted polymers MIP
34 Bisphenol A BPA
35 Glucose oxidase GOx
36 Cholesterol oxidase ChOx
37 Rabbit-immunoglobulin antibodies r-IgGs
38 Bovine serum albumin BSA
39 Improved permeation and retention IPR
40 multidrug resistance MDR
41 Michigan Cancer Foundation-7 MCF-7
42 Doxorubicin DOX
43 Daunorubicin DNR
44 Paclitaxel PTX
45 Polyethylene glycol PEG

REFERENCES

Abd Elkodous, M., G. S. El-Sayyad, I. Y. Abdelrahman, H. S. El-Bastawisy, A. E. Mohamed, F. M. Mosallam, H. A. Nasser, M. Gobara, A. Baraka, M. A. Elsayed, and A. I. El-Batal. 2019. Therapeutic and diagnostic potential of nanomaterials for enhanced biomedical applications. *Colloids and Surfaces B: Biointerfaces* 180: 411–428.

Abdel-Aziz, M. S., M. S. Shaheen, A. A. El-Nekeety, and M. A. Abdel-Wahhab. 2014. Antioxidant and antibacterial activity of silver nanoparticles biosynthesized using chenopodium murale leaf extract. *Journal of Saudi Chemical Society*, 18 (4), 356–363.

Abdulateef, S. A., M. Z. MatJafri, A. F. Omar, N. M. Ahmed, S. A. Azzez, I. M. Ibrahim, and B. E. Al-Jumaili. 2016. Preparation of CuO nanoparticles by laser ablation in liquid. *AIP Conference Proceedings*.

Al-Azawi, M. A., and N. Bidin. 2015. Gold nanoparticles synthesized by laser ablation in deionized water: Influence of liquid layer thickness and defragmentation on the characteristics of gold nanoparticles. *Chinese Journal of Physics* 53 (4): 201–209.

Al-Dahash, G., W. M. Khilkala, and S. N. Abd Alwahid. 2018. Preparation and characterization of ZnO nanoparticles by laser ablation in NaOH aqueous solution. *Iranian Journal of Chemistry and Chemical Engineering (IJCCE)* 37 (1): 11–16.

Algar, W. R., A. J. Tavares, and U. J. Krull. 2010. Beyond labels: A review of the application of quantum dots as integrated components of assays, bioprobes, and biosensors utilizing optical transduction. *Analytica Chimica Acta* 673 (1): 1–25.

Amendola, V. and M. Meneghetti. 2009. Laser ablation synthesis in solution and size manipulation of noble metal nanoparticles. *Physical Chemistry Chemical Physics* 11 (20): 3805.

Basnet, P. and S. Chatterjee. 2020. Structure-directing property and growth mechanism induced by capping agents in nanostructured ZnO during hydrothermal synthesis—A systematic review. *Nano-Structures & Nano-Objects* 22: 100426.

Beig, B., M. B. Niazi, F. Sher, Z. Jahan, U. S. Malik, M. D. Khan, J. H. Américo-Pinheiro, and D.-V.N. Vo. 2022. Nanotechnology-based controlled release of sustainable fertilizers. A Review. *Environmental Chemistry Letters* 20 (4): 2709–2726.

Benyettou, F., R. Rezgui, F. Ravaux, T. Jaber, K. Blumer, M. Jouiad, L. Motte, J.-C. Olsen, C. Platas-Iglesias, M. Magzoub, and A. Trabolsi. 2015. Synthesis of silver nanoparticles for the dual delivery of doxorubicin and alendronate to cancer cells. *Journal of Materials Chemistry B* 3 (36): 7237–7245.

Berry, C. C. and A. S. Curtis. 2003. Functionalisation of magnetic nanoparticles for applications in biomedicine. *Journal of Physics D: Applied Physics* 36 (13).

Biju, V., T. Itoh, and M. Ishikawa. 2010. Delivering quantum dots to cells: Bioconjugated quantum dots for targeted and nonspecific extracellular and intracellular imaging. *Chemical Society Reviews* 39 (8): 3031.

Bokov, D., A. Turki Jalil, S. Chupradit, W. Suksatan, M. Javed Ansari, I. H. Shewael, G. H. Valiev, and E. Kianfar. 2021. Nanomaterial by sol-gel method: Synthesis and application. *Advances in Materials Science and Engineering* 2021: 1–21.

Boutinguiza, M., J. del Val, A. Riveiro, F. Lusquiños, F. Quintero, R. Comesaña, and J. Pou. 2013. Synthesis of titanium oxide nanoparticles by ytterbium fiber laser ablation. *Physics Procedia* 41: 787–793.

Chang, C.-C., H.-L. Wu, C.-H. Kuo, and M. H. Huang. 2008. Hydrothermal synthesis of monodispersed octahedral gold nanocrystals with five different size ranges and their self-assembled structures. *Chemistry of Materials* 20 (24): 7570–7574.

Chen, B. and H. Zhang. 2012. Daunorubicin-Tio_2 nanocomposites as A 'smart' PH-responsive drug delivery system. *International Journal of Nanomedicine* 235.

Chen, G., I. Roy, C. Yang, and P. N. Prasad. 2016. Nanochemistry and nanomedicine for nanoparticle-based diagnostics and therapy. *Chemical Reviews* 116 (5): 2826–2885.

Chiang, P.-C., D.-S. Hung, J.-W. Wang, C.-S. Ho, and Y.-D. Yao. 2007. Engineering water-dispersible FePt nanoparticles for biomedical applications. *IEEE Transactions on Magnetics* 43 (6): 2445–2447.

Chrisey, D. B., and G. K. Hybler. 1994. Pulsed Laser Deposition of Thin Films. *John Willey & Sons.*

Dai, Z., X. Xu, and H. Ju. 2004. Direct electrochemistry and electrocatalysis of myoglobin immobilized on a hexagonal mesoporous silica matrix. *Analytical Biochemistry* 332 (1): 23–31.

De Crozals, G., R. Bonnet, C. Farre, and C. Chaix. 2016. Nanoparticles with multiple properties for biomedical applications: A strategic guide. *Nano Today* 11 (4): 435–463.

De Yoreo, J. J. P. U. Gilbert, N. A. Sommerdijk, R. L. Penn, S. Whitelam, D. Joester, H. Zhang, J. D. Rimer, A. Navrotsky, J. F. Banfield, A. F. Wallace, F. M. Michel, F. C. Meldrum, H. Cölfen, and P. M. Dove, 2015. Crystallization by particle attachment in synthetic, biogenic, and geologic environments. *Science* 349 (6247).

Descalzo, A. B., M. D. Marcos, R. Martínez-Máñez, J. Soto, D. Beltrán, and P. Amorós. 2005. Anthrylmethylamine functionalised mesoporous silica-based materials as hybrid fluorescent chemosensors for ATP. *Journal of Materials Chemistry* 15 (27–28): 2721.

Du, Y., W. Ren, Y. Li, Q. Zhang, L. Zeng, C. Chi, A. Wu, and J. Tian. 2015. The enhanced chemotherapeutic effects of doxorubicin loaded peg coated TiO_2 nanocarriers in an orthotopic breast tumor bearing mouse model. *Journal of Materials Chemistry B* 3 (8): 1518–1528.

Dörner, L., C. Cancellieri, B. Rheingans, M. Walter, R. Kägi, P. Schmutz, M. V. Kovalenko, and L. P. Jeurgens, 2019. Cost-effective sol-gel synthesis of porous CuO nanoparticle aggregates with tunable specific surface area. *Scientific Reports*, 9 (1).

Edison, T. J. and M. G. Sethuraman. 2012. Instant green synthesis of silver nanoparticles using Terminalia Chebula Fruit Extract and evaluation of their catalytic activity on reduction of methylene blue. *Process Biochemistry* 47 (9): 1351–1357.

Elahi, N., M. Kamali, and M. H. Baghersad. 2018. Recent biomedical applications of Gold Nanoparticles: A Review. *Talanta* 184: 537–556.

Elghanian, R., J. J. Storhoff, R. C. Mucic, R. L. Letsinger, and C. A. Mirkin. 1997. Selective colorimetric detection of polynucleotides based on the distance-dependent optical properties of gold nanoparticles. *Science* 277 (5329): 1078–1081.

Elumalai, K., S. Velmurugan, S. Ravi, V. Kathiravan, and S. Ashokkumar. 2015. Retracted: Green synthesis of zinc oxide nanoparticles using moringa oleifera leaf extract and evaluation of its antimicrobial activity. *Spectrochimica Acta Part A: Molecular and Biomolecular Spectroscopy* 143: 158–164.

Fakruddin, M., Z. Hossain, and H. Afroz. 2012. Prospects and applications of nanobiotechnology: A medical perspective. *Journal of Nanobiotechnology* 10 (1): 31.

Fei Yin, Z., L. Wu, H. Gui Yang, and Y. Hua Su. 2013. Recent progress in biomedical applications of titanium dioxide. *Physical Chemistry Chemical Physics* 15 (14): 4844.

Franco-Molina, M. A., E. Mendoza-Gamboa, C. A. Sierra-Rivera, R. A. Gómez-Flores, P. Zapata-Benavides, P. Castillo-Tello, J. M. Alcocer-González, D. F. Miranda-Hernández, R. S. Tamez-Guerra, and C. Rodríguez-Padilla. 2010. Antitumor activity of colloidal silver on MCF-7 Human Breast Cancer Cells. *Journal of Experimental & Clinical Cancer Research* 29 (1).

Freeman, R., J. Girsh, and I. Willner. 2013. Nucleic acid/quantum dots (QDs) hybrid systems for optical and photoelectrochemical sensing. *ACS Applied Materials & Interfaces* 5 (8): 2815–2834.

Fuchigami, T., R. Kawamura, Y. Kitamoto, M. Nakagawa, and Y. Namiki. 2012. A magnetically guided anti-cancer drug delivery system using porous FePt capsules. *Biomaterials* 33 (5): 1682–1687.

Ghorbani, A., R. Rashidi, and R. Shafiee-Nick 2019. Flavonoids for preserving pancreatic beta cell survival and function: A mechanistic review. *Biomedicine & Pharmacotherapy* 111: 947–957.

Ghosh, S., R. Ahmad, Md. Zeyaullah, and S. K. Khare. 2021. Microbial nano-factories: Synthesis and biomedical applications. *Frontiers in Chemistry* 9: 626834.

Giner-Casares, J. J., M. Henriksen-Lacey, M. Coronado-Puchau, and L. M. Liz-Marzán. 2016. Inorganic nanoparticles for biomedicine: Where materials scientists meet medical research. *Materials Today* 19 (1): 19–28.

Gondal, M. A., Q. A. Drmosh, Z. H. Yamani, and T. A. Saleh. 2009. Synthesis of ZnO_2 nanoparticles by laser ablation in liquid and their annealing transformation into ZnO nanoparticles. *Applied Surface Science* 256 (1): 298–304.

Gondal, M. A., T. F. Qahtan, M. A. Dastageer, Y. W. Maganda, and D. H. Anjum. 2013. Synthesis of Cu/Cu_2O nanoparticles by laser ablation in deionized water and their annealing transformation into CuO nanoparticles. *Journal of Nanoscience and Nanotechnology* 13 (8): 5759–5766.

Gorrasi, G. and A. Sorrentino. 2015. Mechanical milling as a technology to produce structural and functional bio-nanocomposites. *Green Chemistry* 17 (5): 2610–2625.

Hajiesmaeilbaigi, F., A. Mohammadalipour, J. Sabbaghzadeh, S. Hoseinkhani, and H. R. Fallah. 2005. Preparation of silver nanoparticles by laser ablation and fragmentation in pure water. *Laser Physics Letters* 3 (5): 252–256.

He, X. and C. Hu. 2011. Building three-dimensional PT catalysts on Tio_2 nanorod arrays for effective ethanol electrooxidation. *Journal of Power Sources* 196 (6): 3119–3123.

Higuita, L. P. and A. F. Vargas. 2017. Effect of addition of calcium ions and hydrothermal treatment on the morphology of calcium phosphates. *Materials Letters* 190: 146–149.

Holzinger, M., A. Le Goff, and S. Cosnier. 2014. Nanomaterials for biosensing applications: A Review. *Frontiers in Chemistry* 2: 63.

Hu, L., C.-C. Fong, X. Zhang, L. L. Chan, P. K. Lam, P. K. Chu, K.-Y. Wong, and M. Yang. 2016. Au nanoparticles decorated TiO_2 nanotube arrays as a recyclable sensor for photoenhanced electrochemical detection of Bisphenol A. *Environmental Science & Technology* 50 (8): 4430–4438.

Huang, J., X. Zhang, S. Liu, Q. Lin, X. He, X. Xing, W. Lian, and D. Tang. 2011. Development of molecularly imprinted electrochemical sensor with titanium oxide and gold nanomaterials enhanced technique for determination of 4-nonylphenol. *Sensors and Actuators B: Chemical* 152 (2): 292–298.

Huang, L., X. Weng, Z. Chen, M. Megharaj, and R. Naidu. 2014. Green synthesis of iron nanoparticles by various tea extracts: Comparative study of the reactivity. *Spectrochimica Acta Part A: Molecular and Biomolecular Spectroscopy* 130: 295–301.

Hussain, I., N. B. Singh, A. Singh, H. Singh, and S. C. Singh. 2015. Green synthesis of nanoparticles and its potential application. *Biotechnology Letters* 38 (4): 545–560.

Idris, N. M., S. S. Lucky, Z. Li, K. Huang, and Y. Zhang. 2014. Photoactivation of core–Shell Titania coated upconversion nanoparticles and their effect on cell death. *Journal of Materials Chemistry B* 2 (40): 7017–7026.

Jaiswal, J. K., H. Mattoussi, J. M. Mauro, and S. M. Simon. 2002. Long-term multiple color imaging of live cells using Quantum Dot bioconjugates. *Nature Biotechnology* 21 (1): 47–51.

Jin, R., Y. Cao, C. A. Mirkin, K. L. Kelly, G. C. Schatz, and J. G. Zheng. 2001. Photoinduced conversion of silver nanospheres to Nanoprisms. *Science* 294 (5548): 1901–1903.

Jo, S. D., G.-H. Nam, G. Kwak, Y. Yang, and I. C. Kwon 2017. Harnessing designed nanoparticles: Current strategies and future perspectives in cancer immunotherapy. *Nano Today* 17: 23–37.

Joseph, M. C., C. Tsotsos, M. A. Baker, P. J. Kench, C. Rebholz, A. Matthews, and A. Leyland. 2005. Characterisation and tribological evaluation of nitrogen-containing molybdenum–copper PVD metallic nanocomposite films. *Surface and Coatings Technology* 190 (2–3): 345–356.

Kafi, A. K. M., G. Wu, P. Benvenuto, and A. Chen. 2011. Highly sensitive amperometric H_2O_2 biosensor based on hemoglobin modified TiO_2 nanotubes. *Journal of Electroanalytical Chemistry* 662 (1): 64–69.

Katz, E., I. Willner, and J. Wang. 2004a. Electroanalytical and bioelectroanalytical systems based on metal and semiconductor nanoparticles. *Electroanalysis* 16 (12): 19–44.

Kaushik, A., P. R. Solanki, A. A. Ansari, S. Ahmad, and B. D. Malhotra. 2009. A nanostructured cerium oxide film-based immunosensor for Mycotoxin Detection. *Nanotechnology* 20 (5): 055105.

Khalil, M. M. H., E. H. Ismail, K. Z. El-Baghdady, and D. Mohamed. 2014. Green synthesis of silver nanoparticles using olive leaf extract and its antibacterial activity. *Arabian Journal of Chemistry* 7 (6): 1131–1139.

Kharissova, O. V., H. V. R. Dias, B. I. Kharisov, B. O. Pérez, and V. M. Pérez. 2013. The greener synthesis of nanoparticles. *Trends in Biotechnology* 31 (4: 240–248.

Khumaeni, A., W. S. Budi, and H. Sutanto. 2017. Synthesis and characterization of high-purity gold nanoparticles by laser ablation method using low-energy Nd:YAG laser 1064 nm. *Journal of Physics: Conference Series* 909: 012037.

Kulkarni, S. K.. 2014. Synthesis of nanomaterials—I (physical methods). *Nanotechnology: Principles and Practices*: 55–76.

Kumar, P. S., J. Sundaramurthy, S. Sundarrajan, V. J. Babu, G. Singh, S. I. Allakhverdiev, and S. Ramakrishna. 2014. Hierarchical electrospun nanofibers for energy harvesting, production and environmental remediation. *Energy & Environmental Science* 7 (10): 3192–3222.

Kunjachan, S. and S. Jose. 2010. Understanding the mechanism of ionic gelation for synthesis of chitosan nanoparticles using qualitative techniques. *Asian Journal of Pharmaceutics* 4 (2): 148.

Kuo, C.-W., J.-Y. Shiu, Y.-H. Cho, and P. Chen. 2003. Fabrication of large-area periodic Nanopillar arrays for nanoimprint lithography using polymer colloid masks. *Advanced Materials* 15 (13): 1065–1068.

Lai, S.-M., T.-Y. Tsai, C.-Y. Hsu, J.-L. Tsai, M.-Y. Liao, and P.-S. Lai. 2012. Bifunctional silica-coated superparamagnetic FePt nanoparticles for fluorescence/mr dual imaging. *Journal of Nanomaterials* 2012: 1–7.

Li, H. and L. Rothberg. 2004. Colorimetric detection of DNA sequences based on electrostatic interactions with unmodified gold nanoparticles. *Proceedings of the National Academy of Sciences* 101 (39): 14036–14039.

Li, H. and L. J. Rothberg. 2004b. DNA sequence detection using selective fluorescence quenching of tagged oligonucleotide probes by gold nanoparticles. *Analytical Chemistry* 76 (18): 5414–5417.

Li, M., Z. Luo, Z. Peng, and K. Cai. 2019. Cascade-amplification of therapeutic efficacy: An emerging opportunity in cancer treatment. *WIREs Nanomedicine and Nanobiotechnology* 11 (5).

Li, S., G. Ye, and G. Chen. 2009. Low-temperature preparation and characterization of nanocrystalline anatase TiO_2. *The Journal of Physical Chemistry C* 113 (10): 4031–4037.

Liang, F., T. Shimizu, M. Tanaka, S. Choi, and T. Watanabe. 2012. Selective preparation of polyhedral graphite particles and multi-wall carbon nanotubes by a transferred arc under atmospheric pressure. *Diamond and Related Materials* 30: 70–76.

Liao, J., S. Lin, Y. Yang, K. Liu, and W. Du. 2015. Highly selective and sensitive glucose sensors based on organic electrochemical transistors using TiO_2 nanotube arrays-based gate electrodes. *Sensors and Actuators B: Chemical* 208: 457–463.

Lieber, C. M. and C.-C. Chen. 1994. Preparation of Fullerenes and Fullerene-based materials. *Solid State Physics*: 109–148.

Liu, B. and H. C. Zeng. 2003. Hydrothermal synthesis of ZnO nanorods in the diameter regime of 50 nm. *Journal of the American Chemical Society* 125 (15): 4430–4431.

Liu, J. and Y. Lu. 2006. Fast colorimetric sensing of adenosine and cocaine based on a general sensor design involving aptamers and nanoparticles. *Angewandte Chemie International Edition* 45 (1): 90–94.

Long, N. V., Y. Yang, T. Teranishi, C. M. Thi, Y. Cao, and M. Nogami. 2015. Biomedical applications of advanced multifunctional magnetic nanoparticles. *Journal of Nanoscience and Nanotechnology* 15 (12): 10091–10107.

Maciulevičius, M., A. Vinčiūnas, M. Brikas, A. Butsen, N. Tarasenka, N. Tarasenko, and G. Račiukaitis. 2013. Online characterization of gold nanoparticles generated by laser ablation in liquids. *Physics Procedia* 41: 531–538.

Mariotti, E., M. Minunni, and M. Mascini. 2002. Surface plasmon resonance biosensor for genetically modified organisms detection. *Analytica Chimica Acta* 453 (2): 165–172.

Matsui, J., K. Akamatsu, N. Hara, D. Miyoshi, H. Nawafune, K. Tamaki, and N. Sugimoto. 2005. SPR sensor chip for detection of small molecules using molecularly imprinted polymer with embedded gold nanoparticles. *Analytical Chemistry* 77 (13): 4282–4285.

Matsumoto, R., S. Adachi, E. H. Sadki, S. Yamamoto, H. Tanaka, H. Takeya, and Y. Takano, 2020. Maskless patterning of gallium-irradiated superconducting silicon using focused ion beam. *ACS Applied Electronic Materials* 2 (3): 677–682.

Millstone, J. E., S. J. Hurst, G. S. Métraux, J. I. Cutler, and C. A. Mirkin. 2009. Colloidal gold and silver triangular Nanoprisms. *Small* 5(6): 646–664.

Mintcheva, N., A. Aljulaih, W. Wunderlich, S. Kulinich, and S. Iwamori. 2018. Laser-ablated Zno nanoparticles and their photocatalytic activity toward organic pollutants. *Materials* 11 (7): 1127.

Mitchell, M. J., M. M. Billingsley, R. M. Haley, M. E. Wechsler, N. A. Peppas, and R. Langer. 2020. Engineering precision nanoparticles for drug delivery. *Nature Reviews Drug Discovery* 20 (2): 101–124.

Mittal, A. K., Y. Chisti, and U. C. Banerjee. 2013. Synthesis of metallic nanoparticles using plant extracts. *Biotechnology Advances* 31 (2): 346–356.

Moghimi, N. and K. T. Leung. 2013. FePt alloy nanoparticles for Biosensing: Enhancement of vitamin C sensor performance and selectivity by nanoalloying. *Analytical Chemistry* 85 (12): 5974–5980.

Murphy, C. J.. 2002. Peer reviewed: Optical Sensing with Quantum Dots. *Analytical Chemistry* 74 (19): 520A–526A.

Murray, C. B., D. J. Norris, and M. G. Bawendi. 1993. Synthesis and characterization of nearly monodisperse CdE (E = sulfur, selenium, tellurium) semiconductor nanocrystallites. *Journal of the American Chemical Society* 115 (19): 8706–8715.

Nesakumar, N., S. Sethuraman, U. M. Krishnan, and J. B. Rayappan. 2013. Fabrication of lactate biosensor based on lactate dehydrogenase immobilized on cerium oxide nanoparticles. *Journal of Colloid and Interface Science* 410: 158–164.

Pankhurst, Q. A., J. Connolly, S. K. Jones, and J. Dobson. 2003. Applications of magnetic nanoparticles in biomedicine. *Journal of Physics D: Applied Physics* 36 (13): R167.

Park, J., J. Joo, S. G. Kwon, Y. Jang, and T. Hyeon. 2007. Synthesis of monodisperse spherical nanocrystals. *Angewandte Chemie International Edition* 46 (25): 4630–4660.

Pastoriza-Santos, I. and L. M. Liz-Marzán. 2008. Colloidal silver nanoplates. state of the art and future challenges. *Journal of Materials Chemistry* 18 (15): 1724.

Patil, D., N. Q. Dung, H. Jung, S. Y. Ahn, D. M. Jang, and D. Kim. 2012. Enzymatic glucose biosensor based on CeO_2 nanorods synthesized by non-isothermal precipitation. *Biosensors and Bioelectronics* 31 (1): 176–181.

Phromma, S., T. Wutikhun, P. Kasamechonchung, T. Eksangsri, and C. Sapcharoenkun. 2020. Effect of calcination temperature on photocatalytic activity of synthesized tio2 nanoparticles via wet ball milling sol-gel method. *Applied Sciences* 10 (3): 993.

Pimpin, A. and W. Srituravanich. 2012. Review on micro- and nanolithography techniques and their applications. *Engineering Journal* 16 (1): 37–56.

Putzbach, W. and N. Ronkainen. 2013. Immobilization techniques in the fabrication of nanomaterial-based electrochemical biosensors: A Review. *Sensors* 13 (4): 4811–4840.

Pyatenko, A., K. Shimokawa, M. Yamaguchi, O. Nishimura, and M. Suzuki. 2004. Synthesis of silver nanoparticles by laser ablation in pure water. *Applied Physics A* 79 (4–6): 803–806.

Qin, Y., L. Sun, X. Li, Q. Cao, H. Wang, X. Tang, and L. Ye 2011. Highly water-dispersible TiO_2 nanoparticles for doxorubicin delivery: Effect of loading mode on therapeutic efficacy. *Journal of Materials Chemistry* 21 (44): 18003.

Rao, C. N. R., P. J. Thomas, and G. U. Kulkarni 2007. Nanocrystals. *Synthesis, properties and applications. Springer, Berlin.*

Reimer, P. and T. Balzer. 2003. Ferucarbotran (Resovist): A new clinically approved RES-specific contrast agent for contrast-enhanced MRI of the liver: Properties, clinical development, and applications. *European Radiology* 13 (6): 1266–1276.

Reiss, P., M. Protière, and L. Li. 2009. Core/Shell Semiconductor nanocrystals. *Small* 5 (2): 154–168.

Renu, G., V. V. Rani, S. V. Nair, K. R. Subramanian, and V.-K. Lakshmanan. 2012. Development of cerium oxide nanoparticles and its cytotoxicity in prostate cancer cells. *Advanced Science Letters* 6 (1): 17–25.

Rubilar, O., M. Rai, G. Tortella, M. C. Diez, A. B. Seabra, and N. Durán. 2013. Biogenic nanoparticles: Copper, copper oxides, copper sulphides, complex copper nanostructures and their applications. *Biotechnology Letters* 35 (9): 1365–1375.

Sanpui, P., A. Chattopadhyay, and S. S. Ghosh. 2011. Induction of apoptosis in cancer cells at low silver nanoparticle concentrations using Chitosan Nanocarrier. *ACS Applied Materials & Interfaces* 3 (2): 218–228.

Savithramma, N., M. L. Rao, K. Rukmini, and P. S. Devi. 2011. Antimicrobial activity of silver nanoparticles synthesized by using medicinal plants. *International Journal of ChemTech Research* 3(3): 1394–1402.

Scheu, M., V. Veefkind, Y. Verbandt, E. M. Galan, R. Absalom, and W. Förster 2006. Mapping nanotechnology patents: The EPO approach. *World Patent Information* 28 (3): 204–211.

Seehra, M. S., V. Singh, P. Dutta, S. Neeleshwar, Y. Y. Chen, C. L. Chen, S. W. Chou, and C. C. Chen. 2010. Size-dependent magnetic parameters of FCC FePt nanoparticles: Applications to magnetic hyperthermia. *Journal of Physics D: Applied Physics* 43 (14): 145002.

Seemann, K. M., M. Luysberg, Z. Révay, P. Kudejova, B. Sanz, N. Cassinelli, A. Loidl, K. Ilicic, G. Multhoff, and T. E. Schmid 2015. Magnetic heating properties and neutron activation of Tungsten-oxide coated biocompatible FePt core–shell nanoparticles. *Journal of Controlled Release* 197: 131–137.

Shi, Y., M. Lin, X. Jiang, and S. Liang. 2015. Recent advances in FePt nanoparticles for Biomedicine. *Journal of Nanomaterials* 2015: 1–13.

Singh, A., J. Vihinen, E. Frankberg, L. Hyvärinen, M. Honkanen, and E. Levänen. 2016. Pulsed laser ablation-induced green synthesis of TiO_2 nanoparticles and application of novel small angle X-ray scattering technique for nanoparticle size and size distribution analysis. *Nanoscale Research Letters* 11 (1).

Soleimani, H., M. K. Baig, N. Yahya, L. Khodapanah, M. Sabet, B. M. Demiral, and M. Burda. 2018. Synthesis of ZnO nanoparticles for oil–water interfacial tension reduction in enhanced oil recovery. *Applied Physics A* 124 (2).

Szabó, Z., J. Volk, E. Fülöp, A. Deák, and I. Bársony. 2013. Regular ZnO nanopillar arrays by nanosphere photolithography. *Photonics and Nanostructures – Fundamentals and Applications* 11 (1): 1–7.

Tajdidzadeh, M., B. Z. Azmi, W. M. Yunus, Z. A. Talib, A. R. Sadrolhosseini, K. Karimzadeh, S. A. Gene, and M. Dorraj. 2014. Synthesis of silver nanoparticles dispersed in various aqueous media using laser ablation. *The Scientific World Journal* 2014: 1–7.

Tartaj, P., M. a Morales, S. Veintemillas-Verdaguer, T. Gonz lez-Carre o, and C. J. Serna. 2003. The preparation of magnetic nanoparticles for applications in biomedicine. *Journal of Physics D: Applied Physics* 36 (13).

Thomas, P. J. and P. O'Brien 2007. Recent developments in the synthesis, properties, and assemblies of Nanocrystals. *Nanomaterials Chemistry*: 1–43.

Tian, L., W. Zhao, L. Li, Y. Tong, G. Peng, and Y. Li. 2017. Multi-talented applications for cell imaging, tumor cells recognition, patterning, staining and temperature sensing by using egg white-encapsulated gold nanoclusters. *Sensors and Actuators B: Chemical* 240: 114–124.

Tsuzuki, T. and P. G. McCormick. 2004. Mechanochemical synthesis of nanoparticles. *Journal of Materials Science* 39 (16/17): 5143–5146.

Tu, W., Y. Dong, J. Lei, and H. Ju. 2010. Low-potential photoelectrochemical biosensing using porphyrin-functionalized TiO_2 nanoparticles. *Analytical Chemistry* 82 (20): 8711–8716.

Tulinski, M. and M. Jurczyk 2017. Nanomaterials synthesis methods. *Metrology and Standardization of Nanotechnology*: 75–98.

Vadhan-Raj, S., W. Strauss, D. Ford, K. Bernard, R. Boccia, J. Li, and L. F. Allen. 2013. Efficacy and safety of IV Ferumoxytol for adults with iron deficiency anemia previously unresponsive to or unable to tolerate oral iron. *American Journal of Hematology* 89 (1): 7–12.

Vallabani, N. V. and S. Singh. 2018. Recent advances and future prospects of iron oxide nanoparticles in Biomedicine and Diagnostics. *3 Biotech* 8 (6).

Venkatasubbu, G. D., S. Ramasamy, G. P. Reddy, and J. Kumar. 2013. In vitro and in vivo anticancer activity of surface modified paclitaxel attached hydroxyapatite and titanium dioxide nanoparticles. *Biomedical Microdevices* 15 (4): 711–726.

Wang, F. and S. Hu. 2009. Electrochemical sensors based on metal and semiconductor nanoparticles. *Microchimica Acta* 165 (1–2): 1–22.

Wang, Q., R. Liu, X. Yang, K. Wang, J. Zhu, L. He, and Q. Li. 2016. Surface plasmon resonance biosensor for enzyme-free amplified microrna detection based on gold nanoparticles and DNA Supersandwich. *Sensors and Actuators B: Chemical* 223: 613–620.

Wang, Y., Q. Zhao, N. Han, L. Bai, J. Li, J. Liu, E. Che, L. Hu, Q. Zhang, T. Jiang, and S. Wang. 2015. Mesoporous silica nanoparticles in drug delivery and biomedical applications. *Nanomedicine: Nanotechnology, Biology and Medicine* 11 (2): 313–327.

Wang, Z. L., Y. Liu, and Z. Zhang 2003. Handbook of Nanophase and Nanostructured Materials, *Kluwer Academic/Plenum Publishers, New York, Chapter 1.*

Wei, Y., H. Dong, J. Xu, and Q. Feng. 2002. Simultaneous immobilization of horseradish peroxidase and glucose oxidase in mesoporous sol–gel host materials. *ChemPhysChem* 3 (9): 802–808.

Wender, H., M. L. Andreazza, R. R. Correia, S. R. Teixeira, and J. Dupont. 2011. Synthesis of gold nanoparticles by laser ablation of an au foil inside and outside Ionic liquids. *Nanoscale* 3 (3): 1240.

Wilchek, M. and E. A. Bayer. 1988. The avidin-biotin complex in bioanalytical applications. *Analytical Biochemistry* 171 (1): 1–32.

Wu, Q., h. Sun, x. Chen, d. Chen, m. Dong, x. Fu, Q. Li, x. Liu, t. Qiu, t. Wan, and s. Li. 2012. Influences of surface coatings and components of FePt nanoparticles on the suppression of glioma cell proliferation. *International Journal of Nanomedicine* 3295.

Wu, Q., S. Liang, Q. Zhou, M. Wang, Y. Zhu, and X. Yang. 2015. Water-soluble L-cysteine-coated FePt nanoparticles as dual MRI/CT imaging contrast agent for Glioma. *International Journal of Nanomedicine* 2325.

Wu, Z., S. Yang, and W. Wu 2016. Shape control of inorganic nanoparticles from solution. *Nanoscale* 8 (3): 1237–1259.

Xie, J., L. Gong, S. Zhu, Y. Yong, Z. Gu, and Y. Zhao. 2018. Emerging strategies of nanomaterial-mediated tumor radiosensitization. *Advanced Materials* 31 (3): 1802244.

Xie, L., Z. Li, L. Sun, B. Dong, Q. Fatima, Z. Wang, Z. Yao, and A. A. Haidry. 2019. Sol-gel synthesis of TiO_2 with P-type response to hydrogen gas at elevated temperature. *Frontiers in Materials* 6.

Yang, L., H. Zhao, S. Fan, B. Li, and C.-P. Li. 2014. A highly sensitive electrochemical sensor for simultaneous determination of hydroquinone and bisphenol A based on the ultrafine PD nanoparticle@ TiO_2 functionalized SiC. *Analytica Chimica Acta* 852: 28–36.

Yang, X., Q. Li, H. Wang, J. Huang, L. Lin, W. Wang, D. Sun, Y. Su, J. B. Opiyo, L. Hong, Y. Wang, N. He, and L. Jia. 2009. Green synthesis of palladium nanoparticles using broth of Cinnamomum camphora leaf. *Journal of Nanoparticle Research*, 12 (5), 1589–1598.

Yeadon, M., J. C. Yang, M. Ghaly, D. L. Olynick, R. S. Averbach, and J. M. Gibson 1997. Nanophase and Nanocomposite Materials 11 (eds S. Komarneni, J. C. Parker, H. J. Wollenberger) *MRS, Pittsburgh*, 179–184.

Zeiri, Y., P. Elia, R. Zach, S. Hazan, S. Kolusheva, and Z. Porat. 2014. Green synthesis of gold nanoparticles using plant extracts as reducing agents. *International Journal of Nanomedicine* 4007.

Zhang, D., K. Ye, Y. Yao, F. Liang, T. Qu, W. Ma, B. Yang, Y. Dai, and T. Watanabe. 2019. Controllable synthesis of carbon nanomaterials by direct current arc discharge from the inner wall of the chamber. *Carbon* 142: 278–284.

Zhang, D., K. Ye, Y. Yao, F. Liang, T. Qu, W. Ma, B. Yang, Y. Dai, and T. Watanabe. 2019a. Controllable synthesis of carbon nanomaterials by direct current arc discharge from the inner wall of the chamber. *Carbon* 142: 278–284.

Zheng, Y., C. Jiang, F. Ren, M. Guo, H. Wang, Z. Bao, Y. Tang, and H. Quan. 2015. FEPT Nanoparticles as a Potential X-Ray Activated Chemotherapy Agent for Hela Cells. *International Journal of Nanomedicine* 6435.

Zweck, A., G. Bachmann, W. Luther, and C. Ploetz 2008. Nanotechnology in Germany: From forecasting to Technological Assessment to Sustainability Studies. *Journal of Cleaner Production* 16 (8–9): 977–987.

7 Recent Advances in Multifunctional Biomacromolecules for Cancer Theranostics

Saroj Saroj, Tatini Rakshit, and Suchetan Pal

7.1 INTRODUCTION

Cancer is a dreadful disease that accounts for many untimely deaths yearly. According to the WHO World Cancer Report 2020, it is anticipated that globally 28.4 million new cancer cases will be reported in the next two decades (Sung et al. 2021). In cancer, uncontrolled cancer cell growth gives rise to a tumor that invades healthy organs and then spreads to other areas of the body to produce secondary tumors known as metastases. Cancer metastasis is believed to be the primary cause of cancer-related death. Various treatments are employed to circumvent this deadly disease. The standard treatments include surgery, immunotherapy, radiotherapy, and chemotherapy. Although a massive advancement has been made in cancer therapy, effective novel treatments and early diagnosis are immediately needed to address the abysmal mortality and morbidity rates. Chemotherapy is one of the most common forms of cancer treatment. Despite significant advancements, patients still experience various side effects, including pain, weight loss, and hair loss (Dickens and Ahmed 2018). Moreover, the cytostatic medicines were found to have insufficient biodistribution and pharmacokinetics profiles, suggesting the urgent need for developing effective methods to tackle the high mortality and morbidity rates associated with cancer. On the other hand, a patient's probability of survival is significantly increased by an early cancer diagnosis. However, current diagnostic techniques (biopsies, histology, medical imaging, and biomarker detection) often involve invasive procedures, have low sensitivity, or only identify cancer in advanced stages, even though novel biomarkers are being researched (Roy Chowdhury et al. 2016). Creating novel therapy approaches and rapid, more accurate, and more sensitive diagnostic technologies is urgently required. Fortunately, advancements in nanotechnology have aided in imaging, real-time monitoring, and diagnosis for cancer management. Currently, one of the primary applications of biomedical nanotechnology is to improve the effectiveness of clinically approved chemotherapeutics by making them safer, increasing their bioavailability, and being able to target cancer cells without harming healthy tissues. The selective targeting of cancer tissue is attained by the "enhanced permeability and retention (EPR)" effect (Shi et al. 2020). In this mechanism, nanomedicines can selectively permeate the leaky vasculature of cancer and accumulate due to poor lymphatic drainage out of cancer tissue due to the optimum size (10–200 nm). Further, the nanomaterial surface can be coated with targeting groups (antibody, nanobody, aptamer, and small molecule) that recognize the cancer cells or cancer microenvironment, also termed active targeting (Bazak et al. 2015). Intriguingly, diagnostic nanomedicines enable the early identification and detection of malignant cells, which is essential to enhancing the disease's prognosis (Shi et al. 2017). It is expected that these nanosystems would achieve substantial benefits, including earlier detection, lessening toxicity, and reducing the cost of cancer treatment (Nicolson et al. 2020).

DOI: 10.1201/9781003305583-7

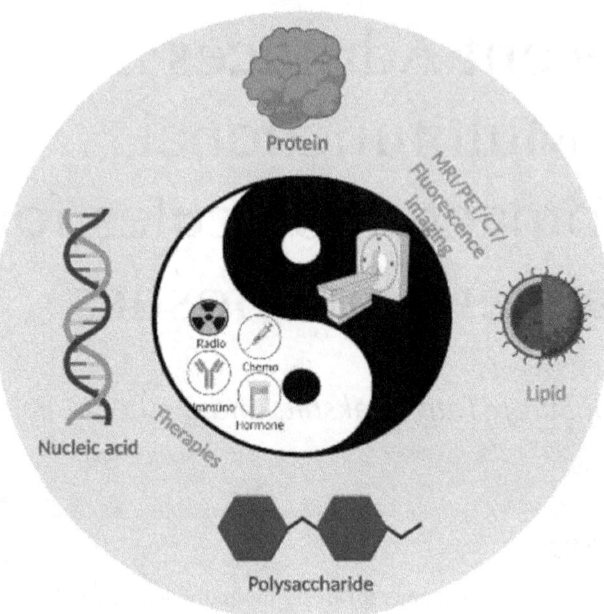

FIGURE 7.1 Biopolymer-based cancer nanotheranostics.

There has been a significant advancement in nanomaterials in oncology medicine, with some nano-based treatments being available on the market and others in various phases of clinical and preclinical research. For example, Myocet, which is a non-PEGylated liposomal doxorubicin nanomedicine carrier used for the therapy of metastatic breast cancer, is approved in Europe and Canada, in combination with cyclophosphamide by the company Sopherion Therapeutics, LLC in North America and Cephalon, Inc. in Europe. Apart from Myocet, several other nanomedicines, such as Doxil, Thermodox, Feridox, and Rexin-G were approved by various regulatory authorities worldwide, and some of them are also being used in clinical trials (Wang et al. 2013). In recent years, scientists have developed nanomedicine-related methods that have drawn the attention of many of the scientific community. It can overcome several current restrictions and increase the capacity of current cancer treatment methods. It involves using nanocarriers based on protein, nucleic acids, and other biological moieties as nano pharmaceutical carriers. This book chapter focuses on the recent developments in cancer-related nanotheranostics based on nanoparticles derived from biomacromolecules, such as proteins, nucleic acids, lipids, and polysaccharides (Figure 7.1).

7.2 STATUS OF CANCER NANOTHERANOSTICS

Traditional cancer management uses personalized medicine to deal with cancer according to the patient's need and to minimize the side effects, resulting in the emergence of the term theranostics. The term "theranostics," coined in 1998 by Funkhouser, refers to a combination of therapy and diagnosis. The novel biosensors have made it possible to diagnose and deliver targeted therapy simultaneously with less side effects, even for recalcitrant pathological illnesses like cancer. A theranostic agent's attributes must enable biological probing and ensure stability during delivery under physiological conditions. The agent must pass through biological barriers on its path, such as epithelial and blood-brain barriers, to the target tissue or organ without producing any detectable adverse side effects (Idée et al. 2013). Additionally, due to decreased lymphatic outflow, nano-sized medicines are maintained in the tumor site and seep preferentially into tumor tissue through permeable capillaries. The enhanced permeability and retention (EPR) effect is the term given to this phenomenon (Nakamura et al. 2016). Although it is widely believed that the EPR effect will improve the delivery of nano-drugs to tumors, it only increases nanodrug delivery by less than a factor of two when compared to important healthy organs, leading to drug concentrations that might be insufficient to treat the majority of cancers. A theranostic agent should ideally include a therapeutic drug (such as nucleic acid, which includes siRNA or mRNA, therapeutic proteins, or chemotherapeutic agent), a carrier of the therapeutic payload, targeting ligands. Further, signal emitters (with characteristic optical, radioactive, or magnetic properties) are conjugated covalently or noncovalently to the delivery platform. Generally, lipid-based carriers, such as liposomes and exosomes, are often used due to their versatility and biocompatibility. A plethora of studies reported that involve the synthetic nanocarrier for cancer theranostics, such as Wang et al., which uses quantum dot-containing micellar formulation, which can decrease the tumor regression by releasing the drug aminoflavone after the intravenous administration by targeting the epidermal growth factor receptor (EGFR)-positive triple-negative breast cancer (TNBC) mouse models (Wang et al. 2017). In another study, Parhi et al. functionalized the lipid-based nanoparticle containing rapamycin (anticancer drug) and quantum dot (bioimaging agent) along with the drug trastuzumab to target the HER+ breast cancer cells (Parhi and Sahoo 2015). Apart from the micelles, theranostics agents are also formulated with various polymers, metal-based, and metal oxide-based nanoparticles for the concurrent imaging and treatment of different types of cancer. Various other types of cells are also present in the cancer microenvironment, which allows the strategies for effectively distributing and penetrating the nanoparticle within the tumor cells. For example, Liu et al. created a gold nanocluster loaded with indocyanine green for tumor fluorescence imaging, the chemotherapeutic medication paclitaxel, and a CD44-targeted tumor-specific hyaluronidase degradable hyaluronic acid (Liu et al. 2018). Subcutaneous injection of these nanoparticles shows the size-reducible properties of these nanoparticles due to the hyaluronidases. Another promising strategy in theranostics includes the use of injectable thermoresponsive hydrogel for local cancer therapy. These examples suggest the advancement of cancer theranostics in biomedicine in the last decades.

7.3 VARIOUS BIONANOCARRIERS FOR CANCER THERANOSTICS

Many cancer theranostics tools have been developed, which use the therapy and the diagnosis in a single platform to provide personalized therapy for cancer management. For cancer management, bypassing the biological barrier can increase the amount of the therapy/drug at target cancer cells. Various types of moieties/nanomaterials are produced naturally, which induces the targeted biodistribution of the nanotheranostics agent, including lipid, protein, nucleic acid, and polysaccharide-based nanotheranostics. Moreover, these nanomaterials are easy to functionalize, have high drug-loading capacity, good biocompatibility, and the capacity to overcome renal excretion. The following sections discuss the different types of nanotheranostics based on biomolecules used as an application for anticancer therapy.

7.3.1 Protein-Based Natural Cancer Nanotheranostics

Proteins derived from natural sources, such as albumin, gliadin, and gelatin, are used for nanotheranostics purposes alone or in combination with biodegradable polymers. The attention on protein-based nanotheranostics compounds has increased recently. Therapeutic and diagnostic substances have been delivered inside engineered protein nanocages due to their proteinaceous structure, and modifications can be made to both their internal and external surfaces. Modifications to the internal surface allow for the loading of molecules (drugs, aptamers, and contrast agents), and the ligand's conjugations can modify external surfaces. Various protein-based nanocarriers are widely used. For example, E2 nanocages multicomplex present in the bacteria *Bacillus stearothermophilus* act as the theranostics agent carrying the supple scaffold, and it is found to be highly thermostable (up to 85 °C). Another example includes ferritin, which enhances the specificity of targeted delivery probability in cancer cells (Elzoghby et al. 2012).

Apoferritin is a self-assembled 24 polypeptide subunit with a hollow protein nanocage containing internal and external dimensions of 8 nm and 12 nm, respectively. The apoferritin goes through a process of disassembly and assembly depending on pH when the iron core in the holo-form is removed. This procedure is widely used to create different nanoparticles for lung cancer theranostics. Li et al. used fluorescence and magnetic resonance imaging to demonstrate the diagnosis of lung cancer in the A549 cell line using apoferritin, a multifunctional ferritin-based nanostructure (Li et al. 2012). Luo et al. utilized apoferritin nanocages conjugated to hyaluronic acid (HA) to control the release of the intracellular prodrug daunomycin in a pH-responsive manner (Luo et al. 2015).

Albumin has emerged as a useful macromolecular carrier for medicinal and diagnostic compounds over the last few decades. It has been demonstrated that albumin is safe, biocompatible, degradable, non-immunogenic, and metabolizable *in vivo*. Commercially, albumins can be made from human serum, bovine serum, egg white, and human serum albumin (HSA). An analysis of albumin-based nanoparticles as medication delivery systems has already been done in various studies. Various methods, such as emulsification, thermal gelation, nano spray drying, desolvation, nab (nanoparticle albumin-bound) technology, and self-assembly, are widely utilized to create albumin nanoparticles. Albumins are widely used in theranostics purposes as they have high binding affinity due to the presence of binding sites in it, which allows the electrostatic absorption of positive and negative charge molecules. The binding of albumin to albumin-binding proteins, such as membrane-associated gp60 and secreted protein, acidic and rich in cysteine (SPARC) and receptor present in the blood vessels of tumors promote the internalization of albumin in cancer tissue. Abraxane® (paclitaxel-albumin nanoparticles prepared by nab-technology) was successfully prepared for the treatment of metastatic breast cancer. HSA nanoparticles allow the delivery of a diversity of drugs across the blood-brain barrier into the brain (Kariduraganavar et al. 2019). Recently, a lactalbumin-based theranostic agent with emissive gold nanoclusters has been developed for breast cancer theranostics (Yang et al. 2020). Along with these, protein-based nanocarriers also include the virus-like particle (VLP). VLPs, or virus-like nanoparticles, are naturally occurring polymer-based nanomaterials that lack viral genomes but imitate viral structures through the hierarchical organization of viral coat proteins. VLPs have drawn significant attention in a variety of nanotechnology-based medical diagnostics and treatments, including theranostics, imaging, and cancer therapy. VLPs have a homogeneous structure and controlled assembly, and they are biocompatible and biodegradable. They can be chemically or genetically modified and include a wide variety of drugs and diagnostic substances. Due to these characteristics, advanced multi-functional theranostic platforms have been developed using VLPs (Kim et al. 2023). A selection of protein-based nanocarriers is summarized in Table 7.1.

7.3.2 Nucleic Acid-Based Nanotheranostics

The conventional concept of nucleic acids, such as deoxyribonucleic acid (DNA) and ribonucleic acid (RNA), as inheritable molecules and genetic components in biological systems. With the rapid development of nanotechnology, nucleic acids have received significant attention and have been extensively investigated as generic materials. Functional nanomaterials based on nucleic acids have been extensively exploited as new drug delivery nanocarriers for cancer therapies. Various nucleic acid-based nanocarriers have been used for cancer treatment by loading Doxorubicin (DOX) as a model drug to DNA origami, DNA tetrahedron, DNA dendrimers, DNA nanoflower, etc. The DNA tetrahedron was extensively researched for theranostics and biosensors as a desirable DNA nanostructure (Yuan et al. 2019). Recently, the DNA tetrahedron was shown to be a viable platform for drug administration due to its high efficiency in cellular absorption, stability to

TABLE 7.1
Examples of Protein-Based Nanocarriers Used in Cancer Theranostics

S.N.	Protein	formulations	Application	Key outcomes	Ref.
1.	HSA	HAS-ICG nanoparticles	Fluorescence and photoacoustic mediated guided dual delivery	Tumor and normal tissue margin were clearly seen via imaging and tumor was completely eradicated	(Sheng et al. 2014)
2.	HSA	HAS-FePC nanoparticles	Photoacoustic imaging-guided Photothermal therapy	After intravenous administration, the NPs demonstrated effective cancer therapy, no weight loss, and minimal long-term harm.	(Jia et al. 2017)
3.	HSA	HSA-coated superparamagnetic iron oxide nanoparticles	MRI and thermoacoustic imaging-guided thermoacoustic therapy	Nanoparticles offer complementary information about 4T1 tumors based on MRI and thermoacoustic imaging. For deep tumor models, the nanoparticle-mediated thermoacoustic treatment demonstrates excellent anti-tumor activity.	(Wen et al. 2017)
4.	BSA	Fe_3O_4-BSA@DOX-PEG	Combination of MRI diagnostics and chemotherapy	The theranostic drug demonstrated comparable cytotoxicity to DOX alone against HEK293 and C6 cells, as well as superparamagnetic and high T2-relaxivity values.	(Semkina et al. 2015)
5.	BSA	Gemcitabine-loaded magnetic BSA nanospheres modified with the drug cetuximab	Pancreatic cancer cell targeting, MRI, and double-targeted thermochemotherapy	The theranostic drug effectively inhibits or kills AsPC-1 cells when combined with antibodies and magnetic targeting.	(Wang et al. 2015)
6.	BSA	Au-BSA-DOX-FA	pH-dependent theranostics agent for CT imaging and targeting therapy	anti-tumor activity shown by the theranostic agent on the MGC-803 tumor and no side effects on normal cell.	(Huang et al. 2017)
7.	Ferritin	CuS-ferritin	PET and photoacoustic imaging-guided Photothermal therapy	After intravenous administration, the theranostic agent showed admirable U87MG tumor imaging ability, significant U87MG tumor inhibition as well as low side effects	(Wang et al. 2016)
8.	Gelatin	DOX- gelatin-EGCG Au nanoparticles	Enzyme-dependent theranostics agent for real-time monitoring and chemotherapy	Shows the significant growth inhibition on PC-3 cells and works as a theranostics agent.	(Tsai et al. 2015)
9.	Transferrin	Docetaxel- and quantum dots-loaded TPGS-Tf	Brain-targeted imaging and application in chemotherapy	The agent shows the crossing of BBB and fluorescence in rat brain	(Sonali et al. 2016)
10.	Silk	DOX-loaded sericin/dextran composite hydrogel	Drug monitoring and therapy	After subcutaneous injection, hydrogel's photoluminescence in C57BL/6 mice is long-term steady and unquenched. The B16-F10 tumor was efficiently inhibited after the hydrogel was injected nearby, and no cytotoxicity was noticed.	(Liu et al. 2016)

HSA: Human serum albumin, ICG: Indocyanine green, FePc: Iron (II) phthalocyanine, EGCG: epigallocatechin gallate, TPGS: D-alpha-tocopheryl PEG 1000 succinate.

resist nuclease degradation, and great biocompatibility. Ahn and colleagues used the DNA tetrahedron (with 26 DOX molecules in each tetrahedron molecule) as a platform for the delivery of anticancer drugs to circumvent the treatment of drug-resistant cancer cells (Kim et al. 2013). Apart from this, various viral nanoparticles obtained from the virus and bacteriophages are also used for theranostics. An insightful review on the combination of conventional chemotherapies with genetically modified oncolytic viruses (OVs) for the treatment of lung cancer was published by Beljanski et al. (Beljanski and Hiscott 2012). Another study by Shu et al. created multifunctional RNA NPs for the imaging and treatment of triple-negative breast cancer (TNBC) (Shu et al. 2015). DNA dendrimers have also been established as a platform for imaging and therapeutic delivery. The nanostructures were constructed utilizing a base pairing assembly method without using any enzymes using three-armed Y-shaped DNA monomers. The resulting structure acted as a method for cancer imaging and drug delivery by including fluorophores, DNA aptamers, and DOX. The structure was a promising option for applications using DNA-based theranostics due to its high biostability, biocompatibility, selectivity, and drug-loading capability (Yuan et al. 2019). A selection of DNA-based nanocarriers is summarized in Table 7.2.

TABLE 7.2
Examples of Nucleic Acid-Based Nanocarriers Used in Cancer Theranostics

S.N.	Agent	Formulations	Application	Key outcomes	Ref.
1.	DNA nanoflower	Aptamer conjugated FRET nanoflower	*in vivo* imaging, pH sensing and targeted drug delivery	Uniform nanoflower having high fluorescence intensity and excellent photostability was obtained that can be tuned for the drug delivery aspect.	(Hu et al. 2014)
2.	DNA origami	Radiolabelled DNA origami nanostructures (DONs) having different shapes	Treatment and imaging of organ failure against acute kidney injury	rectangular shape DONs have renal-protective properties, with efficacy like the N-acetylcysteine—a clinically used drug	(Jiang et al. 2018)
3.	DNA hydrogel-Dox (cargo)	Quantum Dots	Imaging and drug delivery in cancer cells	Delivery via quantum dot DNA hydrogels increases 9-fold against cancer cells.	(Zhang et al. 2017)
4.	DNA origami	DNA origami functionalized with quantum dots and DOX	Imaging and therapy of breast cancer in vivo.	Suggesting the potential applications of DOX delivery and therapeutic efficacy in cancer tissue in vivo.	(Zhang et al. 2014)
5.	DNA nano switch	Aptamer based DNA nano switch to deliver cisplatin	Cancer theranostics	According to tests using live cells, conjugate offers sensitivity to tumor cell identification and high contrast fluorescence imaging, as well as high-efficiency drug administration into tumor cells via an endocytosis mechanism.	(Wang et al. 2015)

7.3.3 Lipid Based Nanotheranostics Carriers

Micelles and lipid-based carriers, such as liposomes, are frequently used because of their adaptability and biocompatibility. Theranostic micelles were recently created by Gregoriou et al. utilizing Pluronic F127 block copolymer and Vitamin ETPGS. They showed promise as a method of administering phytochemical resveratrol to breast cancer patients in a targeted manner. It is possible to integrate the fluorophore coumarin-6 to give the system imaging capabilities (Gregoriou et al. 2021). Exosomes are also appealing for application in medication delivery and molecular diagnosis because of their characteristic. Exosomes can also be coupled to nanoparticles and used for highly accurate imaging. Exosomes play a significant role in liquid biopsy evaluations, which help identify malignancies. Numerous studies are being conducted to figure out how to use exosomes as effective drug delivery systems and to create new types of diagnostic tools (He et al. 2018). Two new nanostructured lipid carriers (NLC) systems were created by Liang et al. and Li et al. for breast cancer treatment. (Liang et al. 2018) (Li et al. 2018). To stop the proliferation of SK-BR-3 cells, Liang et al. employed an NLC functionalized with a HER2 aptamer and loaded with an ATP aptamer linked to EGCG and protamine sulfate (HER2 receptor overexpressed) linked to this. Li et al. created an NLC that was co-loaded with lapachone and DOX, allowing them to boost the DOX concentration in tumor tissues in comparison to free DOX in mice that were receiving MCF-7/ADR tumor xenografts (Liang et al. 2018). A selection of lipid-based nanocarriers is summarized in Table 7.3.

TABLE 7.3
Examples of Lipid-Based Nanocarriers Used in Cancer Theranostics

S.N.	Exosome	formulations	Application	Key outcomes	Ref.
1.	-Exosome derived from mouse macrophage cells	RGD modified exosome	Targeted angiogenesis therapy as well as imaging	Increased targeting of the blood vessel and angiogenesis imaging both *in vivo* and *in vitro*	(Wang et al. 2017)
2.	Exosome	Exosomes loaded with metal-organic salts	Fluorescent/therapeutic drug delivery in Hela cells	No significant toxicity was observed	(Illes et al. 2017)
3.	Macrophages derived Exosome	AuNR@RGD-Exosome was modified with tumor-specific targeting ligand folic acid	Fluorescent Dox molecules were loaded for the targeted delivery and chemo-photothermal synergistic tumor therapy	Engineered exosomes were found to accumulate at the tumor site with high-efficiency	(Wang et al. 2018)
4.	Blood exosome	Dual-functional exosome-based superparamagnetic nanoparticle cluster	Targeted drug delivery vehicle loaded with DOX	Murine hepatoma subcutaneous cancer cells were used in in vivo investigations to demonstrate that drug-loaded exosome-based vehicle delivery improved cancer targeting in the presence of an external magnetic field and inhibited tumor growth.	(Qi et al. 2016)
5.	Macrophage Exosome	Loading of exosome with the drug paclitaxel	To overcome the MDR in cancer cells	The study concludes that modified exosome exhibits significant delivery to MDR cancer cells	(Kim et al. 2016)

7.3.4 Polysaccharide-Based Nanotheranostics Carriers

The most intriguing delivery mechanisms for polymeric drugs in anticancer therapy are nanocarriers. The majority of naturally occurring polysaccharides with complex structures made up of extended chains of monosaccharide or disaccharide units connected by glycosidic connections are carbohydrate-based polymers, also known as polysaccharides or carbohydrates. Their desirable qualities, such as accessibility, biocompatibility and degradability, low toxicity, high chemical reactivity, ease of chemical modification, and low cost, lead to a wide range of applications in the biomedical and pharmaceutical applications, including the creation of nano-vehicles for the delivery of therapeutic agents for treating cancer. Increasing short half-lives while decreasing systemic toxicity and targeting medicines at tumors are the key priorities for effective cancer treatment (Swierczewska et al. 2016). Both therapeutic and imaging agents can be physically incorporated inside polysaccharide by covalently conjugating them at the polysaccharide backbone present in the carbohydrates. The hydroxyl, amino, aldehyde, and carboxylic acid groups are the most frequently employed functional groups in polysaccharides that are available for chemical modification. Recently, amphiphilic polysaccharides were modified with stimuli-labile linkers or hydrophobic moieties to create the stimuli-responsive nanoparticles (NPs) such as dextran was modified and used in theranostics agent against SCC7 cancer cells shown by in vivo study. (Thambi et al. 2014) Some small molecular compounds are also used to conjugate with carbohydrates through various approaches. A selection of carbohydrate-based nanocarriers is summarized in Table 7.4.

TABLE 7.4
Examples of Carbohydrate-Based Nanocarriers Used in Cancer Theranostics

Sl. No	Carbohydrates	formulations	Application	Key outcomes	Ref.
1.	Hyaluronan	Loading of copper sulfide (CuS) into Cy5.5-conjugated hyaluronic acid nanoparticles (HANP)	Cancer theranostics	After intravenous administration of HANP into tumor-bearing mice, high fluorescence was observed in the tumor area over time, and a good tumor inhibition rate (89.74% on day 5) was observed.	(Zhang et al. 2014)
2.	hyaluronic acid	Liposome prepared by amphiphilic hyaluronic acid-ceramide (HACE) and loaded with doxorubicin (DOX) and Magnevist	Targeted cancer therapy and *in vivo* imaging	Liposomes can be employed as tumor-targeting MR imaging probes, and in vivo data shows that the medication circulates in the bloodstream for longer and that the nanohybrid liposomes have higher therapeutic efficacy.	(Park et al. 2014)
3.	hyaluronic acid	Gold nanocages conjugated with hyaluronic acid (AuNCs-HA)	Fluorescence imaging and drug delivery and photothermal therapy	high therapeutic efficacy against cancer cells and minimal side effects, and chemotherapy and photothermal therapy-treated tumors completely inhibited their ability to develop *in vivo*	(Wang et al. 2014)

TABLE 7.4 *(Continued)*
Examples of Carbohydrate-Based Nanocarriers Used in Cancer Theranostics

Sl. No	Carbohydrates	formulations	Application	Key outcomes	Ref.
4.	Dextran	Biocompatible gold nanoparticle-loaded lysozyme–dextran (Au@ Lys–Dex) nanogels	Fluorescence imaging and drug delivery	*In vitro* study suggest that DOX-loaded Au@ Lys–Dex nanogels have the same anti-tumor activity as free DOX. The nanogels can be used as a contrast agent in optical imaging.	(Cai and Yao 2013)
5.	Heparin	Heparin-folic acid-IR-780 nanoparticles (HF-IR-780 NPs)	Theranostic agent for imaging-guided cancer therapy	The HF-IR-780 NPs are selectively targeted to the tumor and can be used for tumor imaging according to *in vivo* biodistribution investigations. According to in vivo photothermal therapy trials and in vitro cell viability assays, MCF-7 cells or MCF-7 xenograft tumors could be destroyed when HF-IR-780 NPs and 808 nm laser radiation were used together.	(Yue et al. 2013)

7.4 CONCLUSIONS

This chapter highlighted applications of biomacromolecular agents for cancer theranostics, which amalgamates detection and treatment in an all-in-one platform to ensure better clinical outcomes. To curb the mortality rate from cancer, early detection as well as treatment are pivotal. Over the past two decades, there have been substantial advancements in cancer diagnosis and treatment. Currently, the focus of cancer treatment is on controlling and treating the disease with increased patient compliance, low patient discomfort, and minimal side effects. Nanomedicine promises targeted medication administration and targeted drug delivery at the tumor site, modifying drug pharmacokinetics, and bioavailability while augmenting tumor permeability and retention and reducing detrimental side effects to healthy tissues. As these versatile platforms can be utilized for the simultaneous identification, treatment, and management of tumors, theranostics has become a crucial tool in personalized medicine. Nanoformulations derived from biomacromolecules are promising and show clear potential. However, before putting them on the market or developing them further, there are a few aspects to be considered. Due to the intricacy of integrating several functionalities into a single platform and preserving their dimensions in the nanoscale range, there are significant challenges in manufacturing, scale-up, and reproducibility. A significant amount of research is to determine the optimal dose that will produce a strong signal for imaging and sustain the ideal medication release kinetics for therapy (Bhushan et al. 2021). Also, the nanodrug uptake by Reticuloendothelial System (RES) organs, such as the liver and spleen, needs to be minimized (Mosquera et al. 2018; Hoshyar et al. 2016).

Further, there should be minimal crosstalk between the nanomedicine and the nearby healthy tissues. While employing imaging modalities with theranostic agents, depth of penetration is a significant hurdle; as a result, imaging agents that obtain high-resolution images at high tissue depth are desired. The materials utilized to create the theranostic system must be tuned to avoid premature release of the therapeutic, encapsulating, and releasing integrated imaging agent. Stimuli-responsive "smart" materials are utilized to deliver therapeutics on demand according to environmental changes (e.g., temperature, pH, light, magnetic field). However, this increases the complexity of the system and may become more challenging to apply the technology in clinical settings. To deliver tailored therapy, the theranostic system must be adjusted against each kind of cancer because different tumor types might upregulate various receptors on their surfaces. However, theranostic nanomedicine has enormous potential for early detection and individualized, targeted cancer treatment. We can anticipate novel biomacromolecular, multifunctional formulations followed by administration and offer targeted and efficient cancer therapy entering clinical trials soon because nanotheranostics is a relatively new field of study.

REFERENCES

Bazak, R., M. Houri, S. El Achy, S. Kamel, and T. Refaat. 2015. "Cancer active targeting by nanoparticles: a comprehensive review of literature." *Journal of Cancer Research and Clinical Oncology* 141 (5): 769–784. doi: 10.1007/s00432-014-1767-3.

Beljanski, Vladimir, and John Hiscott. 2012. "The use of oncolytic viruses to overcome lung cancer drug resistance." *Current Opinion in Virology* 2 (5):629–635. doi: 10.1016/j.coviro.2012.07.006.

Bhushan, Arya, Andrea Gonsalves, and Jyothi U. Menon. 2021. "Current state of breast cancer diagnosis, treatment, and theranostics." *Pharmaceutics* 13 (5):723.

Cai, Huanxin, and Ping Yao. 2013. "In situ preparation of gold nanoparticle-loaded lysozyme–dextran nanogels and applications for cell imaging and drug delivery." *Nanoscale* 5 (7):2892–2900. doi: 10.1039/C3NR00178D.

Dickens, Elena, and Samreen Ahmed. 2018. "Principles of cancer treatment by chemotherapy." *Surgery (Oxford)* 36 (3):134–138. 10.1016/j.mpsur.2017.12.002.

Elzoghby, A. O., W. M. Samy, and N. A. Elgindy. 2012. "Protein-based nanocarriers as promising drug and gene delivery systems." *Journal of Controlled Release* 161 (1):38–49. doi: 10.1016/j.jconrel.2012.04.036.

Gregoriou, Y., G. Gregoriou, V. Yilmaz, K. Kapnisis, M. Prokopi, A. Anayiotos, K. Strati, N. Dietis, A. I. Constantinou, and C. Andreou. 2021. "Resveratrol loaded polymeric micelles for theranostic targeting of breast cancer cells." *Nanotheranostics* 5 (1):113–124. doi: 10.7150/ntno.51955.

He, C., S. Zheng, Y. Luo, and B. Wang. 2018. "Exosome theranostics: Biology and translational medicine." *Theranostics* 8 (1):237–255. doi: 10.7150/thno.21945.

Hoshyar, N., S. Gray, H. Han, and G. Bao. 2016. "The effect of nanoparticle size on in vivo pharmacokinetics and cellular interaction." *Nanomedicine (Lond)* 11 (6):673–692. doi: 10.2217/nnm.16.5.

Hu, Rong, Xiaobing Zhang, Zilong Zhao, Guizhi Zhu, Tao Chen, Ting Fu, and Weihong Tan. 2014. "DNA nanoflowers for multiplexed cellular imaging and traceable targeted drug delivery." *Angewandte Chemie International Edition* 53 (23):5821–5826. doi: 10.1002/anie.201400323.

Huang, H., D. P. Yang, M. Liu, X. Wang, Z. Zhang, G. Zhou, W. Liu, Y. Cao, W. J. Zhang, and X. Wang. 2017. "pH-sensitive Au-BSA-DOX-FA nanocomposites for combined CT imaging and targeted drug delivery." *International Journal of Nanomedicine* 12:2829–2843. doi: 10.2147/ijn.s128270.

Idée, J. M., S. Louguet, S. Ballet, and C. Corot. 2013. "Theranostics and contrast-agents for medical imaging: A pharmaceutical company viewpoint." *Quantitative Imaging in Medicine and Surgery* 3 (6):292–297. doi: 10.3978/j.issn.2223-4292.2013.12.06.

Illes, Bernhard, Patrick Hirschle, Sabine Barnert, Valentina Cauda, Stefan Wuttke, and Hanna Engelke. 2017. "Exosome-coated metal–organic framework nanoparticles: An efficient drug delivery platform." *Chemistry of Materials* 29 (19):8042–8046. doi: 10.1021/acs.chemmater.7b02358.

Jia, Qingyan, Jiechao Ge, Weimin Liu, Xiuli Zheng, Mengqi Wang, Hongyan Zhang, and Pengfei Wang. 2017. "Biocompatible iron phthalocyanine–albumin assemblies as photoacoustic and thermal theranostics in living mice." *ACS Applied Materials & Interfaces* 9 (25):21124–21132. doi: 10.1021/acsami.7b04360.

Jiang, Dawei, Zhilei Ge, Hyung-Jun Im, Christopher G. England, Dalong Ni, Junjun Hou, Luhao Zhang, Christopher J. Kutyreff, Yongjun Yan, Yan Liu, Steve Y. Cho, Jonathan W. Engle, Jiye Shi, Peng Huang, Chunhai Fan, Hao Yan, and Weibo Cai. 2018. "DNA origami nanostructures can exhibit preferential renal uptake and alleviate acute kidney injury." *Nature Biomedical Engineering* 2 (11):865–877. doi: 10.1038/s41551-018-0317-8.

Kariduraganavar, Mahadevappa Y., Geetha B. Heggannavar, Sandra Amado, and Geoffrey R. Mitchell. 2019. "Chapter 6 - Protein Nanocarriers for Targeted Drug Delivery for Cancer Therapy." In *Nanocarriers for Drug Delivery*, edited by Shyam S. Mohapatra, Shivendu Ranjan, Nandita Dasgupta, Raghvendra Kumar Mishra and Sabu Thomas, 173–204. Elsevier.

Kim, Kyeong Rok, Ae Sol Lee, Su Min Kim, Hye Ryoung Heo, and Chang Sup Kim. 2023. "Virus-like nanoparticles as a theranostic platform for cancer." *Frontiers in Bioengineering and Biotechnology* 10. doi: 10.3389/fbioe.2022.1106767.

Kim, Kyoung-Ran, Da-Rae Kim, Taemin Lee, Ji Young Yhee, Byeong-Su Kim, Ick Chan Kwon, and Dae-Ro Ahn. 2013. "Drug delivery by a self-assembled DNA tetrahedron for overcoming drug resistance in breast cancer cells." *Chemical Communications* 49 (20):2010–2012. doi: 10.1039/C3CC38693G.

Kim, M. S., M. J. Haney, Y. Zhao, V. Mahajan, I. Deygen, N. L. Klyachko, E. Inskoe, A. Piroyan, M. Sokolsky, O. Okolie, S. D. Hingtgen, A. V. Kabanov, and E. V. Batrakova. 2016. "Development of exosome-encapsulated paclitaxel to overcome MDR in cancer cells." *Nanomedicine* 12 (3):655–664. doi: 10.1016/j.nano.2015.10.012.

Li, Ke, Zhi-Ping Zhang, Ming Luo, Xiang Yu, Yu Han, Hong-Ping Wei, Zong-Qiang Cui, and Xian-En Zhang. 2012. "Multifunctional ferritin cage nanostructures for fluorescence and MR imaging of tumor cells." *Nanoscale* 4 (1):188–193. doi: 10.1039/C1NR11132A.

Li, X., X. Jia, and H. Niu. 2018. "Nanostructured lipid carriers co-delivering lapachone and doxorubicin for overcoming multidrug resistance in breast cancer therapy." *International Journal of Nanomedicine* 13:4107–4119. doi: 10.2147/ijn.s163929.

Liang, Tingxizi, Zhigang Yao, Jie Ding, Qianhao Min, Liping Jiang, and Jun-Jie Zhu. 2018. "Cascaded aptamers-governed multistage drug-delivery system based on biodegradable envelope-type nanovehicle for targeted therapy of HER2-overexpressing breast cancer." *ACS Applied Materials & Interfaces* 10 (40):34050–34059. doi: 10.1021/acsami.8b14009.

Liu, Jia, Chao Qi, Kaixiong Tao, Jinxiang Zhang, Jian Zhang, Luming Xu, Xulin Jiang, Yunti Zhang, Lei Huang, Qilin Li, Hongjian Xie, Jinbo Gao, Xiaoming Shuai, Guobin Wang, Zheng Wang, and Lin Wang. 2016. "Sericin/dextran injectable hydrogel as an optically trackable drug delivery system for malignant melanoma treatment." *ACS Applied Materials & Interfaces* 8 (10):6411–6422. doi: 10.1021/acsami.6b00959.

Liu, Rui, Wei Xiao, Chuan Hu, Rou Xie, and Huile Gao. 2018. "Theranostic size-reducible and no donor conjugated gold nanocluster fabricated hyaluronic acid nanoparticle with optimal size for combinational treatment of breast cancer and lung metastasis." *Journal of Controlled Release* 278:127–139. 10.1016/j.jconrel.2018.04.005.

Luo, Y., X. Wang, D. Du, and Y. Lin. 2015. "Hyaluronic acid-conjugated apoferritin nanocages for lung cancer targeted drug delivery." *Biomaterials Science* 3 (10):1386–1394. doi: 10.1039/c5bm00067j.

Mosquera, Jesús, Isabel García, and Luis M. Liz-Marzán. 2018. "Cellular uptake of nanoparticles versus small molecules: A matter of size." *Accounts of Chemical Research* 51 (9):2305–2313. doi: 10.1021/acs.accounts.8b00292.

Nakamura, Yuko, Ai Mochida, Peter L. Choyke, and Hisataka Kobayashi. 2016. "Nanodrug delivery: Is the enhanced permeability and retention effect sufficient for curing cancer?" *Bioconjugate Chemistry* 27 (10):2225–2238. doi: 10.1021/acs.bioconjchem.6b00437.

Nicolson, Fay, Akbar Ali, Moritz F. Kircher, and Suchetan Pal. 2020. "DNA nanostructures and DNA-functionalized nanoparticles for cancer theranostics." *Advanced Science* 7 (23): 2001669. doi: 10.1002/advs.202001669.

Parhi, Priyambada, and Sanjeeb Kumar Sahoo. 2015. "Trastuzumab guided nanotheranostics: A lipid based multifunctional nanoformulation for targeted drug delivery and imaging in breast cancer therapy." *Journal of Colloid and Interface Science* 451:198–211. doi: 10.1016/j.jcis.2015.03.049.

Park, J. H., H. J. Cho, H. Y. Yoon, I. S. Yoon, S. H. Ko, J. S. Shim, J. H. Cho, J. H. Park, K. Kim, I. C. Kwon, and D. D. Kim. 2014. "Hyaluronic acid derivative-coated nanohybrid liposomes for cancer imaging and drug delivery." *Journal of Controlled Release* 174:98–108. doi: 10.1016/j.jconrel.2013.11.016.

Qi, Hongzhao, Chaoyong Liu, Lixia Long, Yu Ren, Shanshan Zhang, Xiaodan Chang, Xiaomin Qian, Huanhuan Jia, Jin Zhao, Jinjin Sun, Xin Hou, Xubo Yuan, and Chunsheng Kang. 2016. "Blood exosomes endowed with magnetic and targeting properties for cancer therapy." *ACS Nano* 10 (3):3323–3333. doi: 10.1021/acsnano.5b06939.

Roy Chowdhury, M., C. Schumann, D. Bhakta-Guha, and G. Guha. 2016. "Cancer nanotheranostics: Strategies, promises and impediments." *Biomed Pharmacother* 84:291–304. doi: 10.1016/j.biopha.2016.09.035.

Semkina, A., M. Abakumov, N. Grinenko, A. Abakumov, A. Skorikov, E. Mironova, G. Davydova, A. G. Majouga, N. Nukolova, A. Kabanov, and V. Chekhonin. 2015. "Core–shell–corona doxorubicin-loaded superparamagnetic Fe_3O_4 nanoparticles for cancer theranostics." *Colloids and Surfaces B: Biointerfaces* 136:1073–1080. 10.1016/j.colsurfb.2015.11.009.

Sheng, Zonghai, Dehong Hu, Mingbin Zheng, Pengfei Zhao, Huilong Liu, Duyang Gao, Ping Gong, Guanhui Gao, Pengfei Zhang, Yifan Ma, and Lintao Cai. 2014. "Smart human serum albumin-indocyanine green nanoparticles generated by programmed assembly for dual-modal imaging-guided cancer synergistic phototherapy." *ACS Nano* 8 (12):12310–12322. doi: 10.1021/nn5062386.

Shi, Jinjun, Philip W. Kantoff, Richard Wooster, and Omid C. Farokhzad. 2017. "Cancer nanomedicine: Progress, challenges and opportunities." *Nature Reviews Cancer* 17 (1):20–37. doi: 10.1038/nrc.2016.108.

Shi, Y., R. van der Meel, X. Chen, and T. Lammers. 2020. "The EPR effect and beyond: Strategies to improve tumor targeting and cancer nanomedicine treatment efficacy." *Theranostics* 10 (17):7921–7924. doi: 10.7150/thno.49577.

Shu, Dan, Hui Li, Yi Shu, Gaofeng Xiong, William E. Carson, III, Farzin Haque, Ren Xu, and Peixuan Guo. 2015. "Systemic delivery of anti-mirna for suppression of triple negative breast cancer utilizing RNA nanotechnology." *ACS Nano* 9 (10):9731–9740. doi: 10.1021/acsnano.5b02471.

Sonali, Rahul Pratap Singh, Nitesh Singh, Gunjan Sharma, Mahalingam R. Vijayakumar, Biplob Koch, Sanjay Singh, Usha Singh, Debabrata Dash, Bajarangprasad L. Pandey, and Madaswamy S. Muthu. 2016. "Transferrin liposomes of docetaxel for brain-targeted cancer applications: Formulation and brain theranostics." *Drug Delivery* 23 (4):1261–1271. doi: 10.3109/10717544.2016.1162878.

Sung, H., J. Ferlay, R. L. Siegel, M. Laversanne, I. Soerjomataram, A. Jemal, and F. Bray. 2021. "Global cancer statistics 2020: GLOBOCAN estimates of incidence and mortality worldwide for 36 cancers in 185 countries." *CA: A Cancer Journal for Clinicians* 71 (3):209–249. doi: 10.3322/caac.21660.

Swierczewska, M., H. S. Han, K. Kim, J. H. Park, and S. Lee. 2016. "Polysaccharide-based nanoparticles for theranostic nanomedicine." *Advanced Drug Delivery Reviews* 99 (Pt A):70–84. doi: 10.1016/j.addr.2015.11.015.

Thambi, Thavasyappan, Dong Gil You, Hwa Seung Han, V. G. Deepagan, Sang Min Jeon, Yung Doug Suh, Ki Young Choi, Kwangmeyung Kim, Ick Chan Kwon, Gi-Ra Yi, Jun Young Lee, Doo Sung Lee, and Jae Hyung Park. 2014. "Bioreducible carboxymethyl dextran nanoparticles for tumor-targeted drug delivery." *Advanced Healthcare Materials* 3 (11):1829–1838. doi: 10.1002/adhm.201300691.

Tsai, Li-Chu, Hao-Ying Hsieh, Kun-Ying Lu, Sin-Yu Wang, and Fwu-Long Mi. 2015. "EGCG/gelatin-doxorubicin gold nanoparticles enhance therapeutic efficacy of doxorubicin for prostate cancer treatment." *Nanomedicine* 11 (1):9–30. doi: 10.2217/nnm.15.183.

Wang, Jie, Yue Dong, Yiwei Li, Wei Li, Kai Cheng, Yuan Qian, Guoqiang Xu, Xiaoshuai Zhang, Liang Hu, Peng Chen, Wei Du, Xiaojun Feng, Yuan-Di Zhao, Zhihong Zhang, and Bi-Feng Liu. 2018. "Designer exosomes for active targeted chemo-photothermal synergistic tumor therapy." *Advanced Functional Materials* 28 (18):1707360. 10.1002/adfm.201707360.

Wang, Jie, Wei Li, Zhichao Lu, Leicheng Zhang, Yu Hu, Qiubai Li, Wei Du, Xiaojun Feng, Haibo Jia, and Bi-Feng Liu. 2017. "The use of RGD-engineered exosomes for enhanced targeting ability and synergistic therapy toward angiogenesis." *Nanoscale* 9 (40):15598–15605. doi: 10.1039/C7NR04425A.

Wang, L., Y. An, C. Yuan, H. Zhang, C. Liang, F. Ding, Q. Gao, and D. Zhang. 2015. "GEM-loaded magnetic albumin nanospheres modified with cetuximab for simultaneous targeting, magnetic resonance imaging, and double-targeted thermochemotherapy of pancreatic cancer cells." *International Journal of Nanomedicine* 10:2507–2519. doi: 10.2147/ijn.s77642.

Wang, Ruibing, Paul S. Billone, and Wayne M. Mullett. 2013. "Nanomedicine in action: An overview of cancer nanomedicine on the market and in clinical trials." *Journal of Nanomaterials* 2013:629681. doi: 10.1155/2013/629681.

Wang, Y., Y. Wang, G. Chen, Y. Li, W. Xu, and S. Gong. 2017. "Quantum-dot-based theranostic micelles conjugated with an anti-EGFR nanobody for triple-negative breast cancer therapy." *ACS Applied Materials & Interfaces* 9 (36):30297–30305. doi: 10.1021/acsami.7b05654.

Wang, Yu-Min, Zhan Wu, Si-Jia Liu, and Xia Chu. 2015. "Structure-switching aptamer triggering hybridization chain reaction on the cell surface for activatable theranostics." *Analytical Chemistry* 87 (13):6470–6474. doi: 10.1021/acs.analchem.5b01634.

Wang, Zhantong, Peng Huang, Orit Jacobson, Zhe Wang, Yijing Liu, Lisen Lin, Jing Lin, Nan Lu, Huimin Zhang, Rui Tian, Gang Niu, Gang Liu, and Xiaoyuan Chen. 2016. "Biomineralization-inspired synthesis of copper sulfide–ferritin nanocages as cancer theranostics." *ACS Nano* 10 (3):3453–3460. doi: 10.1021/acsnano.5b07521.

Wang, Zhenzhen, Zhaowei Chen, Zhen Liu, Peng Shi, Kai Dong, Enguo Ju, Jinsong Ren, and Xiaogang Qu. 2014. "A multi-stimuli responsive gold nanocage–hyaluronic platform for targeted photothermal and chemotherapy." *Biomaterials* 35 (36):9678–9688. doi: 10.1016/j.biomaterials.2014.08.013.

Wen, Liewei, Sihua Yang, Junping Zhong, Quan Zhou, and Da Xing. 2017. "Thermoacoustic imaging and therapy guidance based on ultra-short pulsed microwave pumped thermoelastic effect induced with superparamagnetic iron oxide nanoparticles." *Theranostics* 7 (7):1976–1989. doi: 10.7150/thno.17846.

Yang, Jiang, Tai Wang, Lina Zhao, Vinagolu K. Rajasekhar, Suhasini Joshi, Chrysafis Andreou, Suchetan Pal, Hsiao-ting Hsu, Hanwen Zhang, Ivan J. Cohen, Ruimin Huang, Ronald C. Hendrickson, Matthew M. Miele, Wenbo Pei, Matthew B. Brendel, John H. Healey, Gabriela Chiosis, and Moritz F. Kircher. 2020. "Gold/alpha-lactalbumin nanoprobes for the imaging and treatment of breast cancer." *Nature Biomedical Engineering* 4 (7):686–703. doi: 10.1038/s41551-020-0584-z.

Yuan, Ye, Zi Gu, Chi Yao, Dan Luo, and Dayong Yang. 2019. "Nucleic acid–based functional nanomaterials as advanced cancer therapeutics." *Small* 15 (26):1900172. doi: 10.1002/smll.201900172.

Yue, C., P. Liu, M. Zheng, P. Zhao, Y. Wang, Y. Ma, and L. Cai. 2013. "IR-780 dye loaded tumor targeting theranostic nanoparticles for NIR imaging and photothermal therapy." *Biomaterials* 34 (28):6853–6861. doi: 10.1016/j.biomaterials.2013.05.071.

Zhang, Libing, Sae Rin Jean, Sharif Ahmed, Peter M. Aldridge, Xiyan Li, Fengjia Fan, Edward H. Sargent, and Shana O. Kelley. 2017. "Multifunctional quantum dot DNA hydrogels." *Nature Communications* 8 (1):381. doi: 10.1038/s41467-017-00298-w.

Zhang, Liwen, Shi Gao, Fan Zhang, Kai Yang, Qingjie Ma, and Lei Zhu. 2014. "Activatable hyaluronic acid nanoparticle as a theranostic agent for optical/photoacoustic image-guided photothermal therapy." *ACS Nano* 8 (12):12250–12258. doi: 10.1021/nn506130t.

Zhang, Qian, Qiao Jiang, Na Li, Luru Dai, Qing Liu, Linlin Song, Jinye Wang, Yaqian Li, Jie Tian, Baoquan Ding, and Yang Du. 2014. "DNA origami as an in vivo drug delivery vehicle for cancer therapy." *ACS Nano* 8 (7):6633–6643. doi: 10.1021/nn502058j.

8 Applications of Nanotechnology for Preventing Global Infectious Diseases

Rajdeep Saha, Kaberi Chatterjee, Biswatrish Sarkar, and Papiya Mitra Mazumder

8.1 INTRODUCTION

Nanotechnology is a boon that is made up of ideas and concepts from various fields of science and technology. This multidisciplinary field has a combination of different scientific disciplines. The technology finds its application in various daily activities of human life. Its roots go way back to the late 20th century where different scientists explained the hypothesis of different materials which lie at the atomic level (nanoscale) and their potential uses. Moreover, the ability of humankind to alter the properties of such materials and to bring out a desirable activity from them for which it can be later used in different areas of life and can prove to be a game changer (Hulla et al. 2015). This hypothesis was later proven to be correct and there have been commercialization of many such products which has altered activities due to the influence of nanotechnology in them. The major advantages of using such nanoscale objects is that the characters (physical, chemical, and mechanical) of these objects do not match with the bulk material and also with the individual molecules. In most cases, the characters of such objects are enhanced (Farajpour et al. 2018; Nam and Ju 2013; Arya et al. 2012) but, in some cases it can also be weakened and based off of the newer characteristics, the applications are generally designed (Devan et al. 2012).

The inclusion of nanostructures in medicine created "nanomedicine." It helps to optimize treatment and diagnosis to make the process easier, to get better diagnosis, and to obtain greater therapeutic effects. There are various examples where nanostructures are utilized for diagnosis purposes (Pelaz et al. 2017). As in the article by Liang et al., (2020) biosensors were created with the help of nanospheres because they showed better Electrochemiluminescence (ECL). This particular property helped to improve sensitivity and diagnosis ability of the preparation. The authors used the preparation for the detection of MUC1, a glycoprotein that is over-expressed on cancerous cells. Similar proteins along with single stranded DNA were impregnated on the surface of the nanosphere. The space steric effect shown by the MUC1 proteins on the nanosphere as well as on the cancerous cell surface caused a reduction in the ECL signal. Therefore, it could be assumed that the particular cell was cancerous (Liang et al. 2020).

Similarly, when it comes to immunization, nanoparticles (NPs) play an important role. As discussed in the article by Gale et al. (2020) NP can be considered to the best option when it comes to the delivery of poly (I:C), a double stranded RNA structural mimic. Not only in delivery, the utilization of polymeric cationic NPs for encapsulating poly (I: C) helps to stabilize it and improves its therapeutic potential. This mimicking action is essential because it acts as a toll-like receptor (TLR) agonist and ensures optimum immunization. So, use of NPs increases the immune

DOI: 10.1201/9781003305583-8

response against the vaccine (Gale et al. 2020). Just like all aspects of medicine, in the case of nanomedicine too, certain issues like ethical issues and safety issues, need to be kept in the mind. It is expected that the best therapeutic effect should be perceived, and safety maintained thereby enhancing the life expectancy. High efficacy and personalization are also certain expectations from the nanomedicine (Pelaz et al. 2017).

Like all other aspects of medicine, nanotechnology can be utilized effectively in the management, prevention, and treatment of infectious diseases (ID). It can help stop the outbreak and rapid spread of diseases which is the major issue with any ID. There are several well-known ID in the world that affects the global population and annually many individuals lose their lives due to the disease. But there are some emerging infectious diseases (EID) that are newly recognized in a group of people or a population or maybe they are well known but their incidence rate is increasing (McArthur 2019). Such diseases must be stopped before they take the shape of another pandemic like the Coronavirus disease 2019 (COVID-19). Therefore, it is essential to understand the role of nanotechnology clearly. This is expected to help in the accurate application of the nanotechnology in order to control disease spreadability.

The chapter will shed light on the different applications of nanotechnology (especially in the biomedical field), the role of nanotechnology in preventing and treating each of the IDs that are affecting the global population, and the ways for more improvisations so that greater advantages can be obtained from the technology when it comes to the global ID.

8.2 INFECTIOUS DISEASES

Organisms like viruses, fungi, bacteria, or worms are responsible for causing ID. For these organisms to cause the disease and spread it amongst different healthy people there is the necessity of one or multiple hosts. If the immune system of the host is healthy, the organism stays within the body and no symptoms are seen of the disease. On the other hand, if the immune system is compromised, the host becomes sick and the symptoms starts showing as the organism starts taking over the body and the immune system gets overwhelmed. The prions from these organisms can also cross the blood brain barrier (BBB) and can cause neurological disorders (Kotra 2007). Transmission of infective organisms may also occur without the presence of any host. These are known as airborne pathogens. The mode of transmission that takes place through droplets is known as aerosolized transmission. The organisms that are expelled by particles generally settle on another surface very quickly making the surface infected. These droplets are very small in size and they remain suspended in air. The organisms that are transmitted through this route are also known to have a smaller size making their exposure time much longer. This makes the disease more infectious because its ability to spread at a rapid pace and affecting a huge number of people at a very short time span (Fernstrom and Goldblatt 2013).

8.2.1 Emerging Infectious Disease

IDs are constantly emerging. The term "emerging infectious diseases" (EID) can be defined as "those whose incidence in humans has increased within the past two decades or threatens to increase in the near future. Emergence may be due to the spread of a new agent, to the recognition of an infection that has been present in the population but has gone undetected, or to the realization that an established disease has an infectious origin. Emergence may also be used to describe the reappearance (or re-emergence) of a known infection after a decline in incidence" (Oaks et al. 1992; Van Doorn 2014). One of the best examples of an EID is COVID-19 which has taken down the global population since its outbreak from 2019. More detailed discussions on the particular disease have been carried out in the later stages of this chapter.

There are several drivers (biological, social, or environmental) which causes the emergence of ID in the global population. Some of them are as follows (Van Doorn 2014):

a. Breakdown in public health and healthcare system
b. Microbial adaptation, changes, and mutations in genetic material
c. International travel and migration
d. Economic development and land use
e. Intentional biological attacks
f. Inequality in the society and poverty

There are more drivers that can be held responsible for the outbreak of a disease like public ignorance. In some cases, the pathogens evolve from one species to another. It may evolve from animals and after a few generations of the pathogens and a few cycles of the disease, it might permanently reside within the human population. One such classical example is HIV (Human Immunodeficiency Virus) which was originally a disease seen among apes and monkeys. But it got transmitted among human beings and caused an outbreak (Eigen et al. 2002). Generally, it is seen to be spread amongst a certain group of people through networks involving people who remain in close proximity to the ID like essential service people (Oster et al. 2021).

Similarly, many more IDs have been transmitted from apes to humans due to their similar genetic makeup (Keita et al. 2014). So, it can be said that for such kind of emergence, human needs to have an increased amount of contact with the animal host of the disease. This route of emergence is known as zoonotic emergence (Van Doorn 2014).

There is another route of emergence, the non-zoonotic emergence, where humans are responsible for the spread of the disease. This means, the pathogens can get transmitted from one human to another without the need for any non-human species (Van Doorn 2014; Thornhill et al. 2010).

8.2.2 History and Global Burden of Infectious Diseases

Pandemics caused due to the outbreak of ID have been seen through history. They were mostly recorded in western history and are responsible for shaping society. The first well-documented pandemic in history was the Athenian plague which happened within the span of 430–66 B.C. There have been many other pandemics like the Justinian plague and the Antonine plague. Although these pandemics took many people's life, they were restricted to a certain region or with a certain group of people. The first global outbreak was seen when there was an outbreak of the Black Death (bubonic plague). Later, there was an outbreak of the "Spanish Flu." This pandemic occurred after modern medicine was established and was caused by the H1N1 strain of the influenza virus. This was the first pandemic which took the lives of over 50 million people (Chatterjee et al. 2022). Therefore, this was the most deadly pandemic at that time. After this, other outbreaks were recorded that were caused by HIV, Small pox, (Severe acute respiratory syndrome) SARS, H1N1/09 (swine flu), Ebola, and Zika (Huremović 2019). Following the Zika virus outbreak (2015–2016), no more pandemics seen until COVID-19. Different population settings are affected differently by different IDs. It has been represented in Figure 8.1 where different IDs have been indicated by different colors and they have been used to symbolize the country where a particular disease has been reported the most.

After the outbreak of COVID-19 in 2019–2020, it was evident how IDs expanded from a certain population to the global population within a matter of days and became a pandemic. Before this outbreak, there was a steady impact seen within the global burden of ID in almost every country and the health of the global population was becoming better, thanks to the advanced healthcare system and vaccinations. Therefore, the average life expectancy of an individual increased from 66.8 years in 2000 to 73.3 years in 2019 with a steady decrease in the number of people affected by IDs. A graphical representation of the decrease in Disability-Adjusted Life Years or DALYs of almost all IDs has been shown in Figure 8.2. DALYs helps to obtain a clear understanding of the burden of a particular disease concerning a population. This data can be calculated with respect to

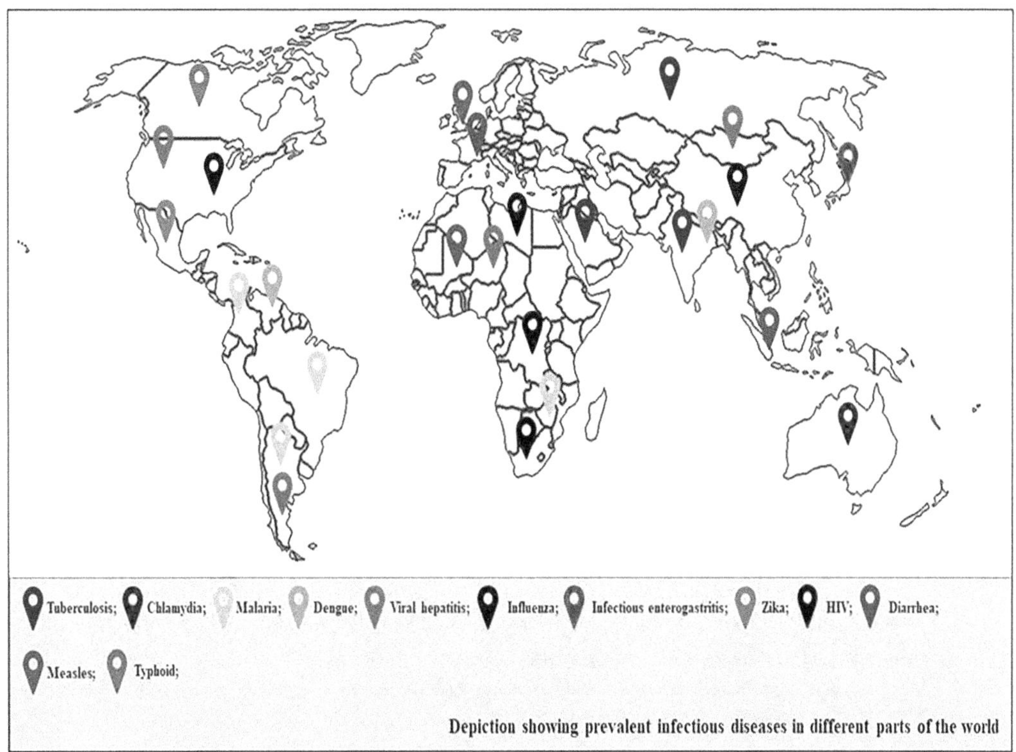

FIGURE 8.1 Prevalent infectious diseases in different parts of the world. Data obtained from https://www.cdc.gov/diseasesconditions/index.html (Prevention).

the different gender (as shown in Figure 8.2 & Figure 8.3) or with respect to different age group in the population (Nurchis et al. 2020). Figure 8.3 is a pie chart depicting the percentage of the burden of different IDs between the years 1990 and 2019. In 2000, around 95% of the global population was infected by such diseases which reduced to 18% by 2019. Although the ID like tuberculosis (TB) and HIV were still the leading causes of deaths but the number of people getting newly infected with the diseases reduced drastically ((WHO) 2021). Global health burden after 2020 changed a lot with a drastic reduction in average life expectancy which had been increasing till 2019.

8.2.3 Pre-Existing Prevention Options of Infectious Diseases, Their Merits and Demerits

Some of the prevention options for ID can be through vaccinations, diagnosis of the disease, tracking the infectivity of the disease, and monitoring to make sure that the infectivity rate of the disease does not become a public health concern and ends up taking the lives of millions of people. There have been various preventive options that were implemented in the past. When it comes to the diagnosis of ID, initially culture-based method was used form which microbial identification was carried out based on the phenotypic characteristics of the microbes. This method is not much reliable and takes a lot of time. After that, nucleic-acid-based detection was introduced. Detection of only targeted microbial DNA could be carried out and the process was comparatively faster and it quickly replaced the culture techniques that were used traditionally. For this detection, a need for the amplification of the DNA material was felt. So, the scientific community introduced the polymerase reaction or PCR (Yang and Rothman 2004). Other than

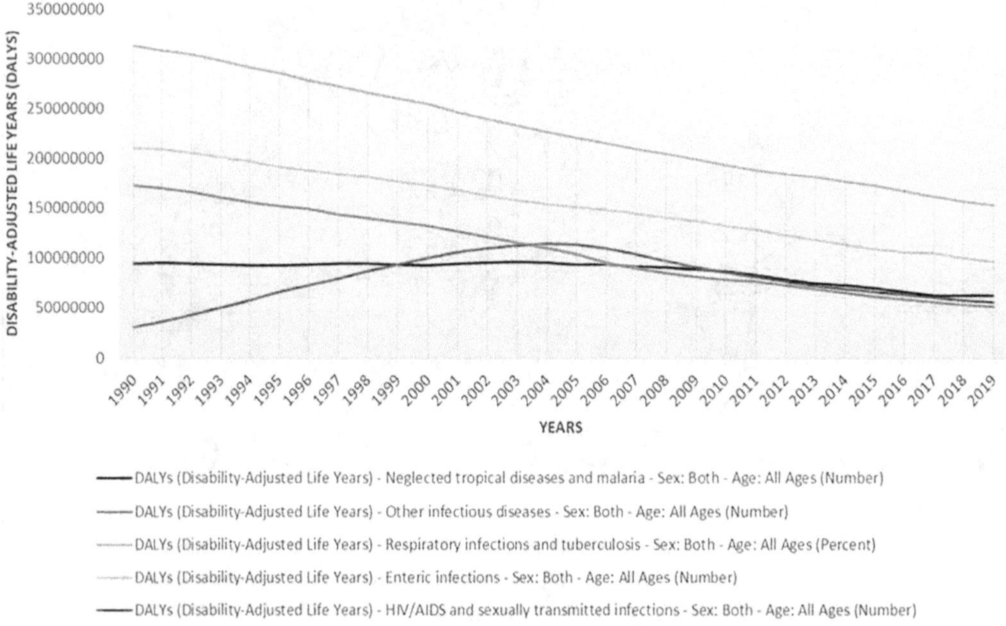

FIGURE 8.2 Disability-Adjusted Life Years (DALYs) of prevailing infectious disease in the world with respect to years for both sexes (male and female). Data obtained from https://www.cdc.gov/diseasesconditions/index.html and https://ourworldindata.org/burden-of-disease (Prevention; H).

PCR, other sequence-based microbial detection methods were used in a clinical setting (Fredricks and Relman 1999). Here enzymes were used to amplify the small regions of microbial DNA. Together these methods of detection are known as "molecular diagnostics." This method is accurate but there are several disadvantages. The time needed to get accurate results is still more. It is faster than the initial method but still it isn't fast enough because when an individual is infected by a disease, the pathogen lives within the body and it changes. Meaning, the condition of the patient can deteriorate within a matter of hours. So, by the time the results are obtained after carrying out the sequence-based microbial detection method, the patient might be in such a stage where it can become very difficult to resuscitate (Yang and Rothman 2004). Inaccurate or time-consuming diagnosis can lead to over 95% of the deaths caused due to the outbreak of an ID. Other than the determination of clinical outcomes, rapid results can also help to determine public health outcomes. According to the WHO, a good diagnostic tool should have the following characteristics: (a) affordability, (b) high sensitivity, (c) user friendly, (d) specific, (e) rapid and robust, (f) will not need any extra equipment, and (g) deliverable to end-users (Su et al. 2015). All these can be summarized as ASSURED (Tay et al. 2016; Lee et al. 2010). The previously mentioned diagnostic tools do not fulfill all the criteria and therefore, there is a need for a new and improved tool.

Tracking of an ID is very essential because they are spread rapidly and efficiently through mass gatherings and traveling (migration). If efficient tracking is done then the probable source of the infection can be identified and the risk groups can also be identified. This makes the management of an ID outbreak easier (Abubakar et al. 2012). The tracking of an ID is a tedious process. But newer technologies can be incorporated to make sure that effective tracking can be done.

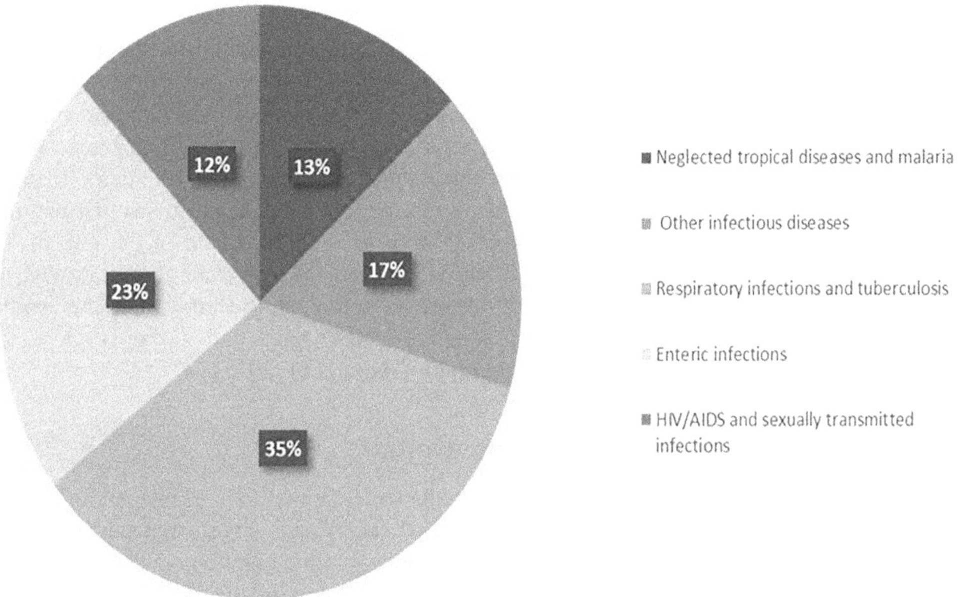

FIGURE 8.3 Pie chart revealing the percentages of the cases reported of the prevailing infectious diseases of the world with respect to Disability-Adjusted Life Years (DALYs) between the years 1990 and 2019 for both sexes (male and female). Data obtained from https://ourworldindata.org/burden-of-disease (H).

Vaccination is a mode of prevention of ID where passive immunization is used. Since its origin in 1888, it has been very successful in the mitigation of ID (Pavia and Wormser 2021). Inactivated vaccines help to incorporate adjuvants within the body which help to improve the quality of the cellular and humoral immune response that is created against the antigens that are present within the vaccines. It has already been proven that the efficiency of vaccines is very high. But this efficiency can be increased much more to see better results (Smith et al. 2013).

8.3 NANOTECHNOLOGY

Nanotechnology can be identified as a technology that evolved from the ideas of the human mind and the creativity within the human species. The word *"nano"* stands for a dwarf in the language, Greek (Boholm 2016). It is a multidisciplinary field which means the involvement of different scientific disciplines (Farokhzad and Langer 2009). As a lot of ideas and tools from different disciplines have been borrowed, it is difficult to provide a clear boundary to the technology (Patra 2013). Just like every form of technology, nanotechnology has also seen many changes. Since the dawn of nanotechnology, different era has seen different versions of it. In other words, it can be said that different eras are marked based on the developmental changes in the technology. All these development helped to carve the path towards the version that is seen in the current ages (Hulla et al. 2015). In the following sections, the history of the technology and later on the characteristic features will be described.

8.3.1 History of Nanotechnology

It is important to understand the beginning of the technology to have a clear picture of how it has developed over the years. History of nanotechnology is closely linked with the Industrial evolution. This is because the exposure of common people to this technology that was initially unheard of; increased dramatically with the industrial evolution. Initially, the concept of "nanometer," proposed by Richard Zsigmondy, was created to properly characterize the small size of gold colloids. Later on, Richard Feynman was responsible for initiating another conversation regarding nanotechnology. He gave a lecture at a famous event which was responsible for the re-introduction of the technology with the help of a new pair of lenses. He was responsible for the development of modern nanotechnology which we see now (Hulla et al. 2015). Till that time, the term "nanotechnology" was not coined. The two scientists just discussed about the idea of nanometer and matter at atomic levels. The term was officially introduced much later. It was first found to be publicly used by Norio Taniguchi in the year 1974 in a conference held in the city of Tokyo (Boholm 2016). This led to the initiation of the golden era of nanotechnology. Slowly, at the initiation of the 21st century, there was an increase in the interest shown in this particular technology. Initially, the United States of America (USA) gave national importance to the technology and gave it the much needed recognition (Hulla et al. 2015).

8.3.2 Characteristic Features of Nanotechnology

Nanotechnology holds specific features which provide the definition to it. The technology exists in the nanoscale range which is 10^{-9} meters (Patra 2013). The size is largely between 1 to 100 nm. Due to this size range and the conversion of bulk products into atomic or molecular dimensions, the properties (both physical and chemical) are altered (Uskokovic 2013). This alteration has enabled a wide range of applications.

8.3.3 Types of Nanostructures

Classification of nanostructures can be done either with respect to their dimensions or based on the composition of the nanostructure (Nasrollahzadeh et al. 2019). Figure 8.4 gives an in-depth representation of the different types of nanostructures in a pictorial format. The classifications have also been enlisted in the following:

8.3.3.1 Classification Based on Dimensions

The different classifications based on the dimensions of the nanostructures are 0D, 1D, 2D, and 3D. These are formed from different elementary blocks like molecules, clusters, grains, etc. (Pokropivny and Skorokhod 2008; Nasrollahzadeh et al. 2019).

- 0D: Fullerenes, rings, particles, Quantum Dots (Li et al. 2019; Reshma and Mohanan 2019)
- 1D (Devan et al. 2012): nanotubes (NT) (hard-matter(Zhang et al. 2020) and soft-matter (Shimizu et al. 2020)) (Vanossi et al. 2020), fibers (Zhang et al. 2021; He et al. 2020), filaments, wires (Zhou et al. 2019), needles (Liu et al. 2021; Meng et al. 2020), belts (Luo et al. 2020), hooks (Fan et al. 2022), cables (Chen et al. 2020), ribbons (Johnson et al. 2020; Watts et al. 2019), helices (Rostamabadi et al. 2019), zigzags (Li et al. 2012; Abid et al. 2017)
- 2D: nanoplates (Cao et al. 2019; Lin et al. 2020; Xie et al. 2022), nanofilms (Sobolev and Kudinov 2021; Wang et al. 2021), nanofiber (Zhang et al. 2021; He et al. 2020)
- 3D: 3D structures are generally made up of combinations of 0D, 1D, and/or 2D structures. Like, the 3D skeleton of heterofibers and nanotubes, etc. (Pokropivny and Skorokhod 2008)

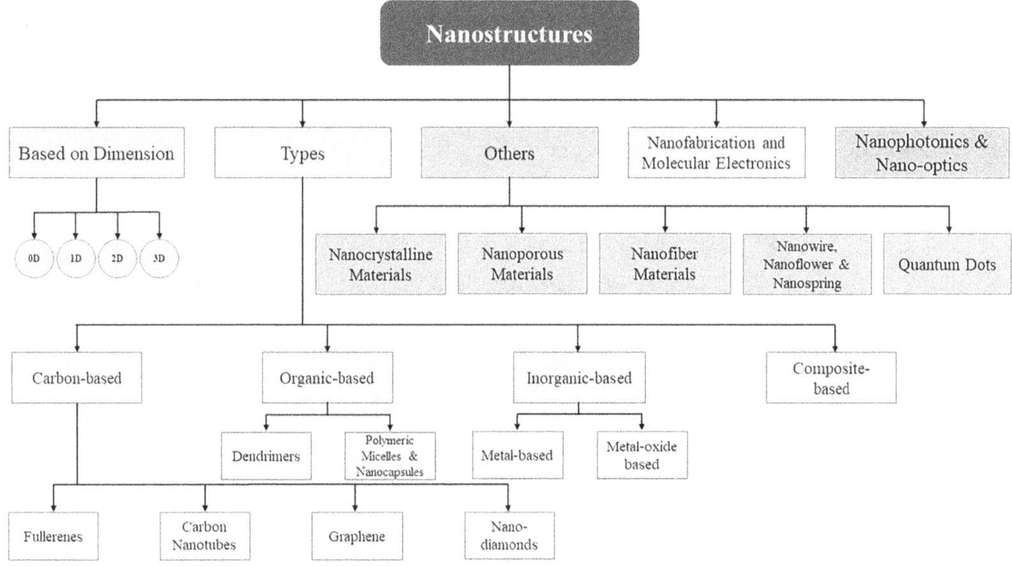

FIGURE 8.4 Pictorial representation of the different nanostructures.

8.3.3.2 Classification Based on the Composition of the Nanostructures

The shape and size of the nanostructures vary greatly. Some of the structures have been mentioned in the previous section. Based on the composition of the nanostructures they can be further classified into the following (Nasrollahzadeh et al. 2019):

- Organic Nanostructures
- Inorganic Nanostructures
- Composite-based Nanostructures (Malaki and Varma 2020; Ali and Ahmed 2018; Liu et al. 2020)
- Carbon-based Nanostructures

8.3.4 Applications of Nanotechnology

Currently, due to the presence of this technology, many novel applications can see the day of light. Some of the applications have been presented in Figure 8.5. There are a very few aspects of the daily life of human life, where the influence of nanotechnology is yet to be felt. For instance, in the article by Qu et al. (2013) they described the usage of nanotechnology in water treatment. The currently established system is not able to provide the entire population of the world with clean drinkable water and the pressure on the naturally available drinkable water sources is increasing with time. So, nanotechnology can be used and it can replace the existing treatment technologies. Like nanoabsorbents can be used to absorb the organic and inorganic contaminants present in water because the by-products of nanotechnology have a high surface area due to their small sizes. Additionally, the authors mentioned nanofilter membranes which help remove micro-sized particles efficiently (Qu et al. 2013). Other than that, the technology finds its use globally in the food industry. For instance, nanocomposites (polymer matrices with entrapped nano fillers like carbon nanotubes, cellulose, etc.) are already used in the packaging section as it is biodegradable, cost-effective, and helps the food to stay fresh for longer. Nanoliposomes can be used as a carrier for different lipids like enzymes, antioxidants, nutraceuticals, etc. (Sahani and Sharma 2021). Additionally, nanocapsulation of the thermolabile compounds and highly reactive compounds can

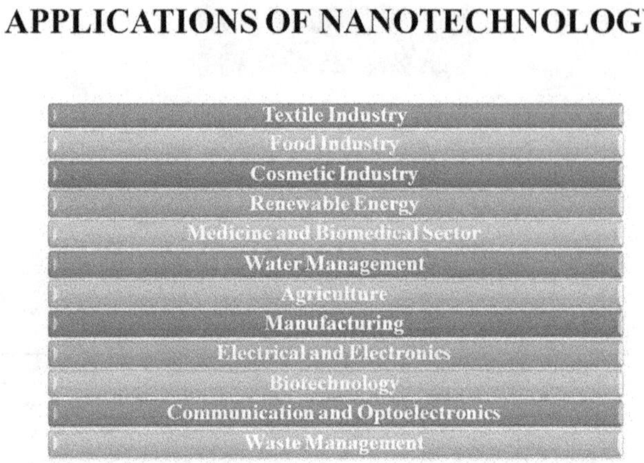

FIGURE 8.5 Applications of nanotechnology in various fields and industrial sectors.

help in better retention of the product which then can be used in the food industry. For example, when encapsulated anthocyanin is a pigment used regularly in the food industry, showing better retention and thus the pigment of the final product remains intact (Zhang et al. 2014). Other nanostructures like nanosponges (DUT-23, MOF-117, etc.) are used for the purpose of storing hydrogen as they have larger pores for which the kinetics and refilling becomes fast (Schlichtenmayer and Hirscher 2012). Perovskite nanowires when incorporated into lasers show better optical properties and these lasers can also be operated at room temperature (Eaton et al. 2016). These are some of the uncountable examples that are present in the world today of the different applications of nanotechnology.

Although this is a developing research field, much more effort is given to commercializing the end products. Given the promising future of the technology, much of the funding generally comes from governments and other educational institutes due to the lack of interest of investors in its early developing stages (Hobson 2009).

8.4 ROLE OF NANOTECHNOLOGY IN THE PREVENTION AND TREATMENT OF INFECTIOUS DISEASES

Nanotechnology has a lot of applications both in the treatment and diagnosis of diseases; similarly, the technology has found its uses in the treatment of ID. Over the years there have been outbreaks of several diseases and there might be future outbreaks. For that reason, newer technologies need to be involved so that humankind can stay prepared and another pandemic like COVID-19 can be avoided. Additionally, the demerits of the pre-existing prevention techniques have been identified to have a clear understanding of the reason nanotechnology needs to be incorporated in the diagnosis, tracking, and monitoring of ID. Discussions focusing on the different IDs and the role of nanotechnology in prevention and treatment are elaborated here.

8.4.1 Role of Nanotechnology in the Treatment of Infectious Diseases

For accuracy in treatment, it is essential to have accurate treatment protocols along with effective drugs which can help the patients to become healthy as soon as possible. For a few IDs like HIV and TB, the span of treatment is very long. An individual, who has been exposed to such disease, has to take medications for years and in some cases till the end of life. Additionally, the dosage regimen is also very complicated which makes it harder for the patient to stick with it. Therefore, formulations designed for the sustained release of drugs can be very useful in this case. It will also help to simplify the dosage regimen (Kirtane et al. 2021). Nanotechnology can help to achieve this. Nanocarriers containing the drug of choice for the purpose of treatment can show good results as they can ensure sustained release (Jain and Thareja 2019). Additionally, it can enhance the quality of the drug by enhancing the therapeutic effects shown by the drug as it travels better to the targeted site. Generally, it can be said that nanocarriers show a much superior transportation method (Huda et al. 2020). The earliest results were seen in the work of Schiffelers et al. (2001) for the treatment of urinary tract infections and infections caused by *Mycobacterium* in the early 2000s by a preparation where aminoglycoside was encapsulated within a liposome. After the study it was concluded that local application of the liposomes showed prolonged release of drugs. Furthermore, on intravenous administration, the liposomes showed prolonged antibiotic effects when compared with free drugs (Schiffelers et al. 2001; Kirtane et al. 2021). Table 8.1 shows some of the nanostructures and their utilization in the treatment of different IDs.

8.4.2 Role of Nanotechnology in the Prevention of Infectious Diseases

Primarily the methods of prevention are diagnosis, tracking, monitoring, and vaccination. While these methods are true for most diseases, but in some cases, there can be more ways of prevention. For instance in the case of malaria, the host (mosquitoes) plays a significant role in the spreading of the disease. So, killing the host can be a prevention technique that can stop the spreading of the disease. Nanotechnology can be used in this aspect also. Silver nanoparticles (AgNP) can act like green nanopesticides which can help to control mosquitoes. These NPs possess intrinsic toxicity against several arthropod vectors including mosquitoes. AgNP synthesized from *Azadirachta indica* (Mehlhorn 2016), *Arstolochia indica* (Murugan et al. 2015), and *Clerodendrum chinense* (Govindarajan et al. 2016) are some of the examples of the NPs which possess mosquitocidal activity. There are several strategies which have been taken up for killing the mosquitoes. For instance, the AgNPs synthesized from *A. cymosa* leaf extract attract the females of different mosquitoes like *An. Stephensi* and show ovicidal property. Therefore, the formulation can be used in different breeding ground of mosquitoes (Benelli and Govindarajan 2017). Other than AgNP, other metal structures also show similar properties. Like zinc nanorods formed from *Myristica fragrans* extract show larvicidal activity against *A. aegypti* mosquitoes (Ashokan et al. 2017) (Benelli et al. 2018). In the upcoming sections, some ways of prevention of IDs will be mentioned.

8.4.2.1 Diagnosis, Tracking, and Monitoring

Diagnosis, tracking, and monitoring of ID are very effective ways of controlling the spread. Additionally, over 95% of the deaths due to ID occur due to the absence of proper diagnosis and treatment facilities (Lee et al. 2010). Due to the advancement of technology, several new techniques are included, which helps in medical diagnosis in resource-limited settings. Technologies like nano/microfluidic technology, droplet microfluidics can be enlisted as some of them. These are powerful tools and they have been development over the years to ensure point-of-care (POC) diagnostics (Wang et al. 2021). Table 8.2 contains different examples where nanostructures have been used for diagnosis, tracking, and monitoring of ID.

8.4.2.2 Vaccine

Vaccination and immunization can be considered an effective way to prevent any ID outbreak and protect public health. Traditional methods have been successful but to get more efficient results newer methods have been introduced of vaccination which also involves nanotechnology. Table 8.3 sheds light on a few of such newer vaccination techniques involving nanotechnology.

TABLE 8.1
Nanostructures Used in the Treatment of Different Infectious Diseases and Their Mechanism of Actions

Infectious Disease	Causative Organism	Nanostructure	Advantages	Reference
Malaria	*Plasmodium* spp.	Dendrimers containing bis-MPA and glycine	It loads the drugs into RBCs that are still not infected by the parasite and therefore, reduces the chance of survival of the parasite.	(Coma-Cros et al. 2019)
		Dihydroartemisinin-loaded solid lipid NP	• The formulation showed sustained release for over 20 hrs. • It was absorbed through the GI tract and showed chemosupression of the parasite (*Plasmodium falciparum*). • The particles containing the positive charge had a higher absorption rate and helped increase the concentration of dihyddroartemisinin in the liver which later went into the systemic circulation.	(Omwoyo et al. 2016)
		Zwitterionic poly(butyl methacrylate-co-morpholinoethyl sulfobetaine methacrylate)-based NP loaded with curumin	• It shows improved solubility of curcumin and therefore reduces the active doses of curcumin. • The encapsulation helps in the entry of the formulation into the systemic circulation through the intestinal epithelium when it has been administered orally. • The formulations targets the parasite-infested RBCs and not the healthy unaffected RBCs.	(Biosca et al. 2021)
		Curcumin incorporated within Eudragit-hyaluronan liposomes and Eudragi-nutrisomes	• Curcumin has poor oral absorption, aqueous solubility, and rapid metabolism by incorporating the drug in liposomes, these issues can be resolved. • The nanocarrier held its integrity through the harsh gastric environment of the mice and deposited the curcumin in the intestine • The Eudragi-nutrisomes had more stability and showed better antimalarial activity.	(Martí Coma-Cros et al. 2018)
		Cyclodextrin nanospheres filled with artemisinin	• The drug has low aqueous solubility, low half-life, and high first-pass metabolism. • Nanocarriers helped to eliminate all the limitations of the drug. • The *in vitro* data showed that the nanospheres had low IC_{50} values than the free drug.	(Yaméogo et al. 2012)

TB	*Mycobacterium tuberculosis*	Niosomes containing D-Cycloserine and Ethionamide	• The drugs were released over a period of 3 days. • The dual-loaded niosomes showed better antibacterial activity than the combination of free drugs according to the MIC study.	(Kulkarni et al. 2019)
		Polymeric micelles containing Rifampicin	• The formulation showed sustained release that lasted for over 24 hrs (by 24 hrs, only 33% (fresh sample) and 38% (lyophilized sample) were released). • On testing the cytotoxicity study it was seen that the formulation had lower cytotoxicity than the free drug. • *In vitro* antibacterial study showed that the efficacy of the nanoformulation has increased than the free drug. • The *in vivo* biodistribution assay shows that the formulation is accurate to be used as an inhaled anti-TB therapy.	(Grotz et al. 2019)
		Aqueous dispersion of nanopolymersomes containing rifampicin	The small size promoted the accumulation of the drug in murine macrophages (RAW 264.7) due to which the formulation can be used as a part of the inhaled therapy.	(Moretton et al. 2015)
Acquired Immunodeficiency Syndrome (AIDS)	HIV	Poly (lactic-co-glycolic acid) NPs loaded with a combination of antiviral drugs, elvitegravir, tenofovir, emtricitabine, and alafenamide	• The nanoformulation showed positive results when tested in the humanized-BLT mice model. • The nanoformulation got deposited in the reservoir organs and the HIV-infection targets present around the body (lymph nodes, colon, etc) which meant the dosing frequency could be reduced to monthly dosing.	(Mandal et al. 2018)
		CD4+ T cell plasma membrane wrapped around PLGA NPs filled with small molecules including peptides, siRNA, Cas9	• This formulation neutralizes HIV-1 and selectively kills the CD4+ T cells and macrophages which are infected by the virus. • The encapsulation helped to reduce the time taken to show the therapeutic action of the formulation. • The cells are killed using autophagy-dependent apoptosis without having any harmful effect on bystander cells.	(Campbell et al. 2021)
Flu (Influenza)	Influenza viruses	Multivalent 6SL-PAMAM dendrimer conjugates	• The nanostructure inhibits the formation of infection caused by the influenza virus. • It provides protection against not only human influenza viruses, but also avian influenza viruses. • It reduces the cytokine and pro-inflammatory mediator levels which helps to reduce the inflammatory lesions caused by the virus in the lungs.	(Kwon et al. 2017)

(Continued)

TABLE 8.1 ***(Continued)***
Nanostructures Used in the Treatment of Different Infectious Diseases and Their Mechanism of Actions

Infectious Disease	Causative Organism	Nanostructure	Advantages	Reference
Genital wart	Human Papillomavirus (HPV)	Bovine lactoferrin formulated in the form of transferosome	• The transferosomes show improved cellular permeability and stability. • They can travel through the narrow skin pores and increase the transdermal flux of the therapeutic agent (lactoferrin)	(Hadidi, Saffari, and Faizi 2018)
Herpes	Herpes Simplex Virus (HSV) 2	Poly(lactic-co-glycolic acid) and chitosan NPs encapsulating propolis	• The formulation helps to overcome the problem regarding the poor water solubility of propolis. • The NPs inactivate the viral particles by showing intracellular activities and increasing the therapeutic efficiency of propolis. • They disrupt the viral entry and release of the virus from the host cell. • The flavonoids and other bioactive compounds derived from propolis interfere with the replication of the virus. They inhibit the viral polymerase and binds to the nucleic acid of the virus which stops the synthesis of viral nucleic acid. • The NPs also interrupt with the viral genes associated with replication (ICP4, gB, and ICP27)	(Sangboonruang et al. 2022)
	HSV 1 or HSV 2	Pseudoboehmite-graphene oxide NPs loaded with acyclovir	Formulating acyclovir in this form helps to stop the interference of gastric fluids during the absorption of the drug. Therefore, it increases the systemic bioavailability of acyclovir.	(Munhoz Jr et al. 2019; Peres et al. 2021)
Polio or poliomyelitis	Poliovirus	Electrochemically synthesized silver NPs	• The NPs, loaded in human rhabdomyosarcoma cell monolayers, to assess the cytotoxicity which was not seen even after 48 hrs post incubation, proving it to be safe for clinical applications. • Also, showed potent antiviral activity against poliovirus at a very low concentration.	(Mehndiratta et al. 2014; Huy et al. 2017)

Viral hepatitis	Hepatitis C virus	Hybrid composite made up of chitosan NPs containing self-assembly of β-cyclodextrin and sofosbuvir	• The composite holds a high loading capacity of 94.54%. • Also, it has a high release efficiency which ensures better activity than the free drug (sofosbuvir).	(El-Shafai et al. 2022)
Meningitis	Bacteria (most frequent), viruses, or fungi *Streptococcus pneumonia* (Pneumococcal meningitis)	PEGylated nano-bacitracin A prepared with the unimers, Pluronic® P85 and RVG_{29}.	• The formulation showed high efficiency to cross the BBB. • The brain performance and brain targeting ability were also excellent. • It showed good brain accumulating ability along with high therapeutic ability with negligible systematic toxicity. • It had an inhibitory effect on the activity of P-glycoprotein and stopped the P-glycoprotein-mediated drug efflux.	((WHO); Hong et al. 2018)
	Cryptococcus neoformans (Fungal meningitis)	Liposome containing Amphotericin B	• The formulation showed less cytotoxicity and systemic side effects than the free drug. • It could cross the BBB more efficiently with higher rates of absorption. • It could be administered in larger doses.	(Joseph et al. 2021)

TABLE 8.2
Nanodiagnostics to Detect, Track, and Monitor Infectious Diseases

Nanostructure	Infectious disease	Strategy	References
Nanoprobes containing mesoporous SiO_2 shells encapsulating gold-platinum nanorods	Mumps virus	• The nanostructure has a peroxidase-like activity which helps in the diagnosis by acting like a reagent. • Also, the structure helps the probe to retain its ability to detect for longer. • It shows sensitivity for mumps-specific IgM antibodies. • Identification was successful at a very low concentration of the antibodies (10 ng/mL).	(Long et al. 2020)
A microfluidic device where AuNPs and silver ions were loaded on an mChip that have been designed in the form of a ELISA-like kit	HIV and *Troponema pallidum* (Syphilis)	• It utilizes blood sample for the detection of both AIDS and syphilis and possesses sensitivity. • The results are obtained within 20 minutes making it fast and efficient. • The surface of the chip contains antigen for HIV (gp41-gp36), syphilis (TpN17), and goat antibody (IgG) through which the sample is run. • Later a flow of AuNP labeled with antibodies is allowed. • The silver ions get reduced that are present on the AuNPs once they interact with the antibodies from the sample which generate the positive signal.	(Chin et al. 2011; Wang et al. 2017)
Nanobioconjugate made up of NPs linked with ssDNA	Zika virus	A electrochemical signal is formed if the sample contains the genetic material of the zika virus	(Cajigas et al. 2020)
Citrate-stabilized AuNPs have been immobilized with the help of electrostatic energy on a glass substrate	Malaria	• It is used as a biosensor of the marker related to malaria (*Plasmodium falciparum* lactate dehydrogenase) • Phytochemical functionalization technique which helps to give rise to fluorescence amplification which acts as a positive signal indicating the presence of the biomarker.	(Minopoli et al. 2021)
QDs encapsulated with modified matrix proteins belonging to HIV-1	Tracking the events of HIV-1 infection	It helped to track the viral entry, disassembly, and interaction of the virus with the mitochondria present within the live primary microphages. This helps to better understand the interaction between the host the virus	(Li et al. 2018)
Carboxylated magnetic NPs encapsulated with hepatitis B surface antigen	Detection of hepatitis B viral infection	By immobilizing the surface antigens, the DNA aptamers were formed which helped in the formation of the chemiluminescence aptasensor. These aptasensors helped to detect the hepatitis B antigens present within the serum samples. Also, they have high sensitivity.	(Xi et al. 2015)

TABLE 8.3
Vaccines Involving Nanotechnology and Their Strategy of Utilization to Prevent the ID

Nanostructure	Disease	Strategy/Mechanism of Action	References
Gel-core nanoliposome for the delivery and controlled release of Pfs25 and CpGODN	Malaria	The nanoliposome contains phospholipid bilayers which help in the slow drug release. Due to this, the drug shows persistence without the need of administering another dose.	(Ahmed et al. 2019)
Proteosomes non-covalently attached with LPS forming a complex structure (Protollin™) which was adjuvanted with H5N1	Avian influenza caused by H5N1 virus	The nasal delivery of the H5N1 vaccine along with the Protollin adjuvant induced the cellular and humoral immune responses.	(Cao et al. 2017)
AuNPs were attached with the highly conserved extracellular region of the M2e protein of the Influenza A virus	Influenza A	The nanostructures were lyophilized and resuspended in water before carrying out intranasal vaccination. This helps to ensure mucosal immunity on top of the overall immune response. The mice were fully protected against the virus with the addition of a CpG adjuvant along with the NPs.	(Tao et al. 2014)
Glycol chitosan NPs containing hepatitis B antigen on the surface	Hepatitis B	The nanoformulation shows nucoadhesive action and has better mucosal uptake than chitosan NPs. For this they are able to showcase increased mucosal immunity along with the humoral immunity.	(Pawar and Jaganathan 2016)
Chitosan NPs coated with alginate which has the ability to entrap Hepatitis B antigen. The NPs have been anchored with lipopolysaccharide	Hepatitis B	This vaccine can be administered via the mucosal route. As lipopolysaccharide was used as an adjuvant, the anchoring ability of the nanoformulation was enhanced and it showed better immunization. It increased the concentration of sIgA and IgG in mucosal secretion and systemic circulation respectively.	(Saraf et al. 2020)
CIA09A composed of cationic liposome, *Quillaja* saponin fraction (QS-21), and de-O-acylated lipoligosaccharide (a TLR4 agonist) which was adjuvanted with VZV glycoprotein E (gE)	Chicken pox caused by VZV	The administered adjuvanted vaccine in mice showed an increase in gE-specific $CD4^+$ T cells and other cytokines expressed by the same, INF–γ, TNF-α, & IL-2. This ensures efficient cell-mediated and humoral immunity shown by the vaccine.	(Wui et al. 2019)
Poly (I:C)- PLGA-PEG copolymer NP adjuvant	TB	The vaccine can be administered intranasally. It increases $CD4^+$ T cells specific to the antigen and other cytokine specific to the $CD4^+$ T cell. Also, there was an increase in the antigen-specific IgG2c, sIgA, IgG1 and IgG which means the nanovaccine ensures both cell-mediated immunity and humoral immune responses.	(Du et al. 2022)

8.4.3 Nanoformulations (Nanomedicines and Nanovaccines) Under Clinical Trial

Due to the advancement in research, several nanomedicines and nanovaccines are in clinical trial to ensure clinical utilization for ID. In the Table 8.4, some of such nanoformulations have been enlisted along with the phase of the clinical trial they are currently in.

8.4.4 Theranostics Involving Nanotechnology for Infectious Diseases

The term "theranostic" is generally used for formulations that have been designed for the purpose of treatment as well as diagnosis. It is generally designed for individualized therapy and it is generally highly effective (Jeelani et al. 2014). Nanomaterials have found their place in this form of medicine especially due to a better understanding of the disease and the biology surrounding it. Also, targeted

TABLE 8.4
Nanoformulations Under Clinical Trial That Can be Used for the Treatment and Prevention of Global Infectious Diseases (Ventola 2017; Sharma et al. 2021; Bobo et al. 2016; Huang et al. 2020; Kumar et al. 2021)

Name of the sponsor/ manufacturer	Particle type (formulation)	Application to be investigated for	Current phase
Arbutus Biopharma (AB-001467 TKM0 HBV)	Lipid particle that targets the HBV genome with the help of three RNAi therapeutics present on the particle	Hepatitis B	Phase II completed
Gilead Sciences (AmBisome)	Liposomal amphotericin B	Cryptococcal meningitis in HIV-infected patients Aspergillus, Candida and/or Cryptococcus species infections (secondary) Visceral leishmaniasis parasite in immunocompromised patients	Approved by the FDA (1997)
Galen (DaunoXome)	Non-PEGylated liposomal daunorubicin	Kaposi's sarcoma associated with HIV	Approved by the FDA (1996)
Janssen (Doxil Caelyx)	PEGylated liposomal daunorubicin	Secondary treatment to chemotherapy in Kaposi's sarcoma associated with HIV	Approved by the FDA (1995) & EMA (1996)
Moderna (mRNA-1944)	The heavy and light chains of anti-Chikungunya antibodies present on two mRNAs that have been formulated in lipid NP technology	Prevention of Chikungunya	Phase I
Novartis Vaccines	Meningococcal ACWY conjugate vaccine containing capsular polysaccharides which were conjugated to a diphtheria toxic mutant CRM197	Bacterial Meningitis (vaccine for infants)	Phase II
AmBIsome (Gilead Sciences)	Liposome NPs (Liposomal amphotericin B)	Fungal/protozoal infections	–
Arrowhead Pharmaceuticals (ARC-520)	A vaccine was created which contained a mixture of polymers and siRNA which targeted the HBV proteins	Hepatitis B	Completed

drug delivery is easier with nanomedicine (Lee et al. 2020). Among the different theranostics applications seen, the ones targeted for IDs stand out as in the case of many IDs like TB, there is drug resistance seen within the patients. The chance for drug resistance for other ID is also very commonly seen. So, theranostic can be an effective strategy. For instance, Liao et al. (2020) designed an aggregation-induced emission carrier loaded with rifampicin. This carrier helped to locate the granulomas and emitted fluorescent signals which helped to diagnose TB at its early stages. Also, it released the rifampicin that helped in the treatment of drug-resistant TB (Liao et al. 2020). In HIV theranostics, there have been great developments. It can be seen in the work presented by Singh et al. (2020). A nanoformulation containing ritonavir, atazanavir, and curcumin was designed. Laser ablation was used in water containing bioactive molecule (Pluronic F-127) fabricate the NPs successfully. These NPs crossed the BBB and accumulated in the brain. Due to the presence of Pluronic F-127, the expression of the viral gene was reduced and the p24 antigen level also reduced drastically. Also, curcumin showed fluorescent activity which helped in tracking the cellular uptake (Singh et al. 2020). Anwar et al. (2018) emphasized the importance of theranostics when it comes to brain-eating amoebae infections which are extremely rare in nature. They mentioned that the drawback of using antiamoebic drugs is that they are unable to cross the BBB and the traditional diagnostic techniques take a long time and are painful. So NPs and QDs are the best option as they can cross the BBB with ease. They can carry out targeted drug delivery and QDs have a long luminescence lifetime for which they be used for various bioanalytical measurements (Anwar et al. 2018).

8.5 COVID-19 AND THE ROLE OF NANOTECHNOLOGY IN COVID-19

8.5.1 COVID-19

COVID-19 is an EID caused by a novel beta coronavirus known as severe acute respiratory syndrome coronavirus two or SARS-CoV-2, belonging to the coronaviridae family (Chatterjee et al. 2022). It was first reported at around the end of 2019, within a few months of 2020, the WHO had identified it as a pandemic and as a serious risk to the global population. By May 2021, the disease had caused 3.2 million individuals to lose their lives ((WHO) 2021). Few regions of the world's population were more affected than the others but this unveiled a great crisis within the healthcare community due to shortages of healthcare professionals, beds in different healthcare institutions, drugs, diagnostic services, etc. Data till the month of November 2021 says that 3.80 billion total SARS-CoV-2 infections around the world. Among this number, 43.9% of individuals have been infected more than once. The global representation has been shown in Figure 8.8. Also, it can be concluded that over 40% of the population around the globe has already been infected by the virus. This data changes all the time due to the high infectivity rate of the virus and the differences in the number of daily and cumulative infections (Barber et al. 2022). The increase in cumulative infections seen every month between January 2020 and June 2022 is represented in Figure 8.6. And the changes in the Case Fatality Rate (CFR) have been represented in Figure 8.7. It must be noted that in both the figures (Figure 8.6 & Figure 8.7) the data has been represented with respect to the global population.

On examination of the genetic material of the virus, it was seen that the virus is much similar to the genetic material of the virus responsible for the SARS outbreak in the year 2003 (Chatterjee et al. 2022). COVID-19 is characterized by fever, cough, and loss of senses of taste and smell, and tiredness. These signs are similar to many other viral infections. Other than that, symptoms like sore throat, headache, skin rash, shortness of breath, etc. are also some of the symptoms that have been reported (Alimohamadi et al. 2020; Chatterjee et al. 2022). As COVID-19 was an EID, there was no available treatment protocol and initially not much information regarding the disease or the virus was available which initiated urgency for research. Initially, it was thought that the disease was a respiratory disorder but over the time this assumption was proven to be wrong. Within a few months of the initial outbreak it could be clearly understood that it is a multi-organ disease (Pal et al. 2022).

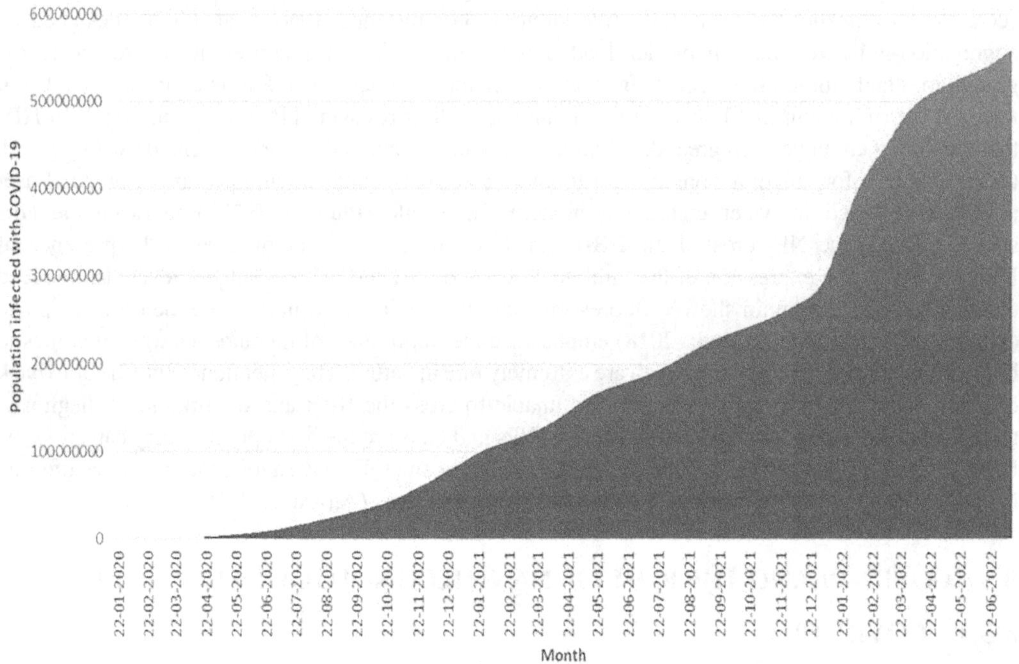

FIGURE 8.6 Representation of the monthly increase in COVID-19 cases between 22-01-2020 and 22-06-2022 in cumulative interval. Data obtained from https://ourworldindata.org/explorers/coronavirus-data-explorer (Data).

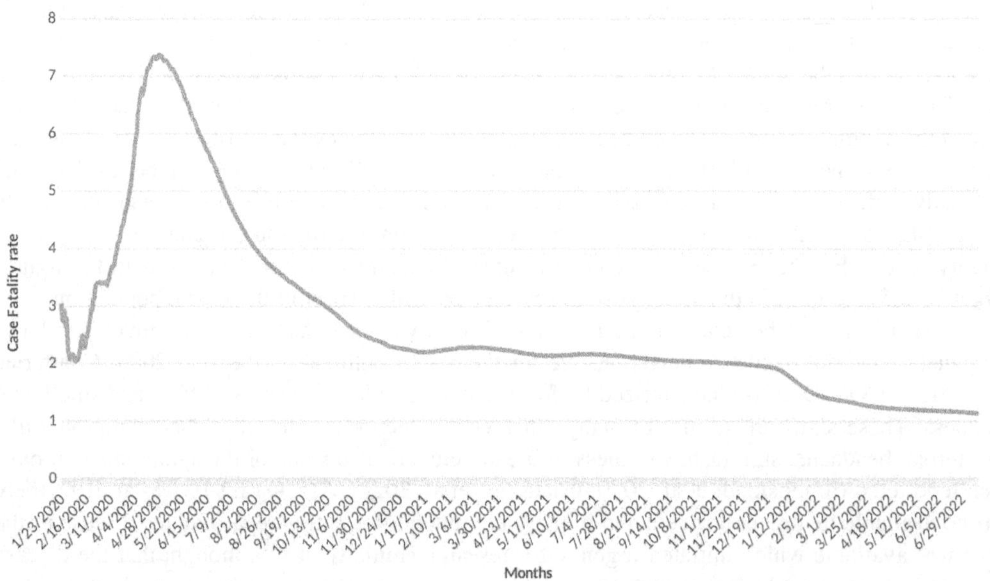

FIGURE 8.7 Graphical representation of the changes in Case Fatality Rate (CFR) due to COVID-19 between 22-01-2020 and 22-06-2022. The data has been represented in cumulative interval and it has been obtained from https://ourworldindata.org/explorers/coronavirus-data-explorer (Data).

FIGURE 8.8 A global view of the different countries and the total number of recorded COVID-19 cases as of 14.07.2022. Data has been obtained from https://ourworldindata.org/explorers/coronavirus-data-explorer (Data).

8.5.2 Role of Nanotechnology in Treatment and Prevention of COVID-19

Nanotechnology has been used intensively in the biomedical field to combat different IDs and it isn't any different when it comes to COVID-19. This utilization is especially important given the huge potential of nanotechnology to show clinical transformation (Tang et al. 2021). Initially, after the outbreak of the disease, many hypotheses were developed involving the utility of nanotechnology in the prevention and treatment of EID. For instance, in the article by Mehta et al. (2020) the authors hypothesized the importance of nanotechnology and the ways it can help to stop the spread of COVID-19. Moreover, it was postulated that nanotechnology and the emerging therapeutic approaches involving the technology could directly target the virus (Mehta et al. 2020). Just like this, various opportunities for technology were created over time. Like the development of surfaces which has self-cleaning abilities. For the development of such surfaces, NPs could be used because of the biological activities of the NPs. For instance, metal NPs have intrinsic antimicrobial properties (Brandelli et al. 2017). Therefore, the usage of metal NPs for the preparation of such surfaces can be possible. Similarly, this property can be used in other aspects like the preparation of fabric which can be later used for the preparation of masks and other personal protective equipment (Campos et al. 2020). Later, these materials (impregnated with nanomaterials) can be used in a wide range of applications which can help to protect an individual from getting infected by the virus. These can be identified as probable preventive measures from the viral disease.

When it comes to the treatment of the IDs, there are several antiviral drugs which can be reused for the said purpose. For instance, Favilavir is a drug approved by the Italian Medicines Agency for experimentation and treatment of COVID-19. Also, Remdesivir was approved by the

TABLE 8.5
Vaccines for COVID-19 Under Clinical Trial (Anselmo and Mitragotri 2021)

Name	Sponsor	Nanostructure	Identifier	Phase
mRNA-1283	Moderna	Lipid NP	NCT04813796	I
DS-5670a	Daiichi Sankyo	Lipid NP	NCT04821674	I/II
CVnCoV	CureVac	Lipid NP	NCT04860258 (III), NCT04838847 (III), NCT04674189 (III), NCT04652102 (II/III), NCT04515147 (II), NCT04449276 (I), NCT04848467 (III)	–
ARCT-021/ LUMAR-COV19	Arcturus	Lipid-enabled nucleomonomer agent containing mRNA	NCT04728347 (II), NCT04668339 (II), NCT04480957 (I/II)	–
SpRN_18-06-PL + ALFQ	U.S. Army Medical Research and Development Command	Spike ferritin NP as the vaccine and a liposomal formulation as the adjuvant	NCT04784767	I
HDT-301	SENAI CIMATEC	Lipid inorganic NP and 15-nm superparamagnetic iron oxide	NCT04844268	I
GBP510	SK Bioscience Co.	Self-assembling NP immunogen	NCT04742738, NCT04740343	I/II
NVX-CoV2373	Novavax	NP containing recombinant spike protein and having saponin based Matrix-M1 as adjuvant	NCT04611802 (III), NCT04368988 (I/II), NCT04533399 (II), NCT04583995 (III)	–

US as well as the Chinese government for the treatment of the viral disease. Several other governments have also approved other drugs including Chloroquine, NP-120, Lopinavir, etc. for treatment (Dube et al. 2021). In this aspect, the nanocarriers can be used for targeted drug delivery. And based on the nature of the drug and the potential target of the drug, the kind of nanocarrier can be chosen (Shah et al. 2021).

In the case of vaccinations, healthcare institutions are currently using vaccines which involve nanotechnology. One of the biggest examples of this is the Moderna vaccine (mRNA-1273) which is a lipid NP containing SM-102 (an ionizable cationic lipid), DSPC, PEG-DMG, and cholesterol. The vaccine was approved by the US FDA on 18th December 2020 after clinical trials were carried out to assure its safety. Another vaccine, BNT162b2 by Pfizer-BioNtech, was also approved on 11th December 2020. This vaccine is also a lipid NP containing ALC-0315 (an ionizable cationic lipid), cholesterol, and PEG-DMA. Both these vaccines did good job when it came to immunization and prevention of the EID (Anselmo and Mitragotri 2021). Other than that, there are some other vaccines which are in clinical trials for approval of clinical use. These vaccines have been mentioned in Table 8.5.

Due to the extensive research work, there have been various instances where groups of scientists have shown excellent utilizations of nanotechnology in case of treatment and prevention of COVID-19. Some examples of such research have been showcased in Table 8.6 where the advantages of using a particular nanoformulation have also been discussed. This helps to shed the limelight on the importance of nanotechnology in combating a deadly disease like COVID-19.

TABLE 8.6
Role of Nanotechnology in the Treatment and Prevention of COVID-19

Nanostructure	Purpose	Strategy and Advantages	Reference
Magnetic NPs acting as a microcarrier of drug	Targeted drug delivery and treatment	• The microcarrier worked under the influence of a magnetic field. • Based on the magnets' positions and inlet velocity, the carriers can be used for delivering drugs to the different corners of the lungs especially in the inflamed parts of the lungs of the patient. • Due to the adhering properties of the microcarriers, they can adhere to the walls of the lungs and efficiently deliver the drugs.	(Ebrahimi et al. 2021)
Halloysite nanoclay and spinel ferrites nanocomposites containing dexamethasome (Dex) was coated with polyethylene glycol (PEG)	Treatment	• The nanocomposite containing Dex shows better cell viability than the free drug. • The formulation can be used for the targeted drug delivery to the lungs of COVID-19 infected patients. • PEG helps to better penetrate respiratory mucus due to its biocompatibility. • Due to the entry of SARS-CoV-2, the pH of the cells gets altered and the nanocomposite releases Dex better at that particular pH (5.0–6.0)	(Jermy et al. 2022)
Separable microneedle patch containing polymer-encapsulated spike protein	Vaccination and immunization	• On the application of the microneedle patch on the skin, the nanovaccines were delivered intradermally. This ensures better immunization as the potency of the vaccine increases. • After the application of the patch, the IFN-γ^+CD4/8$^+$, IL-2$^+$CD4/8$^+$ T, and other cells associated with virus-specific IgG were increased in the blood, ensuring immunization. • Also, according to the *in vivo* tests, the stability of the patch and the therapeutic effects associated with the same, stayed for up to 30 days when stored at room temperature. • The current vaccination techniques call for nucleic acid vaccines where the temperature of storage can hinder the application especially in developing countries. This problem can be solved by the microneedle patch.	(Yin et al. 2021)
Nanovaccine composed of poly (I:C) as immune adjuvant, biomimetic pulmonary surfactant as capsid structure, and receptor binding domains of the virus mimicking the spike	Vaccination and immunization	• The nanovaccine was designed in such a way so that it mimics the structure of the virus. • It can be administered intranasally which ensures mucosal protection. • It increases the secretion of sIgA within the nasal secretions, which helps to neutralize the virus and prevent its entry. • Poly (I: C) helps to activate the cytokines via the TLR pathway.	(Zheng et al. 2021)

(*Continued*)

TABLE 8.6 ***(Continued)***
Role of Nanotechnology in the Treatment and Prevention of COVID-19

Nanostructure	Purpose	Strategy and Advantages	Reference
Plasmonic AuNPs integrated within a colorimetric test which includes a dual-prong approach	POC clinical diagnosis	• The NPs are capped with pre-designed antisense oligonucleotides which help in the detection process. • The RNA of the virus (if present) is extracted from the sample with a lysis buffer which is added to the sample containing NPs and antisense oligonucleotide solution. • The color change of the solution signifies the presence of SARS-CoV-2 virus within the patient. • As AuNP is used here, the chances of occurrence of false positives are reduced in comparison to the processes where pH-responsive dyes are used. • As this test targets the N gene within the virus, it can be used to detect even at the early stages of viral entry when the viral load is very less. • The test gives positive results even when the virus has undergone mutations within its gene.	(Alafeef et al. 2021)
Carbon nanotube field-effect transistor	Diagnosis	• The biosensor utilizes an electronic detection technique which helps in selective diagnosis. • A signal is generated within the biosensor when there is the presence of viral RNA. • It is highly sensitive, cheap, and fast making it a good alternative to pre-existing diagnostic tools.	(Thanihaichelvan et al. 2022)
Smartphone-dependent quantum dot barcode device	Tracking	• There are four quantum dots which are utilized to provide different data including, positive control, negative control, anti-nucleocapsid antibody, and anti-S1 RBD antibodies. • The data can be displayed within a smartphone-based imaging device which ensures more understandability from the patients' end. • The device shows high clinical sensitivity along with analytical sensitivity. • It helps in real-time surveillance and better tracking as the tool was linked with an app which gave data to the physicians as well as the public health agencies.	(Zhang et al. 2021)
Nano-enabled wearable sensors integrated with technologies like IoT, IoNT, etc.	Tracking	This technology helps in understanding and analyzing the data according to health condition and stores in an API-enabled cloud where it is easier to track the disease.	(Singh et al. 2021)

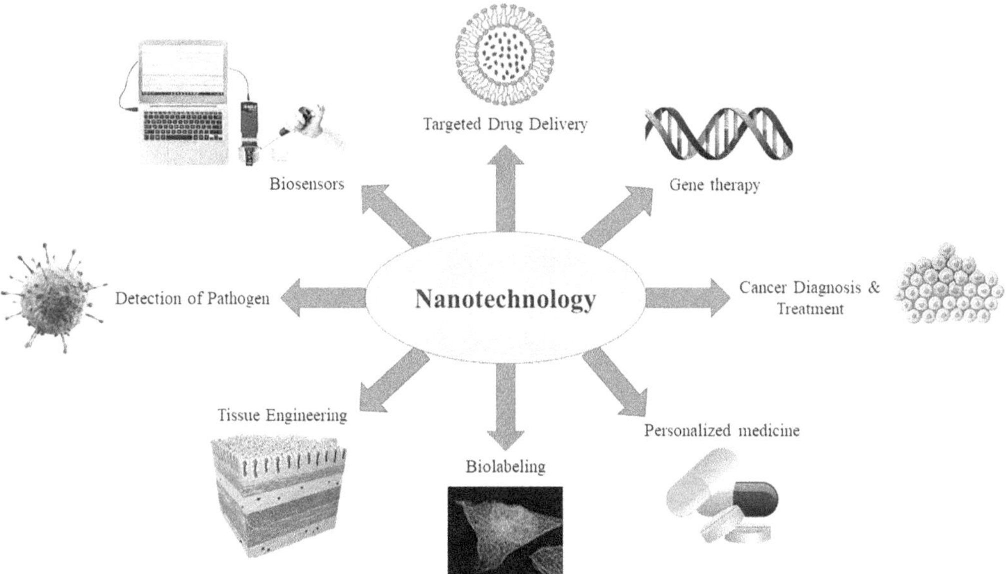

FIGURE 8.9 Different applications of nanotechnology in the field of medicine for treatment, detection, and diagnosis.

8.6 APPLICATION OF NANOTECHNOLOGY IN HEALTHCARE

Nanotechnology has been a boon to medicine and healthcare in general. Modern medicine could take a new turn because of this technology. From diagnostics to therapy, nanotechnology has its influence everywhere (Figure 8.9). As a part of contemporary medicine, healthcare can be approached in three different ways which are: 1) Medical devices, 2) Biological molecules, and 3) Drugs with smaller molecular weight (Alzate-Correa et al. 2022). In most of these approaches, nanotechnology can be applied. Most of the utilization of nanotechnology has been in medical diagnostics and targeted drug delivery (Emerich and Thanos 2003). This utility as a nanocarrier can be used to target any organ of the body. For example, in CNS-related disorders like Parkinson's disease, Alzheimer's disease, etc. nanotechnology can be used. The biggest challenge for any drug before showing any effect in the brain is to cross the BBB. Crossing the BBB can be achieved with the help of nanotechnology. Like cell-penetrating peptide-mediated drug delivery, nanocapsules, liposomes, etc. can be used as a non-invasive way for crossing the BBB and successfully delivering drugs to the CNS (Saeedi et al. 2019). Some other biomedical applications of nanotechnology in medicine have been described in Table 8.7.

8.7 FUTURE IMPROVEMENTS

Nanotechnology is still an active research field where newer ideas, data, and information are added to the knowledge pool on a regular basis. This is also partially the reason why not many examples can be seen in a clinical setting. Also, this is another reason why there is still a lack of proper toxicity and safety information in many cases. Due to the presence of this toxicity issue, the real-life use of the technology gets limited in medicine (Patel et al. 2021). Additionally, nanomaterials, due to their size, can easily enter a cell. This property has been used in a positive aspect and it has been exploited for conducting targeted drug delivery at a cellular level. But on the other hand, it has a negative side as well. Due to such characteristics, the nanomaterials can enter the eukaryotic cell and hinder the biological processes (Rana et al. 2020). This calls for a safety assessment. Similar is the case when it comes to the use of nanotechnology in the treatment and prevention of

TABLE 8.7
Biomedical Applications of Nanotechnology

Application	Nanostructure	Examples	Strategy or Mechanism of Action	References
Cancer Therapy	CNT	P-glycoprotein antibody functionalised CNT loaded with Doxorubicin	The antibody of the P-glycoprotein on K562R cells was present in the preparation and it helped in targeting the drug-resistant cells with the help of near-infrared radiation (NIR), doxorubicin was released	(Li et al. 2010)
	Micelles	Paclitaxel-loaded polymeric micelle (NK105), Doxorubicin-loaded micelles (NK911)	The pre-loaded drugs act as the core of the micelle and help in its stability and they are released due to enzymatic actions	(Cabral and Kataoka 2014)
	Nanosheets	Chitosan-functionalized MoS_2 nanosheets	Photons applied from external sources help to release the drugs and they need to be activated with NIR laser	(Yin et al. 2014; Senapati et al. 2018)
	Nanosphere and nanofiber	Nanoassembly is made up of nanosphere and nanofiber	The nanoassembly had a smaller size and that is why could interact with the cells at a molecular level. The microenvironment is acidic which helps to enhance the reactivity of the nanostructures and the application of laser shows better metastasis inhibition and tumor ablation.	(Jia et al. 2019)
Cancer diagnosis	Inorganic NP	Immunicon, Gold ion imregnated superparamagnetic NPs	These metallic NPs act as contrasting agents during diagnosis and helps in the detection of soft tissues	(Ferreira et al. 2020; Sanvicens and Marco 2008)
		Lipid calcium phosphate NPs	Helps in the imaging of lymph node metastasis by depositing in lymph nodes due to their negative surface charge and lipid surface	(Tseng et al. 2014)
	Quantum dots	Quantum dot-peptide conjugate targeting tumor vasculature, ABC triblock copolymer-coated quantum dots	They are used as imaging probes and helps to recognize specific cell types	(Onoshima et al. 2015; Senapati et al. 2018)

Diabetes (Diabetic nephropathy) treatment		NPs	NP containing Cinaciguat	These drug-loaded NPs efficiently carried the drug and unloaded its contents into the cytosolic compartments present within the target cells after which they activated the sGC pathway and supressed the non-canonical TGF-β pathway	(Fleischmann et al. 2021)
Glucose sensing		NP	Nitrogen-doped carbon dodecahedron embedded with cobalt NP	It helps to sense glucose in human serum without the utilization of any enzymes	(Mo et al. 2021)
Cardio vascular diseases	Myocardial infraction	Liposomal encapsulations	Liposomal encapsulations loaded with berberine	The preparation preserves left ventricular projection and helps to avoid and reduce adverse cardiac remodeling	(Pal et al. 2022)
	Stroke (Diagnosis)	Nanocrystal	Antibodies captured on gold NPs (AuNPs) loaded on MoS_2 nanosheet which was present on Cu_2S structure shaped like a snowflake	The ultrasensitive immunosensor helped in the sensing of NT-proBNP which is a biomarker of stroke with the help of the electrochemiluminescence technique	(Wang et al. 2020)
	Recovery from stroke (through angiogenesis)	Hydrogel	Injectable hydrogel scaffold made up of Konjac glucomannan-derived polysaccharide and heparin	Stimulates the secretion of endogenous pro-angiogenic growth factor which helps to promote angiogenesis	(Liao et al. 2020)
	Hypertension	Nanocapsule	Polycaprolactone encapsulated resveratrol nanocapsule	Shows a reduction in both the systolic and the diastolic blood pressure	(Pal et al. 2022; Hesari et al. 2021)
Kidney diseases		NPs	Resveratrol-loaded NPs	The NPs reduced the concentration of NLRP3 inflammasome and IL-1β and therefore, reduced inflammation of the kidney.	(Ma et al. 2020)
Regenerative medicine		Graphene oxide	-	Graphene oxide showed an increase in the number of tyrosine hydroxylase-positive cells and increased dopamine neuron differentiation of GFP-reporter embryonic stem cells in a dose-dependent manner.	(Yang et al. 2014)

(Continued)

TABLE 8.7 *(Continued)*
Biomedical Applications of Nanotechnology

Application	Nanostructure	Examples	Strategy or Mechanism of Action	References
Tissue engineering	CNT	3D-printed scaffolds with CNT	The 3D-printed scaffolds when coated with single stranded DNA CNT complex show better characteristics like adhesion, differentiation of cells, less toxicity, and improved proliferation of cells due to the altered static electric force of the surface.	(Liu et al. 2020)
Would healing	Nanofiber and hydrogel	Resveratrol-loaded poly(lactic-*co*-glycolic acid) electrospinning nanofiber mat for the upper layer and alginate di-aldehyde–gelatin crosslinking hydrogel as the lower layer	This preparation mimicked the epidermis of the skin and helped the growth of cells (HaCaT cells and human embryonic skin) on it and therefore, increasing the wound healing rate	(Ma et al. 2021)
Diagnosis of Hepatitis B antigen	Nanobeads	Quantum dots-nanobeads	The quantum dot nanobeads helped to amplify the signal indicators of the surface antigen proteins	(Zhang et al. 2014)

ID. So, the future improvisations that can be carried out in this aspect are the development of standardized guidelines and an increase in regulatory requirement. With the help of these, a researcher will be able to better assess the risks associated with the nanomaterial they are working on. Also, another challenge that stands in front of utilization of nanotechnology in ID is the cost. The more developed the technology becomes; the harder it will be for developing countries to use it due to the price. For instance, a currently approved nanoformulation, amphotericin B liposomal formulation, is not cheap making it difficult for many people to opt for this particular nanoformulation. So, researchers need to delve on these issues and accordingly design the nanoformulation so that it can be globally accessible irrespective of the economic situation of the country or the community (Kirtane et al. 2021).

8.8 CONCLUSION

ID does not take much time to convert into a public health emergency if proper treatment and prevention options are absent. To have an effective response against the ID, involvement of multiple disciplines is quintessential. Proper prevention strategies are also an important aspect to be taken care of to avoid one ID to spread and becoming a public health concern. Over the years, the globe has experienced the outbreak of several diseases and the global health institutions have figured out a way to tackle them effectively. But after the outbreak of COVID-19, everyone was taken aback due to the unknown nature of the disease resulting in numerous deaths around the globe. Before the latest outbreak, several treatment protocols were present for effective prevention and treatment the IDs. Nanotechnology has been slowly introduced as a possible option to the pre-existing protocols. Due to the nature of the nanomaterials, they can be used effectively as a carrier (of vaccines as well as drugs), diagnostic and tracking tools. Now after the COVID-19 outbreak there have been more additions and the involvement of nanotechnology has increased drastically. Two of the vaccines (the Moderna vaccine and BNT162b2) that have been approved by the FDA for emergency use involve nanotechnology. This can be served as the best example of nanotechnology and its involvement in the preventive measures of IDs which in this case is COVID-19. All these have been previously discussed in the chapter and with the understanding that has been developed, it can be stated that nanotechnology has immense potential when it comes to the prevention of IDs. But some challenges do exist where improvisations are necessary to make sure more use of the technology is seen at a clinical setting.

LIST OF ABBREVIATIONS

0D	0 Dimensional
1D	1 Dimensional
2D	2 Dimensional
3D	3 Dimensional
AgNP	Silver nanoparticle
AIDS	Acquired immunodeficiency syndrome
ASSUERD	Criteria for the designing of ideal diagnostic tool by WHO
AuNP	Gold nanoparticles
BBB	Blood brain barrier
CNS	Central nervous system
CNT	Carbon nanotube
COVID-19	Coronavirus disease 2019
DALY	Disability-Adjusted Life Year
Dex	Dexamethasome
DNA	Deoxyribonucleic acid

ECL	Electrochemiluminescence
EID	Emerging infectious disease
GI tract	Gastrointestinal tract
H1N1	Strain of Influenza causing virus in human
HIV	Human Immunodeficiency Virus
HSV	Herpes simplex virus
IC_{50}	Value depicting the half of the maximum inhibitory concentration
ID	Infectious disease
MIC	Minimum inhibitory concentration
NIR	Near infrared radiation
NP	Nanoparticle
NT	Nanotube
PCR	Polymerase chain reaction
PEG	Polyethylene glycol
POC	Point-of-care
QD	Quantum dot
RBC	Red blood cell
RNA	Ribonucleic acid
SARS-CoV-2	Severe acute respiratory syndrome coronavirus 2
TB	Tuberculosis
TLR	Toll-like receptor
VZV	Varicella zoster virus
WHO	World Health Organization

REFERENCES

Abid, M., Anwer Shoaib, M Hassan Farooq, Hongbo Wu, Dashuai Ma, and Botao Fu. 2017. "Edge magnetism and electronic structure properties of zigzag nanoribbons of arsenene and antimonene." *Journal of Physics and Chemistry of Solids* 110:167–172.

Abubakar, Ibrahim, Philippe Gautret, Gary W. Brunette, Lucille Blumberg, David Johnson, Gilles Poumerol, Ziad A. Memish, Maurizio Barbeschi, and Ali S. Khan. 2012. "Global perspectives for prevention of infectious diseases associated with mass gatherings." *The Lancet infectious diseases* 12 (1):66–74.

Ahmed, Toqeer, Muhammad Zeeshan Hyder, Irfan Liaqat, and Miklas Scholz. 2019. "Climatic conditions: conventional and nanotechnology-based methods for the control of mosquito vectors causing human health issues." *International Journal of Environmental Research and Public Health* 16 (17):3165.

Alafeef, Maha, Parikshit Moitra, Ketan Dighe, and Dipanjan Pan. 2021. "RNA-extraction-free nano-amplified colorimetric test for point-of-care clinical diagnosis of COVID-19." *Nature Protocols* 16 (6): 3141–3162.

Ali, Akbar, and Shakeel Ahmed. 2018. "A review on chitosan and its nanocomposites in drug delivery." *International Journal of Biological Macromolecules* 109:273–286.

Alimohamadi, Yousef, Mojtaba Sepandi, Maryam Taghdir, and Hadiseh Hosamirudsari. 2020. "Determine the most common clinical symptoms in COVID-19 patients: a systematic review and meta-analysis." *Journal of Preventive Medicine and Hygiene* 61 (3):E304.

Alzate-Correa, D., W. R. Lawrence, A. Salazar-Puerta, N. Higuita-Castro, and D. Gallego-Perez. 2022. "Nanotechnology-driven cell-based therapies in regenerative medicine." *The AAPS Journal* 24 (2):1–15.

Anselmo, Aaron C., and Samir Mitragotri. 2021. "Nanoparticles in the clinic: An update post COVID-19 vaccines." *Bioengineering & Translational Medicine* 6 (3):e10246.

Anwar, Ayaz, Ruqaiyyah Siddiqui, and Naveed Ahmed Khan. 2018. "Importance of theranostics in rare brain-eating amoebae infections." *ACS Chemical Neuroscience* 10 (1):6–12.

Arya, Sunil K., Shibu Saha, Jaime E. Ramirez-Vick, Vinay Gupta, Shekhar Bhansali, and Surinder P. Singh. 2012. "Recent advances in ZnO nanostructures and thin films for biosensor applications." *Analytica Chimica Acta* 737:1–21.

Ashokan, Anila P., Manickam Paulpandi, Devakumar Dinesh, Kadarkarai Murugan, Chithravel Vadivalagan, and Giovanni Benelli. 2017. "Toxicity on dengue mosquito vectors through Myristica fragrans-synthesized zinc oxide nanorods, and their cytotoxic effects on liver cancer cells (HepG2)." *Journal of Cluster Science* 28 (1):205–226.

Barber, Ryan M., Reed J. D. Sorensen, David M. Pigott, Catherine Bisignano, Austin Carter, Joanne O. Amlag, James K. Collins, Cristiana Abbafati, Christopher Adolph, and Adrien Allorant. 2022. "Estimating global, regional, and national daily and cumulative infections with SARS-CoV-2 through Nov 14, 2021: a statistical analysis." *The Lancet*.

Benelli, Giovanni, and Marimuthu Govindarajan. 2017. "Green-synthesized mosquito oviposition attractants and ovicides: towards a nanoparticle-based "lure and kill" approach?" *Journal of Cluster Science* 28 (1):287–308.

Benelli, Giovanni, Filippo Maggi, Roman Pavela, Kadarkarai Murugan, Marimuthu Govindarajan, Baskaralingam Vaseeharan, Riccardo Petrelli, Loredana Cappellacci, Suresh Kumar, and Anders Hofer. 2018. "Mosquito control with green nanopesticides: towards the One Health approach? A review of non-target effects." *Environmental Science and Pollution Research* 25 (11): 10184–10206.

Biosca, Arnau, Pol Cabanach, Muthanna Abdulkarim, Mark Gumbleton, Cristian Gómez-Canela, Miriam Ramírez, Inés Bouzón-Arnáiz, Yunuen Avalos-Padilla, Salvador Borros, and Xavier Fernàndez-Busquets. 2021. "Zwitterionic self-assembled nanoparticles as carriers for Plasmodium targeting in malaria oral treatment." *Journal of Controlled Release* 331:364–375.

Bobo, Daniel, Kye J. Robinson, Jiaul Islam, Kristofer J. Thurecht, and Simon R. Corrie. 2016. "Nanoparticle-based medicines: a review of FDA-approved materials and clinical trials to date." *Pharmaceutical research* 33 (10):2373–2387.

Boholm, Max. 2016. "The use and meaning of nano in American English: Towards a systematic description." *Ampersand* 3:163–173.

Brandelli, Adriano, Ana Carolina Ritter, and Flávio Fonseca Veras. 2017. "Antimicrobial activities of metal nanoparticles." In *Metal Nanoparticles in Pharma*, 337–363. Springer.

Cabral, Horacio, and Kazunori Kataoka. 2014. "Progress of drug-loaded polymeric micelles into clinical studies." *Journal of Controlled Release* 190:465–476.

Cajigas, Sebastian, Daniel Alzate, and Jahir Orozco. 2020. "Gold nanoparticle/DNA-based nanobioconjugate for electrochemical detection of Zika virus." *Microchimica Acta* 187 (11):1–10.

Campbell, Grant R., Jia Zhuang, Gang Zhang, Igor Landa, Luke J. Kubiatowicz, Diana Dehaini, Ronnie H. Fang, Liangfang Zhang, and Stephen A. Spector. 2021. "CD4+ T cell-mimicking nanoparticles encapsulating DIABLO/SMAC mimetics broadly neutralize HIV-1 and selectively kill HIV-1-infected cells." *Theranostics* 11 (18):9009.

Campos, Estefânia VR, Anderson E. S. Pereira, Jhones Luiz De Oliveira, Lucas Bragança Carvalho, Mariana Guilger-Casagrande, Renata De Lima, and Leonardo Fernandes Fraceto. 2020. "How can nano-technology help to combat COVID-19? Opportunities and urgent need." *Journal of Nanobiotechnology* 18 (1):1–23.

Cao, Kangzhe, Huiqiao Liu, Wangyang Li, Qingqing Han, Zhang Zhang, Kejing Huang, Qiangshan Jing, and Lifang Jiao. 2019. "CuO nanoplates for high-performance potassium-ion batteries." *Small* 15 (36):1901775.

Cao, Weiping, Jin Hyang Kim, Adrian J. Reber, Mary Hoelscher, Jessica A. Belser, Xiuhua Lu, Jacqueline M. Katz, Shivaprakash Gangappa, Martin Plante, and David S. Burt. 2017. "Nasal delivery of Protollin-adjuvanted H5N1 vaccine induces enhanced systemic as well as mucosal immunity in mice." *Vaccine* 35 (25):3318–3325.

Chatterjee, Amrita, Rajdeep Saha, Arpita Mishra, Deepak Shilkar, Venkatesan Jayaprakash, Pawan Sharma, and Biswatrish Sarkar. 2022. "Molecular determinants, clinical manifestations and effects of immunization on cardiovascular health during COVID-19 pandemic era-A review." *Current Problems in Cardiology*:101250.

Chen, Yijun, Muhammad Yousaf, Yunsong Wang, Zhipeng Wang, Shuaifeng Lou, Ray P. S. Han, Yuan Yang, and Anyuan Cao. 2020. "Nanocable with thick active intermediate layer for stable and high-areal-capacity sodium storage." *Nano Energy* 78:105265.

Chin, Curtis D., Tassaneewan Laksanasopin, Yuk Kee Cheung, David Steinmiller, Vincent Linder, Hesam Parsa, Jennifer Wang, Hannah Moore, Robert Rouse, and Gisele Umviligihozo. 2011. "Microfluidics-based diagnostics of infectious diseases in the developing world." *Nature medicine* 17 (8):1015–1019.

Coma-Cros, Elisabet Martí, Alexandre Lancelot, María San Anselmo, Livia Neves Borgheti-Cardoso, Juan José Valle-Delgado, José Luis Serrano, Xavier Fernàndez-Busquets, and Teresa Sierra. 2019. "Micelle carriers based on dendritic macromolecules containing bis-MPA and glycine for antimalarial drug delivery." *Biomaterials science* 7 (4):1661–1674.

Data, Our World in. COVID-19 Data Explorer. https://ourworldindata.org/explorers/coronavirus-data-explorer. Last accessed on 21-05-2023.

Devan, Rupesh S., Ranjit A. Patil, Jin-Han Lin, and Yuan-Ron Ma. 2012. "One-dimensional metal-oxide nanostructures: recent developments in synthesis, characterization, and applications." *Advanced Functional Materials* 22 (16):3326–3370.

Du, Xiufen, Daquan Tan, Yang Gong, Yifan Zhang, Jiangyuan Han, Wei Lv, Tao Xie, Pu He, Zongjie Hou, and Kun Xu. 2022. "A new poly (I: C)-decorated PLGA-PEG nanoparticle promotes Mycobacterium tuberculosis fusion protein to induce comprehensive immune responses in mice intranasally." *Microbial Pathogenesis* 162:105335.

Dube, Taru, Amrito Ghosh, Jibanananda Mishra, Uday B. Kompella, and Jiban Jyoti Panda. 2021. "Repurposed drugs, molecular vaccines, immune-modulators, and nanotherapeutics to treat and prevent COVID-19 associated with SARS-CoV-2, a deadly nanovector." *Advanced Therapeutics* 4 (2): 2000172.

Eaton, Samuel W., Anthony Fu, Andrew B. Wong, Cun-Zheng Ning, and Peidong Yang. 2016. "Semiconductor nanowire lasers." *Nature Reviews Materials* 1 (6):1–11.

Ebrahimi, Sina, Amir Shamloo, Mojgan Alishiri, Yasaman Mozhdehbakhsh Mofrad, and Fatemeh Akherati. 2021. "Targeted pulmonary drug delivery in coronavirus disease (COVID-19) therapy: A patient-specific in silico study based on magnetic nanoparticles-coated microcarriers adhesion." *International Journal of Pharmaceutics* 609:121133.

Eigen, Manfred, Werner J. Kloft, and Gerhard Brandner. 2002. "Transferability of HIV by arthropods supports the hypothesis about transmission of the virus from apes to man." *Naturwissenschaften* 89 (4):185–186.

El-Shafai, Nagi M, Mamdouh S. Masoud, Mohamed M. Ibrahim, Mohamed S. Ramadan, Gaber A. M. Mersal, and Ibrahim M. El-Mehasseb. 2022. "Drug delivery of sofosbuvir drug capsulated with the β-cyclodextrin basket loaded on chitosan nanoparticle surface for anti-hepatitis C virus (HCV)." *International Journal of Biological Macromolecules* 207:402–413.

Emerich, Dwaine F., and Christopher G. Thanos. 2003. "Nanotechnology and medicine." *Expert opinion on biological therapy* 3 (4):655–663.

Fan, Xin, Xizheng Wu, Fan Yang, Lei Wang, Kai Ludwig, Lang Ma, Andrej Trampuz, Chong Cheng, and Rainer Haag. 2022. "A nanohook-equipped bionanocatalyst for localized near-infrared-enhanced catalytic bacterial disinfection." *Angewandte Chemie* 134 (8):e202113833.

Farajpour, Ali, Mergen H. Ghayesh, and Hamed Farokhi. 2018. "A review on the mechanics of nanostructures." *International Journal of Engineering Science* 133:231–263.

Farokhzad, Omid C., and Robert Langer. 2009. "Impact of nanotechnology on drug delivery." *ACS Nano* 3 (1):16–20.

Fernstrom, Aaron, and Michael Goldblatt. 2013. "Aerobiology and its role in the transmission of infectious diseases." *Journal of pathogens* 2013:1–14.

Ferreira, Maria, João Sousa, Alberto Pais, and Carla Vitorino. 2020. "The role of magnetic nanoparticles in cancer nanotheranostics." *Materials* 13 (2):266.

Fleischmann, Daniel, Manuela Harloff, Sara Maslanka Figueroa, Jens Schlossmann, and Achim Goepferich. 2021. "Targeted delivery of soluble guanylate cyclase (sGC) activator cinaciguat to renal mesangial cells via virus-mimetic nanoparticles potentiates anti-fibrotic effects by cGMP-mediated suppression of the TGF-β pathway." *International journal of molecular sciences* 22 (5):2557.

Fredricks, David N., and David A. Relman. 1999. "Application of polymerase chain reaction to the diagnosis of infectious diseases." *Clinical Infectious Diseases*:475–486.

Gale, Emily C., Gillie A. Roth, Anton A. A. Smith, Marcela Alcántara-Hernández, Juliana Idoyaga, and Eric A. Appel. 2020. "A nanoparticle platform for improved potency, stability, and adjuvanticity of poly (I: C)." *Advanced Therapeutics* 3 (1):1900174.

Govindarajan, Marimuthu, Mohan Rajeswary, S. L. Hoti, Kadarkarai Murugan, Kalimuthu Kovendan, Subramanian Arivoli, and Giovanni Benelli. 2016. "Clerodendrum chinense-mediated biofabrication of silver nanoparticles: Mosquitocidal potential and acute toxicity against non-target aquatic organisms." *Journal of Asia-Pacific Entomology* 19 (1):51–58.

Grotz, Estefanía, Nancy L. Tateosian, Jimena Salgueiro, Ezequiel Bernabeu, Lorena Gonzalez, Maria Letizia Manca, Nicolas Amiano, Donatella Valenti, Maria Manconi, and Verónica García. 2019. "Pulmonary delivery of rifampicin-loaded soluplus micelles against Mycobacterium tuberculosis." *Journal of Drug Delivery Science and Technology* 53:101170.

Hadidi, Naghmeh, Mostafa Saffari, and Mehrdad Faizi. 2018. "Optimized transferosomal bovine lactoferrin (BLF) as a promising novel non-invasive topical treatment for genital warts caused by human papiluma virus (HPV)." *Iranian Journal of Pharmaceutical Research: IJPR* 17 (Suppl2):12.

He, Yanghua, Hui Guo, Sooyeon Hwang, Xiaoxuan Yang, Zizhou He, Jonathan Braaten, Stavros Karakalos, Weitao Shan, Maoyu Wang, and Hua Zhou. 2020. "Single cobalt sites dispersed in hierarchically porous nanofiber networks for durable and high-power PGM-free cathodes in fuel cells." *Advanced Materials* 32 (46):2003577.

Hesari, Mahvash, Pantea Mohammadi, Fatemeh Khademi, Dareuosh Shackebaei, Saeideh Momtaz, Narges Moasefi, Mohammad Hosein Farzaei, and Mohammad Abdollahi. 2021. "Current Advances in the use of nanophytomedicine therapies for human cardiovascular diseases." *International journal of nanomedicine* 16:3293.

Hobson, David W. 2009. "Commercialization of nanotechnology." *Wiley Interdisciplinary Reviews: Nanomedicine and Nanobiotechnology* 1 (2):189–202.

Hong, Wei, Zehui Zhang, Lipeng Liu, Yining Zhao, Dexian Zhang, and Mingchun Liu. 2018. "Brain-targeted delivery of PEGylated nano-bacitracin A against Penicillin-sensitive and-resistant Pneumococcal meningitis: Formulated with RVG29 and Pluronic® P85 unimers." *Drug delivery* 25 (1):1886–1897.

Huang, Hui, Wei Feng, Yu Chen, and Jianlin Shi. 2020. "Inorganic nanoparticles in clinical trials and translations." *Nano Today* 35:100972.

Huda, Shamsul, Md Aftab Alam, and Pramod Kumar Sharma. 2020. "Smart nanocarriers-based drug delivery for cancer therapy: An innovative and developing strategy." *Journal of Drug Delivery Science and Technology* 60:102018.

Hulla, J. E., S. C. Sahu, and A. W. Hayes. 2015. "Nanotechnology: History and future." *Human & experimental toxicology* 34 (12):1318–1321.

Huremović, Damir. 2019. "Brief history of pandemics (pandemics throughout history)." In *Psychiatry of Pandemics*, 7–35. Springer.

Huy, Tran Quang, Nguyen Thi Hien Thanh, Nguyen Thanh Thuy, Pham Van Chung, Pham Ngoc Hung, Anh-Tuan Le, and Nguyen Thi Hong Hanh. 2017. "Cytotoxicity and antiviral activity of electrochemical–synthesized silver nanoparticles against poliovirus." *Journal of Virological Methods* 241:52–57.

Jain, Akhlesh K., and Suresh Thareja. 2019. "In vitro and in vivo characterization of pharmaceutical nanocarriers used for drug delivery." *Artificial cells, Nanomedicine, and Biotechnology* 47 (1): 524–539.

Jeelani, S., RC Jagat Reddy, Thangadurai Maheswaran, G. S. Asokan, A. Dany, and B. Anand. 2014. "Theranostics: A treasured tailor for tomorrow." *Journal of Pharmacy & Bioallied Sciences* 6 (Suppl 1):S6.

Jermy, B Rabindran, Vijaya Ravinayagam, D. Almohazey, W. A. Alamoudi, H. Dafalla, Sultan Akhtar, and Gazali Tanimu. 2022. "PEGylated green halloysite/spinel ferrite nanocomposites for pH sensitive delivery of dexamethasone: A potential pulmonary drug delivery treatment option for COVID-19." *Applied Clay Science* 216:106333.

Jia, Hao-Ran, Ya-Xuan Zhu, Xiaoyang Liu, Guang-Yu Pan, Ge Gao, Wei Sun, Xiaodong Zhang, Yao-Wen Jiang, and Fu-Gen Wu. 2019. "Construction of dually responsive nanotransformers with nanosphere–nanofiber–nanosphere transition for overcoming the size paradox of anticancer nanodrugs." *ACS Nano* 13 (10):11781–11792.

Johnson, Asha P., H. V. Gangadharappa, and K. Pramod. 2020. "Graphene nanoribbons: A promising nanomaterial for biomedical applications." *Journal of Controlled Release* 325:141–162.

Joseph, Sharon K., M. A. Arya, Sachin Thomas, and Sreeja C. Nair. 2021. "Nanomedicine as a future therapeutic approach for treating meningitis." *Journal of Drug Delivery Science and Technology* 67:102968.

Keita, Mamadou B., Ibrahim Hamad, and Fadi Bittar. 2014. "Looking in apes as a source of human pathogens." *Microbial Pathogenesis* 77:149–154.

Kirtane, Ameya R., Malvika Verma, Paramesh Karandikar, Jennifer Furin, Robert Langer, and Giovanni Traverso. 2021. "Nanotechnology approaches for global infectious diseases." *Nature Nanotechnology* 16 (4):369–384.

Kotra, Lakshmi P. 2007. "Viral Disease." *xPharm: The Comprehensive Pharmacology Reference*:1.

Kulkarni, Pratik, Deepak Rawtani, and Tejas Barot. 2019. "Formulation and optimization of long acting dual niosomes using box-Behnken experimental design method for combinative delivery of ethionamide and D-cycloserine in tuberculosis treatment." *Colloids and Surfaces A: Physicochemical and Engineering Aspects* 565:131–142.

Kumar, Ramya, Cristiam F Santa Chalarca, Matthew R. Bockman, Craig Van Bruggen, Christian J. Grimme, Rishad J. Dalal, Mckenna G. Hanson, Joseph K. Hexum, and Theresa M. Reineke. 2021. "Polymeric delivery of therapeutic nucleic acids." *Chemical Reviews* 121 (18):11527–11652.

Kwon, Seok-Joon, Dong Hee Na, Jong Hwan Kwak, Marc Douaisi, Fuming Zhang, Eun Ji Park, Jong-Hwan Park, Hana Youn, Chang-Seon Song, and Ravi S. Kane. 2017. "Nanostructured glycan architecture is important in the inhibition of influenza A virus infection." *Nature Nanotechnology* 12 (1):48–54.

Lee, Songyi, Thanh Chung Pham, Chaeeon Bae, Yeonghwan Choi, Yong Kyun Kim, and Juyoung Yoon. 2020. "Nano theranostics platforms that utilize proteins." *Coordination Chemistry Reviews* 412: 213258.

Lee, Won Gu, Yun-Gon Kim, Bong Geun Chung, Utkan Demirci, and Ali Khademhosseini. 2010. "Nano/ Microfluidics for diagnosis of infectious diseases in developing countries." *Advanced Drug Delivery Reviews* 62 (4-5):449–457.

Li, Meixiu, Tao Chen, J Justin Gooding, and Jingquan Liu. 2019. "Review of carbon and graphene quantum dots for sensing." *ACS Sensors* 4 (7):1732–1748.

Li, Qin, Wen Yin, Wei Li, Zhiping Zhang, Xiaowei Zhang, Xian-En Zhang, and Zongqiang Cui. 2018. "Encapsulating quantum dots within HIV-1 Virions through site-specific decoration of the matrix protein enables single virus tracking in live primary macrophages." *Nano Letters* 18 (12): 7457–7468.

Li, Ruibin, Ren'an Wu, Liang Zhao, Minghuo Wu, Ling Yang, and Hanfa Zou. 2010. "P-glycoprotein antibody functionalized carbon nanotube overcomes the multidrug resistance of human leukemia cells." *ACS Nano* 4 (3):1399–1408.

Li, Yafei, Dihua Wu, Zhen Zhou, Carlos R. Cabrera, and Zhongfang Chen. 2012. "Enhanced Li adsorption and diffusion on MoS_2 zigzag nanoribbons by edge effects: a computational study." *The Journal of Physical Chemistry Letters* 3 (16):2221–2227.

Liang, Zihui, Yang Liu, Qian Zhang, Yupeng Guo, and Qiang Ma. 2020. "The high luminescent polydopamine nanosphere-based ECL biosensor with steric effect for MUC1 detection." *Chemical Engineering Journal* 385:123825.

Liao, Xiaoshan, Xushan Yang, Hong Deng, Yuting Hao, Lianzhi Mao, Rongjun Zhang, Wenzhen Liao, and Miaomiao Yuan. 2020. "Injectable hydrogel-based nanocomposites for cardiovascular diseases." *Frontiers in Bioengineering and Biotechnology* 8:251.

Liao, Yuhui, Bin Li, Zheng Zhao, Yu Fu, Qingqin Tan, Xingyu Li, Wei Wang, Jialing Yin, Hong Shan, and Ben Zhong Tang. 2020. "Targeted theranostics for tuberculosis: a rifampicin-loaded aggregation-induced emission carrier for granulomas tracking and anti-infection." *ACS Nano* 14 (7):8046–8058.

Lin, Yifan, Hao Wan, Dan Wu, Gen Chen, Ning Zhang, Xiaohe Liu, Junhui Li, Yijun Cao, Guanzhou Qiu, and Renzhi Ma. 2020. "Metal–organic framework hexagonal nanoplates: bottom-up synthesis, topotactic transformation, and efficient oxygen evolution reaction." *Journal of the American Chemical Society* 142 (16):7317–7321.

Liu, Hongbin, Jianchen Li, Lin Ye, Lijun Zhao, and Yongfu Zhu. 2021. "Co_{1-x-y} Ni $_x$ Zn $_y$ (CO_3) 0.5 (OH)· 0.11 H_2O nanoneedles–NiCo-layered double hydroxide nanosheet composites on vulcanized Ni foams for supercapacitors." *ACS Applied Nano Materials* 4 (2):1743–1753.

Liu, Xifeng, Matthew N. George, Sungjo Park, A Lee Miller II, Bipin Gaihre, Linli Li, Brian E. Waletzki, Andre Terzic, Michael J. Yaszemski, and Lichun Lu. 2020. "3D-printed scaffolds with carbon nanotubes for bone tissue engineering: Fast and homogeneous one-step functionalization." *Acta Biomaterialia* 111:129–140.

Liu, Yang, Tiannan Yang, Bing Zhang, Teague Williams, Yen-Ting Lin, Li Li, Yao Zhou, Wenchang Lu, Seong H. Kim, and Long-Qing Chen. 2020. "Structural insight in the interfacial effect in ferroelectric polymer nanocomposites." *Advanced Materials* 32 (49):2005431.

Long, Lin, Rui Cai, Jianbo Liu, and Xiaochun Wu. 2020. "A novel nanoprobe based on core–shell Au@ Pt@ mesoporous SiO_2 nanozyme with enhanced activity and stability for mumps virus diagnosis." *Frontiers in Chemistry* 8:463.

Luo, Jingting, Xiaoying Feng, Hao Kan, Hui Li, and Chen Fu. 2020. "One-dimensional Bi_2S_3 nanobelts-based surface acoustic wave sensor for NO_2 detection at room temperature." *IEEE Sensors Journal* 21 (2):1404–1408.

Ma, Wendi, Mingjuan Zhou, Wenying Dong, Shanshan Zhao, Yilong Wang, Jihang Yao, Zhewen Liu, Hongshuang Han, Dahui Sun, and Mei Zhang. 2021. "A bi-layered scaffold of a poly (lactic-co-glycolic acid) nanofiber mat and an alginate–gelatin hydrogel for wound healing." *Journal of Materials Chemistry B* 9 (36):7492–7505.

Ma, Yanhong, Fanghao Cai, Yangyang Li, Jianghua Chen, Fei Han, and Weiqiang Lin. 2020. "A review of the application of nanoparticles in the diagnosis and treatment of chronic kidney disease." *Bioactive Materials* 5 (3):732–743.

Malaki, Massoud, and Rajender S. Varma. 2020. "Mechanotribological aspects of MXene-reinforced nanocomposites." *Advanced Materials* 32 (38):2003154.

Mandal, Subhra, Guobin Kang, Pavan Kumar Prathipati, Wenjin Fan, Qingsheng Li, and Christopher J. Destache. 2018. "Long-acting parenteral combination antiretroviral loaded nano-drug delivery system to treat chronic HIV-1 infection: A humanized mouse model study." *Antiviral Research* 156:85–91.

Martí Coma-Cros, Elisabet, Arnau Biosca, Elena Lantero, Maria Letizia Manca, Carla Caddeo, Lucía Gutiérrez, Miriam Ramírez, Livia Neves Borgheti-Cardoso, Maria Manconi, and Xavier Fernàndez-Busquets. 2018. "Antimalarial activity of orally administered curcumin incorporated in Eudragit®-containing liposomes." *International Journal of Molecular Sciences* 19 (5):1361.

McArthur, Donna Behler. 2019. "Emerging infectious diseases." *Nursing Clinics* 54 (2):297–311.

Mehlhorn, Heinz. 2016. *Nanoparticles in the Fight Against Parasites*. Vol. 8: Springer.

Mehndiratta, Man Mohan, Prachi Mehndiratta, and Renuka Pande. 2014. "Poliomyelitis: historical facts, epidemiology, and current challenges in eradication." *The Neurohospitalist* 4 (4):223–229.

Mehta, Meenu, Parteek Prasher, Mousmee Sharma, Madhur D. Shastri, Navneet Khurana, Manish Vyas, Harish Dureja, Gaurav Gupta, Krishnan Anand, and Saurabh Satija. 2020. "Advanced drug delivery systems can assist in targeting coronavirus disease (COVID-19): a hypothesis." *Medical hypotheses* 144:110254.

Meng, Xiangchun, Zeyu Zhang, and Linlin Li. 2020. "Micro/nano needles for advanced drug delivery." *Progress in Natural Science: Materials International* 30 (5):589–596.

Minopoli, Antonio, Bartolomeo Della Ventura, Raffaele Campanile, Julian A. Tanner, Andreas Offenhäusser, Dirk Mayer, and Raffaele Velotta. 2021. "Randomly positioned gold nanoparticles as fluorescence enhancers in apta-immunosensor for malaria test." *Microchimica Acta* 188 (3):1–9.

Mo, Guangquan, Xinru Zheng, Naobei Ye, and Zhixiong Ruan. 2021. "Nitrogen-doped carbon dodecahedron embedded with cobalt nanoparticles for the direct electro-oxidation of glucose and efficient nonenzymatic glucose sensing." *Talanta* 225:121954.

Moretton, Marcela A., Maximiliano Cagel, Ezequiel Bernabeu, Lorena Gonzalez, and Diego A. Chiappetta. 2015. "Nanopolymersomes as potential carriers for rifampicin pulmonary delivery." *Colloids and Surfaces B: Biointerfaces* 136:1017–1025.

Munhoz Jr, Antônio Hortêncio, Marcos Romero Filho, Henrique de Arruda Kleist, Guilherme Dias Moreno, Mariana Oliva de Oliveira, Renato Meneghetti Peres, Maura Vincenza Rossi, Leila Figueiredo de Miranda, Roberto Rodrigues Ribeiro, and Bruno Filipe Carmelino Cardoso Sarmento. 2019. "Synthesis of pseudoboehmite-graphene oxide for drug delivery system." *Materials Today: Proceedings* 14:700–707.

Murugan, Kadarkarai, Mohammed Aamina Labeeba, Chellasamy Panneerselvam, Devakumar Dinesh, Udaiyan Suresh, Jayapal Subramaniam, Pari Madhiyazhagan, Jiang-Shiou Hwang, Lan Wang, and Marcello Nicoletti. 2015. "Aristolochia indica green-synthesized silver nanoparticles: a sustainable control tool against the malaria vector Anopheles stephensi?" *Research in Veterinary Science* 102:127–135.

Nam, Youngsuk, and Y Sungtaek Ju. 2013. "A comparative study of the morphology and wetting characteristics of micro/nanostructured Cu surfaces for phase change heat transfer applications." *Journal of Adhesion Science and Technology* 27 (20):2163–2176.

Nasrollahzadeh, Mahmoud, Zahra Issaabadi, Mohaddeseh Sajjadi, S Mohammad Sajadi, and Monireh Atarod. 2019. "Types of nanostructures." *Interface Science and Technology* 28:29–80.

Nurchis, Mario Cesare, Domenico Pascucci, Martina Sapienza, Leonardo Villani, Floriana D'ambrosio, Francesco Castrini, Maria Lucia Specchia, Patrizia Laurenti, and Gianfranco Damiani. 2020. "Impact of the burden of COVID-19 in Italy: results of disability-adjusted life years (DALYs) and productivity loss." *International Journal of Environmental Research and Public Health* 17 (12):4233.

Oaks Jr, Stanley C., Robert E. Shope, and Joshua Lederberg. 1992. *Emerging infections: microbial threats to health in the United States*. National Academies Press.

Omwoyo, Wesley N., Paula Melariri, Jeremiah W. Gathirwa, Florence Oloo, Geoffrey M. Mahanga, Lonji Kalombo, Bernhards Ogutu, and Hulda Swai. 2016. "Development, characterization and antimalarial efficacy of dihydroartemisinin loaded solid lipid nanoparticles." *Nanomedicine: Nanotechnology, Biology and Medicine* 12 (3):801–809.

Onoshima, Daisuke, Hiroshi Yukawa, and Yoshinobu Baba. 2015. "Multifunctional quantum dots-based cancer diagnostics and stem cell therapeutics for regenerative medicine." *Advanced Drug Delivery Reviews* 95:2–14.

Oster, Alexandra M., Sheryl B. Lyss, R Paul McClung, Meg Watson, Nivedha Panneer, Angela L. Hernandez, Kate Buchacz, Susan E. Robilotto, Kathryn G. Curran, and Rashida Hassan. 2021. "HIV cluster and outbreak detection and response: the science and experience." *American Journal of Preventive Medicine* 61 (5):S130–S142.

Pal, Shreyashi, Rajdeep Saha, Shivesh Jha, and Biswatrish Sarkar. 2022. "Nanotherapeutics: A way to cure cardiac complications associated with COVID-19." Presented at the the 2nd International Electronic Conference on Healthcare.

Patel, Priya, Naimish Vyas, and Mihir Raval. 2021. "Safety and toxicity issues of polymeric nanoparticles: A serious concern." *Nanotechnology in Medicine: Toxicity and Safety*:156–173.

Patra, Debasmita. 2013. "Nanoscience, nanotechnology, or nanotechnoscience: Perceptions of Indian nanoresearchers." *Public Understanding of Science* 22 (5):590–605.

Pavia, Charles S., and Gary P. Wormser. 2021. "Passive immunization and its rebirth in the era of the COVID-19 pandemic." *International Journal of Antimicrobial Agents* 57 (3):106275.

Pawar, Dilip, and K. S. Jaganathan. 2016. "Mucoadhesive glycol chitosan nanoparticles for intranasal delivery of hepatitis B vaccine: enhancement of mucosal and systemic immune response." *Drug Delivery* 23 (1):185–194.

Pelaz, Beatriz, Christoph Alexiou, Ramon A. Alvarez-Puebla, Frauke Alves, Anne M. Andrews, Sumaira Ashraf, Lajos P. Balogh, Laura Ballerini, Alessandra Bestetti, and Cornelia Brendel. 2017. "Diverse applications of nanomedicine." *ACS Nano* 11 (3):2313–2381.

Peres, Renato Meneghetti, Jéssica Maiara Leme Sousa, Mariana Oliva de Oliveira, Maura Vincenza Rossi, Rene Ramos de Oliveira, Nelson Batista de Lima, Ayrton Bernussi, Juliusz Warzywoda, Bruno Sarmento, and Antonio Hortencio Munhoz. 2021. "Pseudoboehmite as a drug delivery system for acyclovir." *Scientific Reports* 11 (1):1–12.

Pokropivny, V. V., and V. V. Skorokhod. 2008. "New dimensionality classifications of nanostructures." *Physica E: Low-Dimensional Systems and Nanostructures* 40 (7):2521–2525.

Prevention, Centers for Disease Control and. Disease & Conditions. https://www.cdc.gov/index.htm. Last accessed on 21-05-2023.

Qu, Xiaolei, Pedro J. J. Alvarez, and Qilin Li. 2013. "Applications of nanotechnology in water and wastewater treatment." *Water Research* 47 (12):3931–3946.

Rana, Anu, Krishna Yadav, and Sheeja Jagadevan. 2020. "A comprehensive review on green synthesis of nature-inspired metal nanoparticles: Mechanism, application and toxicity." *Journal of Cleaner Production* 272:122880.

Reshma, V. G., and P. V. Mohanan. 2019. "Quantum dots: Applications and safety consequences." *Journal of Luminescence* 205:287–298.

Roser M. H. and Ritchie. Burden of Disease. https://ourworldindata.org/burden-of-disease. Last accessed on 21-05-2023.

Rostamabadi, Hadis, Seid Reza Falsafi, and Seid Mahdi Jafari. 2019. "Nano-helices of amylose for encapsulation of food ingredients." In *Biopolymer Nanostructures for Food Encapsulation Purposes*, 463–491. Elsevier.

Saeedi, Majid, Masoumeh Eslamifar, Khadijeh Khezri, and Solmaz Maleki Dizaj. 2019. "Applications of nanotechnology in drug delivery to the central nervous system." *Biomedicine & Pharmacotherapy* 111:666–675.

Sahani, Shalini, and Yogesh Chandra Sharma. 2021. "Advancements in applications of nanotechnology in global food industry." *Food Chemistry* 342:128318.

Sangboonruang, Sirikwan, Natthawat Semakul, Sanonthinee Sookkree, Jiraporn Kantapan, Nicole Ngo-Giang-Huong, Woottichai Khamduang, Natedao Kongyai, and Khajornsak Tragoolpua. 2022. "Activity of propolis nanoparticles against HSV-2: Promising approach to inhibiting infection and replication." *Molecules* 27 (8):2560.

Sanvicens, Nuria, and M Pilar Marco. 2008. "Multifunctional nanoparticles–properties and prospects for their use in human medicine." *Trends in Biotechnology* 26 (8):425–433.

Saraf, Surendra, Shailesh Jain, Rudra Narayan Sahoo, and Subrata Mallick. 2020. "Lipopolysaccharide derived alginate coated Hepatitis B antigen loaded chitosan nanoparticles for oral mucosal immunization." *International Journal of Biological Macromolecules* 154:466–476.

Schiffelers, Raymond, Gert Storm, and Irma Bakker-Woudenberg. 2001. "Liposome-encapsulated aminoglycosides in pre-clinical and clinical studies." *Journal of Antimicrobial Chemotherapy* 48 (3):333–344.

Schlichtenmayer, Maurice, and Michael Hirscher. 2012. "Nanosponges for hydrogen storage." *Journal of Materials Chemistry* 22 (20):10134–10143.

Senapati, Sudipta, Arun Kumar Mahanta, Sunil Kumar, and Pralay Maiti. 2018. "Controlled drug delivery vehicles for cancer treatment and their performance." *Signal Transduction and Targeted Therapy* 3 (1):1–19.

Shah, Afzal, Saima Aftab, Jan Nisar, Muhammad Naeem Ashiq, and Faiza Jan Iftikhar. 2021. "Nanocarriers for targeted drug delivery." *Journal of Drug Delivery Science and Technology* 62:102426.

Sharma, Neelam, Ishrat Zahoor, Monika Sachdeva, Vetriselvan Subramaniyan, Shivkanya Fuloria, Neeraj Kumar Fuloria, Tanveer Naved, Saurabh Bhatia, Ahmed Al-Harrasi, and Lotfi Aleya. 2021. "Deciphering the role of nanoparticles for management of bacterial meningitis: an update on recent studies." *Environmental Science and Pollution Research* 28 (43):60459–60476.

Shimizu, Toshimi, Wuxiao Ding, and Naohiro Kameta. 2020. "Soft-matter nanotubes: a platform for diverse functions and applications." *Chemical Reviews* 120 (4):2347–2407.

Singh, Ajay, Hilliard L. Kutscher, Julia C. Bulmahn, Supriya D. Mahajan, Guang S. He, and Paras N. Prasad. 2020. "Laser ablation for pharmaceutical nanoformulations: Multi-drug nanoencapsulation and theranostics for HIV." *Nanomedicine: Nanotechnology, Biology and Medicine* 25:102172.

Singh, Kshitij R. B., Vanya Nayak, Jay Singh, and Ravindra Pratap Singh. 2021. "Nano-enabled wearable sensors for the internet of things (IoT)." *Materials Letters* 304:130614.

Smith, Douglas M., Jakub K. Simon, and James R. Baker Jr. 2013. "Applications of nanotechnology for immunology." *Nature Reviews Immunology* 13 (8):592–605.

Sobolev, S. L., and I. V. Kudinov. 2021. "Heat conduction across 1D nano film: Local thermal conductivity and extrapolation length." *International Journal of Thermal Sciences* 159:106632.

Su, Wentao, Xinghua Gao, Lei Jiang, and Jianhua Qin. 2015. "Microfluidic platform towards point-of-care diagnostics in infectious diseases." *Journal of Chromatography A* 1377:13–26.

Tang, Zhongmin, Xingcai Zhang, Yiqing Shu, Ming Guo, Han Zhang, and Wei Tao. 2021. "Insights from nanotechnology in COVID-19 treatment." *Nano Today* 36:101019.

Tao, Wenqian, Katherine S. Ziemer, and Harvinder S. Gill. 2014. "Gold nanoparticle–M2e conjugate coformulated with CpG induces protective immunity against influenza A virus." *Nanomedicine* 9 (2):237–251.

Tay, Andy, Andrea Pavesi, Saeed Rismani Yazdi, Chwee Teck Lim, and Majid Ebrahimi Warkiani. 2016. "Advances in microfluidics in combating infectious diseases." *Biotechnology Advances* 34 (4): 404–421.

Thanihaichelvan, M., S. N. Surendran, T. Kumanan, U. Sutharsini, P. Ravirajan, R. Valluvan, and T. Tharsika. 2022. "Selective and electronic detection of COVID-19 (Coronavirus) using carbon nanotube field effect transistor-based biosensor: A proof-of-concept study." *Materials Today: Proceedings* 49:2546–2549.

Thornhill, Randy, Corey L. Fincher, Damian R. Murray, and Mark Schaller. 2010. "Zoonotic and non-zoonotic diseases in relation to human personality and societal values: Support for the parasite-stress model." *Evolutionary Psychology* 8 (2):147470491000800201.

Tseng, Yu-Cheng, Zhenghong Xu, Kevin Guley, Hong Yuan, and Leaf Huang. 2014. "Lipid–calcium phosphate nanoparticles for delivery to the lymphatic system and SPECT/CT imaging of lymph node metastases." *Biomaterials* 35 (16):4688–4698.

Uskokovic, Vuk. 2013. "Entering the era of nanoscience: time to be so small." *Journal of Biomedical Nanotechnology* 9 (9):1441–1470.

Van Doorn, H Rogier. 2014. "Emerging infectious diseases." *Medicine* 42 (1):60–63.

Vanossi, Andrea, Clemens Bechinger, and Michael Urbakh. 2020. "Structural lubricity in soft and hard matter systems." *Nature Communications* 11 (1):1–11.

Ventola, C Lee. 2017. "Progress in nanomedicine: approved and investigational nanodrugs." *Pharmacy and Therapeutics* 42 (12):742.

Wang, Chao, Lei Liu, Xuejing Liu, Yu Chen, Xueying Wang, Dawei Fan, Xuan Kuang, Xu Sun, Qin Wei, and Huangxian Ju. 2020. "Highly-sensitive electrochemiluminescence biosensor for NT-proBNP using MoS_2@ Cu_2S as signal-enhancer and multinary nanocrystals loaded in mesoporous UiO-66-NH_2 as novel luminophore." *Sensors and Actuators B: Chemical* 307:127619.

Wang, Chao, Mei Liu, Zhifei Wang, Song Li, Yan Deng, and Nongyue He. 2021. "Point-of-care diagnostics for infectious diseases: From methods to devices." *Nano Today* 37:101092.

Wang, Ting, Jianmin Yun, Yu Zhang, Yang Bi, Fengyun Zhao, and Yaoxing Niu. 2021. "Effects of ozone fumigation combined with nano-film packaging on the postharvest storage quality and antioxidant capacity of button mushrooms (Agaricus bisporus)." *Postharvest Biology and Technology* 176:111501.

Wang, Yongzhong, Li Yu, Xiaowei Kong, and Leming Sun. 2017. "Application of nanodiagnostics in point-of-care tests for infectious diseases." *International Journal of Nanomedicine* 12:4789.

Watts, Mitchell C., Loren Picco, Freddie S. Russell-Pavier, Patrick L. Cullen, Thomas S. Miller, Szymon P. Bartuś, Oliver D. Payton, Neal T. Skipper, Vasiliki Tileli, and Christopher A. Howard. 2019. "Production of phosphorene nanoribbons." *Nature* 568 (7751):216–220.

(WHO), World Health Organisation. Meningitis. https://www.who.int/health-topics/meningitis#tab=tab_1. Last accessed on 21-05-2023.

(WHO), World Health Organisation. 2021. *World Health Statistics 2021: Monitoring Health for the SDGs, Sustainable Development Goals.* World Health Organisation.

Wui, Seo Ri, Kwang Sung Kim, Ji In Ryu, Ara Ko, Hien Thi Thu Do, Yeon Jung Lee, Hark Jun Kim, Soo Jeong Lim, Shin Ae Park, and Yang Je Cho. 2019. "Efficient induction of cell-mediated immunity to varicella-zoster virus glycoprotein E co-lyophilized with a cationic liposome-based adjuvant in mice." *Vaccine* 37 (15):2131–2141.

Xi, Zhijiang, Rongrong Huang, Zhiyang Li, Nongyue He, Ting Wang, Enben Su, and Yan Deng. 2015. "Selection of HBsAg-specific DNA aptamers based on carboxylated magnetic nanoparticles and their application in the rapid and simple detection of hepatitis B virus infection." *ACS Applied Materials & Interfaces* 7 (21):11215–11223.

Xie, Zhongshuai, Xiaolong Tang, Jiafeng Shi, Yaojin Wang, Guoliang Yuan, and Jun-Ming Liu. 2022. "Excellent piezo-photocatalytic performance of Bi4Ti3O12 nanoplates synthesized by molten-salt method." *Nano Energy* 98:107247.

Yaméogo, Josias B. G., Annabelle Gèze, Luc Choisnard, Jean-Luc Putaux, Adama Gansané, Sodiomon B. Sirima, Rasmané Semdé, and Denis Wouessidjewe. 2012. "Self-assembled biotransesterified cyclodextrins as Artemisinin nanocarriers–I: Formulation, lyoavailability and in vitro antimalarial activity assessment." *European Journal of Pharmaceutics and Biopharmaceutics* 80 (3):508–517.

Yang, Dehua, Ting Li, Minghan Xu, Feng Gao, Juan Yang, Zhi Yang, and Weidong Le. 2014. "Graphene oxide promotes the differentiation of mouse embryonic stem cells to dopamine neurons." *Nanomedicine* 9 (16):2445–2455.

Yang, Samuel, and Richard E. Rothman. 2004. "PCR-based diagnostics for infectious diseases: uses, limitations, and future applications in acute-care settings." *The Lancet infectious diseases* 4 (6): 337–348.

Yin, Wenyan, Liang Yan, Jie Yu, Gan Tian, Liangjun Zhou, Xiaopeng Zheng, Xiao Zhang, Yuan Yong, Juan Li, and Zhanjun Gu. 2014. "High-throughput synthesis of single-layer MoS2 nanosheets as a near-infrared photothermal-triggered drug delivery for effective cancer therapy." *ACS Nano* 8 (7): 6922–6933.

Yin, Yue, Wen Su, Jie Zhang, Wenping Huang, Xiaoyang Li, Haixia Ma, Mixiao Tan, Haohao Song, Guoliang Cao, and Shengji Yu. 2021. "Separable microneedle patch to protect and deliver DNA nanovaccines against COVID-19." *ACS Nano* 15 (9):14347–14359.

Zhang, Min, Chen Han, Wen-Qiang Cao, Mao-Sheng Cao, Hui-Jing Yang, and Jie Yuan. 2021. "A nano-micro engineering nanofiber for electromagnetic absorber, green shielding and sensor." *Nano-Micro Letters* 13 (1):1–12.

Zhang, Pengfei, Huiqi Lu, Jia Chen, Huanxing Han, and Wei Ma. 2014. "Simple and sensitive detection of HBsAg by using a quantum dots nanobeads based dot-blot immunoassay." *Theranostics* 4 (3):307.

Zhang, Tuo, Chenyan Lv, Lingli Chen, Guangling Bai, Guanghua Zhao, and Chuanshan Xu. 2014. "Encapsulation of anthocyanin molecules within a ferritin nanocage increases their stability and cell uptake efficiency." *Food Research International* 62:183–192.

Zhang, Wanli, Hanxue Sun, Zhaoqi Zhu, Rui Jiao, Peng Mu, Weidong Liang, and An Li. 2020. "N-doped hard carbon nanotubes derived from conjugated microporous polymer for electrocatalytic oxygen reduction reaction." *Renewable Energy* 146:2270–2280.

Zhang, Yuwei, Ayden Malekjahani, Buddhisha N. Udugama, Pranav Kadhiresan, Hongmin Chen, Matthew Osborne, Max Franz, Mike Kucera, Simon Plenderleith, and Lily Yip. 2021. "Surveilling and tracking COVID-19 patients using a portable quantum dot smartphone device." *Nano Letters* 21 (12): 5209–5216.

Zheng, Bin, Wenchang Peng, Mingming Guo, Mengqian Huang, Yuxuan Gu, Tao Wang, Guangjian Ni, and Dong Ming. 2021. "Inhalable nanovaccine with biomimetic coronavirus structure to trigger mucosal immunity of respiratory tract against COVID-19." *Chemical Engineering Journal* 418:129392.

Zhou, Guangmin, Lin Xu, Guangwu Hu, Liqiang Mai, and Yi Cui. 2019. "Nanowires for electrochemical energy storage." *Chemical Reviews* 119 (20):11042–11109.

9 Nanomedicine for Cancer Therapy
Current Status and Challenges

P. S. Rajalakshmi, Kalyani Eswar, Paloma Patra, and Aravind Kumar Rengan

9.1 INTRODUCTION

Significant research progress has been made towards comprehending the functioning of cancer biology in previous two decades. The World Health Organization (WHO) reported 8.2 million related cancer deaths in the year 2012, which is expected to touch 22 million over the subsequent two decades. Although development of safe and effective cancer therapeutics is perhaps the major area of research in today's era, certain challenges are still to be resolved, such as off-target systemic toxicity and multidrug resistance. To circumvent these challenges, nanomedicine has been especially advantageous as it helps in efficient delivery of the chemotherapeutic to the target tissue (Landis and Kola 2007).

Developing nanoscale therapeutic products with desirable qualities like rapid disintegration and low toxicity is the heart of nanomedicine (Krown et al. 2004). Nanomedicine has been exploited to design therapies as well as diagnose diseases for the betterment of human health. Biomedical applications have been actively explored nanoparticles (NPs) in the 1–100 nm size range. They are effective tools for imaging, treatment, as well as drug carriers due to their large surface-to-volume ratio, increased load - carrying ability, capability to generate NPs of various sizes, morphology, and surface characteristics, and ease of functionalization by diverse biomolecules (Navya et al. 2019). This remarkably high specific surface area of NPs supports their desirable physicochemical features, including magnetic, optical, and electrical properties. The explosive growth of nanotechnology usage across biomedicine, particularly for timely and accurate biosensing as well as effective treatment of multiple diseases, has prompted the birth of a novel scientific discipline termed as "nanomedicine" (Madamsetty et al. 2019). This fresh and extremely interesting field has enormous potential for prevention, diagnosing, tracking, and treatment of diseases, most importantly cancer. Diagnostics, biosensing, biomedical imaging, biopharmaceuticals, and theranostics are active areas of nanomedicine, which rely on the novel attributes of nanoparticles and nanoplatforms (Navya et al. 2019).

There are currently around 50 cancer nanomedicines available in the market that have been approved by the FDA, demonstrating that nanomedicine's primary focus has been on cancer treatment in the recent past (Aulic et al. 2020). The advent of smart nanocarriers having the flexibility to administer hydrophobic pharmaceuticals in a directed and regulated manner has transformed the cancer treatment scenario (Guan et al. 2020). Recent research studies have also examined functionalized nanoplatforms for intracellular delivery of various payloads (Morshedi Rad et al. 2021). Numerous innovative nanostructures, including polymeric nanoparticles, nanoshells, liposomes, carbon nanotubes, dendrimers, superparamagnetic nanoparticles, metal organic framework NPs, and nucleic acid-based nanoparticles, have been implemented to deliver specific chemotherapeutic agents, including drugs of low molecular weight as well as

DOI: 10.1201/9781003305583-9

biomacromolecules (proteins, peptides, or genes). These nanostructures may be carved out of organic, inorganic, protein, lipid, and synthetic polymer materials. Nanoparticulate carriers can enhance anti-cancer treatment by altering the pharmacokinetics and enhancing the tissue distribution of the encapsulated drug. Additionally, nanomedicine can also eliminate multidrug resistant cancers as indicated by many reports. Nanomedicines usually take up either passive or active targeting to reach the cancer tissue. While passive targeting relies on aberrant basement membranes, expanding endothelial cells, and even the absence of pericytes, active targeting depends on the receptor-ligand interaction. In 1995, liposomal doxorubicin was the first anti-cancer nanomedicine with increased permeability and retention (EPR) to be licensed by FDA (Gabizon et al. 2008). Despite the advancements, nanomedicine in this field appears to have a long road ahead before reaching its full potential (Havel et al. 2016). As nanomedicine is critical in developing cancer nanotherapeutics, the periodical evaluation of latest innovations and accomplishments in this domain is essential. This book chapter thus provides a comprehensive insight into the status and challenges in nanomedicine as a platform for cancer therapy.

9.2 CURRENT TREATMENT MODALITIES

9.2.1 Chemotherapy

Chemotherapy, the golden standard, uses cytotoxic small-molecule drugs to inhibit the replicative stage of the cell cycle and cause apoptosis or necrosis of tumor cells in clinics for both primary and metastatic tumors (Mills et al. 2018). Chemotherapy was first used to treat cancer in the 1940s with nitrogen mustard, a cytotoxic substance. In the 1960s, chemotherapeutics with natural origins, like anthracyclines and vinca alkaloids, were developed (Falzone et al. 2018; Morrison 2010). These drugs are used to treat cancer today. Since then, 50 distinct types of chemotherapeutic drugs have been formulated to treat 200 different types of tumors. The FDA has approved various chemotherapeutic drugs that have been used exclusively for cancer and is currently one of the most clinically prevalent therapies due to its ease. Nevertheless, these drugs have several unavoidable adverse effects, such as myelosuppression, baldness, appetite loss, organ damage, nausea, and vomiting brought on by chemotherapy (Cheng et al. 2021). The harsh tumor microenvironment, low bioavailability of the drugs and multidrug resistance cause the failure of the chemotherapeutic drug in clinics (Alfarouk et al. 2015). The development of efficient clinical chemotherapeutic combinations and procedures have been constrained by the cytotoxicity of currently available anti-cancer agents and the ineffectiveness of chemotherapy, particularly in the advanced stages of cancer. Cardiotoxicity is a frequent chemotherapeutic side effect and the main factor in morbidity and mortality in cancer patients (Broder et al. 2008; Florescu et al. 2013). Two of the most frequent CIC adverse effects of systemic cancer treatment are arrhythmias and heart failure (Senkus and Jassem 2011). Multidrug resistance (MDR) is one of the significant barrier in chemotherapy, resulting in the generation of cancer cells insensitive to administered therapy (Emran et al. 2022). MDR causes genetic and epigenetic alteration in cells, which results in drug resistance and decreases the bioavailability of chemotherapeutic agents to the cancer site (Gillet and Gottesman 2009)(Fu et al. 2019).

Given their distinct physicochemical characteristics, nanomedicines hold considerable therapeutic potential to be used for the treatment of cancer, especially in terms of enhancing the effectiveness and reducing the adverse side effect of conventional chemotherapy. By manipulating the intrinsic features like shape and size as well as the surface functions like charge, the nanosystem can be designed to build a multifunctional platform responding to the induced stimulations like pH, temperature, etc. (Patra et al. 2018; M. K. Yu et al. 2012). As a result, encapsulating chemotherapeutic agents within the nano-architecture has been shown to improve the solubility, bioavailability, drug targeting ability, and sustainable release of drug (Wicki et al. 2015). Nanomedicine enables the sustainable release of medicines through biodegradation and self-regulation of nanomaterials *in vitro* and *in vivo*. As a result of their unique material characteristics, nanotechnologies are distinguished by excellent drug encapsulation, controlled self-assembly, specificity, and biocompatibility (Singh et al. 2018). Due to the specific nature of the nanoscale and the different bioeffects of nanomaterials, nanotechnology holds the potential to overcome the present chemotherapeutic hurdles in the treatment of cancer. Multidrug resistance and the issues with conventional chemotherapy may be resolved with the aid of nanotechnology. These promising technologies regulates the required dose load of hydrophobic chemotherapeutics to facilitate the design of safe, controlled, and effective administration, precisely at tumor microenvironment.

The clinical application of nanotechnology in diagnosing cancer and its treatment is still in its infancy. Although various nano-based therapy modalities are now in clinical trials, only a few nanocarrier-based drugs are approved in clinics. The use of precisely engineered materials to develop innovative technologies and therapies that could improve efficacy and target while reducing the adverse side effect is one of the prominent examples of how nanotechnology is being employed in medicine. As a result, using the application of nanotechnology for diagnosis, treatment, and prevention of cancer can lead to a number of advancements. Liposomes are highly desirable category of nanoparticle systems for the application of sustainable drug delivery due to their simplicity of synthesis, drug loading, and surface modification, and are biocompatible (Ali et al. 2021; Al-Zoubi and Al-Zoubi 2022). These are lipid-based vesicles with an aqueous interior and a lipid bilayer composed mostly of amphipathic phospholipids (Akbarzadeh et al. 2013; Nsairat et al. 2022). The first FDA-approved nanosystem for Cancer is Doxil®, a liposomal formulation with cytotoxic drug doxorubicin in 1995 (Y. (Chezy) Barenholz 2012). The formulation was designed on three independent principles, such as prolonged blood circulation time, loading efficacy and passive targeting to tumor region because of the EPR effect (Y. Barenholz 2012). The prolonged circulation is due to the use of PEGylation. Ammonium sulfate gradient method helps in increasing the loading efficacy of the doxorubicin to the liposome and also enables the pH-based drug release. Additionally, Doxil significantly reduces the cardiotoxicity when compared with free doxorubicin in all studied conditions for cardiac function. (Y. Barenholz 2012). Next year, a liposomal formulation DaunoXome® of daunorubicin was approved for HIV-associated Kaposi sarcoma. There are a total of seven FDA-approved liposomal formulations for the treatment of Cancer including Myocet® (DOX),Mepact® (Mifamurtide), Depocyt® (Cytarabine/Ara-C) and Onivyde® (Irinotecan) (van Hoogevest et al. 2021; Bulbake et al. 2017). Genexol-PM® is a PTX polymer based micellar formulation approved in South Koreain 2007 (He et al. 2016a, 2016b; Lee et al. 2008a). Nanoxel® is used in Indian clinics from 2006 and the clinical trials are undergoing for the FDA approval (Miguel et al. 2022; Parodi et al. 2022; Lee et al. 2008b).

9.2.2 Phototherapy

9.2.2.1 Photodynamic Therapy

The use of light in therapeutics date back centuries ago. Light is used for the treatment of psoriasis, vitiligo, and skin cancer by Egyptians, Chinese, and Indians. Niels Finsen was awarded the Nobel Prize in 1903 for his work in treating lupus vulgaris with phototherapy (carbon arc) (Grzybowski and Pietrzak 2012). Cremer et al. pioneered phototherapy, which is now routinely used to treat infant jaundice (Maisels 2015). Von Tappeiner and Jodlbauer introduced photodynamic treatment for oxygen-dependent photosensitization and employed eosin as a photosensitizer in clinical trials for skin cancer in 1904. Scherer produced hematoporphyrin for the first time from blood in 1841 by removing the iron content using sulfuric acid. Hausmann investigated the biological features of the photosensitizer in 1908. Policard, a Frenchman, discovered porphyrin buildup in the tumor location in 1924, and Auler and Banzer investigated the photodynamic effect (Nakamura et al. 2016; Josefsen and Boyle 2012; Y. Zhang and Lovell 2012). The light stimulates the photosensitizer and transfers its energy to cellular oxygen resulting in the formation of a particular ROS, known as singlet oxygen, causing the cancer cell death by peroxidation of nucleic acid (RNA and DNA), proteins and lipids (Z. J. Zhang et al. 2020; Castano et al. 2005). PDT can stimulate the patient's immune system by boosting the amount of cancer cell-derived antigen presented to T-cells and cause indirect damage by blocking tumor blood arteries (Nath et al. 2019). Photofrin and hemato-porphyrin derivatives, the first-generation photosensitizer for PDT uses visible red light for treatment (Abrahamse and Hamblin 2016; Ormond and Freeman 2013). They have been supplanted by second-generation photosensitizers, including their derivatives, due to their poor targeting abilities, complicated architectures, and cytotoxicity. Motexafin lutetium, Palladium bacteriopheophorbide, Porfimer sodium (Photofrin®), Purlytin®, Temoporfin (Foscan®), Talaporfin (Laserphyrin®), and Verteporfin (Visudyne®) are the second-generation PS under clinical trials or approved in clinics (Niculescu and Grumezescu 2021; Baskaran et al. 2018). Photosensitizers have many disadvantages, including poor solubility leading to the aggregation of dye, hostile biodistribution, uncontrollable photoactivity, and a delayed elimination from the body resulting in a variety of post-treatment adverse effects. The limitations to classic PDT are their poor water solubility, ineffective tumor targeting and difficult scheduling, the requirement for several procedures, patient monitoring following the treatment, and photosensitivity problems in the initial few weeks following therapy. Therefore, the progress of PDT in cancer treatment and its standardization in clinical research depend on the development of third generation photosensitizers. Photosensitizers have been combined with nanotechnology to decrease these side effects and improve the effectiveness of the treatment.

Nanotechnology protects the photosensitizer from enzymatic degradation, enables sustained release to achieve a constant and uniform concentration in the targeted area, and can be combined with various therapeutic molecules for synergistic impact while avoiding traditional drug resistance mechanisms (Misra et al. 2010; Calixto et al. 2014; Bovis et al. 2012). Nanosystems like polymeric systems, liposomes, and metallic nanosystems, can be combined with traditional photosensitizers to create a unique and stable system with minimal toxicity and great efficiency. The benefits over conventional photosensitizers include compact size (below 200 nm), an increase in surface area, high catalytic efficiency, a larger number of active centers, and good adsorption capability. Aminolevulinic acid (ALA), a photosensitizer for PDT, was encapsulated in liposomes and the *in vitro* results showed enhanced uptake and photocytotoxic efficiency compared to the prodrug (Hong et al. 2016; Wachowska et al. 2011). The Ce6, the photosensitizer, is encapsulated in a lipopolymeric hybrid nanosystem functionalized with folic acid to achieve active tumor targeting (John et al. 2016). Both pH and the temperature responsiveness of the nanosystem can trigger the release of the photosensitizer. The functionalization of the nanosystem with folic acid helps target the tumor. The dual responsiveness of the nanosystem helps in the sustained release of the dye and increases the efficiency of the therapy compared to free Ce6. Fan et al. developed a nanosystem composed of iodine-enriched material (heavy atom) to elevate the $^1O_2$ generation, thereby improving the efficacy of the photodynamic therapy (W. Zhou et al. 2020; Hu et al. 2020). The nanoparticle composed of NIR light absorbing semiconductor-based polymer serves as photodynamic agent and an iodine-grafted PEG-PHEMA (amphiphilic copolymer) enhances $^1O_2$ generation. This improves the circulation time and dual imaging CT and fluorescence. The nanosystem was used for dual-modal imaging and the generation of the $^1O_2$ helps in enhancing the efficacy of the photodynamic therapy.

9.2.2.2 Photothermal Therapy

Photothermal therapy (PTT) is a potential cancer treatment strategy involving the nonradiative conversion of light energy (especially near infrared light) into heat energy, which induces cancer cell death. Tumor regression in patients due to high fever sparked a foundation for studies into cancer hyperthermia during the 19th century. Several advances have been made in this sector to the point that non-invasive localized hyperthermic treatments in cancer cells using electromagnetic radiation are now available. Honda et al. published the first study of its sort in 1994, demonstrating the efficacy of cancer hyperthermia by injecting a dosage of hot saline percutaneously directly into the tumor mass to elicit hyperthermia-induced necrosis (Honda et al. 1994). Hyperpyrexia, a temperature of 40–46 °C causes irreversible damage to the cancer cell membranes and promotes protein denaturation (Krenacs et al. 2020; Roti 2008). NIR light has a low absorption coefficient and deeper penetrating power in tissues due to its longer wavelength. As a result, NIR light is referred to as the "biological window" and is regarded as an important tool for diagnosis and treatment. Indocyanine green (ICG), a carbocyanine dye andis an FDA-approved dye in 1959 is used as a contrast agent for bioimaging and photosensitizer for photothermal therapy (Jogdand et al. 2020). Most photosensitizers have a poor fluorescence quantum yield and are vulnerable to photobleaching, limiting their application in bioimaging and photothermal therapy (S. B. Alvi et al. 2022; Jogdand et al. 2020). Many researchers have noted that the photosensitizer is oxidized in the aqueous medium, leading to decreased light absorption, reduced fluorescence, interaction with the blood plasma proteins, and reduced photothermal activity. Encapsulation of the photosensitizer in a nanosystem composed of polymeric nanosystem, liposomes, protein-based nanosystem, and metallic nanosystem helps improve the thermal stability, increases the circulation time, and decreases the toxicity of the chemical dyes (Appidi et al. 2020).

Nanotechnology improves the photothermal agent by increasing thermal stability, thereby decreasing the number of dosages for the treatment. The optical absorption of NIR light was employed to enhance PTT by using nanostructures like carbon-based nanoparticles, noble metal-based nanoparticles, semiconductor nanosystem like copper-based nanosystem, graphene oxide-based nanoparticles, polypyrrole, and polydopamine nanosystem (Oldenburg et al. 1998; Niidome et al. 2006; R et al. 2021; Syed Baseeruddin Alvi et al. 2019; P S et al. 2021; Shihao Li and Zhang 2020; D. Wang et al. 2018; Jin et al. 2020; Ramalingam et al. 2018; J. Xu et al. 2017a). These nanosystems were functionalized for targeting tumor tissue for selective treatment of tumor tissues. The accumulation of photothermal agents in tumor lesions due to passive targeting or active targeting may further enhance the outcome of the therapy, thereby reducing the adverse effect in normal tissues. Noble-metal-based nanomaterials like gold, palladium, platinum, and silver used in PTT have made significant progress in the past. The photostability of these particles is high, thereby not causing rapid photobleaching and they have good light absorption and high heat conversion efficiency. Gold-based nanoparticles (Au NPs) are an excellent photothermal agent because of their surface plasmon resonance, which enables photothermal conversion (SPR). The gold nanoshells and related nanomaterials are currently gaining a lot of attention, even though most of the photothermal agents are only restricted to preclinical models. The main problem with gold nanoparticles is the biodegradation and renal clearance from the body. A liposome coated with nanoparticles was used as the photothermal agent (Rengan et al. 2015). The gold coated over the liposome will degrade into 5–10 nm gold seeds when irradiated with NIR light (Figure 9.1). The size of the disintegrated gold seeds aids in renal clearance, thereby reducing the side effects by preventing the accumulation of gold in different organs. In 2019, a phase-I clinical trial of prostate tumors in 16 patients employing a silicon core and a gold shell proved the efficiency of photothermal therapy (NCT04240639). According to the findings, 94% of the patients' tumors were thermally ablated, and no significant adverse effects were noticed following the treatment. In addition, a pilot study of the PTT platform was also investigated in patients with head and neck cancer (NCT00848042).

9.3 CANCER IMMUNOTHERAPY

The immunotherapy has emerged as a popular and novel approach for treating various disease condition by modifying the patient's immune system. It can be divided depending on

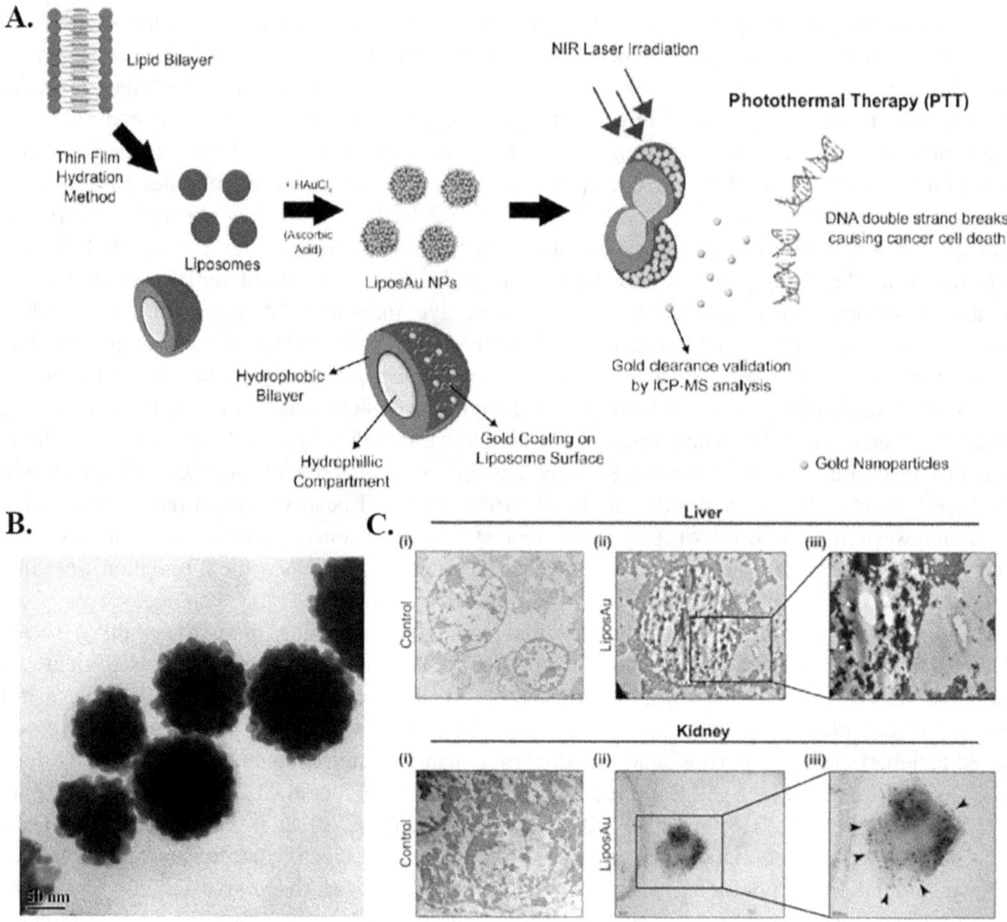

FIGURE 9.1 A. Schematic representation of the synthesis of gold coated liposome (LiposAu NPs) and the mode of action of LiposAu NPs while irradiation with the NIR Light. B. TEM image of LiposAu NPs. C. TEM image of liver tissue (i) control and (ii), (iii) aggregated state of LiposAu NPs in hepatocyte and kidney tissue (i) control (ii), (iii) tissue section containing dissociated LiposAu NP with less than 5 nm sized gold seeds (Rengan et al. 2015).

its effect on the immune system into immunosuppressive and immunostimulatory (Bhaskar and Singh 2014). The cancer immunomodulation involves the stimulation of the immune system of the host to combat the disease as opposed to the traditional treatment modalities like radiotherapy, chemotherapy, and surgery (Zhaogang Yang et al. 2020). The immune system of the human body is bestowed with the ability to differentiate between normal and cancerous cells by identifying the antigens that are of two different types, namely, tumor-specific antigens (TSAs) and tumor-associated antigens (TAAs) (Koury et al. 2018). However, the immune suppressive characteristic of the tumor microenvironment (TME) inhibits the host immune system from eliciting a response against the tumors (Musetti and Huang 2018). Several strategies, like monoclonal antibodies, can bring about the activation of the immune system of the host, cell-based therapies, monoclonal antibodies, and checkpoint blockers (Ventola 2017). Cancer immunotherapy confers better treatment owing to its ability to induce a robust immune response against the tumor and control metastasis (Kroemer and Zitvogel 2018; Hodi et al. 2010).

9.3.1 Adoptive Cellular Immunotherapy

Stimulating the self-immune system of the host enables the suppressed immune system to detect and remove tumors. Adoptive cellular immunotherapy (ACI) involves the injection of genetically amplified and modified effector cells. This field of therapeutic modality has gained huge attention due to its specificity, absence of drug resistance, and ease of preservation. ACI can be specific or non-specific. The specific ACI involves the induction of immune cells by Chimeric Antigen Receptor T-Cell Immunotherapy (CAR-T), TIL therapy (tumor-infiltrating lymphocyte-therapy), and T-cell receptor-T (TCR-T), which are specific factors and specific tumor antigens that are stimulated. In contrast, non-specific ACI stimulates the immune cells through cytokines or lymphocytes in the peripheral blood. These immune cells exhibit the ability to kill tumor cells. Cytokine-induced killer cells (CIK), dendritic cells (DC), lymphocyte activated killer cells (LAK), natural killer cells (NKC), tumor-infiltrating lymphocyte (TIL), macrophages-activated killer cells (MAK), etc., are the effector immune cells participating in non-specific ACI (Marei et al. 2019). $CD8^+$ T-cells and $CD4^+$ T-cells are the primary effector cells. This type of therapy offers strong specificity, no drug resistance, high lethality, few side effects, and strong targeting and can be employed in some advanced patients or patients who have not responded to other therapeutic effects. Due to the challenges of separating and collecting TIL, melanoma patients only receive it momentarily, and CAR-T therapy is mostly used to treat blood tumors (Panda et al. 2017).

9.3.2 NK Cells

Natural killer cells (NK cells) are the immune cells and play a key role in the immune system. Their surface is endowed with two types of receptors, namely, killer inhibitory receptors (KIRs) and killer activation receptors (KARs). The inhibitory receptors on the surface of the NK cells recognize the self-cells and prevent their killing. But NK cells can rapidly react to viruses or tumor occurrence without any prior antibody exposure and can be understood that prior activation of NK cells is not required for their action and immunological activities. In the case of metastatic tumors and blood cancers, NK cell-mediated killing was brought about when they interacted with tumor cells without MHC-1 receptors and tumor cells with enhanced expression of activated ligands (Shevtsov and Multhoff 2016). Additionally, cytokines, such as IL-5 and IL-2, also activate NK cells and antibody-dependent cell-mediated cytotoxicity (ADCC) to bring about the killing of antibody-coated tumor cells. The efficiency of ADCC might be enhanced by newly created BiKE and TriKE molecules (Davis et al. 2017). Despite this, tumor cells display the potential to evade the immune system through various strategies. 1) The MICA and MICB proteins secreted by the tumor cells lead to the development of the relevant antibodies to achieve the target, for instance, mAb 7C6 (Ferrari de Andrade et al. 2018). 2) Increased HLA-G expression on the tumor cells binds the NK cell inhibitory receptors and inhibits their killing by NK cells (Yan 2011; Chapman et al. 1999). 3) NK cells bind to the tumor cells via NKG2D ligand, and tumor cells shed and secrete the ligand thus preventing recognition by NK cells. This also leads to the constant activation of the NK cells around the tumor and their reduction in sensitivity of recognition (Saito et al. 2012). 4) The anti-tumor activity of NK cells can be inhibited by the secretion of cytokines or direct contact with certain inhibitory immune cells like regulatory T-cells (Treg) and myeloid-derived suppressor cells (MDSCs) (Mellman et al. 2011). NK cells-based tumor therapy with limited side effects can be achieved by inhibiting the inhibitory receptors, activating the activation receptors on the NK cells, or targeting drugs to cue NK cells toward the tumor site (Morvan and Lanier 2016). In the current state of medical research, NK cells paired with immune checkpoint inhibitors, NK cells mixed with CAR, and NK cell-based immunotherapy have all become popular (Siegler et al. 2018).

9.3.3 Chimeric Antigen Receptor T-Cell Immunotherapy (CAR-T)

CAR-T immunotherapy involves the isolation of the patient-derived T-cells by leukocyte reduction process and genetically modifying them to express CAR-t on their surface. These engineered cells are then injected into the tumors for targeted tumor depletion. The remission rate associated with this treatment is elevated, especially in tumors with CD19 protein expression, like in B-cell acute leukemia and large B-cell lymphoma. FDA-approved drugs include Kite/Gilead's axicabagene ciloleucel (Yescarta) and tisagenlecleucel-T (Kymriah) by Novartis. The latter is associated with treating relapsed B-cell acute lymphoblastic leukemia (Wei et al. 2017). While the former is used to treat relapsed adult large B-cell lymphoma. However, the side effects like neurotoxicity and cytokine release syndrome were also observed (Kroschinsky et al. 2017). Though CAR-T is efficient in treating leukemia and lymphoma, T-cell depletion and suppression of tumor environment are also associated.

Research has been directed toward the reduction of T-cell depletion. The primary objective is to prevent the T-cells population depletion, which is essential for tumor targeting, thereby increasing the benefits to patients from this treatment modality (Wherry and Kurachi 2015). The levels of perforation and granulation enzyme were increased by enhanced muscle memory T-cells. Further, this prolonged the time of T-cells in the central memory state in the absence of TET-2 protein to achieve effective CAR-T therapy. Further, this will reduce the cost of production associated with processing TET-2 proteins/genes through drug-mediation or gene editing technologies (Fraietta et al. 2018). In the case of cancer or viral infections, the $CD8^+$ T-cells exhibit enhanced expression of the transcription factor Nr4a. Downregulation of this transcription factor could improve the efficiency of CAR-T treatment. Experiments have shown a reduction in tumor size and enhanced survival in mice treated with CAR-T-cells which were devoid of Nr4a. Therefore, inhibition of Nr4a could be an excellent strategy to prevent T-cell depletion (J. Chen et al. 2019). Epigenetic drug decitabine, an inhibitor of DNA methylation, can reverse the T-cells from depletion states into long-lasting memory T-cells, thereby aiding infection and tumor control and reducing the effects of immunosuppressants (Mardiana et al. 2019; Ghoneim et al. 2017; Pauken et al. 2016; Sen et al. 2016; Xia et al. 2019). Additionally, immune checkpoint inhibitors play an important role in tumor inhibition.

9.3.4 Immune Checkpoint Inhibitors (ICIs)

Checkpoint proteins, under normal circumstances, keep the immune cells under control from undesired proliferation and activation to prevent autoimmune diseases. However, the same proteins are unsuitable for the tumor, where they prevent the T-cells from getting activated and thus reduce the potential of the immune system to combat the tumor (Ramsay 2013). Immune checkpoint inhibitors (ICIs) help the T-cells to activate and regain the anti-tumor property. Several studies have exploited the checkpoint proteins for immunotherapy. However, the number of target immune checkpoints is vast, and attempts are made to identify specific sites. The efficiency of the ICIs depends mainly on the expression of biomarkers, such as the amount/extent of PD-L1 expression, microsatellite instability (MSI), and tumor mutation burden (TMB) (Buder-Bakhaya and Hassel 2018).

Antigen-stimulated T-cells have PD-1 receptors all over their surface, which are connected to both T-cells and tumor cells and can be coupled with PDL-1 and PDL-2 ligands to prevent T-cell activation (Pardoll 2012). In addition to improving the body's immune response of T-cells against tumors, the administration of PD-1/PD-L1 receptor antagonists also acts on tumor-associated macrophages (TAMs), restores their capacity to phagocytose tumors, reduces tumor growth, and lengthens patient survival (KATOH 2016). The PD-1/PD-L1 inhibitors are very safe and selective compared to CTLA-4 inhibitors. Several cancers, such as urinary epithelial cancer, melanoma, gastric cancer, Hodgkin's lymphoma, non-small cell lung cancer, kidney cancer, liver cancer, head and neck cancer, and so on, are being treated with PD-L1 monoclonal antibodies (Y. Pan et al. 2019; Mohan et al. 2019; Kang et al. 2019; Bian et al. 2019; Angell et al. 2019).

PD-1 receptor-targeted antagonists include PD-1 and PD-L1 monoclonal antibodies. These antagonists are now commercially available as drug formulations. These include Keytruda (Pembrolizumab), Tecentriq (Atezolizumab), and Opdivo (Nivolumab). The most recent target identified is called Siglec-15, and the research study on this subject is growing. Siglec-15 can decrease the immunological activity of T-cells, which allows tumor cells to escape the body's defense against them. Siglec-15 expression is elevated in several tumor cell types, including those found in liver, bladder, lung, and kidney cancer. Eliminating siglec-15 or utilizing a siglec-15 inhibitor can alleviate immunosuppression and result in the killing of tumor specifically. According to several research studies, the combination of PD-1 / PD-L1 and siglec-15 inhibitors can remove the tumors' resistance to the latter, increasing the likelihood that the treatment would be effective (Cao et al. 2019; J. Wang et al. 2019). See Figure 9.2.

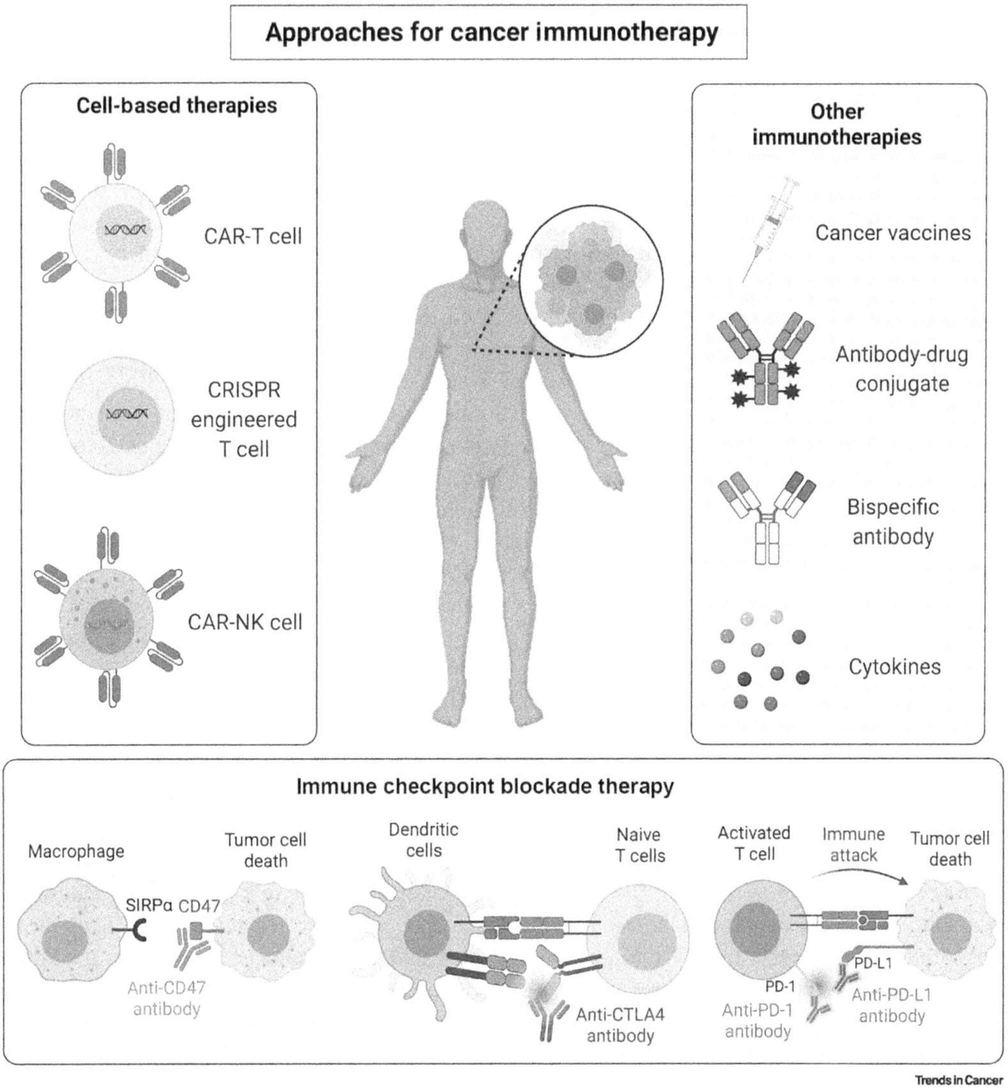

FIGURE 9.2 Various approaches of cancer immunotherapy (10.1016/j.trecan.2022.01.008).

9.3.5 Nanoparticles as Delivery Platforms for Immunomodulatory Agents

Novel physical and chemical properties of synthetic or natural-derived nanoparticles render excellent drug delivery characteristics. Conventional chemotherapeutic drugs when made into nanoformulations were found to be less cytotoxic. Their pharmacokinetic and pharmacodynamic properties could also be tailored with their tumorogenic potential not getting affected (Blanco et al. 2015). Various materials can be used for the synthesis of nanoparticles, and these nanoparticles can act as a ferry agent owing to their hollow core where several therapeutic moieties can be loaded. Surface modification of the synthesized nanoparticles with antibodies, recombinant proteins, peptides etc., aids targeted delivery and enhanced accumulation of the nanoparticles at the tumor site (Wilhelm et al. 2016a). This novel property of nanoparticles has been exploited for several immune-oncology therapeutic purposes. Viral peptides and antigens could be presented to the antigen-presenting cells for activating T-cell memory by loading them into lipid-based and synthetic nanoparticles (Moon et al. 2011; Kuai et al. 2017). Self-assembled nanoparticles, including those produced from viruses enhances the synthesis of inflammatory cytokines like interleukin (IL-2) and interferon (IFN-γ) inside active leukocytes. This increased their potential to induce robust immune responses against low immunogenic malignancies (44). Further, several cocktails of immune stimulants, growth factors, and cytokines can be delivered via nanoparticles to enhance the activity of immune cells (Song et al. 2017). In the recent past, nanoparticles have not only been used for the delivery of conventional drugs, growth factors, and cytokines but also for the delivery of nucleic acids. Owing to the developments in gene editing technologies, research groups across the globe are showing increasing interest in the delivery of nucleic acids, such as siRNA and mRNA. While the latter is associated with the repair of specific disease-related genes *in vivo*, the former plays an important role in transcriptional modification. For instance, circulating T-cells were re-programmed into anti-tumor phenotype by the insertion of leukemia-targeting CAR receptors (chimeric antigen receptor), into the nucleus of T-cells by synthetic DNA carriers. This technology would help to overcome the drawbacks associated with CAR-T immunotherapy like difficulty in *ex vivo* expansion of the isolated T-cells (Smith et al. 2017). The RNA sequences loaded into the lipid nanoparticles with specified compositions were found to be protected from degradation by ribonucleases. Enhanced uptake by professional antigen-presenting cells to express tailor-made antigenic peptides was also observed *in vivo* (Oberli et al. 2017). These nanoparticle structures produce potent anti-tumor effector and memory T-cell responses through powerful IFN activation by using RNA sequences that encode viral or mutant neo-antigens. Surprisingly, preliminary clinical results from the first three melanoma patients treated with these lipid nanoparticle-RNA complexes demonstrated antigen-specific anti-tumor responses and potent IFNα, providing substantial hope that such a method can be quickly applied to the clinic (Kranz et al. 2016).

Furthermore, nanoparticles not only aid in ferrying multiple moieties but also brings about immune targeting as they are readily taken up by the cells of the innate immune system, namely, dendritic cells, monocytes, and macrophages (Blanco et al. 2015; Wilhelm et al. 2016a). The surface of the synthetic nanoparticles acts as an excellent surface for binding serum proteins, such as apolipoproteins, albumin, and complements. These interactions lead to the formation of biological corona that interacts with phagocyte surface receptors (e.g., complement and scavenger receptors) (Wilhelm et al. 2016a). Such non-specific ingestion by phagocytic cells is undesirable for traditional cancer nanomedicine because it lowers the availability of encapsulated medicines to tumor tissues. However, this physiological process might be advantageous for immunological nanomedicine. Synthesis of nanoparticles of novel composition and surface charge properties would render the ability of immunomodulation by homing at the lymphoid organs, such as the spleen. The previouslymentioned properties confer the nanoparticles with the ability to be a promising candidate for cancer vaccine delivery. Adjuvants for anti-cancer vaccines can be delivered by nanoformulation to increase their efficacy while lowering side effects by limiting systemic distribution and extending their activity in draining lymph nodes (Lynn et al. 2015). The natural lymph node-targeting of endogenous proteins can also help with the delivery of adjuvants and antigens. For instance, lipid micellar nanoparticles designed with amphiphilic adjuvant components readily dissociated and bound to albumin when administered *in vivo*. This idea of "hitchhiking" elicit T-cells response against tumor by causing enhanced drug delivery to the antigen-presenting cells (H. Liu et al. 2014).

Nanomaterials mimicking endogenous proteins have been exploited for applications in cancer immunotherapy. Strong polyfunctional, neo-antigen-specific CD8 and CD4 T-cell immunity produced by high-density lipoprotein-mimicking nanodiscs "personalized" with tumor-specific neo-antigens have been developed. These nanodiscs improved lymph-node-targeting of neo-antigens and eliminated established tumors when combined with immune checkpoint blockade (Kuai et al. 2017). These results opened the door for tailored nanomedicine to enhance cancer immunotherapy.

9.3.6 Protein-Based Immunomodulators Delivery

In the past decade, the use of immune checkpoint inhibitors in the form of monoclonal antibodies (mAbs) has gained tremendous attention for the immunotherapeutic treatment of cancers. Monocytes, B-cells, and activated T-cells express PD-1, a key immune checkpoint inhibitor protein. Interaction of PD-1 with its ligand PD-L1 prevents T-cell activation, leading to an immune-suppressed state favoring cancer progression. Antibodies against PD-1 prevent the interaction between PD-1 and PD-L1, thus reversing the immune-suppressed state and activating anti-tumor responses. Though the administration of mAbs as a therapeutic module in clinics has shown great success, early degradation and off-targeted toxicity demand better delivery strategies. In this perspective, researchers have developed nanoparticles of PLGA (N. Zhang et al. 2019; Ordikhani et al. 2018) or upconversion nanoparticles made of mesoporous silica (Z. Wang et al. 2021). Co-delivery of PD-1 antibodies with other small molecule inhibitors or adjuvants improved the efficiency of the treatment modality. The development of biodegradable nanocarriers would increase the safety and release of the entrapped PD-1 that would be desired in the clinical aspect. Triggered release systems that are responsive to reactive oxygen species (ROS), matrix metalloproteinase (MMP), and pH of the tumor microenvironment are widely researched in the recent past. Novel nanosystems sensitive to MMP, made of DNA-nano cocoon, loaded with PD-1 and CpG-ODNs (CpG oligonucleotides), were tested for their efficiency in preventing tumor relapse post-surgery (C. Wang et al. 2016). Liu and his team developed nanogels encapsulated with PD-1 antibodies and IMD-0354 containing lipid nanoparticles. This nanosystem was MMP-2 responsive, and they blocked PD-1 and increased the TAM (tumor-associated macrophages) repolarization. IMD-0354 downregulated the PD-1 expression on T-cells by inhibiting the NF-κβ pathway. Owing to this combinational therapeutic intervention, a reduction in tumor growth and increased survival was observed in mice with B16 tumors. Not only PD-1 but also other checkpoint inhibitors, such as CD47 (Y. Xu et al. 2021) and CTLA-4 (C. Wang et al. 2014; J. Xu et al. 2017b) have also been delivered via nanomaterials to improve their therapeutic efficiency.

9.3.7 Nucleic Acid-Based Immunomodulators Delivery

Delivery of nucleic acids, such as micro-RNA, plasmid-DNA (pDNA), and short interfering-RNA (siRNA), is another strategy to modulate the immune system. Unlike mAbs with heterogenicity in their pattern, the negative charge of the nucleic acids enables their ready encapsulation inside the cationic nanosystems. Lipid (Shobaki et al. 2020), inorganic (Y. Wu et al. 2019b), and polymeric (Kwak et al. 2019) nanosystems have been used for the delivery of siRNA to modulate immune cells. RNA-based therapeutics hold an advantage over DNA-based therapeutics, as the former can start their activity in the cytosol, whereas the latter must be targeted to the nucleus for their effective activity. This created a significant barrier for non-viral plasmid-DNA delivery into immune cells because several immune cells, such as primary T-cells, macrophages, DCs, and NK cells, are resistant to pDNA transfection utilizing chemical-based techniques. Therefore, RNA as a therapeutic cargo is preferred over DNA for nanosystem-based nucleic acid delivery. Huang et al. developed PD-1 siRNA encapsulated calcium phosphate-lipid nanoparticles (LCP NPs) for siRNA delivery due to the synergistic effect obtained from membrane fusion mediated by lipid and proton sponge effect arising from calcium phosphate degradation leading to endosomal escape (J. Li et al. 2010). The biocompatibility and biodegradability of calcium phosphate under acidic conditions favor their use in this system. The nucleic acids are encapsulated in these tiny nanoparticles (30 nm) owing to the electrostatic interaction between the positively charged calcium group in the nanoparticles and the negatively charged phosphate group in the nucleic acids. The siRNA delivered by this method effectively knocked down the PD-1 in TILs, resulting in enhanced cytokine production and killing efficacy (Y. Wu et al. 2019a). Tumor-infiltrating macrophages and monocytes play an important role in tumor progression, metastasis, and invasion. Lipid nanoparticles with PEG linkers that are acid-labile have been synthesized by Hanafyand and co-workers to encapsulate PD-1 siRNA for the downregulation of PD-1. The PD-1 in TAMs has been targeted in this study rather than the PD-1 in lymphocytes. The results showed an increased uptake of these nanoparticles by the J774A.1 macrophages. Furthermore, a reduction in tumor size and PD-1 expression was observed in B16-F10 tumor model-bearing mice. The authors have claimed that these results could be associated with the M2 to M1 repolarization of the macrophages after checkpoint blockage. Other than the knockdown of immune checkpoint proteins, strategies to knock down other critical genes involved in the pro-tumorigenic activities of TAM have also shown promising results. For instance, lipid nanoparticles were loaded with siRNA to block/silence the hypoxia-inducible factor 1α (HIF-1α) and activator of transcription 3 (STAT3) that plays a significant role in tumor angiogenesis and inhibits tumor suppression (Werno et al. 2010; Y. Chen et al. 2019). The developed nanoparticles were readily uptaken by the TAMs without any specific targeting ligands. HIF-1α and STAT3 silencing resulted in an increase in the CD169+ macrophages (M1) and CD11+ macrophage infiltration. Furthermore, a reduction in CD31, TGF-β, and an increase in TNF-α, and INF-γ with a reduction in tumor size was also observed by quantitative PCR. CD163 and CD206 have been effectively studied for the increased targeting of TAMs. Generally, CD206 or MRC1 (the macrophage mannose receptor 1) has been exploited as a target for delivery using nanoparticles (Zhu et al. 2013; Y. Zhou et al. 2020; F. Zhang et al. 2019).

9.4 CLINICAL TRIALS OF VARIOUS NANOFORMULATIONS FOR CANCER

Several nanoformulations like liposomes, polymeric conjugates, polymeric microspheres, protein conjugates, and many other unique nanomaterials have been extensively tested in clinical trials for their efficacy in drug delivery and targeting ability (Craig et al. 2021). Targeted delivery is the future of cancer therapeutics as it reduces the risk of off-target-related toxicity. Moreover, nanomedicine-based therapeutics are effective therapy owing to their small size and increased drug loading, and targeted delivery. Liposomes are lipid bilayered-vesicles made up of a hydrophobic outer layer and hydrophilic inner layer and are mainly composed of amphipathic phospholipids. The advantages of liposomes are 1) ease of synthesis, 2) simple drug loading process, 3) biocompatible components, and 4) possibility for surface modification (Piffoux et al. 2018). Owing to the previouslymentioned advantages, liposomes are widely used as nanomedicine in clinical trials.

The first liposomal formulation was Doxil, liposomes loaded with the toxic chemotherapeutic drug doxorubicin. It was approved by FDA in 1995. A year after this approval, another drug named

DaunoXome® was approved by FDA for the treatment of Kaposis sarcoma. The formulation contained the drug daunorubicin encapsulated into the liposomes (Beltrán-Gracia et al. 2019). Vincristine sulfate loaded onto a liposomal formulation of sphingomyelin/cholesterol was approved in 2012 by FDA and were exhibiting better PK/PD properties as opposed to plain vincristine. They were marketed in the name of Marqibo®. Depocyt® (Cytarabine/Ara-C), Mepact® (Mifamurtide), Onivyde® (Irinotecan), and Myocet®(DOX) are the other liposomal formulations that are approved by the FDA for the treatment of cancer. However, Depocyt was discontinued as it was on the microscale. Due to its effectiveness against a variety of cancer types, cisplatin is one of the most frequently employed chemotherapies. Nonetheless, it has serious side effects, highlighting the urgent need for re-formulation and specificity (Ciarimboli 2014). A nanoformulation was developed with trigger-sensitive carriers that would release the drug cisplatin in response to tumor environmental conditions. The liposomal formulation LiPlaCis was hydrolyzed by the phospholipase A2–IIA isoenzyme expressed in many solid tumors including breast, colorectal, pancreatic, gastric, and prostate cancers (Cummings 2007). Compared to cisplatin, LiPlaCis offers a wider therapeutic window due to its improved PK characteristics, higher efficacy, and higher maximum tolerated dose (ClinicalTrials.gov Identifier: NCT01861496). The probability of success in clinical trials has been considerably enhanced with the aid of Drug Response Prediction (DRP®). The patients are subjected to genetic screening and are selected for trials based on the results that show the probability of patients' response to the treatment. This has led to the reduced risk and cost associated with the treatment modality.

Glioblastomas are one of the most common forms of brain tumor with a very poor prognosis. The currently available therapeutic modalities of glioblastoma are not effective owing to the reduced activity, resistance to the drug, damage to the brain from the therapeutic modalities, and restricted access to the tumor sites in the intracranial regions (Grek et al. 2018). Liposomal formulations with irinotecan and Gd are currently under phase 1 investigation and are administered by convection-enhanced delivery (CED) (ClinicalTrials.gov Identifier: NCT02022644). The components of the formulation enable effective and targeted drug delivery. Liposomal components aid breaching of the blood-brain barrier, whereas Gd enabled real-time imaging. CED assisted in the targeted delivery by virtue of fluid convention (Mehta et al. 2017). A nanoformulation, E7389-LF, made of liposomes loaded with eribulin (microtubule dynamics inhibitor), has been approved for the treatment of metastatic breast cancer and liposarcoma (Clinical-Trials.gov Identifier: NCT04078295) (Y. Yu et al. 2013). E7389-LF is under phase 1/2 clinical trials along with nivolumab in phase 1b part to understand the effectiveness, tolerability, and effective dosage in phase 2 trials. Liposomal formulations triggered by heat for the release of the encapsulated drug have been developed. ThermoDox releases the encapsulated drug when heated to 40–45 °C (Dou et al. 2017). This drug improved the survival in liver cancer patients by 2.1 years, and the drug is in multiple trials now (ClinicalTrials.gov Identifiers: NCT02181075, NCT04852367, NCT02112656, NCT04791228). Compared with plain doxorubicin and Doxil, ThermoDox delivered 25X and 5X drugs to tumors, respectively.

Paclitaxel is one of the most used drugs to treat cancer, derived from natural components. They are effective in inhibiting cancer progression by blocking mitosis. Despite its effective cancer inhibition, the side effects, such as neuropathy, hepatotoxicity, and cardiotoxicity, severely affect the patient's quality of life. Additionally, paclitaxel is hydrophobic, so to increase the solubility Cremaphor® was used, but this caused an increase in toxicity of the drug in several nanoformulations (L. Kiss et al. 2013). A polymeric micellar formulation, Genexol-PM® made from PEG-b-poly (D,L-lactic acid) (PEG-PLA) loaded with paclitaxel, was approved in Korea for breast cancer and non-small cell lung cancer (NSCLC) in 2007. The developed nanoformulation exhibited lower toxicity and improved maximum tolerable dose compared with Taxol (Paclitaxel with Cremaphor®). A nanoformulation of poly-(L-glutamic acid) conjugated to paclitaxel, Apealea, was approved by Europen Union in 2018. Nanoxel® was approved by India in 2006 and is under clinical trials currently for approval by the FDA (ClinicalTrials. gov Identifier: NCT04066335).

Despite many efforts and new formulations, the occurrence of peripheral neuropathy is still a challenge, thus demanding the need for further optimization of the formulations (Riedel et al. 2021). Poly (ethylene glycol)-block-poly(aspartic acid) (PEG-b-pAsp) loaded paclitaxel micelles (NK105) underwent a clinical study recently, and the results showed that the incidence of peripheral sensory neuropathy was 1.4% in NK105. In contrast, it was 7.5% in paclitaxel alone, thus showing the efficiency of the formulation in inhibiting neuropathy (Clinical-Trials.gov Identifier: NCT01644890). NC-6300, a micellar formulation of doxorubicin, sensitive to pH (pH<5), is under phase 2 clinical trial (ClinicalTrials.gov Identifier: NCT03168061). Cisplatin encapsulated in poly(ethylene glycol)-block-poly (glutamic acid) (PEG-b-pGlu) micelle has gone through many trials. It is now under investigation to understand its effectiveness when used along with Pembrolizumab in the treatment of head and neck cancer (ClinicalTrials.gov Identifier: NCT03771820).

To prevent plasma clearance, decrease immunogenicity, inhibit degradation, and to PK profile, polymer-drug conjugates have been developed, mainly made of PEG (polyethylene glycol) (Ekladious et al. 2019). With the advent of development in technology, several polymers like chitosan, hyaluronic acid, polysaccharides, polypeptides, and polysialic acid have also been used in the development of polymer-drug conjugate due to their biocompatibility and biodegradability than PEG (Girase et al. 2020). A polyglutamate-paclitaxel conjugate, Opaxio® (previously-Xyotax®) that was named an orphan drug by FDA, is now under phase 3 trials on patients with stage III/IV ovarian, fallopian or peritoneal cancers (ClinicalTrials.gov Identifier: NCT00108745). JK-1201I, a PEG-Irinotecan complex, is under phase-I trials to treat solid malignant tumors (ClinicalTrials.gov Identifier: NCT04366648). Recombinant human arginase loaded onto the PEGylated complex, called PEG-BCT-100, is used to starve the tumor cells by depleting the arginine. This drug is now entering phase 2 clinical trials (ClinicalTrials.gov Identifier: NCT03455140).

EP0057 (previously - CLX101/IT-101) is made up of a cyclodextrin-based polymer backbone and conjugated with camptothecin (CPT) (topoisomerase inhibitor) and underwent phase 1b/2 trials for its use in epithelial ovarian cancer treatment (ClinicalTrials.gov Identifier: NCT02389985). Furthermore, it is currently in phase 1/2 for lung cancer in combination with Olaparib (ClinicalTrials.gov Identifier: NCT02769962). Colorectal cancers show the enhanced expression of somatostatin receptors (SSR) and a novel nano-drug formulation of ethylcellulose loaded with Cetuximab is under clinical trial at stage 1. These nanoparticles have been designed for targeted delivery to colorectal cancer by decorating them with an SSR agonist octreotide (Reubi et al. 1996). The novelty of the formulation is that it is stable at pH 1.5 and hence would not cause a toxic reaction in the stomach and it releases the drug at pH 6.8 and showed an overall reduction in toxicity (ClinicalTrials.gov Identifier: NCT03774680).

Proteins have been exploited widely for drug delivery applications owing to their biocompatible and biodegradable properties. Further, specific interactions of proteins can also be utilized for targeted delivery and uptake of the particles. Albumin-based nanoparticles loaded with paclitaxel, Abraxane®, have been approved by FDA to be used as the first line of treatment modality for NSCLC, metastatic breast cancer, and pancreatic cancer in the advanced stage. Another albumin-based formulation, ABI-009 conjugated with Rapamycin, is being under investigation in combination with mFOLFOX6 and Bevacizumab for treating patients with advanced colorectal cancer (ClinicalTrials.gov Identifier: NCT04390399).

9.4.1 Nanocarriers for Gene Delivery

Gene therapy has opened new avenues for the treatment of several types of cancers as the delivered nucleic acids can express pro-apoptotic proteins, silence oncogenic genes, synthesize anti-tumor cytokines, replace mutated genes, and induce the immune system. However, one major concern in gene delivery is the successful targeting of the nucleic acids to the target site before their degradation. Patisiran (ONPATTRO®), the first liposomal formulation encapsulated with siRNA encoding the genes for transthyretin protein expression, that is responsible for hereditary transthyretin amyloidosis. The formulation was the first in-kind nanoformulation to deliver siRNA that was approved by FDA in 2019. Recombinant viral vectors are preferred over non-viral vectors for gene delivery, but there are limitations associated with it, such as narrow cell tropism, immune reaction, size of the gene limitation, inability to modify the surface, and up-scaling issues (Lugin et al. 2020). Non-viral vectors dynamic, elicits low immunogenicity, and large-scale synthesis is possible, but they are associated with low transfection efficiency.

Polo-like kinase 1 (PLK1) is highly expressed in several human cancers and blocking their expression can lead to apoptosis by inhibiting mitosis. Stable nucleic acid-lipid particles (SNALPs) are comprised of PEGylated lipids, high transition temperature phospholipids, and ionizable cationic phospholipids (Judge et al. 2009). The SNALP formulation containing siRNA (TKM-080301) to silence PLK-1 is under clinical trials for liver cancer treatment (ClinicalTrials.gov Identifier: NCT01437007). TKM-080301 showed preliminary anti-cancer activity in earlier clinical studies and was generally well tolerated by patients with solid tumors (ClinicalTrials.gov Identifier: NCT02191878). Eph receptor A2 (EphA2) is a receptor of the tyrosine kinase family highly expressed in several cancers and is responsible for cell proliferation, differentiation, and survival (Huang et al. 2014). 1,2-Dioleoyl-sn-glycero-3-phosphocholine (DOPC) liposomes with siRNAs for EphA2 are under phase 1 trials for treating patients with relapsing solid tumors (Clinical- Trials.gov Identifier: NCT01591356). Transforming growth factor-β (TGF-β) is responsible for cell differentiation, migration, epithelial-to-mesenchymal transition, and apoptosis (Fabregat and Caballero-Díaz 2018). STP705, a polypeptide nanoparticle carrying siRNA against both cyclooxygenase-2 and TGF-β1 is under clinical trials for hepatocellular carcinoma, cutaneous squamous cell carcinoma, and cell carcinoma (ClinicalTrials.gov Identifier: NCT04844983, NCT04676633, NCT04669808). A lipid nanoparticle for the delivery of the gene TUSC2, which is responsible for anti-tumor effects, has been developed for treating NSCLCs. GPX-001 (quaratusugene ozeplasmid) was named a fast-track drug by FDA in 2020 for treating NSCLCs along with Osimertinib for combinational therapy (ClinicalTrials.gov Identifier: NCT04486833).

Poly(lactic-co-glycolic acid) (PLGA) nanoparticles loaded with tumor antigen NY-ESO-1 (New York esophageal squamous cell carcinoma 1) and iNKT (invariant natural killer T-cell) cell activator IMM60 are being tested in phase 1 trials for treating cancer patients and checking their anti-cancer responses (ClinicalTrials.gov Identifier: NCT04751786). mRNA vaccines are the emerging trend in cancer therapy as they show simple integration, ease of synthesis, and increased protein expression levels (Jahanafrooz et al. 2020). Moderna has developed a personalized cancer vaccine obtained from sequencing the patient's tumor to elicit a better anti-cancer response. Lipid nanoparticles coating the mRNA 4157 are under clinical trial and are given to patients with solid tumors resected. They are given along with Pembrolizumab to patients with unresectable solid tumors (ClinicalTrials.gov Identifier: NCT03313778). A liposomal formulation containing mRNA against four different malignancy-associated antigens has been developed. Lipo-MERIT is the first mRNA vaccine to be tested in humans (BNT111/Melanoma FixVac). The vaccine carries mRNA against tyrosinase, NY-ESO-1, melanoma-associated antigen A3, and transmembrane phosphatase with tensin homology (ClinicalTrials.gov Identifier: NCT02410733). The interim analysis showed that the vaccine when used along with immune checkpoint inhibitor PD-1 induced of strong CD4+ and $CD8^+$ T-cell immunity. The vaccine is currently under phase 1 trial for HPV16-positive cancers (BNT113) (Clinicaltrials.gov Identifier NCT03418480), ovarian cancer (BNT115) (Clinicaltrials.gov Identifier NCT04163094), prostate cancer (BNT112) (Clinicaltrials.gov Identifier NCT04382898), and triple-negative breast cancer (BNT114) (Clinicaltrials.gov Identifier NCT02316457). Liposoaml-mRNA vaccine formulation with three different TSA RNA for ovarian cancer patients has been developed and is in open-labeled, clinical trial phase 1. The patients will be treated with the W_ova1 vaccine before and during the adjuvant chemotherapy (ClinicalTrials.gov Identifier: NCT04163094).

9.5 PERSONALIZED CANCER NANOMEDICINE

Precision or personalized medicine comprises of customized medical diagnosis, treatments, and therapeutic products for each patient. In selecting pharmacological molecules, customized therapy takes patients' omics data, such as genomics, proteomics, and metabolomics into account. Leveraging nanomedicine as the platform for targeted drug delivery, the specific receptor and metabolizing enzyme are the predominant biological components which are anticipated to determine the efficacy of a tailored therapy. Nanomedicine can be designed with an appropriate targeting ligand that possesses a high affinity and selectivity towards the target receptor. Profiling metabolizing enzymes permits the selection of suitable matrices and targeting ligand that aren't unduly destroyed prior to reaching their therapeutic capability (Hood and Friend 2011) (Bragazzi 2013).

Trastuzumab is one of the gold standards in the field of precision medicine, and it has been widely used in the treatment of metastatic breast cancer overexpressing HER2 (Vogel et al. 2002). It has been documented that the probable molecular targets and subsequent fate of a drug molecule could be ascertained by analyzing the proteomics and genomics profile of an individual's cancerous and healthy tissues. Additionally, a drug-cocktail could be administered depending on the individual needs and multiple targets (Minko et al. 2013). Upregulated receptors, including estrogen receptor (ER), HER2, and EGFR, are some of the extensively studied cellular targets for personalized cancer nanomedicine therapy. Estrone conjugated noscapine loaded gelatin nanoparticles were more accumulated in ER expressing breast cancer cell lines rather than in ER negative breast cancer cells (Madan et al. 2014). In another study, poly (lactide-co-glycolide) core-hydrated polyethylene glycol shell type NP loaded with docetaxel and surface decorated with anti-HER2/neu peptide conjugated modified HIV-1 Tat was reported. Rather than caveolae-mediated endocytosis, it activated several endocytic routes for its uptake, increasing the drug load available to HER2-overexpressing tumors (Zhe Yang et al. 2015). Similarly, cisplatin carrying gelatin NPs decked with biotinylated EGF tumor-specific ligand could penetrate EGFR-overexpressing lung tumor cells with minimal toxicity and increased anti-tumor efficacy (Tseng et al. 2009).

Cellular uptake of a nano medication is critically influenced by its surface chemistry, size, shape, and other physical properties (Yameen et al. 2014). Nanoparticles composed of polylactide-co-polyethylene glycol or poly(L-lysine)-surface complexed poly(lactide-co-glycolide) are predominantly internalized by clathrin associated endocytosis and degradable by lysosomes. If it is ingested by caveolae dependent endocytosis, nanomedicine becomes less sensitive to lysosomal breakdown. Folic acid is an example of a ligand that facilitates receptor-mediated endocytosis (Musalli et al. 2020). Folate decorated PEG coated NPs engage caveolae folate receptors, stimulating endocytic uptake mechanism (Bareford and Swaan 2007; Zwicke et al. 2012; A. L. Kiss and Botos 2009; Xie et al. 2013; Cheung et al. 2016).

In recent years, cancer cell membrane coated NPs have garnered a lot of attention with respect to effective tumor tissue targeting. To be precise, personalized cancer nanomedicine could benefit enormously from employing an individual's own cancer membrane as the targeting moiety, as the former has remarkable homologous binding abilities. Numerous studies have reported that these cloaked NPs were able to locate their autologous tumor, owing to the presence of tumor associated antigens (TAAs) and surface proteins on the surface of these engineered NPs (Fang et al. 2014; Fan et al. 2021; Rao et al. 2019). Rather than having to discover precise targeting moieties, the direct utilization of cancer cell membrane permits the manufacturing of tailored therapy. This avoids the wasted effort and resources associated in identifying the targeted moiety. The cellular membrane nanotechnology also facilitates the provision of tailored nanomedicine to each individual patient without categorizing them or excluding some of them. In the following table (Table 9.1) we have summarized some recent cancer cell membrane coated NPs in various cancer therapy modalities.

TABLE 9.1
Various Cancer Cell Membrane Coated NPs Applied for Cancer Nanomedicine

Cell Source	Core Material	Cargo	Applications	Ref.
4T1	Dendritic large-pore mesoporous silica nanoparticles (DLMSNs)	Copper sulfide (CuS) nanoparticles, Resiquimod (R848)	Photothermal ablation, immune remodeling of metastatic triple-negative breast cancer	(Y. Cheng et al. 2020)
B16-F10 – bacterial OMV	Hollow polydopamine (HPDA) NPs	N/A	Photothermal therapy of melanoma	(D. Wang et al. 2020)
HeLa	Chitosan NPs	Ca^{2+} channel siRNA, DOX	Anti-multidrug resistance	(Zhao et al. 2020)
CT26 and RAW 264.7	ZGGO PLNPs coated with mesoporous silica	IR825, Irinotecan	Imaging-Guided Photothermal Therapy of Colorectal Cancer	(Z. H. Wang et al. 2020)
MCF-7	Carbon, silica semi-yolk-spiky-shell nanomotor	DOX	Synergistic photothermal and chemotherapy of breast Cancer	(M. Zhou et al. 2020)
HepG2	ZIF-8 MOF NPs	Dihydroartemisinin	Targeted chemotherapy of hepatocellular carcinoma	(Xiao et al. 2020)
KB	PEG coated gold nanorods	N/A	Combined photothermal and radiotherapy of oral squamous cancer	(Q. Sun et al. 2020)
MDA-MB-231	Rare earth doped NPs	N/A	Tumor imaging in NIR II window	(X. Zhang et al. 2020)
MCF-7 and erythrocyte	Gold nanocages	DOX	Photo-radio-chemotherapy of breast cancer	(M. Sun et al. 2020)
A549	PLGA	Perfluoro-15-crown-5-ether (PFCE), ICG	19F MR/PA/FL imaging-guided targeted photothermal therapy	(Sha Li et al. 2020)
B16-F10	Hollow CuS NPs	ICG, DOX	Targeted photoacoustic (PA) imaging and photothermal-chemotherapy	(M. Wu et al. 2020)
B16-F10	Aluminum phosphate NPs	CpG	Anti-cancer vaccination	(Gan et al. 2020)
4T1	PEG modified EGaIn liquid metal NPs	N/A	Tumor nanovaccine	(Yu Zhang et al. 2020)
4T1	PEG-PLGA NPs	Ce6, MnO_2 NPs	Photodynamic and immunotherapy	(He et al. 2020)
4T1	PAMAM dendrimer	DOX	Targeted chemotherapy	(Pei et al. 2021)
C1498	PLGA NPs	CpG	Anti-cancer vaccination	(Johnson et al. 2021)

(*Continued*)

TABLE 9.1 ***(Continued)***
Various Cancer Cell Membrane Coated NPs Applied for Cancer Nanomedicine

Cell Source	Core Material	Cargo	Applications	Ref.
4T1	Catalase immobilized hollow mesoporous silica	Nitrogen-doped graphene quantum dots (N-GQDs), protoporphyrin IX (PpIX).	Targeted photothermal and photodynamic therapy	(Chen et al. 2021)
4T1	Poly(allylamine hydrochloride)-formylphenylboronic acid-sodium alginate nanogel	DOX	Stimuli-responsive targeted chemotherapy	(Xu et al. 2021)
HeLa	Cerium oxide nanocubes	Glucose oxidase	Tumor selective therapy	(Online et al. 2021)
4T1	Manganese dioxide (MnO2) nanosheet coated gold nanorod (GNR) core	DOX	Targeted photothermal-chemotherapy	(D. Zhang et al. 2021)
HepG2	Magnetic iron oxide (Fe3O4) NPs with silicon dioxide shell	N/A	Immunotherapy	(D. Wu et al. 2021)
HCT116	PVP microneedles containing poloxamer 407 (F127) NPs	Imiquimod	Immunotherapy	(Park et al. 2021)
4T1 and bacterial OMV	N/A	N/A	Personalized cancer immunotherapy	MENDELEY CITATION PLACEHOLDER 49
TE10	PLGA NPs	DOX, curcumin	Targeted chemotherapy of multidrug resistant esophageal cancer	(Gao et al. 2021)
Transferrin conjugated HepG2	N/A	Hypocrellin B NPs	Photodynamic therapy of hepatocellular carcinoma	(Z. Zhang et al. 2021)
MCF-7	PLGA NPs	Curcumin, chlorin e6	Photothermal and chemotherapy	(Yongtai Zhang et al. 2021)
MDA-MB-231	PVP-NMOF of Fe-TCPP	Tirapazamine (TPZ)	Sequential ferroptosis, photodynamic and chemotherapy	(W.-L. Pan et al. 2022)
SMMC-7721	Paclitaxel nanocrystals	N/A	Chemotherapy of hepatocellular carcinoma	(Shen et al. 2022)
K7M2	Alendronate modified hollow MnO_2 NPs	Ginsenosides Rh2	MRI guided chemodynamic-immunotherapy of osteosarcoma	(L. Fu et al. 2022)

9.6 CHALLENGES

The research on application of nanomaterials with high functional adaptability in diagnostic and therapeutic applications drives ongoing research and optimization as cancer nanomedicine, however, significant problems still need to be resolved. Nevertheless, despite advances in preclinical research, a few have made it into clinics. One of the primary concerns about nanomaterials is their toxicity. These even stimulate the immune response, and their incredibly small size allows penetration of physiological barriers that may cause side effects. Many research groups used polyethylene glycol (PEG), a hydrophilic polymer, as one of the main coating or surface modifiers to nanosystem. This decreases the interaction with the serum protein and increase the half-life of circulation time, thereby enhancing the therapeutic outcome (B.-M. Chen et al. 2021). PEG molecules with regularly used contour dimensions vary from 12.5 nm for PEG2000 to 253 nm for PEG40,000 (linear polymer) (B.-M. Chen et al. 2021). According to studies, some PEG-related nanosystems generate anti PEG antibodies, which trigger an immune response during successive treatments and reduce the effectiveness of the therapy (Hershfield et al. 2014; Armstrong et al. 2007; Y. Liu et al. 2019). Many factors influence nanomaterial toxicity, interaction with the serum proteins, and half-life of circulation time. Therefore, optimizing the fabrication process is essential for the application of nanotechnology to clinics.

A central paradigm in the application of nanomedicine for the cancer treatment is the concept of nanoparticle delivery through the enhanced permeation and retention effect (EPR effect) (inter-endothelial gaps in the tumor vascular system) (Hobbs et al. 1998; Jain 1999). These gaps were found to be up to 2,000 nm in size (Jain 1999; Hashizume et al. 2000; Hobbs et al. 1998). The size of inter-endothelial gaps in the tumor vascular system justifies the nanoparticles for the treatment of solid tumors since their dimensions are small enough to extravasate and reach the tumor microenvironment (Jain 1999). The EPR effect has long been explored as the major passive delivery mechanism utilizing the nanosystem (Wilhelm et al. 2016b). However, most engineered nanomaterials hardly reach clinical stage. Some researchers attempted to reexamine the concept of EPR and investigated the potential efficacy of this "royal gate" towards cancer treatment (Nel et al. 2017; Huynh and Zheng 2015). The EPR effect in small animals differs from that in humans. Sindhwani et al. indicate that these inter-endothelial gaps do not mediate the trafficking of nanoparticles into solid tumors (Sindhwani et al. 2020). The study was conducted in four distinct mouse models with three different types of human tumors. Mathematical simulation and modeling, and two different kinds of imaging techniques were used to evaluate the trafficking of nanoparticles to the tumor. Instead, the study concluded that endothelial cells play a direct role in the active process by which approximately 97% of nanoparticles infiltrate tumors. More investigations are required to examine diverse characteristics and efficiency of EPR effect and understand the mechanism of the transport of nanocarriers in order to effectively utilize the EPR effect in cancer therapy.

Clinical translation is a challenging hurdle in the application of nanomaterials for cancer therapies. Although there has been a considerable progress of research on nanocarriers for cancer therapy, most of it employs primary or secondary cell lines and different animal models that are not guaranteed to reflect the behaviors in real human organs. No proper models exist to study the advanced cancer stage, especially metastasis. Metastasis is only the main reason for the increase in the mortality rate due to cancer worldwide. Biomimetic 'organ/tumor-on-a-chip' technologies and organoid-based model systems should be the future models to mimic the in vivo condition of nanocarriers employed in cancer patients.

9.7 CONCLUSIONS

Nanomedicine is among the rapidly expanding research fields and one of the most effective cancer treatments at the cutting edge. Numerous nanomedicine technologies have been designed, and several are utilized in clinical cancer treatment. However, nanostructured materials characterization, safety issues, and legal and production challenges impede the broad use of cancer nanomedicine therapies. To anticipate the efficacy and safety of nanomedicines in human, a thorough and reproducible characterization is required. To forecast the efficacy of nanomedicine therapies in the clinic, standards for models in vitro and in vivo are necessary. Additionally, well-defined parameters and thorough guidelines are required, including both regulatory authorization and major manufacturing. In this potential domain of cancer nanomedicine, collaboration among research labs, government regulators, and industries is essential to expedite the development of innovative, safe, and effective therapeutic alternatives for patients.

ABBREVIATIONS

ACI	Adoptive cellular immunotherapy
ALA	Aminolaevulinic acid
ADCC	Antibody-dependent cell-mediated cytotoxicity
Au	NPs Gold nanoparticles
Ce6	Chlorin e6
CAR-T	Chimeric Antigen Receptor T-Cell
CIK	Cytokine-induced killer cells
CED	Convection-enhanced delivery
°C	Degree Celsius
DC	Dendritic cell
DNA	Deoxyribonucleic acid
DRP	Drug Response Prediction
Doxil	Liposomal formulation of Doxorubicin
e.g	Example
EPR	Enhanced Permeability and Retention Effect
FDA	Food and Drug Administration
HIV	Human Immunodeficiency Virus
HLA-G	Human leukocyte antigen G
HIF-1alpha	Hypoxia-inducible factor 1α
IFN	Interferon
ICIs	Immune checkpoint inhibitors
IL	Interleukins
ICG	Indocyanine green
KIRs	Killer inhibitory receptors
KARs	Killer Activation receptors
LCP	NPs Calcium phosphate-lipid nanoparticles
LAK	Lymphocyte activated killer cells
MDR	Multidrug resistance
mRNA	Messenger RNA
MMP	Matrix metalloproteinase
MDSCs	Myeloid-derived suppressor cells
MSI	Microsatellite instability
MHC-1	Major histocompatibility complex class 1
MICA	MHC Class I Polypeptide-Related Sequence A
MICB	MHC class I polypeptide-related sequence B
mAb	Monoclonal antibody
MRC1	Macrophage mannose receptor 1
MAK	Macrophages-activated killer cells
NKC	Natural killer cells

NIR	Near Infra Red
NPs	Nanoparticles
nm	Nanometer
PDT	Photodynamic Therapy
PTT	Photothermal therapy
PEG	Poly (ethylene glycol)
PLGA	Poly(lactic-co-glycolic acid)
pDNA	Plasmid-DNA
ROS	Reactive Oxygen Species
RNA	Ribonucleic acid
siRNA	Short interfering Ribonucleic acid
SPR	Surface Plasmon Resonance
STAT3	Activator of transcription 3
TSAs	Tumor-specific antigens
TAAs	Tumor-associated antigens
TME	Tumor microenvironment
TIL	Tumor-infiltrating lymphocyte
Treg	Regulatory T-cells
TMB	Tumor mutation burden
TAMs	Tumor-associated macrophages
TGF-beta	Transforming growth factor beta
WHO	World Health Organization

REFERENCES

Abrahamse, Heidi, and Michael R. Hamblin. 2016. "New Photosensitizers for Photodynamic Therapy." *Biochemical Journal*. Portland Press Ltd. doi:10.1042/BJ20150942.

Akbarzadeh, Abolfazl, Rogaie Rezaei-Sadabady, Soodabeh Davaran, Sang Woo Joo, Nosratollah Zarghami, Younes Hanifehpour, Mohammad Samiei, Mohammad Kouhi, and Kazem Nejati-Koshki. 2013. "Liposome: Classification, Preparation, and Applications." *Nanoscale Research Letters* 8 (1). United States: 102. doi:10.1186/1556-276X-8-102.

Alfarouk, Khalid O., Christian Martin Stock, Sophie Taylor, Megan Walsh, Abdel Khalig Muddathir, Daniel Verduzco, Adil H. H. Bashir, et al. 2015. "Resistance to Cancer Chemotherapy: Failure in Drug Response from ADME to P-Gp." *Cancer Cell International*. BioMed Central Ltd. doi:10.1186/s12935-015-0221-1.

Ali, Eunus S., Shazid Md Sharker, Muhammad Torequl Islam, Ishaq N. Khan, Subrata Shaw, Md Atiqur Rahman, Shaikh Jamal Uddin, et al. 2021. "Targeting Cancer Cells with Nanotherapeutics and Nanodiagnostics: Current Status and Future Perspectives." *Seminars in Cancer Biology* 69 (January 2020). Elsevier Ltd: 52–68. doi:10.1016/j.semcancer.2020.01.011.

Alvi, S. B., P. S. Rajalakshmi, A. B. Jogdand, B. Nazia, V. Bantal, and A. K. Rengan. 2022. "Chitosan IR806 Dye-Based Polyelectrolyte Complex Nanoparticles with Mitoxantrone Combination for Effective Chemo-Photothermal Therapy of Metastatic Triple-Negative Breast Cancer." *International Journal of Biological Macromolecules* 216 (September). Elsevier B.V.: 558–570. doi:10.1016/j.ijbiomac.2022.07.018.

Alvi, Syed Baseeruddin, Tejaswini Appidi, B. Pemmaraju Deepak, P. S. Rajalakshmi, Gillipsie Minhas, Surya Prakash Singh, Afreen Begum, et al. 2019. "The 'Nano to Micro' Transition of Hydrophobic Curcumin Crystals Leading to in Situ Adjuvant Depots for Au-Liposome Nanoparticle Mediated Enhanced Photothermal Therapy." *Biomaterials Science* 7 (9): 3866–3875. doi:10.1039/c9bm00932a.

Al-Zoubi, Mazhar Salim, and Raed M. Al-Zoubi. 2022. "Nanomedicine Tactics in Cancer Treatment: Challenge and Hope." *Critical Reviews in Oncology/Hematology* 174 (April). Elsevier B.V.: 103677. doi:10.1016/j.critrevonc.2022.103677.

Angell, H. K., J. Lee, K-M. Kim, K. Kim, S-T. Kim, S. H. Park, W. K. Kang, et al. 2019. "PD-L1 and Immune Infiltrates Are Differentially Expressed in Distinct Subgroups of Gastric Cancer." *OncoImmunology* 8 (2): e1544442. doi:10.1080/2162402X.2018.1544442.

Appidi, Tejaswini, Deepak Bharadwaj Pemmaraju, Rafiq Ahmad Khan, Syed Baseeruddin Alvi, Rohit Srivastava, Mahadeb Pal, Nooruddin Khan, and Aravind Kumar Rengan. 2020. "Light-Triggered Selective ROS-Dependent Autophagy by Bioactive Nanoliposomes for Efficient Cancer Theranostics." *Nanoscale* 12(3). Royal Society of Chemistry: 2028–2039. doi:10.1039/c9nr05211a.

Armstrong, Jonathan K., Georg Hempel, Susanne Koling, Linda S. Chan, Timothy Fisher, Herbert J. Meiselman, and George Garratty. 2007. "Antibody against Poly(Ethylene Glycol) Adversely Affects PEG-Asparaginase Therapy in Acute Lymphoblastic Leukemia Patients." *Cancer* 110 (1). United States: 103–111. doi:10.1002/cncr.22739.

Aulic, Suzana, Domenico Marson, Erik Laurini, Maurizio Fermeglia, and Sabrina Pricl. 2020. "Breast Cancer Nanomedicine Market Update and Other Industrial Perspectives of Nanomedicine." *Nanomedicines for Breast Cancer Theranostics*, January. 371–404. Elsevier. doi:10.1016/B978-0-12-820016-2.00016-1.

Bareford, Lisa M., and Peter W. Swaan. 2007. "Endocytic Mechanisms for Targeted Drug Delivery." *Advanced Drug Delivery Reviews* 59 (8): 748–758. doi:10.1016/j.addr.2007.06.008.

Barenholz, Yechezkel. 2012. "Doxil® - The First FDA-Approved Nano-Drug: Lessons Learned." *Journal of Controlled Release*. doi:10.1016/j.jconrel.2012.03.020.

Barenholz, Yechezkel (Chezy). 2012. "Doxil® — The First FDA-Approved Nano-Drug: Lessons Learned." *Journal of Controlled Release* 160 (2): 117–134. doi:10.1016/j.jconrel.2012.03.020.

Baskaran, Rengarajan, Junghan Lee, and Su Geun Yang. 2018. "Clinical Development of Photodynamic Agents and Therapeutic Applications." *Biomaterials Research*. BioMed Central Ltd. doi:10.1186/s40824-018-0140-z.

Beltrán-Gracia, Esteban, Adolfo López-Camacho, Inocencio Higuera-Ciapara, Jesús B Velázquez-Fernández, and Alba A. Vallejo-Cardona. 2019. "Nanomedicine Review: Clinical Developments in Liposomal Applications." *Cancer Nanotechnology* 10 (1): 11. doi:10.1186/s12645-019-0055-y.

Bhaskar, Sangeeta, and Manu Smriti Singh. 2014. "Nanocarrier-Based Immunotherapy in Cancer Management and Research." *ImmunoTargets and Therapy*, June, 121. doi:10.2147/ITT.S62471.

Bian, Benjamin, Daniele Fanale, Nelson Dusetti, Julie Roque, Sonia Pastor, Anne-Sophie Chretien, Lorena Incorvaia, Antonio Russo, Daniel Olive, and Juan Iovanna. 2019. "Prognostic Significance of Circulating PD-1, PD-L1, Pan-BTN3As, BTN3A1 and BTLA in Patients with Pancreatic Adenocarcinoma." *OncoImmunology* 8 (4): e1561120. doi:10.1080/2162402X.2018.1561120.

Blanco, Elvin, Haifa Shen, and Mauro Ferrari. 2015. "Principles of Nanoparticle Design for Overcoming Biological Barriers to Drug Delivery." *Nature Biotechnology* 33 (9): 941–951. doi:10.1038/nbt.3330.

Bovis, Melissa J., Josephine H. Woodhams, Marilena Loizidou, Dietrich Scheglmann, Stephen G. Bown, and Alexander J. Macrobert. 2012. "Improved in Vivo Delivery of M-THPC via Pegylated Liposomes for Use in Photodynamic Therapy." *Journal of Controlled Release: Official Journal of the Controlled Release Society* 157 (2). Netherlands: 196–205. doi:10.1016/j.jconrel.2011.09.085.

Bragazzi, Nicola Luigi. 2013. "From P0 to P6 Medicine, a Model of Highly Participatory, Narrative, Interactive, and 'Augmented' Medicine: Some Considerations on Salvatore Iaconesi's Clinical Story." *Patient Preference and Adherence* 7 (April): 353–359. doi:10.2147/PPA.S38578.

Broder, Howard, Roberta A. Gottlieb, and Norman E. Lepor. 2008. "Chemotherapy and Cardiotoxicity." *Reviews in cardiovascular medicine* 9 (2), 75.

Buder-Bakhaya, Kristina, and Jessica C. Hassel. 2018. "Biomarkers for Clinical Benefit of Immune Checkpoint Inhibitor Treatment—A Review From the Melanoma Perspective and Beyond." *Frontiers in Immunology* 9 (June). doi:10.3389/fimmu.2018.01474.

Bulbake, Upendra, Sindhu Doppalapudi, Nagavendra Kommineni, and Wahid Khan. 2017. "Liposomal Formulations in Clinical Use: An Updated Review." *Pharmaceutics*. MDPI AG. doi:10.3390/pharmaceutics9020012.

Calixto, Giovana, Jéssica Bernegossi, Bruno Fonseca-Santos, and Marlus Chorilli. 2014. "Nanotechnology-Based Drug Delivery Systems for Treatment of Oral Cancer: A Review." *International Journal of Nanomedicine* 9. New Zealand: 3719–3735. doi:10.2147/IJN.S61670.

Cao, Guangchao, Zhiqiang Xiao, and Zhinan Yin. 2019. "Normalization Cancer Immunotherapy: Blocking Siglec-15!" *Signal Transduction and Targeted Therapy* 4 (1): 10. doi:10.1038/s41392-019-0045-x.

Castano, Ana P., Tatiana N. Demidova, and Michael R. Hamblin. 2005. "Mechanisms in Photodynamic Therapy: Part Two - Cellular Signaling, Cell Metabolism and Modes of Cell Death." *Photodiagnosis and Photodynamic Therapy*. Elsevier. doi:10.1016/S1572-1000(05)00030-X.

Chapman, Tara L., Astrid P. Heikema, and Pamela J. Bjorkman. 1999. "The Inhibitory Receptor LIR-1 Uses a Common Binding Interaction to Recognize Class I MHC Molecules and the Viral Homolog UL18." *Immunity* 11 (5): 603–613. doi:10.1016/S1074-7613(00)80135-1.

Chen, Bing-Mae, Tian-Lu Cheng, and Steve R. Roffler. 2021. "Polyethylene Glycol Immunogenicity: Theoretical, Clinical, and Practical Aspects of Anti-Polyethylene Glycol Antibodies." *ACS Nano* 15 (9). American Chemical Society: 14022–14048. doi:10.1021/acsnano.1c05922.

Chen, Hongliang, Donghui Zheng, Wenzhen Pan, Xiang Li, Bin Lv, Wenxiang Gu, Jeremiah Ong, et al. 2021. "Biomimetic Nanotheranostics Camou Fl Aged with Cancer Cell Membranes Integrating Persistent Oxygen Supply and Homotypic Targeting for Hypoxic Tumor Elimination." doi:10.1021/acsami.1c03010.

Chen, Joyce, Isaac F. López-Moyado, Hyungseok Seo, Chan-Wang J. Lio, Laura J. Hempleman, Takashi Sekiya, Akihiko Yoshimura, James P. Scott-Browne, and Anjana Rao. 2019. "NR4A Transcription Factors Limit CAR T Cell Function in Solid Tumours." *Nature* 567 (7749): 530–534. doi:10.1038/s41586-019-0985-x.

Chen, Yibing, Yucen Song, Wei Du, Longlong Gong, Haocai Chang, and Zhengzhi Zou. 2019. "Tumor-Associated Macrophages: An Accomplice in Solid Tumor Progression." *Journal of Biomedical Science* 26 (1): 78. doi:10.1186/s12929-019-0568-z.

Cheng, Yuanyuan, Qian Chen, Zhaoyang Guo, Mengwen Li, Xiaoying Yang, Guoyun Wan, Hongli Chen, Qiqing Zhang, and Yinsong Wang. 2020. "An Intelligent Biomimetic Nanoplatform for Holistic Treatment of Metastatic Triple-Negative Breast Cancer via Photothermal Ablation and Immune Remodeling." *ACS Nano* 14 (11): 15161–15181. doi:10.1021/acsnano.0c05392.

Cheng, Zhe, Maoyu Li, Raja Dey, and Yongheng Chen. 2021. "Nanomaterials for Cancer Therapy: Current Progress and Perspectives." *Journal of Hematology & Oncology* 14 (1): 85. doi:10.1186/s13045-021-01096-0.

Cheung, Anthony, Heather J. Bax, Debra H. Josephs, Kristina M. Ilieva, Giulia Pellizzari, James Opzoomer, Jacinta Bloomfield, et al. 2016. "Targeting Folate Receptor Alpha for Cancer Treatment." *Oncotarget* 7 (32): 52553–52574. doi:10.18632/oncotarget.9651.

Ciarimboli, Giuliano. 2014. "Membrane Transporters as Mediators of Cisplatin Side-Effects." *Anticancer Research* 34 (1): 547–550.

Craig, Morgan, Adrianne L. Jenner, Bumseok Namgung, Luke P. Lee, and Aaron Goldman. 2021. "Engineering in Medicine To Address the Challenge of Cancer Drug Resistance: From Micro- and Nanotechnologies to Computational and Mathematical Modeling." *Chemical Reviews* 121 (6): 3352–3389. doi:10.1021/acs.chemrev.0c00356.

Cummings, Brian S. 2007. "Phospholipase A2 as Targets for Anti-Cancer Drugs." *Biochemical Pharmacology* 74 (7): 949–959. doi:10.1016/j.bcp.2007.04.021.

Davis, Zachary B., Daniel A. Vallera, Jeffrey S. Miller, and Martin Felices. 2017. "Natural Killer Cells Unleashed: Checkpoint Receptor Blockade and BiKE/TriKE Utilization in NK-Mediated Anti-Tumor Immunotherapy." *Seminars in Immunology* 31 (June): 64–75. doi:10.1016/j.smim.2017.07.011.

Dou, Yannan, Kullervo Hynynen, and Christine Allen. 2017. "To Heat or Not to Heat: Challenges with Clinical Translation of Thermosensitive Liposomes." *Journal of Controlled Release* 249 (March): 63–73. doi:10.1016/j.jconrel.2017.01.025.

Ekladious, Iriny, Yolonda L. Colson, and Mark W. Grinstaff. 2019. "Polymer–Drug Conjugate Therapeutics: Advances, Insights and Prospects." *Nature Reviews Drug Discovery* 18 (4): 273–294. doi:10.1038/s41573-018-0005-0.

Emran, Talha bin, Asif Shahriar, Aar Rafi Mahmud, Tanjilur Rahman, Mehedy Hasan Abir, Mohd Faijanur Rob Siddiquee, Hossain Ahmed, et al. 2022. "Multidrug Resistance in Cancer: Understanding Molecular Mechanisms, Immunoprevention and Therapeutic Approaches." *Frontiers in Oncology*. Frontiers Media S.A. doi:10.3389/fonc.2022.891652.

Fabregat, Isabel, and Daniel Caballero-Díaz. 2018. "Transforming Growth Factor-β-Induced Cell Plasticity in Liver Fibrosis and Hepatocarcinogenesis." *Frontiers in Oncology* 8 (September). doi:10.3389/fonc.2018.00357.

Falzone, Luca, Salvatore Salomone, and Massimo Libra. 2018. "Evolution of Cancer Pharmacological Treatments at the Turn of the Third Millennium." *Frontiers in Pharmacology* 9. Switzerland: 1300. doi:10.3389/fphar.2018.01300.

Fan, Yueyue, Yuexin Cui, Wenyan Hao, Mengyu Chen, Qianqian Liu, Yuli Wang, Meiyan Yang, et al. 2021. "Carrier-Free Highly Drug-Loaded Biomimetic Nanosuspensions Encapsulated by Cancer Cell Membrane Based on Homology and Active Targeting for the Treatment of Glioma." *Bioactive Materials* 6 (12). Elsevier: 4402–4414. doi:10.1016/J.BIOACTMAT.2021.04.027.

Fang, Ronnie H., Che Ming J. Hu, Brian T. Luk, Weiwei Gao, Jonathan A. Copp, Yiyin Tai, Derek E. O'Connor, and Liangfang Zhang. 2014. "Cancer Cell Membrane-Coated Nanoparticles for Anticancer Vaccination and Drug Delivery." *Nano Letters* 14 (4). American Chemical Society: 2181–2188. doi:10.1021/nl500618u.

Ferrari de Andrade, Lucas, Rong En Tay, Deng Pan, Adrienne M. Luoma, Yoshinaga Ito, Soumya Badrinath, Daphne Tsoucas, et al. 2018. "Antibody-Mediated Inhibition of MICA and MICB Shedding Promotes NK Cell–Driven Tumor Immunity." *Science* 359 (6383): 1537–1542. doi:10.1126/science.aao0505.

Florescu, Maria, Mircea Cinteza, Dragos Vinereanu, and " Carol. 2013. "Maedica-a Journal of Clinical Medicine Chemotherapy-Induced Cardiotoxicity." *Maedica A Journal of Clinical Medicine* 8, 59.

Fraietta, Joseph A., Christopher L. Nobles, Morgan A. Sammons, Stefan Lundh, Shannon A. Carty, Tyler J. Reich, Alexandria P. Cogdill, et al. 2018. "Disruption of TET_2 Promotes the Therapeutic Efficacy of CD19-Targeted T Cells." *Nature* 558 (7709): 307–312. doi:10.1038/s41586-018-0178-z.

Fu, Jijun, Bo Wu, Minyan Wei, Yugang Huang, and Yi Zhou. 2019. "Prussian Blue Nanosphere-Embedded in Situ Hydrogel for Photothermal Therapy by Peritumoral Administration" 9 (3): 604–614. doi:10.1016/j.apsb.2018.12.005.

Fu, Liwen, Weiying Zhang, Xiaojun Zhou, Jingzhong Fu, and Chuanglong He. 2022. "Tumor Cell Membrane-Camouflaged Responsive Nanoparticles Enable MRI-Guided Immuno-Chemodynamic Therapy of Orthotopic Osteosarcoma." *Bioactive Materials* 17 (November). Elsevier: 221–233. doi:10.1016/J.BIOACTMAT.2022.01.035.

Gabizon, Alberto, Rut Isacson, Ora Rosengarten, Dina Tzemach, Hilary Shmeeda, and Rama Sapir. 2008. "An Open-Label Study to Evaluate Dose and Cycle Dependence of the Pharmacokinetics of Pegylated Liposomal Doxorubicin." *Cancer Chemotherapy and Pharmacology* 61 (4). Springer: 695–702. doi:10.1007/S00280-007-0525-5/METRICS.

Gan, Jingyao, Guangsheng Du, Chunting He, Min Jiang, Xingyue Mou, Jiao Xue, and Xun Sun. 2020. "Tumor Cell Membrane Enveloped Aluminum Phosphate Nanoparticles for Enhanced Cancer Vaccination." *Journal of Controlled Release* 326 (July). Elsevier: 297–309. doi:10.1016/j.jconrel.2020.07.008.

Gao, Yi, Yue Zhu, Xiaopeng Xu, Fangjun Wang, Weidong Shen, Xia Leng, and Jiyi Zhao. 2021. "Surface PEGylated Cancer Cell Membrane-Coated Nanoparticles for Codelivery of Curcumin and Doxorubicin for the Treatment of Multidrug Resistant Esophageal Carcinoma" Frontiers in Cell and Developmental Biology 9 (July): 1–13. doi:10.3389/fcell.2021.688070.

Ghoneim, Hazem E., Yiping Fan, Ardiana Moustaki, Hossam A. Abdelsamed, Pradyot Dash, Pranay Dogra, Robert Carter, et al. 2017. "De Novo Epigenetic Programs Inhibit PD-1 Blockade-Mediated T Cell Rejuvenation." *Cell* 170 (1): 142–157.e19. doi:10.1016/j.cell.2017.06.007.

Gillet, Jean-pierre, and Michael M. Gottesman. 2009. "Chapter 4 Mechanisms of Multidrug Resistance in Cancer." *Gillet and Gottesman* 596. doi:10.1007/978-1-60761-416-6.

Girase, Monika Lotansing, Priyanka Ganeshrao Patil, and Pradum Pundlikrao Ige. 2020. "Polymer-Drug Conjugates as Nanomedicine: A Review." *International Journal of Polymeric Materials and Polymeric Biomaterials* 69 (15): 990–1014. doi:10.1080/00914037.2019.1655745.

Grek, Christina L, Zhi Sheng, Christian C. Naus, Wun Chey Sin, Robert G. Gourdie, and Gautam G. Ghatnekar. 2018. "Novel Approach to Temozolomide Resistance in Malignant Glioma: Connexin43-Directed Therapeutics." *Current Opinion in Pharmacology* 41 (August): 79–88. doi:10.1016/j.coph.2018.05.002.

Grzybowski, Andrzej, and Krzysztof Pietrzak. 2012. "From Patient to Discoverer-Niels Ryberg Finsen (1860-1904)-the Founder of Phototherapy in Dermatology." *Clinics in Dermatology* 30 (4): 451–455. doi:10.1016/j.clindermatol.2011.11.019.

Guan, Guijian, Mingda Wu, and Ming Yong Han. 2020. "Stimuli-Responsive Hybridized Nanostructures." *Advanced Functional Materials* 30 (2). John Wiley & Sons, Ltd: 1903439. doi:10.1002/ADFM.201903439.

Hashizume, Hiroya, Peter Baluk, Shunichi Morikawa, John W. McLean, Gavin Thurston, Sylvie Roberge, Rakesh K. Jain, and Donald M. McDonald. 2000. "Openings between Defective Endothelial Cells Explain Tumor Vessel Leakiness." *The American Journal of Pathology* 156 (4): 1363–1380. doi:10.1016/S0002-9440(10)65006-7.

Havel, Henry, Gregory Finch, Pamela Strode, Marc Wolfgang, Stephen Zale, Iulian Bobe, Hagop Youssoufian, Matthew Peterson, and Maggie Liu. 2016. "Nanomedicines: From Bench to Bedside and Beyond." *AAPS Journal* 18 (6). Springer New York LLC: 1373–1378. doi:10.1208/S12248-016-9961-7/METRICS.

He, Qianyuan, Huiping Hu, Qian Zhang, Tingting Wu, Yu Zhang, Ke Li, and Chen Shi. 2020. "Ultra-Dispersed Biomimetic Nanoplatform Fabricated by Controlled Etching Agglomerated MnO2 for Enhanced Photodynamic Therapy and Immune Activation." *Chemical Engineering Journal* 397 (May). doi:10.1016/j.cej.2020.125478.

He, Zhijian, Xiaomeng Wan, Anita Schulz, Herdis Bludau, Marina A. Dobrovolskaia, Stephan T. Stern, Stephanie A. Montgomery, et al. 2016a. "A High Capacity Polymeric Micelle of Paclitaxel: Implication of High Dose Drug Therapy to Safety and in Vivo Anti-Cancer Activity." *Biomaterials* 101 (September). Elsevier Ltd: 296–309. doi:10.1016/j.biomaterials.2016.06.002.

He, Zhijian, Xiaomeng Wan, Anita Schulz, Herdis Bludau, Marina A. Dobrovolskaia, Stephan T. Stern, Stephanie A. Montgomery, et al. 2016b. "A High Capacity Polymeric Micelle of Paclitaxel: Implication of High Dose Drug Therapy to Safety and in Vivo Anti-Cancer Activity." *Biomaterials* 101 (September). Elsevier Ltd: 296–309. doi:10.1016/j.biomaterials.2016.06.002.

Hershfield, Michael S., Nancy J. Ganson, Susan J. Kelly, Edna L. Scarlett, Denise A. Jaggers, and John S. Sundy. 2014. "Induced and Pre-Existing Anti-Polyethylene Glycol Antibody in a Trial of Every 3-Week Dosing of Pegloticase for Refractory Gout, Including in Organ Transplant Recipients." *Arthritis Research & Therapy* 16 (2). England: R63. doi:10.1186/ar4500.

Hobbs, Susan K., Wayne L. Monsky, Fan Yuan, W Gregory Roberts, Linda Griffith, Vladimir P. Torchilin, and Rakesh K. Jain. 1998. "Regulation of Transport Pathways in Tumor Vessels: Role of Tumor Type and Microenvironment." *Proceedings of the National Academy of Sciences* 95(8): 4607–4612. doi:10.1073/pnas.95.8.4607.

Hodi, F. Stephen, Steven J. O'Day, David F. McDermott, Robert W. Weber, Jeffrey A. Sosman, John B. Haanen, Rene Gonzalez, et al. 2010. "Improved Survival with Ipilimumab in Patients with Metastatic Melanoma." *New England Journal of Medicine* 363 (8): 711–723. doi:10.1056/NEJMoa1003466.

Honda, N., Q. Guo, H. Uchida, H. Ohishi, and Y. Hiasa. 1994. "Percutaneous Hot Saline Injection Therapy for Hepatic Tumors: An Alternative to Percutaneous Ethanol Injection Therapy." *Radiology* 190 (1): 53–57.

Hong, Eun Ji, Dae Gun Choi, and Min Suk Shim. 2016. "Targeted and Effective Photodynamic Therapy for Cancer Using Functionalized Nanomaterials." *Acta Pharmaceutica Sinica B*. Chinese Academy of Medical Sciences. doi:10.1016/j.apsb.2016.01.007.

Hood, Leroy, and Stephen H. Friend. 2011. "Predictive, Personalized, Preventive, Participatory (P4) Cancer Medicine." *Nature Reviews Clinical Oncology 2011 8:3* 8 (3). Nature Publishing Group: 184–187. doi:10.1038/nrclinonc.2010.227.

Hoogevest, Peter van, Harry Tiemessen, Josbert M. Metselaar, Simon Drescher, and Alfred Fahr. 2021. "The Use of Phospholipids to Make Pharmaceutical Form Line Extensions." *European Journal of Lipid Science and Technology*. John Wiley and Sons Inc. doi:10.1002/ejlt.202000297.

Hu, DanRong, Meng Pan, Yan Yu, Ao Sun, Kun Shi, Ying Qu, and ZhiYong Qian. 2020. "Application of Nanotechnology for Enhancing Photodynamic Therapy via Ameliorating, Neglecting, or Exploiting Tumor Hypoxia." *VIEW* 1 (1). John Wiley & Sons, Ltd: e6. doi:10.1002/viw2.6.

Huang, J., D. Xiao, G. Li, J. Ma, P. Chen, W. Yuan, F. Hou, et al. 2014. "EphA2 Promotes Epithelial–Mesenchymal Transition through the Wnt/β-Catenin Pathway in Gastric Cancer Cells." *Oncogene* 33 (21): 2737–2747. doi:10.1038/onc.2013.238.

Huynh, Elizabeth, and Gang Zheng. 2015. "Cancer Nanomedicine: Addressing the Dark Side of the Enhanced Permeability and Retention Effect." *Nanomedicine* 10 (13). Future Medicine: 1993–1995. doi:10.2217/nnm.15.86.

Jahanafrooz, Zohreh, Behzad Baradaran, Jafar Mosafer, Mahmoud Hashemzaei, Tayebeh Rezaei, Ahad Mokhtarzadeh, and Michael R. Hamblin. 2020. "Comparison of DNA and MRNA Vaccines against Cancer." *Drug Discovery Today* 25 (3): 552–560. doi:10.1016/j.drudis.2019.12.003.

Jain, Rakesh K. 1999. "Transport of Molecules, Particles, and Cells in Solid Tumors." *Annual Review of Biomedical Engineering* 1 (1): 241–263. doi:10.1146/annurev.bioeng.1.1.241.

Jin, Anting, Yitong Wang, Kaili Lin, and Lingyong Jiang. 2020. "Nanoparticles Modified by Polydopamine: Working as 'Drug' Carriers." *Bioactive Materials* 5 (3): 522–541. doi:10.1016/j.bioactmat.2020.04.003.

Jogdand, Anil, Syed Baseeruddin Alvi, P. S. Rajalakshmi, and Aravind Kumar Rengan. 2020. "NIR-Dye Based Mucoadhesive Nanosystem for Photothermal Therapy in Breast Cancer Cells." *Journal of Photochemistry and Photobiology B: Biology* 208 (July). Elsevier B.V. doi:10.1016/j.jphotobiol.2020.111901.

John, Johnson V., Chung-Wook Chung, Renjith P. Johnson, Young-Il Jeong, Kyu-Don Chung, Dae Hwan Kang, Hongsuk Suh, Hongyu Chen, and Il Kim. 2016. "Dual Stimuli-Responsive Vesicular Nanospheres Fabricated by Lipopolymer Hybrids for Tumor-Targeted Photodynamic Therapy." *Biomacromolecules* 17 (1). American Chemical Society: 20–31. doi:10.1021/acs.biomac.5b01474.

Johnson, Daniel T., Jiarong Zhou, Ashley V. Kroll, Ronnie H. Fang, Ming Yan, Crystal Xiao, Xiufen Chen, and Justin Kline. 2021. "Acute Myeloid Leukemia Cell Membrane-Coated Nanoparticles for Cancer Vaccination Immunotherapy," January. Springer US. doi:10.1038/s41375-021-01432-w.

Josefsen, Leanne B., and Ross W. Boyle. 2012. "Unique Diagnostic and Therapeutic Roles of Porphyrins and Phthalocyanines in Photodynamic Therapy, Imaging and Theranostics." *Theranostics*. doi:10.7150/thno.4571.

Judge, Adam D., Marjorie Robbins, Iran Tavakoli, Jasna Levi, Lina Hu, Anna Fronda, Ellen Ambegia, Kevin McClintock, and Ian MacLachlan. 2009. "Confirming the RNAi-Mediated Mechanism of Action of SiRNA-Based Cancer Therapeutics in Mice." *Journal of Clinical Investigation* 119 (3): 661–673. doi:10.1172/JCI37515.

Kang, Seong-Ho, Bhumsuk Keam, Yong-Oon Ahn, Ha-Ram Park, Miso Kim, Tae Min Kim, Dong-Wan Kim, and Dae Seog Heo. 2019. "Inhibition of MEK with Trametinib Enhances the Efficacy of Anti-PD-L1 Inhibitor by Regulating Anti-Tumor Immunity in Head and Neck Squamous Cell Carcinoma." *OncoImmunology* 8 (1): e1515057. doi:10.1080/2162402X.2018.1515057.

Katoh, Masaru. 2016. "FGFR Inhibitors: Effects on Cancer Cells, Tumor Microenvironment and Whole-Body Homeostasis (Review)." *International Journal of Molecular Medicine* 38 (1): 3–15. doi:10.3892/ijmm.2016.2620.

Kiss, Anna L., and Erzsébet Botos. 2009. "Endocytosis via Caveolae: Alternative Pathway with Distinct Cellular Compartments to Avoid Lysosomal Degradation?" *Journal of Cellular and Molecular Medicine* 13 (7). John Wiley & Sons, Ltd: 1228–1237. doi:10.1111/J.1582-4934.2009.00754.X.

Kiss, Lóránd, Fruzsina R. Walter, Alexandra Bocsik, Szilvia Veszelka, Béla Ózsvári, László G. Puskás, Piroska Szabó-Révész, and Mária A. Deli. 2013. "Kinetic Analysis of the Toxicity of Pharmaceutical Excipients Cremophor EL and RH40 on Endothelial and Epithelial Cells." *Journal of Pharmaceutical Sciences* 102 (4): 1173–1181. doi:10.1002/jps.23458.

Koury, Jeffrey, Mariana Lucero, Caleb Cato, Lawrence Chang, Joseph Geiger, Denise Henry, Jennifer Hernandez, et al. 2018. "Immunotherapies: Exploiting the Immune System for Cancer Treatment." *Journal of Immunology Research* 2018: 1–16. doi:10.1155/2018/9585614.

Kranz, Lena M., Mustafa Diken, Heinrich Haas, Sebastian Kreiter, Carmen Loquai, Kerstin C. Reuter, Martin Meng, et al. 2016. "Systemic RNA Delivery to Dendritic Cells Exploits Antiviral Defence for Cancer Immunotherapy." *Nature* 534 (7607): 396–401. doi:10.1038/nature18300.

Krenacs, Tibor, Nora Meggyeshazi, Gertrud Forika, Eva Kiss, Peter Hamar, Tamas Szekely, and Tamas Vancsik. 2020. "Modulated Electro-Hyperthermia-Induced Tumor Damage Mechanisms Revealed in Cancer Models." *International Journal of Molecular Sciences* 21 (17). MDPI AG: 1–25. doi:10.3390/ijms21176270.

Kroemer, Guido, and Laurence Zitvogel. 2018. "The Breakthrough of the Microbiota." *Nature Reviews Immunology* 18 (2): 87–88. doi:10.1038/nri.2018.4.

Kroschinsky, Frank, Friedrich Stölzel, Simone von Bonin, Gernot Beutel, Matthias Kochanek, Michael Kiehl, and Peter Schellongowski. 2017. "New Drugs, New Toxicities: Severe Side Effects of Modern Targeted and Immunotherapy of Cancer and Their Management." *Critical Care* 21 (1): 89. doi:10.1186/s13054-017-1678-1.

Krown, Susan E., Donald W. Northfelt, David Osoba, and J. Simon Stewart. 2004. "Use of Liposomal Anthracyclines in Kaposi's Sarcoma." *Seminars in Oncology* 31 (SUPPL. 13): 36–52. doi:10.1053/j.seminoncol.2004.08.003.

Kuai, Rui, Lukasz J. Ochyl, Keith S. Bahjat, Anna Schwendeman, and James J. Moon. 2017. "Designer Vaccine Nanodiscs for Personalized Cancer Immunotherapy." *Nature Materials* 16 (4): 489–496. doi:10.1038/nmat4822.

Kwak, Seo Young, Seonmin Lee, Hee Dong Han, Suhwan Chang, Kyu-pyo Kim, and Hyung Jun Ahn. 2019. "PLGA Nanoparticles Codelivering SiRNAs against Programmed Cell Death Protein-1 and Its Ligand Gene for Suppression of Colon Tumor Growth." *Molecular Pharmaceutics* 16 (12): 4940–4953. doi:10.1021/acs.molpharmaceut.9b00826.

Landis, John, and Ismail Kola. 2007. "Can the Pharmaceutical Industry Reduce Attrition Rates." *Architectural Record* 195 (3): 153. doi:10.1038/news070604-1.

Lee, Keun Seok, Hyun Cheol Chung, Seock Ah Im, Yeon Hee Park, Chul Soo Kim, Sung Bae Kim, Sun Young Rha, Min Young Lee, and Jungsil Ro. 2008a. "Multicenter Phase II Trial of Genexol-PM, a Cremophor-Free, Polymeric Micelle Formulation of Paclitaxel, in Patients with Metastatic Breast Cancer." *Breast Cancer Research and Treatment* 108 (2): 241–250. doi:10.1007/s10549-007-9591-y.

Lee, Keun Seok, Hyun Cheol Chung, Seock Ah Im, Yeon Hee Park, Chul Soo Kim, Sung Bae Kim, Sun Young Rha, Min Young Lee, and Jungsil Ro. 2008b. "Multicenter Phase II Trial of Genexol-PM, a Cremophor-Free, Polymeric Micelle Formulation of Paclitaxel, in Patients with Metastatic Breast Cancer." *Breast Cancer Research and Treatment* 108 (2): 241–250. doi:10.1007/s10549-007-9591-y.

Li, Jun, Yun-Ching Chen, Yu-Cheng Tseng, Subho Mozumdar, and Leaf Huang. 2010. "Biodegradable Calcium Phosphate Nanoparticle with Lipid Coating for Systemic SiRNA Delivery." *Journal of Controlled Release* 142 (3): 416–421. doi:10.1016/j.jconrel.2009.11.008.

Li, Sha, Weiping Jiang, Yaping Yuan, Meiju Sui, Yuqi Yang, Liqun Huang, Ling Jiang, Maili Liu, Shizhen Chen, and Xin Zhou. 2020. "Delicately Designed Cancer Cell Membrane-Camouflaged Nanoparticles for Targeted 19F MR/PA/FL Imaging-Guided Photothermal Therapy." *ACS Applied Materials and Interfaces* 12 (51): 57290–57301. doi:10.1021/acsami.0c13865.

Li, Shihao, and Lin Zhang. 2020. "Erythrocyte Membrane Nano-Capsules: Biomimetic Delivery and Controlled Release of Photothermal–Photochemical Coupling Agents for Cancer Cell Therapy." *Dalton Transactions* 49 (8). Royal Society of Chemistry: 2645–2651. doi:10.1039/C9DT04335G.

Liu, Haipeng, Kelly D. Moynihan, Yiran Zheng, Gregory L. Szeto, Adrienne v. Li, Bonnie Huang, Debra S. van Egeren, Clara Park, and Darrell J. Irvine. 2014. "Structure-Based Programming of Lymph-Node Targeting in Molecular Vaccines." *Nature* 507 (7493): 519–522. doi:10.1038/nature12978.

Liu, Yiwei, Colton A. Smith, John C. Panetta, Wenjian Yang, Lauren E. Thompson, Jacob P. Counts, Alejandro R. Molinelli, et al. 2019. "Antibodies Predict Pegaspargase Allergic Reactions and Failure of Rechallenge." *Journal of Clinical Oncology: Official Journal of the American Society of Clinical Oncology* 37 (23). United States: 2051–2061. doi:10.1200/JCO.18.02439.

Lugin, Margaret L., Rebecca T. Lee, and Young Jik Kwon. 2020. "Synthetically Engineered Adeno-Associated Virus for Efficient, Safe, and Versatile Gene Therapy Applications." *ACS Nano* 14 (11): 14262–14283. doi:10.1021/acsnano.0c03850.

Lynn, Geoffrey M., Richard Laga, Patricia A. Darrah, Andrew S. Ishizuka, Alexandra J. Balaci, Andrés E Dulcey, Michal Pechar, et al. 2015. "In Vivo Characterization of the Physicochemical Properties of Polymer-Linked TLR Agonists That Enhance Vaccine Immunogenicity." *Nature Biotechnology* 33 (11): 1201–1210. doi:10.1038/nbt.3371.

Madamsetty, Vijay Sagar, Anubhab Mukherjee, and Sudip Mukherjee. 2019. "Recent Trends of the Bio-Inspired Nanoparticles in Cancer Theranostics." *Frontiers in Pharmacology* 10 (October). Frontiers Media S.A.: 1264. doi:10.3389/FPHAR.2019.01264/BIBTEX.

Madan, Jitender, Sushma R. Gundala, Yoganjaneyulu Kasetti, Prasad V. Bharatam, Ritu Aneja, Anju Katyal, and Upendra K. Jain. 2014. "Enhanced Noscapine Delivery Using Estrogen-Receptor-Targeted Nanoparticles for Breast Cancer Therapy." *Anti-Cancer Drugs* 25 (6): 704–716. doi:10.1097/CAD.0000000000000098.

Maisels, M. J. 2015. "Sister Jean Ward, Phototherapy, and Jaundice: A Unique Human and Photochemical Interaction." *Journal of Perinatology* 35 (9). Nature Publishing Group: 671–675. doi:10.1038/jp.2015.56.

Mardiana, Sherly, Benjamin J. Solomon, Phillip K. Darcy, and Paul A. Beavis. 2019. "Supercharging Adoptive T Cell Therapy to Overcome Solid Tumor–Induced Immunosuppression." *Science Translational Medicine* 11 (495). doi:10.1126/scitranslmed.aaw2293.

Marei, Hany E., Asma Althani, Thomas Caceci, Roberto Arriga, Tommaso Sconocchia, Alessio Ottaviani, Giulia Lanzilli, et al. 2019. "Recent Perspective on CAR and Fcγ-CR T Cell Immunotherapy for Cancers: Preclinical Evidence versus Clinical Outcomes." *Biochemical Pharmacology* 166 (August): 335–346. doi:10.1016/j.bcp.2019.06.002.

Mehta, A. M., A. M. Sonabend, and J. N. Bruce. 2017. "Convection-Enhanced Delivery." *Neurotherapeutics* 14 (2): 358–371. doi:10.1007/s13311-017-0520-4.

Mellman, Ira, George Coukos, and Glenn Dranoff. 2011. "Cancer Immunotherapy Comes of Age." *Nature* 480 (7378): 480–489. doi:10.1038/nature10673.

Miguel, Rodrigo dos A., Amanda S. Hirata, Paula C. Jimenez, Luciana B. Lopes, and Leticia v. Costa-Lotufo. 2022. "Beyond Formulation: Contributions of Nanotechnology for Translation of Anticancer Natural Products into New Drugs." *Pharmaceutics*. MDPI. doi:10.3390/pharmaceutics14081722.

Mills, Christopher C., E. A. Kolb, and Valerie B. Sampson. 2018. "Development of Chemotherapy with Cell-Cycle Inhibitors for Adult and Pediatric Cancer Therapy." *Cancer Research* 78 (2). United States: 320–325. doi:10.1158/0008-5472.CAN-17-2782.

Minko, Tamara, Lorna Rodriguez-Rodriguez, and Vitaly Pozharov. 2013. "Nanotechnology Approaches for Personalized Treatment of Multidrug Resistant Cancers." *Advanced Drug Delivery Reviews* 65 (13–14). Elsevier: 1880–1895. doi:10.1016/J.ADDR.2013.09.017.

Misra, Ranjita, Sarbari Acharya, and Sanjeeb K. Sahoo. 2010. "Cancer Nanotechnology: Application of Nanotechnology in Cancer Therapy." *Drug Discovery Today* 15 (19–20). England: 842–850. doi:10.1016/j.drudis.2010.08.006.

Mohan, Nishant, Salman Hosain, Jun Zhao, Yi Shen, Xiao Luo, Jiangsong Jiang, Yukinori Endo, and Wen Jin Wu. 2019. "Atezolizumab Potentiates Tcell-Mediated Cytotoxicity and Coordinates with FAK to Suppress Cell Invasion and Motility in PD-L1 $^{+}$ Triple Negative Breast Cancer Cells." *OncoImmunology* 8 (9): e1624128. doi:10.1080/2162402X.2019.1624128.

Moon, James J., Heikyung Suh, Anna Bershteyn, Matthias T. Stephan, Haipeng Liu, Bonnie Huang, Mashaal Sohail, et al. 2011. "Interbilayer-Crosslinked Multilamellar Vesicles as Synthetic Vaccines for Potent Humoral and Cellular Immune Responses." *Nature Materials* 10 (3): 243–251. doi:10.1038/nmat2960.

Morrison, W. B. 2010. "Cancer Chemotherapy: An Annotated History." *Journal of Veterinary Internal Medicine* 24 (6). John Wiley & Sons, Ltd: 1249–1262. doi:10.1111/j.1939-1676.2010.0590.x.

Morshedi Rad, Dorsa, Maryam Alsadat Rad, Sajad Razavi Bazaz, Navid Kashaninejad, Dayong Jin, and Majid Ebrahimi Warkiani. 2021. "A Comprehensive Review on Intracellular Delivery." *Advanced Materials* 33 (13). John Wiley & Sons, Ltd: 2005363. doi:10.1002/ADMA.202005363.

Morvan, Maelig G., and Lewis L. Lanier. 2016. "NK Cells and Cancer: You Can Teach Innate Cells New Tricks." *Nature Reviews Cancer* 16 (1): 7–19. doi:10.1038/nrc.2015.5.

Musalli, Abdul Hadi, Priyanka Dey Talukdar, Partha Roy, Pradeep Kumar, and Tin Wui Wong. 2020. "Folate-Induced Nanostructural Changes of Oligochitosan Nanoparticles and Their Fate of Cellular Internalization by Melanoma." *Carbohydrate Polymers* 244 (September). Elsevier: 116488. doi:10.1016/J.CARBPOL.2020.116488.

Musetti, Sara, and Leaf Huang. 2018. "Nanoparticle-Mediated Remodeling of the Tumor Microenvironment to Enhance Immunotherapy." *ACS Nano* 12 (12): 11740–11755. doi:10.1021/acsnano.8b05893.

Nakamura, Yuko, Ai Mochida, Peter L. Choyke, and Hisataka Kobayashi. 2016. "Nanodrug Delivery: Is the Enhanced Permeability and Retention Effect Sufficient for Curing Cancer?" *Bioconjugate Chemistry*. American Chemical Society. doi:10.1021/acs.bioconjchem.6b00437.

Nath, Shubhankar, Girgis Obaid, and Tayyaba Hasan. 2019. "The Course of Immune Stimulation by Photodynamic Therapy: Bridging Fundamentals of Photochemically Induced Immunogenic Cell Death to the Enrichment of T-Cell Repertoire." *Photochemistry and Photobiology*. Blackwell Publishing Inc. doi:10.1111/php.13173.

Navya, P. N., Anubhav Kaphle, S. P. Srinivas, Suresh Kumar Bhargava, Vincent M. Rotello, and Hemant Kumar Daima. 2019. "Current Trends and Challenges in Cancer Management and Therapy Using Designer Nanomaterials." *Nano Convergence 2019 6:1* 6 (1). SpringerOpen: 1–30. doi:10.1186/S40580-019-0193-2.

Nel, Andre, Erkki Ruoslahti, and Huan Meng. 2017. "New Insights into 'Permeability' as in the Enhanced Permeability and Retention Effect of Cancer Nanotherapeutics." *ACS Nano* 11 (10). American Chemical Society: 9567–9569. doi:10.1021/acsnano.7b07214.

Niculescu, Adelina Gabriela, and Alexandru Mihai Grumezescu. 2021. "Photodynamic Therapy—an up-to-Date Review." *Applied Sciences (Switzerland)*. MDPI AG. doi:10.3390/app11083626.

Niidome, Takuro, Masato Yamagata, Yuri Okamoto, Yasuyuki Akiyama, Hironobu Takahashi, Takahito Kawano, Yoshiki Katayama, and Yasuro Niidome. 2006. "PEG-Modified Gold Nanorods with a Stealth Character for in Vivo Applications." *Journal of Controlled Release: Official Journal of the Controlled Release Society* 114 (3). Netherlands: 343–347. doi:10.1016/j.jconrel.2006.06.017.

Nsairat, Hamdi, Dima Khater, Usama Sayed, Fadwa Odeh, Abeer Al Bawab, and Walhan Alshaer. 2022. "Liposomes: Structure, Composition, Types, and Clinical Applications." *Heliyon* 8 (5). England: e09394. doi:10.1016/j.heliyon.2022.e09394.

Oberli, Matthias A., Andreas M. Reichmuth, J. Robert Dorkin, Michael J. Mitchell, Owen S. Fenton, Ana Jaklenec, Daniel G. Anderson, Robert Langer, and Daniel Blankschtein. 2017. "Lipid Nanoparticle Assisted MRNA Delivery for Potent Cancer Immunotherapy." *Nano Letters* 17 (3): 1326–1335. doi:10.1021/acs.nanolett.6b03329.

Oldenburg, S. J., R. D. Averitt, S. L. Westcott, and N. J. Halas. 1998. "Nanoengineering of Optical Resonances." *Chemical Physics Letters* 288 (2): 243–247. doi:10.1016/S0009-2614(98)00277-2.

Online, View ArticleZongjun Liu, Peng Wan, Mingxin Yang, Fang Han, Tianran Wang, You Wang, and Yu Li. 2021. "Cell Membrane Camouflaged Cerium Oxide Nanocubes for Targeting Enhanced," 9524–9532. doi:10.1039/d1tb01685g.

Ordikhani, Farideh, Mayuko Uehara, Vivek Kasinath, Li Dai, Siawosh K. Eskandari, Baharak Bahmani, Merve Yonar, et al. 2018. "Targeting Antigen-Presenting Cells by Anti–PD-1 Nanoparticles Augments Antitumor Immunity." *JCI Insight* 3 (20). doi:10.1172/jci.insight.122700.

Ormond, Alexandra B., and Harold S. Freeman. 2013. "Dye Sensitizers for Photodynamic Therapy." *Materials*. doi:10.3390/ma6030817.

P. S., Rajalakshmi, Syed Baseeruddin Alvi, Nazia Begum, Bantal Veeresh, and Aravind Kumar Rengan. 2021. "Self-Assembled Fluorosome-Polydopamine Complex for Efficient Tumor Targeting and Commingled Photodynamic/Photothermal Therapy of Triple-Negative Breast Cancer." *Biomacromolecules* 22 (9): 3926–3940. doi:10.1021/acs.biomac.1c00744.

Pan, Wei-Lun, Yong Tan, Wei Meng, Nai-Han Huang, Yi-Bang Zhao, Zhi-Qiang Yu, Zhong Huang, Wen-Hua Zhang, Bin Sun, and Jin-Xiang Chen. 2022. "Microenvironment-Driven Sequential Ferroptosis, Photodynamic Therapy, and Chemotherapy for Targeted Breast Cancer Therapy by a Cancer-Cell-Membrane-Coated Nanoscale Metal-Organic Framework." *Biomaterials* 283 (April). Elsevier: 121449. doi:10.1016/J.BIOMATERIALS.2022.121449.

Pan, Yu, Qinglin Fei, Ping Xiong, Jianyang Yang, Zheyang Zhang, Xianchao Lin, Minggui Pan, Fengchun Lu, and Heguang Huang. 2019. "Synergistic Inhibition of Pancreatic Cancer with Anti-PD-L1 and c-Myc Inhibitor JQ1." *OncoImmunology* 8 (5): e1581529. doi:10.1080/2162402X.2019.1581529.

Panda, Anshuman, Anil Betigeri, Kalyanasundaram Subramanian, Jeffrey S. Ross, Dean C. Pavlick, Siraj Ali, Paul Markowski, et al. 2017. "Identifying a Clinically Applicable Mutational Burden Threshold as a Potential Biomarker of Response to Immune Checkpoint Therapy in Solid Tumors." *JCO Precision Oncology*, no. 1 (November): 1–13. doi:10.1200/PO.17.00146.

Pardoll, Drew M. 2012. "The Blockade of Immune Checkpoints in Cancer Immunotherapy." *Nature Reviews Cancer* 12 (4): 252–264. doi:10.1038/nrc3239.

Park, Wonchan, Keum Yong Seong, Hye Hyeon Han, Seung Yun Yang, and Sei Kwang Hahn. 2021. "Dissolving Microneedles Delivering Cancer Cell Membrane Coated Nanoparticles for Cancer Immunotherapy." *RSC Advances* 11 (17). Royal Society of Chemistry: 10393–10399. doi:10.1039/d1ra00747e.

Parodi, Alessandro, Ekaterina P. Kolesova, Maya v. Voronina, Anastasia S. Frolova, Dmitry Kostyushev, Daria B. Trushina, Roman Akasov, Tatiana Pallaeva, and Andrey A. Zamyatnin. 2022. "Anticancer Nanotherapeutics in Clinical Trials: The Work behind Clinical Translation of Nanomedicine." *International Journal of Molecular Sciences* 23 (21). MDPI AG: 13368. doi:10.3390/ijms232113368.

Patra, Jayanta Kumar, Gitishree Das, Leonardo Fernandes Fraceto, Estefania Vangelie Ramos Campos, Maria Del Pilar Rodriguez-Torres, Laura Susana Acosta-Torres, Luis Armando Diaz-Torres, et al. 2018. "Nano Based Drug Delivery Systems: Recent Developments and Future Prospects 10 Technology 1007 Nanotechnology 03 Chemical Sciences 0306 Physical Chemistry (Incl. Structural) 03 Chemical Sciences 0303 Macromolecular and Materials Chemistry 11 Medical and Health Sciences 1115 Pharmacology and Pharmaceutical Sciences 09 Engineering 0903 Biomedical Engineering Prof Ueli Aebi, Prof Peter Gehr." *Journal of Nanobiotechnology*. BioMed Central Ltd. doi:10.1186/s12951-018-0392-8.

Pauken, Kristen E., Morgan A. Sammons, Pamela M. Odorizzi, Sasikanth Manne, Jernej Godec, Omar Khan, Adam M. Drake, et al. 2016. "Epigenetic Stability of Exhausted T Cells Limits Durability of Reinvigoration by PD-1 Blockade." *Science* 354 (6316): 1160–1165. doi:10.1126/science.aaf2807.

Pei, Xiaochen, Xiuhua Pan, Xiaoyi Xu, Xiang Xu, Haiqin Huang, Zhenghong Wu, and Xiaole Qi. 2021. "4T1 Cell Membrane Fragment Reunited PAMAM Polymer Units Disguised as Tumor Cell Clusters for Tumor Homotypic Targeting and Anti-Metastasis Treatment." *Biomaterials Science* 9 (4). Royal Society of Chemistry: 1325–1333. doi:10.1039/D0BM01731K.

Piffoux, Max, Amanda K. A. Silva, Claire Wilhelm, Florence Gazeau, and David Tareste. 2018. "Modification of Extracellular Vesicles by Fusion with Liposomes for the Design of Personalized Biogenic Drug Delivery Systems." *ACS Nano* 12 (7): 6830–6842. doi:10.1021/acsnano.8b02053.

R., Gayathri, Rajalakshmi P. S., Aswathi Thomas, and Aravind Kumar Rengan. 2021. "Doxorubicin Loaded Polyvinylpyrrolidone-Copper Sulfide Nanoparticles Enabling Mucoadhesiveness and Chemo-Photothermal Synergism for Effective Killing of Breast Cancer Cells." *Materialia* 19 (August). Elsevier B.V.: 101195. doi:10.1016/j.mtla.2021.101195.

Ramalingam, Vaikundamoorthy, Krishnamoorthy Varunkumar, and Vilwanathan Ravikumar. 2018. "Target Delivery of Doxorubicin Tethered with PVP Stabilized Gold Nanoparticles for Effective Treatment of Lung Cancer." *Scientific Reports*, no. February. Springer US: 1–12. doi:10.1038/s41598-018-22172-5.

Ramsay, Alan G. 2013. "Immune Checkpoint Blockade Immunotherapy to Activate Anti-Tumour T-Cell Immunity." *British Journal of Haematology* 162 (3): 313–325. doi:10.1111/bjh.12380.

Rao, Lang, Guang Tao Yu, Qian Fang Meng, Lin Lin Bu, Rui Tian, Li Sen Lin, Hongzhang Deng, et al. 2019. "Cancer Cell Membrane-Coated Nanoparticles for Personalized Therapy in Patient-Derived Xenograft Models." *Advanced Functional Materials* 29 (51). Wiley-VCH Verlag. doi:10.1002/ADFM.201905671.

Rengan, Aravind Kumar, Amirali B. Bukhari, Arpan Pradhan, Renu Malhotra, Rinti Banerjee, Rohit Srivastava, and Abhijit De. 2015. "In Vivo Analysis of Biodegradable Liposome Gold Nanoparticles as Efficient Agents for Photothermal Therapy of Cancer." *Nano Letters* 15 (2): 842–848. doi:10.1021/nl5045378.

Reubi, J. C., L. Mazzucchelli, I. Hennig, and J. A. Laissue. 1996. "Local Up-Regulation of Neuropeptide Receptors in Host Blood Vessels around Human Colorectal Cancers." *Gastroenterology* 110 (6): 1719–1726. doi:10.1053/gast.1996.v110.pm8964396.

Riedel, Jennifer, Maria Natalia Calienni, Ezequiel Bernabeu, Valeria Calabro, Juan Manuel Lázaro-Martinez, Maria Jimena Prieto, Lorena Gonzalez, et al. 2021. "Paclitaxel and Curcumin Co-Loaded Mixed Micelles: Improving in Vitro Efficacy and Reducing Toxicity against Abraxane®." *Journal of Drug Delivery Science and Technology* 62 (April): 102343. doi:10.1016/j.jddst.2021.102343.

Roti, Joseph L. 2008. "Cellular Responses to Hyperthermia (40-46°C): Cell Killing and Molecular Events." In *International Journal of Hyperthermia*, 24:3–15. doi:10.1080/02656730701769841.

Saito, Hiroaki, Tomohiro Osaki, and Masahide Ikeguchi. 2012. "Decreased NKG2D Expression on NK Cells Correlates with Impaired NK Cell Function in Patients with Gastric Cancer." *Gastric Cancer* 15 (1): 27–33. doi:10.1007/s10120-011-0059-8.

Sen, Debattama R., James Kaminski, R. Anthony Barnitz, Makoto Kurachi, Ulrike Gerdemann, Kathleen B. Yates, Hsiao-Wei Tsao, et al. 2016. "The Epigenetic Landscape of T Cell Exhaustion." *Science* 354 (6316): 1165–1169. doi:10.1126/science.aae0491.

Senkus, Elzbieta, and Jacek Jassem. 2011. "Cardiovascular Effects of Systemic Cancer Treatment." *Cancer Treatment Reviews*. doi:10.1016/j.ctrv.2010.11.001.

Shen, Wenwen, Shuke Ge, Xiaoyao Liu, Qi Yu, Xue Jiang, Qian Wu, Yu Chen Tian, Yu Gao, Ying Liu, and Chao Wu. 2022. "Folate-Functionalized SMMC-7721 Liver Cancer Cell Membrane-Cloaked Paclitaxel Nanocrystals for Targeted Chemotherapy of Hepatoma." *Drug Delivery* 29 (1). Taylor and Francis Ltd.: 31–42. doi:10.1080/10717544.2021.2015481/SUPPL_FILE/IDRD_A_2015481_SM0912.DOCX.

Shevtsov, Maxim, and Gabriele Multhoff. 2016. "Immunological and Translational Aspects of NK Cell-Based Antitumor Immunotherapies." *Frontiers in Immunology* 7 (November). doi:10.3389/fimmu.2016.00492.

Shobaki, Nour, Yusuke Sato, Yuichi Suzuki, Nana Okabe, and Hideyoshi Harashima. 2020. "Manipulating the Function of Tumor-Associated Macrophages by SiRNA-Loaded Lipid Nanoparticles for Cancer Immunotherapy." *Journal of Controlled Release* 325 (September): 235–248. doi:10.1016/j.jconrel.2020.07.001.

Siegler, Elizabeth L., Yanni Zhu, Pin Wang, and Lili Yang. 2018. "Off-the-Shelf CAR-NK Cells for Cancer Immunotherapy." *Cell Stem Cell* 23 (2): 160–161. doi:10.1016/j.stem.2018.07.007.

Sindhwani, Shrey, Abdullah Muhammad Syed, Jessica Ngai, Benjamin R. Kingston, Laura Maiorino, Jeremy Rothschild, Presley MacMillan, et al. 2020. "The Entry of Nanoparticles into Solid Tumours." *Nature Materials* 19 (5): 566–575. doi:10.1038/s41563-019-0566-2.

Singh, Surya Prakash, Syed Baseeruddin Alvi, Deepak Bharadwaj Pemmaraju, Anula Divyash Singh, Sasidhar Venkata Manda, Rohit Srivastava, and Aravind Kumar Rengan. 2018. "NIR Triggered Liposome Gold Nanoparticles Entrapping Curcumin as in Situ Adjuvant for Photothermal Treatment of Skin Cancer." *International Journal of Biological Macromolecules* 110 (April). Elsevier B.V.: 375–382. doi:10.1016/j.ijbiomac.2017.11.163.

Smith, Tyrel T., Sirkka B. Stephan, Howell F. Moffett, Laura E. McKnight, Weihang Ji, Diana Reiman, Emmy Bonagofski, Martin E. Wohlfahrt, Smitha P. S. Pillai, and Matthias T. Stephan. 2017. "In Situ Programming of Leukaemia-Specific T Cells Using Synthetic DNA Nanocarriers." *Nature Nanotechnology* 12 (8): 813–820. doi:10.1038/nnano.2017.57.

Song, Qingle, Yijia Yin, Lihuan Shang, Tingting Wu, Dan Zhang, Miao Kong, Yongdan Zhao, et al. 2017. "Tumor Microenvironment Responsive Nanogel for the Combinatorial Antitumor Effect of Chemotherapy and Immunotherapy." *Nano Letters* 17 (10): 6366–6375. doi:10.1021/acs.nanolett.7b03186.

Sun, Mengqi, Yuchen Duan, Yumeng Ma, and Qingyuan Zhang. 2020. "Cancer Cell-Erythrocyte Hybrid Membrane Coated Gold Nanocages for near Infrared Light-Activated Photothermal/Radio/Chemotherapy of Breast Cancer." *International Journal of Nanomedicine* 15: 6749–6760. doi:10.2147/IJN.S266405.

Sun, Qiang, Jinggen Wu, Lulu Jin, Liangjie Hong, Fang Wang, Zhengwei Mao, and Mengjie Wu. 2020. "Cancer Cell Membrane-Coated Gold Nanorods for Photothermal Therapy and Radiotherapy on Oral Squamous Cancer." *Journal of Materials Chemistry B* 8 (32). Royal Society of Chemistry: 7253–7263. doi:10.1039/d0tb01063d.

Tseng, Ching Li, Yueh Hsiu Wu, Su Wen Yu, Kai Chiang Yang, and Feng Huei Lin. 2009. "Development of EGFR-Targeting Nanomedicine for Effectively and Noninvasively Treats Lung Cancer Patients by Aerosol Delivery." *IFMBE Proceedings* 25 (8): 347–350. doi:10.1007/978-3-642-03887-7_100.

Ventola, C Lee. 2017. "Cancer Immunotherapy, Part 1: Current Strategies and Agents." *P & T: A Peer-Reviewed Journal for Formulary Management* 42 (6): 375–383.

Vogel, Charles L., Melody A. Cobleigh, Debu Tripathy, John C. Gutheil, Lyndsay N. Harris, Louis Fehrenbacher, Dennis J. Slamon, et al. 2002. "Efficacy and Safety of Trastuzumab as a Single Agent in First-Line Treatment of HER2-Overexpressing Metastatic Breast Cancer." *Journal of Clinical Oncology* 20 (3): 719–726. doi:10.1200/JCO.20.3.719.

Wachowska, Malgorzata, Angelika Muchowicz, Malgorzata Firczuk, Magdalena Gabrysiak, Magdalena Winiarska, Malgorzata Wańczyk, Kamil Bojarczuk, and Jakub Golab. 2011. "Aminolevulinic Acid (Ala) as a Prodrug in Photodynamic Therapy of Cancer." *Molecules*. doi:10.3390/molecules16054140.

Wang, Chao, Wujin Sun, Grace Wright, Andrew Z. Wang, and Zhen Gu. 2016. "Inflammation-Triggered Cancer Immunotherapy by Programmed Delivery of CpG and Anti-PD1 Antibody." *Advanced Materials* 28 (40): 8912–8920. doi:10.1002/adma.201506312.

Wang, Chao, Ligeng Xu, Chao Liang, Jian Xiang, Rui Peng, and Zhuang Liu. 2014. "Immunological Responses Triggered by Photothermal Therapy with Carbon Nanotubes in Combination with Anti-CTLA-4 Therapy to Inhibit Cancer Metastasis." *Advanced Materials* 26 (48): 8154–8162. doi:10.1002/adma.201402996.

Wang, Dongdong, Conghui Liu, Shiquan You, Kai Zhang, Meng Li, Yu Cao, Changtao Wang, Haifeng Dong, and Xueji Zhang. 2020. "Bacterial Vesicle-Cancer Cell Hybrid Membrane-Coated Nanoparticles for Tumor Specific Immune Activation and Photothermal Therapy." *ACS Applied Materials and Interfaces* 12 (37): 41138–41147. doi:10.1021/acsami.0c13169.

Wang, Dongdong, Huihui Wu, Jiajia Zhou, Pengping Xu, Changlai Wang, Ruohong Shi, Haibao Wang, Hui Wang, Zhen Guo, and Qianwang Chen. 2018. "In Situ One-Pot Synthesis of MOF-Polydopamine Hybrid Nanogels with Enhanced Photothermal Effect for Targeted Cancer Therapy." *Advanced Science* 5 (6): 1800287. doi:10.1002/advs.201800287.

Wang, Jun, Jingwei Sun, Linda N. Liu, Dallas B. Flies, Xinxin Nie, Maria Toki, Jianping Zhang, et al. 2019. "Siglec-15 as an Immune Suppressor and Potential Target for Normalization Cancer Immunotherapy." *Nature Medicine* 25 (4): 656–666. doi:10.1038/s41591-019-0374-x.

Wang, Zhenqing, Liang Chen, Yiqun Ma, Xilei Li, Annan Hu, Huiren Wang, Wenxing Wang, Xiaomin Li, Bo Tian, and Jian Dong. 2021. "Peptide Vaccine-Conjugated Mesoporous Carriers Synergize with Immunogenic Cell Death and PD-L1 Blockade for Amplified Immunotherapy of Metastatic Spinal." *Journal of Nanobiotechnology* 19 (1): 243. doi:10.1186/s12951-021-00975-5.

Wang, Zhi Hao, Jing Min Liu, Ning Zhao, Chun Yang Li, Shi Wen Lv, Yaozhong Hu, Huan Lv, Di Wang, and Shuo Wang. 2020. "Cancer Cell Macrophage Membrane Camouflaged Persistent Luminescent Nanoparticles for Imaging-Guided Photothermal Therapy of Colorectal Cancer." *ACS Applied Nano Materials* 3 (7): 7105–7118. doi:10.1021/acsanm.0c01433.

Wei, Guoqing, Jiasheng Wang, He Huang, and Yanmin Zhao. 2017. "Novel Immunotherapies for Adult Patients with B-Lineage Acute Lymphoblastic Leukemia." *Journal of Hematology & Oncology* 10 (1): 150. doi:10.1186/s13045-017-0516-x.

Werno, C., H. Menrad, A. Weigert, N. Dehne, S. Goerdt, K. Schledzewski, J. Kzhyshkowska, and B. Brune. 2010. "Knockout of HIF-1 in Tumor-Associated Macrophages Enhances M2 Polarization and Attenuates Their pro-Angiogenic Responses." *Carcinogenesis* 31 (10): 1863–1872. doi:10.1093/carcin/bgq088.

Wherry, E. John, and Makoto Kurachi. 2015. "Molecular and Cellular Insights into T Cell Exhaustion." *Nature Reviews Immunology* 15 (8): 486–499. doi:10.1038/nri3862.

Wicki, Andreas, Dominik Witzigmann, Vimalkumar Balasubramanian, and Jörg Huwyler. 2015. "Nanomedicine in Cancer Therapy: Challenges, Opportunities, and Clinical Applications." *Journal of Controlled Release: Official Journal of the Controlled Release Society* 200 (February). Netherlands: 138–157. doi:10.1016/j.jconrel.2014.12.030.

Wilhelm, Stefan, Anthony J. Tavares, Qin Dai, Seiichi Ohta, Julie Audet, Harold F. Dvorak, and Warren C. W. Chan. 2016a. "Analysis of Nanoparticle Delivery to Tumours." *Nature Reviews Materials* 1 (5): 16014. doi:10.1038/natrevmats.2016.14.

Wilhelm, Stefan, Anthony J. Tavares, Qin Dai, Seiichi Ohta, Julie Audet, Harold F. Dvorak, and Warren C. W. Chan. 2016b. "Analysis of Nanoparticle Delivery to Tumours." *Nature Reviews Materials* 1 (5): 16014. doi:10.1038/natrevmats.2016.14.

Wu, Dan, Xin Shou, Yalan Zhang, Zihan Li, Guohua Wu, Di Wu, Jianguo Wu, Shengyu Shi, and Shuqi Wang. 2021. "Cell Membrane-Encapsulated Magnetic Nanoparticles for Enhancing Natural Killer Cell-Mediated Cancer Immunotherapy." *Nanomedicine: Nanotechnology, Biology, and Medicine* 32. doi:10.1016/j.nano.2020.102333.

Wu, Minliang, Tianxiao Mei, Chenyu Lin, Yuchong Wang, Jingyao Chen, Wenjun Le, Mengyan Sun, et al. 2020. "Melanoma Cell Membrane Biomimetic Versatile CuS Nanoprobes for Homologous Targeting Photoacoustic Imaging and Photothermal Chemotherapy." *ACS Applied Materials and Interfaces* 12 (14): 16031–16039. doi:10.1021/acsami.9b23177.

Wu, Yanheng, Wenyi Gu, Jiang Li, Chen Chen, and Zhi Ping Xu. 2019a. "Silencing PD-1 and PD-L1 with Nanoparticle-Delivered Small Interfering RNA Increases Cytotoxicity of Tumor-Infiltrating Lymphocytes." *Nanomedicine* 14 (8): 955–967. doi:10.2217/nnm-2018-0237.

Wu, Yanheng, Wenyi Gu, Li Li, Chen Chen, and Zhi Xu. 2019b. "Enhancing PD-1 Gene Silence in T Lymphocytes by Comparing the Delivery Performance of Two Inorganic Nanoparticle Platforms." *Nanomaterials* 9 (2): 159. doi:10.3390/nano9020159.

Xia, An-Liang, Jin-Cheng Wang, Kun Yang, Dong Ji, Zheng-Ming Huang, and Yong Xu. 2019. "Genomic and Epigenomic Perspectives of T-Cell Exhaustion in Cancer." *Briefings in Functional Genomics* 18 (2): 113–118. doi:10.1093/bfgp/ely005.

Xiao, Yusha, Wei Huang, Daoming Zhu, Quanxiong Wang, Baiyang Chen, Zhisu Liu, Yang Wang, and Quanyan Liu. 2020. "Cancer Cell Membrane-Camouflaged MOF Nanoparticles for a Potent Dihydroartemisinin-Based Hepatocellular Carcinoma Therapy." *RSC Advances* 10 (12). Royal Society of Chemistry: 7194–7205. doi:10.1039/c9ra09233a.

Xie, Minqiang, H. Zhang, Y. Xu, T. Liu, S. Chen, J. Wang, and T. Zhang. 2013. "Expression of Folate Receptors in Nasopharyngeal and Laryngeal Carcinoma and Folate Receptor-Mediated Endocytosis by Molecular Targeted Nanomedicine." *International Journal of Nanomedicine* 8. Dove Medical Press Ltd.: 2443–2451. doi:10.2147/IJN.S46327.

Xu, Jun, Ligeng Xu, Chenya Wang, Rong Yang, Qi Zhuang, Xiao Han, Ziliang Dong, Wenwen Zhu, Rui Peng, and Zhuang Liu. 2017a. "Near-Infrared-Triggered Photodynamic Therapy with Multitasking Upconversion Nanoparticles in Combination with Checkpoint Blockade for Immunotherapy of Colorectal Cancer." *ACS Nano* 11 (5): 4463–4474. doi:10.1021/acsnano.7b00715.

Xu, Jun, Ligeng Xu, Chenya Wang, Rong Yang, Qi Zhuang, Xiao Han, Ziliang Dong, Wenwen Zhu, Rui Peng, and Zhuang Liu. 2017b. "Near-Infrared-Triggered Photodynamic Therapy with Multitasking Upconversion Nanoparticles in Combination with Checkpoint Blockade for Immunotherapy of Colorectal Cancer." *ACS Nano* 11 (5): 4463–4474. doi:10.1021/acsnano.7b00715.

Xu, Weide, Jilong Wang, Qinghua Li, Chenghu Wu, Lingling Wu, Kaiqiang Li, Qin Li, et al. 2021. "Cancer Cell Membrane-Coated Nanogels as a Redox/PH Dual-Responsive Drug Carrier for Tumor-Targeted Therapy." *Journal of Materials Chemistry B* 9 (38). Royal Society of Chemistry: 8031–8037. doi:10.1039/d1tb00788b.

Xu, Yanan, Bin Zheng, Mengqian Huang, Xianhuang Li, Zhiyun Wang, Jin Chang, and Tao Wang. 2021. "Sendai Virus Acts as a Nano-Booster to Excite Dendritic Cells for Enhancing the Efficacy of CD47-Directed Immune Checkpoint Inhibitors against Breast Carcinoma." *Materials Chemistry Frontiers* 5 (1): 223–237. doi:10.1039/D0QM00393J.

Yameen, Basit, Won Il Choi, Cristian Vilos, Archana Swami, Jinjun Shi, and Omid C. Farokhzad. 2014. "Insight into Nanoparticle Cellular Uptake and Intracellular Targeting." *Journal of Controlled Release* 190. Elsevier B.V.: 485–499. doi:10.1016/j.jconrel.2014.06.038.

Yan, Wei-Hua. 2011. "HLA-G Expression in Cancers: Potential Role in Diagnosis, Prognosis and Therapy." *Endocrine, Metabolic & Immune Disorders - Drug Targets* 11 (1): 76–89. doi:10.2174/187153011794982059.

Yang, Zhaogang, Yifan Ma, Hai Zhao, Yuan Yuan, and Betty Y. S. Kim. 2020. "Nanotechnology Platforms for Cancer Immunotherapy." *WIREs Nanomedicine and Nanobiotechnology* 12 (2). doi:10.1002/wnan.1590.

Yang, Zhe, Wenxin Tang, Xingen Luo, Xiaofang Zhang, Chao Zhang, Hao Li, Di Gao, Huiyan Luo, Qing Jiang, and Jie Liu. 2015. "Dual-Ligand Modified Polymer-Lipid Hybrid Nanoparticles for Docetaxel Targeting Delivery to Her2/Neu Overexpressed Human Breast Cancer Cells." *Journal of Biomedical Nanotechnology* 11 (8): 1401–1417. doi:10.1166/jbn.2015.2086.

Yu, Mi Kyung, Jinho Park, and Sangyong Jon. 2012. "Targeting Strategies for Multifunctional Nanoparticles in Cancer Imaging and Therapy." *Theranostics*. doi:10.7150/thno.3463.

Yu, Yanke, Christopher DesJardins, Phil Saxton, George Lai, Edgar Schuck, and Y. Nancy Wong. 2013. "Characterization of the Pharmacokinetics of a Liposomal Formulation of Eribulin Mesylate (E7389) in Mice." *International Journal of Pharmaceutics* 443 (1–2): 9–16. doi:10.1016/j.ijpharm.2013.01.010.

Zhang, Di, Zhongju Ye, Hua Liu, Xin Wang, Jianhao Hua, Yunyun Ling, Lin Wei, Yunsheng Xia, Shaokai Sun, and Lehui Xiao. 2021. "Cell Membrane Coated Smart Two-Dimensional Supraparticle for in Vivo Homotypic Cancer Targeting and Enhanced Combinational Theranostics." *Nanotheranostics* 5 (3): 275–287. doi:10.7150/ntno.57657.

Zhang, F., N. N. Parayath, C. I. Ene, S. B. Stephan, A. L. Koehne, M. E. Coon, E. C. Holland, and M. T. Stephan. 2019. "Genetic Programming of Macrophages to Perform Anti-Tumor Functions Using Targeted MRNA Nanocarriers." *Nature Communications* 10 (1): 3974. doi:10.1038/s41467-019-11911-5.

Zhang, Ni, Jiao Song, Yi Liu, Mingzhu Liu, Liang Zhang, Danli Sheng, Liming Deng, et al. 2019. "Photothermal Therapy Mediated by Phase-Transformation Nanoparticles Facilitates Delivery of Anti-PD1 Antibody and Synergizes with Antitumor Immunotherapy for Melanoma." *Journal of Controlled Release* 306 (July): 15–28. doi:10.1016/j.jconrel.2019.05.036.

Zhang, Xiao, Shuqing He, Bingbing Ding, Chunrong Qu, Qing Zhang, Hao Chen, Yu Sun, et al. 2020. "Cancer Cell Membrane-Coated Rare Earth Doped Nanoparticles for Tumor Surgery Navigation in NIR-II Imaging Window." *Chemical Engineering Journal* 385 (December 2019). Elsevier: 123959. doi:10.1016/j.cej.2019.123959.

Zhang, Yongtai, Zehui He, Yanyan Li, Qing Xia, Zhe Li, Xuefeng Hou, and Nianping Feng. 2021. "Materials Science & Engineering C Tumor Cell Membrane-Derived Nano-Trojan Horses Encapsulating Phototherapy and Chemotherapy Are Accepted by Homologous Tumor Cells." *Materials Science & Engineering C* 120 (June 2020). Elsevier B.V.: 111670. doi:10.1016/j.msec.2020.111670.

Zhang, Yu, Miao Deng Liu, Chu Xin Li, Bin Li, and Xian Zheng Zhang. 2020. "Tumor Cell Membrane-Coated Liquid Metal Nanovaccine for Tumor Prevention‡." *Chinese Journal of Chemistry* 38 (6): 595–600. doi:10.1002/cjoc.201900557.

Zhang, Yumiao, and Jonathan F. Lovell. 2012. "Porphyrins as Theranostic Agents from Prehistoric to Modern Times." *Theranostics*. doi:10.7150/thno.4908.

Zhang, Zhiqiang, Xianghui Li, Kai Zhao, Zhenhuan Guo, Yonglu Liu, Dai Zeng, and Xueting Ban. 2021. "Medicine in Drug Discovery Transferrin-Modi Fi Ed Cancer Cell Member Coating Hypocrellin B-Derived Nanomaterials for Enhanced Photodynamic Therapy Ef Fi Cacy in Hepatocellular Carcinoma." *Medicine in Drug Discovery* 11. The Authors: 100088. doi:10.1016/j.medidd.2021.100088.

Zhang, Zi Jian, Kun Peng Wang, Jing Gang Mo, Li Xiong, and Yu Wen. 2020. "Photodynamic Therapy Regulates Fate of Cancer Stem Cells through Reactive Oxygen Species." *World Journal of Stem Cells* 12 (7). Baishideng Publishing Group Co: 562–584. doi:10.4252/wjsc.v12.i7.562.

Zhao, Zhao, Mei Ji, Qianqing Wang, Nannan He, and Yue Li. 2020. "Ca^{2+} Signaling Modulation Using Cancer Cell Membrane Coated Chitosan Nanoparticles to Combat Multidrug Resistance of Cancer." *Carbohydrate Polymers* 238 (November 2019). Elsevier: 116073. doi:10.1016/j.carbpol.2020.116073.

Zhou, Mengyun, Yi Xing, Xiaoyu Li, Xin Du, Tailin Xu, and Xueji Zhang. 2020. "Cancer Cell Membrane Camouflaged Semi-Yolk@Spiky-Shell Nanomotor for Enhanced Cell Adhesion and Synergistic Therapy." *Small* 16 (39): 1–11. doi:10.1002/smll.202003834.

Zhou, Wen, Ying Chen, Yutao Zhang, Xiaoyan Xin, Rutian Li, Chen Xie, and Quli Fan. 2020. "Iodine-Rich Semiconducting Polymer Nanoparticles for CT/Fluorescence Dual-Modal Imaging-Guided Enhanced Photodynamic Therapy." *Small* 16 (5). John Wiley & Sons, Ltd: 1905641. doi:10.1002/smll.201905641.

Zhou, Yun, Ke-Ting Que, Hua-Ming Tang, Peng Zhang, Qian-Mei Fu, and Zuo-Jin Liu. 2020. "Anti-CD206 Antibody-conjugated Fe_3O_4-based PLGA Nanoparticles Selectively Promote Tumor-associated Macrophages to Polarize to the Pro-inflammatory Subtype." *Oncology Letters* 20 (6): 1–1. doi:10.3892/ol.2020.12161.

Zhu, Saijie, Mengmeng Niu, Hannah O'Mary, and Zhengrong Cui. 2013. "Targeting of Tumor-Associated Macrophages Made Possible by PEG-Sheddable, Mannose-Modified Nanoparticles." *Molecular Pharmaceutics* 10 (9): 3525–3530. doi:10.1021/mp400216r.

Zou, Mei-zhen, Zi-hao Li, Xue-feng Bai, Chuan-jun Liu, and Xian-zheng Zhang. 2021. "Hybrid Vesicles Based on Autologous Tumor Cell Membrane and Bacterial Outer Membrane To Enhance Innate Immune Response and Personalized Tumor Immunotherapy." Nano Letters 21 (20):8609–8618. doi:10.1021/acs.nanolett.1c02482.

Zwicke, Grant L., G. Ali Mansoori, and Constance J. Jeffery. 2012. "Utilizing the Folate Receptor for Active Targeting of Cancer Nanotherapeutics." Nano Reviews 3 (1). Taylor & Francis: 18496. doi:10.3402/NANO.V3I0.18496.

10 Developing Nanoparticles for Brain Delivery by Receptor-mediated Targeting and Transport across the Blood-brain Barrier

Iason Papademetriou

10.1 INTRODUCTION

Improving methods for drug delivery to the brain has historically been a major challenge, and this remains the case today. Physiological and pathological barriers to efficient drug transport greatly limit the efficacy of many existing therapeutics and prevent use of agents with therapeutic potential. Systemic administration often results in limited therapeutic delivery, particularly in the case of biologics. Nanoparticles made of various materials and architectures have been explored as vehicles for brain drug delivery. Yet after decades of exploration and optimization of design parameters (e.g., size, shape), it remains a challenge to deliver beyond even 1% of the injected dose into brain tissue from systemically administered nanoparticles (Papademetriou and Porter 2015; Wong et al. 2013). Rapid clearance from the circulation by the liver remains a primary barrier (Zhang et al. 2016). Accumulation in other peripheral organs also removes nanoparticles from the circulation, but generally to a lesser extent. Nanoparticles which do not distribute to the liver or other peripheral organs must then cross the blood-brain barrier (BBB) present in brain capillaries in order to reach therapeutic targets in the parenchyma.

The BBB presents a formidable physical barrier to nanoparticle transport due to the expression of tight junctions which restrict transport between brain microvascular endothelial cells (BMECs) (Kadry et al. 2020). Nanoparticle transport across BMECs is limited as well by lack of fenestrae present in other vascular beds and downregulated basal transcytosis. Transport across BMECs is regulated by transporters and receptors which recognize specific substrates. In this chapter, targeting receptors capable of receptor-mediated transcytosis (RMT) will be reviewed as a viable strategy for facilitating BBB transport of nanoparticles. The complexity inherent in this approach will be discussed as well as the use of next-generation *in vitro* BBB models to aid design of RMT-targeted nanoparticles. Targeting nanoparticles to transporters (e.g., glucose transporter 1), tailoring nanoparticle surface properties to favor adsorption to the plasma membrane, transient BBB disruption, or utilizing alternative routes of administration also represent viable strategies in certain contexts, but will not be covered in this chapter. The reader is referred to excellent articles on these topics (Battaglia et al. 2018; Fowler et al. 2020; Lamson et al. 2020; Arora et al. 2020; Lu 2012).

DOI: 10.1201/9781003305583-10

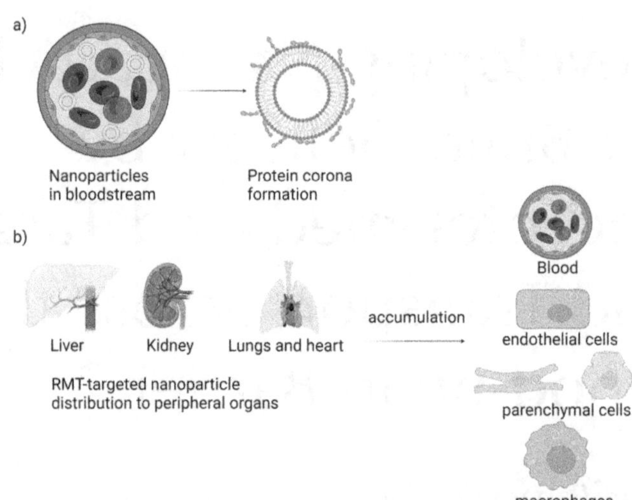

FIGURE 10.1 Distribution of RMT-targeted nanoparticles to peripheral organs.

10.2 BARRIERS TO SYSTEMIC ADMINISTRATION OF NANOPARTICLES TARGETED TO BRAIN RMT RECEPTORS

Systemic administration of nanoparticles by injection into the bloodstream results in rapid adsorption of circulating proteins to the nanoparticle surface (e.g., opsonins, apolipoproteins), a surface coating known as the protein corona (Figure 10.1a.) (Poon et al. 2019; Zhang et al. 2016). This results in rapid clearance of nanoparticles to peripheral organs from the circulation, primarily by the liver where the endothelium is discontinuous, fenestrated, and lacks a basement membrane. Nanoparticles are freely transported to the parenchyma where uptake occurs by resident phagocytes and hepatocytes (Figure 10.1b). Generally, 30–99% of nanoparticles injected into the bloodstream are distributed to the liver (Poon et al. 2019; Zhang et al. 2016). Nanoparticle size and surface properties can influence the protein corona and by extension the degree of liver clearance. Secondary to the liver, nanoparticles >6 nm distribute to the spleen due to uptake by resident immune cells (Poon et al. 2019; Zhang et al. 2016). Nanoparticles <6 nm in size are subject to rapid elimination by filtration in the kidneys. Peripheral organs beyond the liver, spleen, and kidneys accumulate nanoparticles to a significant extent in the blood compartment, although this can be mitigated by active targeting strategies (Figure 10.1b). As such, accumulation in these organs (e.g., lungs, heart) is influenced by organ vascularity. For example, the lungs contain ~30% of the total surface area of endothelium in the body, which can increase accumulation of nanoparticles compared with other organs beyond spleen, liver, kidneys. Other physiological factors influence nanoparticle clearance from the circulation, albeit to a lesser extent. Fluid shear stress (FSS), for example, can alter the frequency of nanoparticle collisions with endothelial cells, or potentially transendothelial transport (Papademetriou et al. 2018; Swaminathan et al. 2011). Nanoparticle size, charge, and other physicochemical properties influence interactions in the bloodstream and the endothelium, which can significantly impact the biodistribution (Muro et al. 2008; Walkey et al. 2012).

10.3 RMT AT THE BLOOD-BRAIN BARRIER

Nanoparticles which distribute to the brain vasculature will then need to reach the brain parenchyma to interact with therapeutic targets for various diseases (e.g., diseased neurons). Capillaries are the primary site for transport in the vasculature, but transport to the brain is limited

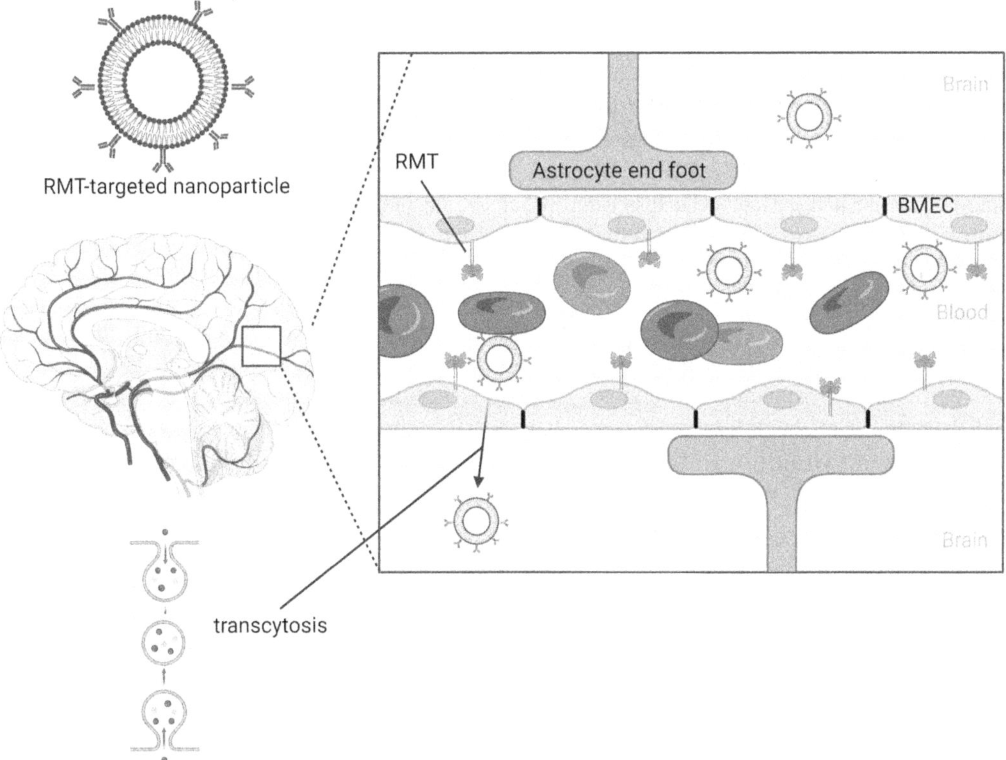

FIGURE 10.2 Transcytosis of RMT-targeted nanoparticles across the BBB.

by the presence of the BBB (Figure 10.2). Targeting nanoparticles to RMT receptors can help circumvent this obstacle. Drug efflux transporters and degrading enzymes are also expressed in the BBB. Loading therapeutics in the interior of nanoparticles can also mitigate these barriers.

RMT is an endogenous transport process of the BBB which is mediated by endocytic receptors expressed on the luminal cell surface of brain microvascular endothelial cells (BMECs). These receptors are internalized by BMECs and transported to the abluminal surface by vesicular trafficking (Georgieva et al. 2020; Papademetriou and Porter 2015). Nanoparticles which follow the transcellular route must then be released at the abluminal surface of BMECs.

10.3.1 Targeting Nanoparticles to RMT

To engage receptors enabling RMT, nanoparticles are functionalized with ligands (e.g., antibodies, peptides) displayed on the nanoparticle surface. These ligands must be designed to enable efficient binding to the RMT receptor (Muro 2012). For example, the number of RMT receptors expressed per BMEC influences the number of nanoparticles which can bind to the cell surface. The RMT receptor must also be localized in a domain of the luminal plasma membrane which is accessible. RMT receptors expressed in caveolae, for example ganglioside GM1, may not be accessible to larger nanoparticles since caveolae are ~70 nm in diameter. In addition, the location of the target epitope within the RMT receptor must also be accessible. For example, epitopes proximal to the plasma membrane may be difficult to target relative to more distal epitopes (Garnacho et al. 2008). Affinity for the target epitope should be high enough that the nanoparticle can bind and remain attached in the presence of FSS in the capillary lumen. The overall strength of the nanoparticle binding interaction with RMT receptors also depends on the number of targeting ligands per nanoparticle (i.e., valency). Multivalent engagement of targeted receptors strengthens the binding interaction (Muro et al. 2006), and avidity is the term used to describe binding strength from multivalent engagement. Multivalency can also impact internalization of nanoparticles by endocytosis, particularly for RMT receptors which are activated by receptor clustering. For example, monovalent engagement of Intercellular Adhesion Molecule 1 (ICAM-1) results in little to no internalization of antibodies, whereas multivalent engagement with nanoparticles results is highly efficient (Muro et al. 2003). With respect to transcytosis, high avidity and low avidity binding has been reported to be less effective than intermediate avidity (Manthe et al. 2020). A proposed hypothesis for this result is that high avidity reduces the ability of nanoparticles to detach from RMT receptors at the abluminal surface and increases trafficking to lysosomes from the abluminal plasma membrane. Low avidity increases detachment from the abluminal surface, but reduces binding of the RMT receptor so that fewer nanoparticles are internalized. Intermediate binding avidity therefore achieves optimal transcytosis (in the case of ICAM-1 targeted nanoparticles) by balancing the opposing effects of avidity on binding at the luminal surface and detachment at the abluminal surface. It is possible that ligand valency impacts transcytosis through other mechanisms as well (e.g., altered internalization and trafficking rates).

10.3.2 Internalization and Transcytosis of RMT-targeted Nanoparticles

Vesicular trafficking of the RMT receptor can also impact the efficiency of transport across the BBB by RMT-targeted nanoparticles. Internalized nanoparticles bound to RMT receptors are, to a large extent, destined for degradation via the endolysosomal pathway. The basal level of transcytosis at the BBB is low (Wood et al. 2021), and is consistent with studies where only a smaller fraction of nanoparticles is routed transcellularly. Design criteria which can increase transcellular transport are not well established. However, several studies inform relationships between endothelial cell (EC) physiology and nanoparticle targeting and transport in ECs. For example, FSS in the lumen of blood vessels can increase binding to ECs by increasing the frequency of collisions between nanoparticles and the endothelium (Swaminathan et al. 2011). On the other hand, binding could be reduced by reducing the time nanoparticles are proximal to the endothelium or by detachment of bound nanoparticles. FSS also is known to alter EC morphological and biophysical characteristics in part by inducing cytoskeletal rearrangement. For example, flow-adapted ECs in cell culture form actin stress fibers and internalization of nanoparticles targeting PECAM-1 is inhibited compared to non-flow-adapted ECs (Han et al. 2012). This observation was consistent with observations *in vivo*. Nanoparticles targeted to ICAM-1, which like platelet endothelial cell adhesion molecule 1 (PECAM-1) utilizes cell adhesion mediated (CAM-mediated) endocytosis upon receptor clustering, exhibited relatively slow internalization in the pulmonary vasculature which was slightly faster in areas of lower FSS (Bhowmick et al. 2012). Interestingly, the internalization rate of nanoparticles in the pulmonary vasculature increased with exposure to an inflammatory insult, suggesting that internalization rate can be modulated by pathological cell status.

Acute changes in flow have also been shown to impact nanoparticle internalization. FSS exposure in non-flow-adapted ECs stimulated internalization in the absence of actin stress fibers (Han et al. 2012). Interestingly, transport in ECs can also be modulated by acute FSS exposure, as observed for various solutes (Papademetriou et al. 2018). Nanoparticles targeting RMT receptor, low density lipoprotein receptor-related protein 1 (LRP-1) were transported across a microfluidic BBB model more efficiently in the presence of acute FSS compared with static incubation, although mechanistic analysis of the paracellular vs transcellular contribution was not definitive. These observations are consistent with the presence of mechanosensitive endocytic receptors in ECs. For example, ECs have been observed to respond to FSS by increasing expression of CAV-1 and the number of caveolae on the cell surface (Filippini and D'Alessio 2020). In contrast with endothelium of peripheral organs, brain ECs resist elongation from exposure to FSS (Ye et al. 2014). One potential explanation is that the function of tight junctions would be reduced in elongated ECs. In an interesting computational study, the hypothesis that caveolae enable activation of shear-sensitive receptors by sequestration from chronic FSS was examined. The authors reported that FSS in caveolae is reduced relative to the planar plasma membrane, whereas solutes entered caveolae efficiently (Shin et al. 2019). This mechanism may help explain how shear-sensitive receptors in ECs are able to respond to changes in FSS in the presence of continuous blood flow.

10.3.3 RMT of Nanoparticles in Healthy vs Diseased BBB

Changes associated with disease states can alter multiple parameters influencing RMT. This necessitates empirical evaluation of RMT-targeted nanoparticles models which replicate critical features of the diseased BBB in humans. For example, nanoparticles targeted to RMT receptors (transferrin receptor, ganglioside GM1, or ICAM-1 respectively) which utilize clathrin-mediated, caveolae-mediated, or CAM-mediated endocytosis were evaluated for transport across BBB in healthy or acid sphingomyelinase (ASM)-deficient models of Niemann Pick Disease (Solomon et al. 2022). The authors found that ASM deficiency reduced expression of the transferrin receptor and clathrin, resulting in reduced transcytosis and brain accumulation relative to healthy control. GM1 levels were increased, but caveolin-1 expression and transcytosis rate decreased, resulting in similar brain accumulation in ASM-deficient mice and healthy control. ICAM-1 expression increased, while NHE-1 expression decreased and transcytosis rate increased, resulting in increased transcytosis rate and brain accumulation related to healthy control. Additionally, transcytotic vesicles have been observed with increased frequency in pathological states involving BBB disruption (Arvanitis et al. 2018), suggesting that nanoparticles targeting RMT receptors may be utilized to increase therapeutic delivery to sites of disease. These results highlight the significant impact that changes in disease can have on the performance of nanoparticles targeted to RMT and imply the necessity of evaluation in models which replicate relevant features of human disease.

10.3.4 RMT Expression in Brain Vasculature

Some RMT (e.g., transferrin receptor) are expressed throughout the brain at the BBB. It is therefore possible to-deliver RMT-targeted nanoparticles to the whole brain which is useful for diseases where the entire brain is affected (e.g., lysosomal storage disorders). Interestingly, a recent study provides evidence that transcytosis by targeting nanoparticles to the transferrin receptor occurs more readily in post-capillary venules than in capillaries of the brain (Kucharz et al. 2021). This intriguing result suggests that RMT receptors expressed in post-capillary venules merit increased scrutiny. It will be also interesting for future studies to examine whether post-capillary venules are also preferred over capillaries for other RMT utilizing clathrin-mediated endocytosis, RMT utilizing alternative endocytic pathways, and in disease models.

10.4 NANOPARTICLE TRANSPORT IN THE EXTRACELLULAR SPACE OF THE BRAIN PARENCHYMA

Following transport across the BBB, nanoparticles will need to penetrate the extracellular space to reach target cells in the brain parenchyma. Recent work in this area has helped to establish design criteria in this regard. Nanoparticle size is a primary consideration, as most pores in the extracellular space are <100 nm in diameter. Efficient brain tissue penetration of dextran and quantum dots up to 35 nm in diameter was observed *in vivo*, and a normal width of 38–64 nm was predicted by computational modeling (Thorne and Nicholson 2006). Interestingly, polymer nanoparticles up to 114 nm in size penetrated human brain tissue *ex vivo* when coated with a dense layer of PEG (Nance et al. 2012). This result was confirmed *in vivo* as 40 nm and 100 nm nanoparticles rapidly penetrated mouse brain tissue, but not 200 nm nanoparticles. In a subsequent study, the effect of PEG coating was characterized in paclitaxel-loaded PLGA nanoparticles and found to improve the diffusion rate by 100-fold in comparison to uncoated PLGA nanoparticles (Nance et al. 2014). A separate study showed that 60 nm PEG coated nanoparticles efficiently penetrated brain tissue following BBB disruption with FUS and microbubbles (Nance et al. 2014). A later study found that PEG coating improved the colloidal stability of polystyrene nanoparticles when tested with varying CSF ion concentration, ion composition, pH, and temperature (Curtis et al. 2018). The authors concluded that improved colloidal stability was key to the improved diffusive capability imparted by the PEG coating. Most recently, modifying PEG nanoparticles with surfactants to increase transport across the BBB and internalization by neurons was reported (Joseph et al. 2021). While surfactants did so successfully, they also reduced diffusivity in brain tissue, illustrating the complexity inherent in designing nanoparticles to cross multiple biological barriers. Although relatively unexplored, the effects of nanoparticle shape and rigidity, as well as functionalization with targeting ligands may also impact brain tissue penetration.

10.5 MICROPHYSIOLOGICAL SYSTEMS AND HUMAN iPSCS TO AID DESIGN OF NANOPARTICLES FOR BRAIN DELIVERY

As is evident from the previous section, design of nanoparticles for this purpose is complex in part because a design parameter can have multiple effects (e.g., increasing nanoparticle size can increase clearance rate from circulation, but also reduce penetration of target tissue). It is also well established that traditional, macroscale, 2D cell culture does not recapitulate the (patho) physiological environment present *in vivo*. Additionally, species differences can significantly alter outcomes observed in animal models compared to humans. These factors have made a strong case for the adoption of microphysiological systems (MPS) incorporating human induced pluripotent stem cells (iPSCs) in translational studies, with recent legislation no longer requiring animal studies before trials in humans. Here, we review the advantages and limitations of these models for design of RMT-targeted nanoparticles.

10.5.1 Microphysiological Systems (MPS)

MPS are attractive for analyzing the effect of nanoparticle design parameters on targeting and transport across the BBB. As discussed in the previous section, FSS on the luminal surface of BMECs is of primary importance. Most MPS BBB models utilize overlapping microchannels separated by a porous membrane (Figure 10.3). BMECs are cultured on the membrane in the presence of FSS imparted by flow in the apical channel. Flow can be driven by syringe or peristaltic pumps, or by pumpless methods (Azizgolshani et al. 2021; Wang et al. 2017; Pediaditakis et al. 2021). The membrane pore size and topography, as well as ECM coating can impact the function of cultured cells (Frohlich et al. 2013). Importantly, MPS enable precise control of flow rates so that effects of FSS can be investigated systematically. The two-compartment architecture recreates the luminal and abluminal compartments where perivascular cells (e.g., pericytes) can be cultured. Another significant advantage of MPS is that cell culture is done at the microscale which more closely replicates physiological parameters (e.g., ratio of fluid to cells, distance between BMECs and perivascular cells (Montanez-Sauri et al. 2015)). MPS have been shown to improve tissue differentiation and function relative to macroscale 2D cultures (Booth and Kim 2012). On the other hand, cells in MPS are often cultured in 2D rather than 3D matrices, incorporate artificial components (e.g., porous membrane and microchannels), and do not fully replicate ratios of cell types or cell to fluid ratio.

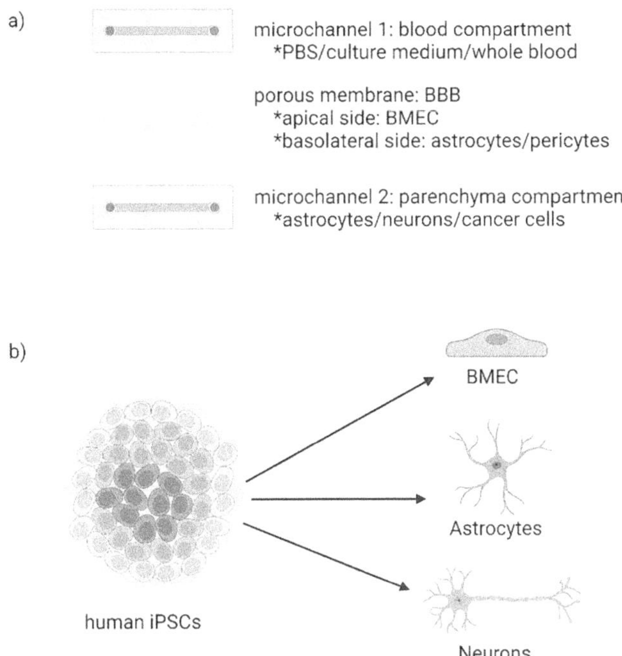

FIGURE 10.3 Components of a canonical MPS for BBB modeling with iPSC-derived cells.

10.5.2 Human Induced Pluripotent Stem Cells (iPSCs)

Human iPSCs are also critical to development of *in vitro* models which better represent aspects of *in vivo* physiology relevant to RMT-targeted nanoparticles. Species differences in RMT receptors at the genetic, transcriptional, or translational level can potentially alter RMT expression, accessibility and affinity for the binding epitope, subcellular localization, trafficking and other characteristics that impact binding to BMECs and transport of RMT-targeted nanoparticles. Human iPSCs can be differentiated to BMEC-like cells with barrier functionality that replicates current benchmarks of *in vivo* functionality (DeStefano et al. 2018). iPSC-derived BMEC-like cells can be co-cultured with iPSC-derived perivascular cells to further improve BBB functionality and to provide target cells for therapeutic delivery (e.g., neurons, astrocytes, Figure 10.3). Recently the precise identity of iPSC-derived BMEC-like cells has been questioned, with several reports of epithelial signatures in the differentiated cells (Lippmann et al. 2020; Lu et al. 2021). iPSC-derived BMEC-like cells expressed a fusion protein of VE-Cadherin and GFP, whereas undifferentiated cells did not. However, expression of the VE-CAD-GFP in iPSC-derived BMEC-like cells was lower than in primary BMEC. Nevertheless, iPSC-derived BMEC-like cells remain a reproducible source of human cells which replicate barrier functionality *in vivo*. It remains to be seen how closely iPSC-derived BMEC-like cells replicate aspects of human RMT receptor biology, and *in vitro* to *in vivo* correlations for the technology need to be established. Incorporation of iPSC-derived BMEC-like cells into MPS has been achieved (Wang et al. 2017).

10.6 CONCLUSIONS

Nanoparticles targeted to RMT receptors remain challenging to design. However, improved understanding of the interaction of RMT-targeted nanoparticles with the BBB, coupled with more predictive *in vitro* models, may very well enable improvements in therapeutic brain delivery over the next decade.

REFERENCES

Arora, S., D. Sharma, and J. Singh. 2020. "GLUT-1: An effective target To deliver brain-derived neurotrophic factor gene across the blood brain barrier." *ACS Chem Neurosci* 11 (11):1620–1633. doi: 10.1021/acschemneuro.0c00076.

Arvanitis, C. D., V. Askoxylakis, Y. Guo, M. Datta, J. Kloepper, G. B. Ferraro, M. O. Bernabeu, D. Fukumura, N. McDannold, and R. K. Jain. 2018. "Mechanisms of enhanced drug delivery in brain metastases with focused ultrasound-induced blood-tumor barrier disruption." *Proc Natl Acad Sci U S A* 115 (37):E8717–E8726. doi: 10.1073/pnas.1807105115.

Azizgolshani, H., J. R. Coppeta, E. M. Vedula, E. E. Marr, B. P. Cain, R. J. Luu, M. P. Lech, S. H. Kann, T. J. Mulhern, V. Tandon, K. Tan, N. J. Haroutunian, P. Keegan, M. Rogers, A. L. Gard, K. B. Baldwin, J. C. de Souza, B. C. Hoefler, S. S. Bale, L. B. Kratchman, A. Zorn, A. Patterson, E. S. Kim, T. A. Petrie, E. L. Wiellette, C. Williams, B. C. Isenberg, and J. L. Charest. 2021. "High-throughput organ-on-chip platform with integrated programmable fluid flow and real-time sensing for complex tissue models in drug development workflows." *Lab Chip* 21 (8):1454–1474. doi: 10.1039/d1lc00067e.

Battaglia, L., P. P. Panciani, E. Muntoni, M. T. Capucchio, E. Biasibetti, P. De Bonis, S. Mioletti, M. Fontanella, and S. Swaminathan. 2018. "Lipid nanoparticles for intranasal administration: application to nose-to-brain delivery." *Expert Opin Drug Deliv* 15 (4):369–378. doi: 10.1080/17425247.2018.1429401.

Bhowmick, T., E. Berk, X. Cui, V. R. Muzykantov, and S. Muro. 2012. "Effect of flow on endothelial endocytosis of nanocarriers targeted to ICAM-1." *J Control Release* 157 (3):485–492. doi: 10.1016/j.jconrel.2011.09.067.

Booth, R., and H. Kim. 2012. "Characterization of a microfluidic in vitro model of the blood-brain barrier (muBBB)." *Lab Chip* 12 (10):1784–1792. doi: 10.1039/c2lc40094d.

Curtis, C., D. Toghani, B. Wong, and E. Nance. 2018. "Colloidal stability as a determinant of nanoparticle behavior in the brain." *Colloids Surf B Biointerfaces* 170:673–682. doi: 10.1016/j.colsurfb.2018.06.050.

DeStefano, J. G., J. J. Jamieson, R. M. Linville, and P. C. Searson. 2018. "Benchmarking in vitro tissue-engineered blood-brain barrier models." *Fluids Barriers CNS* 15 (1):32. doi: 10.1186/s12987-018-0117-2.

Filippini, A., and A. D'Alessio. 2020. "Caveolae and lipid rafts in endothelium: Valuable organelles for multiple functions." *Biomolecules* 10 (9). doi: 10.3390/biom10091218.

Fowler, M. J., J. D. Cotter, B. E. Knight, E. M. Sevick-Muraca, D. I. Sandberg, and R. W. Sirianni. 2020. "Intrathecal drug delivery in the era of nanomedicine." *Adv Drug Deliv Rev* 165–166:77–95. doi: 10.1016/j.addr.2020.02.006.

Frohlich, E. M., J. L. Alonso, J. T. Borenstein, X. Zhang, M. A. Arnaout, and J. L. Charest. 2013. "Topographically-patterned porous membranes in a microfluidic device as an in vitro model of renal reabsorptive barriers." *Lab Chip* 13 (12):2311–2319. doi: 10.1039/c3lc50199j.

Garnacho, C., S. M. Albelda, V. R. Muzykantov, and S. Muro. 2008. "Differential intra-endothelial delivery of polymer nanocarriers targeted to distinct PECAM-1 epitopes." *J Control Release* 130 (3):226–233. doi: 10.1016/j.jconrel.2008.06.007.

Georgieva, J. V., L. I. Goulatis, C. C. Stutz, S. G. Canfield, H. W. Song, B. D. Gastfriend, and E. V. Shusta. 2020. "Antibody screening using a human iPSC-based blood-brain barrier model identifies antibodies that accumulate in the CNS." *FASEB J* 34 (9):12549–12564. doi: 10.1096/fj.202000851R.

Han, J., B. J. Zern, V. V. Shuvaev, P. F. Davies, S. Muro, and V. Muzykantov. 2012. "Acute and chronic shear stress differently regulate endothelial internalization of nanocarriers targeted to platelet-endothelial cell adhesion molecule-1." *ACS Nano* 6 (10):8824–8836. doi:10.1021/nn302687n.

Joseph, A., G. M. Simo, T. Gao, N. Alhindi, N. Xu, D. J. Graham, L. J. Gamble, and E. Nance. 2021. "Surfactants influence polymer nanoparticle fate within the brain." *Biomaterials* 277:121086. doi: 10.1016/j.biomaterials.2021.121086.

Kadry, H., B. Noorani, and L. Cucullo. 2020. "A blood-brain barrier overview on structure, function, impairment, and biomarkers of integrity." *Fluids Barriers CNS* 17 (1):69. doi: 10.1186/s12987-020-00230-3.

Kucharz, K., K. Kristensen, K. B. Johnsen, M. A. Lund, M. Lonstrup, T. Moos, T. L. Andresen, and M. J. Lauritzen. 2021. "Post-capillary venules are the key locus for transcytosis-mediated brain delivery of therapeutic nanoparticles." *Nat Commun* 12 (1):4121. doi: 10.1038/s41467-021-24323-1.

Lamson, N. G., A. Berger, K. C. Fein, and K. A. Whitehead. 2020. "Anionic nanoparticles enable the oral delivery of proteins by enhancing intestinal permeability." *Nat Biomed Eng* 4 (1):84–96. doi: 10.1038/s41551-019-0465-5.

Lippmann, E. S., S. M. Azarin, S. P. Palecek, and E. V. Shusta. 2020. "Commentary on human pluripotent stem cell-based blood-brain barrier models." *Fluids Barriers CNS* 17 (1):64. doi: 10.1186/s12987-020-00222-3.

Lu, T. M., S. Houghton, T. Magdeldin, J. G. B. Duran, A. P. Minotti, A. Snead, A. Sproul, D. T. Nguyen, J. Xiang, H. A. Fine, Z. Rosenwaks, L. Studer, S. Rafii, D. Agalliu, D. Redmond, and R. Lis. 2021. "Pluripotent stem cell-derived epithelium misidentified as brain microvascular endothelium requires ETS factors to acquire vascular fate." *Proc Natl Acad Sci U S A* 118 (8). doi: 10.1073/pnas.2016950118.

Lu, W. 2012. "Adsorptive-mediated brain delivery systems." *Curr Pharm Biotechnol* 13 (12):2340–2348. doi: 10.2174/138920112803341851.

Manthe, R. L., M. Loeck, T. Bhowmick, M. Solomon, and S. Muro. 2020. "Intertwined mechanisms define transport of anti-ICAM nanocarriers across the endothelium and brain delivery of a therapeutic enzyme." *J Control Release* 324:181–193. doi: 10.1016/j.jconrel.2020.05.009.

Montanez-Sauri, S. I., D. J. Beebe, and K. E. Sung. 2015. "Microscale screening systems for 3D cellular microenvironments: platforms, advances, and challenges." *Cell Mol Life Sci* 72 (2):237–249. doi: 10.1007/s00018-014-1738-5.

Muro, S. 2012. "Challenges in design and characterization of ligand-targeted drug delivery systems." *J Control Release* 164 (2):125–137. doi: 10.1016/j.jconrel.2012.05.052.

Muro, S., T. Dziubla, W. Qiu, J. Leferovich, X. Cui, E. Berk, and V. R. Muzykantov. 2006. "Endothelial targeting of high-affinity multivalent polymer nanocarriers directed to intercellular adhesion molecule 1." *J Pharmacol Exp Ther* 317 (3):1161–1169. doi: 10.1124/jpet.105.098970.

Muro, S., C. Garnacho, J. A. Champion, J. Leferovich, C. Gajewski, E. H. Schuchman, S. Mitragotri, and V. R. Muzykantov. 2008. "Control of endothelial targeting and intracellular delivery of therapeutic enzymes by modulating the size and shape of ICAM-1-targeted carriers." *Mol Ther* 16 (8):1450–1458. doi: 10.1038/mt.2008.127.

Muro, S., R. Wiewrodt, A. Thomas, L. Koniaris, S. M. Albelda, V. R. Muzykantov, and M. Koval. 2003. "A novel endocytic pathway induced by clustering endothelial ICAM-1 or PECAM-1." *J Cell Sci* 116 (Pt 8):1599–1609. doi: 10.1242/jcs.00367.

Nance, E. A., G. F. Woodworth, K. A. Sailor, T. Y. Shih, Q. Xu, G. Swaminathan, D. Xiang, C. Eberhart, and J. Hanes. 2012. "A dense poly(ethylene glycol) coating improves penetration of large polymeric nanoparticles within brain tissue." *Sci Transl Med* 4 (149):149ra119. doi:10.1126/scitranslmed.3003594.

Nance, E., K. Timbie, G. W. Miller, J. Song, C. Louttit, A. L. Klibanov, T. Y. Shih, G. Swaminathan, R. J. Tamargo, G. F. Woodworth, J. Hanes, and R. J. Price. 2014. "Non-invasive delivery of stealth, brain-penetrating nanoparticles across the blood-brain barrier using MRI-guided focused ultrasound." *J Control Release* 189:123–132. doi: 10.1016/j.jconrel.2014.06.031.

Nance, E., C. Zhang, T. Y. Shih, Q. Xu, B. S. Schuster, and J. Hanes. 2014. "Brain-penetrating nanoparticles improve paclitaxel efficacy in malignant glioma following local administration." *ACS Nano* 8 (10):10655–10664. doi: 10.1021/nn504210g.

Papademetriou, I. T., and T. Porter. 2015. "Promising approaches to circumvent the blood-brain barrier: progress, pitfalls and clinical prospects in brain cancer." *Ther Deliv* 6 (8):989–1016. doi: 10.4155/tde.15.48.

Papademetriou, I., E. Vedula, J. Charest, and T. Porter. 2018. "Effect of flow on targeting and penetration of angiopep-decorated nanoparticles in a microfluidic model blood-brain barrier." *PLoS One* 13 (10): e0205158. doi: 10.1371/journal.pone.0205158.

Pediaditakis, I., K. R. Kodella, D. V. Manatakis, C. Y. Le, C. D. Hinojosa, W. Tien-Street, E. S. Manolakos, K. Vekrellis, G. A. Hamilton, L. Ewart, L. L. Rubin, and K. Karalis. 2021. "Modeling alpha-synuclein pathology in a human brain-chip to assess blood-brain barrier disruption." *Nat Commun* 12 (1):5907. doi: 10.1038/s41467-021-26066-5.

Poon, W., Y. N. Zhang, B. Ouyang, B. R. Kingston, J. L. Y. Wu, S. Wilhelm, and W. C. W. Chan. 2019. "Elimination pathways of nanoparticles." *ACS Nano* 13 (5):5785–5798. doi: 10.1021/acsnano.9b01383.

Shin, H., J. H. Haga, T. Kosawada, K. Kimura, Y. S. Li, S. Chien, and G. W. Schmid-Schonbein. 2019. "Fine control of endothelial VEGFR-2 activation: caveolae as fluid shear stress shelters for membrane receptors." *Biomech Model Mechanobiol* 18 (1):5–16. doi: 10.1007/s10237-018-1063-2.

Solomon, M., M. Loeck, M. Silva-Abreu, R. Moscoso, R. Bautista, M. Vigo, and S. Muro. 2022. "Altered blood-brain barrier transport of nanotherapeutics in lysosomal storage diseases." *J Control Release* 349:1031–1044. doi: 10.1016/j.jconrel.2022.07.022.

Swaminathan, T. N., J. Liu, U. Balakrishnan, P. S. Ayyaswamy, R. Radhakrishnan, and D. M. Eckmann. 2011. "Dynamic factors controlling carrier anchoring on vascular cells." *IUBMB Life* 63 (8):640–647. doi: 10.1002/iub.475.

Thorne, R. G., and C. Nicholson. 2006. "In vivo diffusion analysis with quantum dots and dextrans predicts the width of brain extracellular space." *Proc Natl Acad Sci U S A* 103 (14):5567–5572. doi: 10.1073/pnas.0509425103.

Walkey, C. D., J. B. Olsen, H. Guo, A. Emili, and W. C. Chan. 2012. "Nanoparticle size and surface chemistry determine serum protein adsorption and macrophage uptake." *J Am Chem Soc* 134 (4):2139–2147. doi: 10.1021/ja2084338.

Wang, Y. I., H. E. Abaci, and M. L. Shuler. 2017. "Microfluidic blood-brain barrier model provides in vivo-like barrier properties for drug permeability screening." *Biotechnol Bioeng* 114 (1):184–194. doi: 10.1002/bit.26045.

Wong, A. D., M. Ye, A. F. Levy, J. D. Rothstein, D. E. Bergles, and P. C. Searson. 2013. "The blood-brain barrier: an engineering perspective." *Front Neuroeng* 6:7. doi: 10.3389/fneng.2013.00007.

Wood, C. A. P., J. Zhang, D. Aydin, Y. Xu, B. J. Andreone, U. H. Langen, R. O. Dror, C. Gu, and L. Feng. 2021. "Structure and mechanism of blood-brain-barrier lipid transporter MFSD2A." *Nature* 596 (7872): 444–448. doi: 10.1038/s41586-021-03782-y.

Ye, M., H. M. Sanchez, M. Hultz, Z. Yang, M. Bogorad, A. D. Wong, and P. C. Searson. 2014. "Brain microvascular endothelial cells resist elongation due to curvature and shear stress." *Sci Rep* 4:4681. doi: 10.1038/srep04681.

Zhang, Y. N., W. Poon, A. J. Tavares, I. D. McGilvray, and W. C. W. Chan. 2016. "Nanoparticle-liver interactions: Cellular uptake and hepatobiliary elimination." *J Control Release* 240:332–348. doi: 10.1016/j.jconrel.2016.01.020.

11 Nanotechnology-Based Advances in Stem Cell Research

Exploring Interactions and Applications

Shreya Pande, Nimeet Desai, Poonam Dogra, and Jyotsnendu Giri

11.1 INTRODUCTION

Nanotechnology advancements have inspired the creation of novel biomaterials with tailored combinations of attributes. A plethora of biomedical nanomaterials that replicate tissue complexity and modulate stem cell behaviors have been designed, resulting in numerous medicinal benefits (Singla et al. 2019). This book chapter focuses on tissue engineering based on nanomaterials with at least one physical dimension in the nanoscale (between 1 and 100 nm). Nanomaterials are not a simple downsizing of their macroscopic counterparts; their high specific surface area lends them unique physical, chemical, optical, and mechanical properties. The field of biomedical engineering can benefit from the incorporation of nanomaterials because of their distinctive features. A great majority of biological components, such as structural protein, deoxyribonucleic acid (DNA), signaling molecules, antibodies, and extracellular matrix (ECM), function at a scale similar to nanomaterials. Technically, a cell can be considered a functional micro-sized particle made up of various nano-constituents (like cell membranes, surface proteins, cytoskeletons, etc.). Owing to their similar size, nanomaterials can interact extensively with the physiological environment and allow the modulation of cellular-level activities, hence serving as a platform that can imitate the structural complexity of cell constituents (Zhao et al. 2021).

One of the important thrust areas for the clinical application of nanomaterials is stem cell research. Stem cells are primordial cells that can self-renew and differentiate into several specialized functional cells (Rajabzadeh et al. 2019). Embryonic stem cells (ESCs) and somatic stem cells (SSCs) are two primary types of stem cells based on their developmental stage. ESCs are produced from the blastocyst's inner cell mass. Induced pluripotent stem cells (iPSCs) are derived from somatic cells that have been genetically reprogrammed to an ESCs-like state by introducing the expression of specific genes (Zakrzewski et al. 2019). iPSCs possess comparable properties to ESCs. Both are pluripotent stem cells with the highest potential for differentiation and a limitless ability for self-renewal. SSCs, which are produced from adult tissues, are more accessible than ESCs and iPSCs but less potent (Ramalingam and Rana 2015). In recent years, stem cell research has enabled the isolation of an increasing variety of SSCs from bone marrow, adipose tissues, cord blood, and brain tissues. Despite the associated ethical difficulties, ESCs, alongside mesenchymal stem cells (MSCs)

DOI: 10.1201/9781003305583-11

and adipose-derived stem cells (ADSCs), have become appealing stem cell sources for tissue regeneration and engineering (Kwon et al. 2018; Wang et al. 2019).

With the convergence of nanomaterials and stem cells, the repair and regeneration of damaged cells/tissues are becoming a therapeutic reality, which will lead to the development of treatments for otherwise incurable diseases. A number of obstacles must be overcome before stem cells may be employed extensively in medical therapies. These include regulating stem cell self-renewal mechanisms, proliferation, and differentiation (Hiew et al. 2018). Throughout the last decade, several attempts have been undertaken to influence the differentiation of stem cells into distinct cell types. However, the poor efficiency and success rate of differentiation limit the development of stem cell differentiation for stem cell treatment. Additionally, undifferentiated ESCs after implantation in vivo are a risk factor for teratoma; therefore, it is essential to facilitate committed differentiation of ESCs into particular lineages prior to implantation for safe use in cell-based treatments (Aly 2020). In this context, the bioactive characteristics of nanomaterials allow excellent control over stem cell behavior. The chemical composition, surface topography, mechanical attributes, electrical properties, and morphology of nanomaterials considerably affect stem cell responses. Numerous research findings have demonstrated that nanomaterials can be engineered to offer an inductive milieu to controllably drive stem cell development. While a variety of multifunctional nanomaterials for tissue engineering applications have been produced and studied, their interactions with stem cells are not thoroughly understood (Kerativitayanan et al. 2015).

With a focus on tissue engineering applications, we critically evaluate the interactions between various kinds of nanomaterials and stem cells in this chapter. Stem cells respond to integrated nanomaterials by a variety of biological pathways, depending chiefly on the material type, structure, chemical functionality, and topographical properties. These biological interactions can be used as tools to control stem cell function and accomplish therapeutic objectives. The subsequent sections will focus on understanding different classes of nanomaterials, their interactions with different stem cells, and how such interactions can be tuned to develop stem cell-based regenerative medicine. It is anticipated that a more profound comprehension of nanomaterial–stem cell interactions will aid in developing the next generation of therapeutics.

11.2 NANOMATERIALS IN STEM CELL RESEARCH

Biomedical nanomaterials provide several therapeutic advantages in stem cell research as they possess the potential to accurately simulate tissue complexity and thereby modulate the stem cell fate. Owing to their small size, nanomaterials can widely interact with the physiological environment and enable the development of systems that mimic the structural complexity and functions of extracellular matrices. The small size also facilitates interactions and alterations of important biological functions at a cellular level. Hence, nanomaterials are highly touted to provide a new framework for therapeutic intervention based on stem cells. Since the nanoscale size ultimately governs the physicochemical properties and nature of interaction with stem cells, for ease of understanding, nanomaterials in stem cell research can be organized into four distinct categories based on their dimensions. Zero-dimensional (0-D) nanomaterials have all dimensions (x, y, z) at the nanoscale, i.e., no dimensions are greater than 100 nm. One-dimensional (1-D) nanomaterials have two dimensions (x, y) at the nanoscale, while two-dimensional (2-D) nanomaterials possess only a single dimension (x) at the nanoscale. Three-dimensional (3-D) nanostructures are not confined to the nanoscale but are composed of constituents with nanoscale features. Figure 11.1 highlights the dimension-based classification of nanomaterials in stem cell research with examples. The subsequent section discusses the unique attributes and properties of individual nanomaterials and how these materials can be extended to modulate the stem cell fate for diverse biomedical applications.

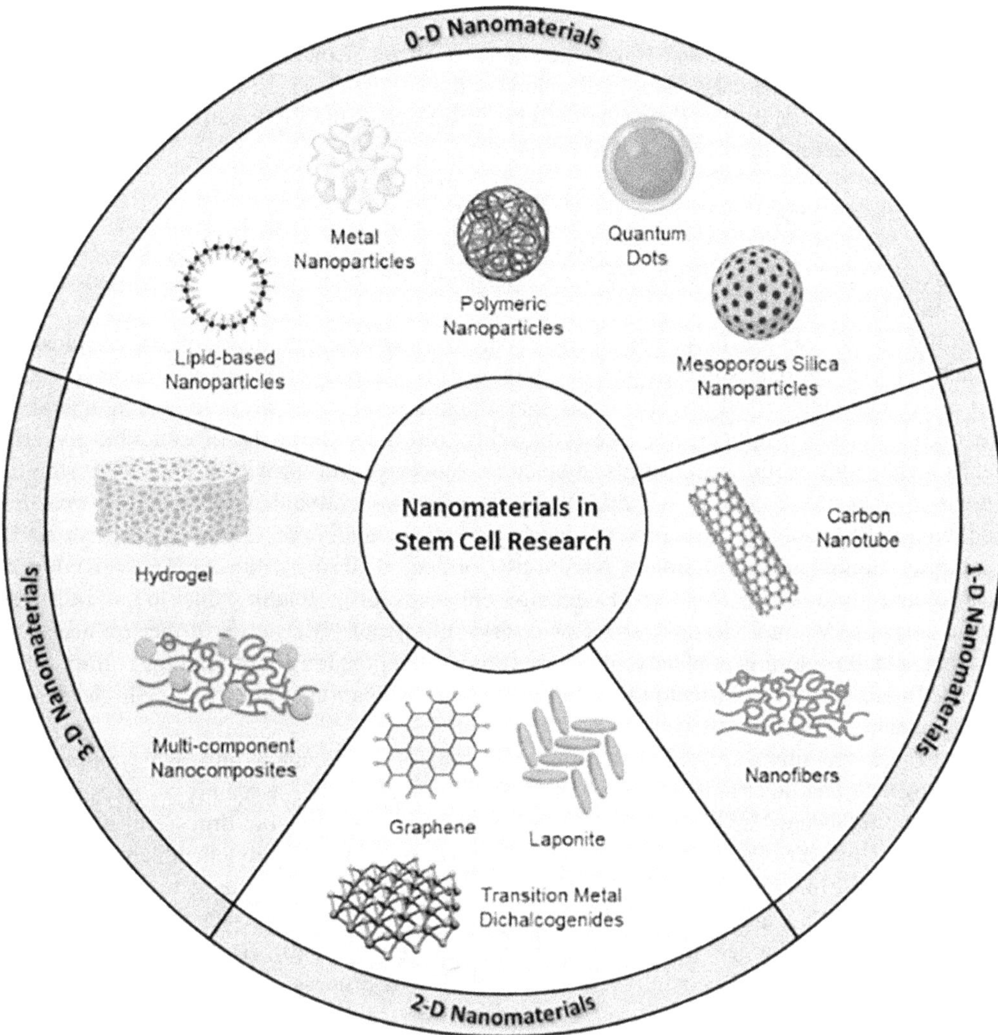

FIGURE 11.1 Dimension-based classification of prominent nanomaterials that are explored for stem cell research.

11.3 0-D NANOMATERIALS

11.3.1 Organic Nanoparticles

11.3.1.1 Lipid-based Nanoparticles

Lipids are vital components of plant and animal cells, along with carbohydrates and proteins. They play a crucial role in maintaining physiological balance by regulating cell structure, energy storage, and cellular signaling. Recently, lipids have gained attention as promising substances for developing drug and biomolecule delivery systems. Biomedically significant lipids can be categorized into different types based on their chemical structure, including fatty acids (such as propionic acid, decanoic acid, and α-linolenic acid), glycerolipids (including monocaprylate, 1,3-diacylglycerol, and tricaprilin), glycerophospholipids (like phosphatidylcholine, phosphatidylethanolamine, and phosphatidylserine), sphingolipids (such as ceramide, sphingomyelin, and glycosphingolipid), and saccharolipids (like lipopolysaccharide and mono-phosphoryl lipid A) (Park et al. 2021). Lipids are easily available in large quantities and can be chemically modified to change their properties. Contrary to popular belief, many lipids have both hydrophilic and hydrophobic characteristics. This enables lipids to form bilayer structures, with the hydrophobic part inside and the hydrophilic part facing the surrounding aqueous environment (Apolinario et al. 2021). Lipid-based nanosystems are widely used in tissue engineering and regenerative medicine due to their versatility, flexibility, and capacity to carry various types of cargo. These nanosystems can be classified into solid lipid nanoparticles (SLNs), nanostructured lipid carriers (NLCs), liposomes (and their derivatives), and lipid nanocapsules (LNCs), based on their composition, structure, and fabrication method.

Briefly, SLNs can be described as colloidal nanocarriers composed of solid lipids and surfactants that form a spherical nanostructure with a solid core and a monolayer shell. Typical constituents of SLNs are lipids that remain solid at a physiological temperature, like highly purified triglycerides, free fatty acids, free fatty alcohols, or complex glyceride blends. The rigid solid core limits molecular mobility and provides improved stability (versus free form) (Duan et al. 2020). NLCs offer key advantages, including excellent biocompatibility, solvent-free production, high cell uptake, and extended shelf life. By combining liquid and solid lipids, NLCs are formed, differing from SLNs by having an unstructured lipid core. This unique feature enhances drug loading due to an imperfect crystal interior (Mu and Holm 2018). Lower crystallinity of NLCs core further reduces drug leaching and the gelling tendency of SLNs (observed upon long-term storage). Both SLNs and NLCs are prepared via techniques like micro emulsification, sonication, or high-pressure homogenization (Kumar 2019).

Liposomes are tiny spherical structures made up of lipids, including phospholipids and cholesterol, with an inner aqueous core. They form through the self-assembly of lipids in water. Hydrophilic molecules are trapped inside the core, while hydrophobic drugs are incorporated into the lipid bilayer. Cationic phospholipids like Dioleoylphosphatidylethanolamine (DOPE) and Dioleoyl-3-trimethylammonium propane (DOTAP) are commonly used to carry biomolecules such as miRNA and siRNA, utilizing the charged interaction between their cationic head and the negatively charged phosphate backbone (Liu et al. 2020). The covalent linking of polyethylene glycol (PEG) chains with phospholipids (a process called PEGylation) is also used to improve stability and systemic circulation time (Mohamed et al. 2019). Replacement of phospholipids with non-ionic surfactants yields niosomes (possess improved stability and longer shelf life) (Chen et al. 2019), while the addition of edge activators (imparts flexibility which promotes tissue penetration) to conventional liposome constituents yield transfersomes (Opatha et al. 2020). While both liposome derivates have distinct benefits, their use pertaining to stem cell research is limited. Conventional liposomes, owing to their ease of preparation and broad scope of functionalization, are widely explored.

LNCs, the latest generation of lipid-based nanoparticles, are composed of an oily core surrounded by a solid lipid shell and emulsifying agents. They exhibit biodegradability and exceptional physical stability. The phase inversion temperature technique is commonly used to create LNCs, but new low-temperature methods have emerged for thermolabile molecules, offering precise size control (20–100 nm) (Yadava et al. 2020; Basu et al. 2021). Initially, the use of LNCs was confined to the delivery of hydrophobic payload (loaded within the oily core), but over recent years, hybrid LNCs with charge-baring lipid surface (for loading nucleic acid) and aqueous core derivatives (for hydrophilic payload) have also been reported (Lagarce and Passirani 2016; Vrignaud et al. 2013).

Since all lipid-based nanoparticles have low mechanical properties, any direct effect of physical interaction with stem cells is insignificant. Despite this, lipid-based nanoparticles have enjoyed great success as carriers of drugs/genes for direct stem cell modulation (alone or as part of a scaffold-like system) and tracking.

11.3.1.2 Polymeric Nanoparticles

Polymers are chain-like structures formed by linking repeating building blocks (monomeric subunits). By using different monomers, polymer-based nanosystems can be created with specific properties for various applications. Over the years, advancements in polymer chemistry have discovered natural and synthetic polymers suitable for preparing polymeric nanoparticles. These nanoparticles enable controlled release of therapeutic or diagnostic agents, offering protection against degradation and off-target delivery (Desai et al. 2021). Natural polymers derived from animals or plants are biocompatible and non-toxic, but they lack consistency in molecular weight and properties, resulting in poor repeatability for nanoparticle preparation. Examples include hyaluronic acid, cellulose, alginate, chitosan, albumin, collagen, and silk fibroin. On the other hand, synthetic polymers produced from petroleum-based materials provide tunable physical and chemical properties at the expense of overall biocompatibility. Common synthetic polymers used for nanoparticle fabrication include poly(lactic acid), poly (lactic-co-glycolic acid), poly(ε-caprolactone), and poly(ethylenimine) (Jarai et al. 2020; Desai et al. 2023).

In biomedical applications, common polymer-based morphologies include nanoshells, nanospheres, polyplexes, polymersomes, and dendrimers. Nanoshells are polymeric vesicles with a core-shell structure. The core contains the drug, trapped within a liquid (oil or water), and is surrounded by a solid polymeric shell. Diblock polymers that undergo self-assembly are preferred for preparing nanoshells (Ahmadi and Arami 2014). If the polymer forms a continuous spherical matrix wherein the drugs are loaded via physical entrapment or covalent bonding, the system is called a nanosphere. Nanoparticles of unmodified natural polymers usually fall under this category (Begines et al. 2020). Polyplexes are a self-assembled layered complex of negatively charged nucleic acid (mostly siRNA) that is condensed within a polymer having cationic functional groups. This arrangement protects nucleic acids from enzymatic degradation when administered systemically and further facilitates cargo release to tumor sites via "endosomal escape" (Uchida and Kataoka 2019). Poly(ethyleneimine) and its modifications are commonly used for the formation of polyplexes. Polymersomes can be considered polymer-based counterparts to liposomes. Instead of amphiphilic lipids, they are formed using diblock- or triblock-copolymers that form vesicles in an effort to minimize their aqueous interaction (Pawar et al. 2013). Dendrimers are more stable than liposomes and offer superior protection for sensitive payloads. They are easily modified for targeted delivery or triggered release. Dendrimers are well-defined globular polymers with a hyperbranched structure, characterized by their small size (1–100 nm) and low polydispersity index. They consist of an initiator core, hyperbranched dendron layers called "generations," a terminal functional group, and void spaces (Patel et al. 2020). As branching increases, the spherical shape and ability to accommodate hydrophobic substances also increase. This is interesting because the hydrophilic surface of dendrimers allows interaction with water-soluble substances and improves solubility. Dendrimers like polyamidoamine, poly(propylene imine), poly-L-lysine, melamine, and poly (ester amine) are commonly used in biomedical applications as carriers for drugs, therapeutic nucleic acids, and imaging agents (Nikzamir et al. 2021).

Since polymeric particles can be easily functionalized to modulate their immunogenicity, biodistribution, and cell internalization, they have been widely explored in stem cell research to deliver drugs, genes, biological factors, and cell-tracking agents.

11.3.2 Inorganic Nanoparticles

11.3.2.1 Metal Nanoparticles

Colloidal science advancements have enabled the creation of diverse metal and metal oxide nanoparticles. These nanostructures offer flexibility in regulating essential properties such as shape, size, structure, assembly, and optics through different synthesis methods. Metallic nanoparticles are prepared using chemical, electrochemical, and photochemical techniques (Jamkhande et al. 2019). Nanoparticles made from metals like gold (Au), and silver (Ag), as well as metal oxides like iron oxide, zinc oxide (ZnO), titanium dioxide (TiO_2), copper oxide (CuO), cerium oxide (CeO_2) have been reported as a carrier to deliver therapeutic agents (especially nucleic acids) using appropriate surface modification strategies (coating with surfactants or polymers) (Illath et al. 2021).

The major biomedical application of metal-based nanoparticles in stem cell research is their utilization for labeling/tracking. After internalization by stem cells, unique optical and magnetic properties allow their use as "external labels or probes" owing to the phenomenon of magnetic resonance (MR) (Liang et al. 2013). MR imaging (MRI) allows for non-invasive visualization of cellular structures in great detail. It can be used to study the behavior and distribution of transplanted stem cells in the local microenvironment. Metal-based nanoparticles used in MRI need to be magnetic. These nanoparticles can be classified as T1 or T2 agents based on their relaxation process, providing contrasts in T1-weighted or T2-weighted MRI, respectively. Gadolinium (Gd) and Gd oxide nanoparticles are commonly used as T1 contrast agents. Enhancements in functionality have been achieved through chelation with lipid nanoparticles and dextran (Han et al. 2019a). T2 contrast agents include superparamagnetic iron oxide nanoparticles (SPIONs) and bimetallic ferrite nanoparticles. SPIONs has received regulatory approval for clinical applications. Approaches like surface coating and modification of surface functional groups are used to deal with in vivo toxicity (Dulińska-Litewka et al. 2019). Metal nanoclusters are a new type of luminescent nanomaterial used for diagnostics. They consist of a few hundred atoms and have similar properties to molecules, approaching the fermi-wavelength of electrons in size. These nanoclusters are being extensively studied as biological labels due to their photoluminescence, photostability, and biocompatibility, either on their own or when incorporated into biofunctionalized scaffolds (Zhang and Wang 2014).

Multiple independent studies have reported that metal nanoparticles can regulate the fate of stem cells. Gold (Au), silver (Ag), and zinc oxide (ZnO) nanoparticles have been found to facilitate phenotypic differentiation in stem cells by modifying signaling pathways following their internalization (Abdal Dayem et al. 2018). Like other nanomaterials, the biological nature of stem cell interaction is directly dependent on overall size, morphology, surface chemistry, and concentration of metal nanoparticles.

11.3.2.2 Mesoporous Silica Nanoparticles

Mesoporous silica nanoparticles (MSNs) are silica-based nanostructures with a well-defined arrangement of mesopores, resulting in a large surface area (>1000 m^2/g). In recent years, extensive research has focused on exploring the biomedical potential of MSNs, which exhibit superior biocompatibility, high stability, biodegradability, and tunable pore size (2 to 50 nm). MSNs can be synthesized using template-directed, sol-gel, microwave-assisted, or chemical etching methods. Particle size and pore volume are key variables that can be controlled by adjusting the silica source and reaction parameters (pH, temperature, surfactant concentration) (Narayan et al. 2018).

MSNs have been successfully employed as carriers for loading various cargo ranging from drugs to macromolecules such as proteins, DNA, and RNA. The pore size can be tuned depending on the type of cargo and the desired release profile. Generally, MSNs with large pore diameters (up to 50 nm) are preferred to facilitate the delivery of macromolecules, whereas small pore size is preferred to achieve a controlled release profile of drug molecules (Tu et al. 2016). Additionally, different pore morphologies (hexagonal, cubic, concentric, radial) and textures have been reported. Controlling pore morphology offers an opportunity to influence release kinetics. Due to the silicon dioxide composition of MSNs, they can undergo hydrolytic breakdown when exposed to physiological fluids, resulting in the formation of ortho-silicic acid, which is eliminated through urine. Surface modification is unnecessary to enhance cytocompatibility, but functionalization strategies can be used to achieve targeted nanoparticle delivery or drug release triggered by external stimuli (Lin et al. 2012).

Stem cell research utilizes MSNs for controlled delivery and imaging. MSNs have a unique role as regulators of morphological and physicochemical characteristics in stem cells. They interact with the cell membrane and are taken up through endocytosis. Studies have demonstrated that internalized MSNs promote temporary osteogenesis by triggering actin polymerization and activating RhoA, a GTP-bound protein (Huang et al. 2008). MSNs are also used as contrast agents for stem cell imaging. While blank MSNs can serve as an effective tool, the incorporation of suitable dyes or smaller magnetic nanoparticles within porous matrix grants the ability to use them in optical, ultrasound, and MR-based imaging (Cha and Kim 2019).

11.3.2.3 Quantum Dots

Quantum dots (QDs) can be described as radiant nanosized particles or nanocrystals composed of semiconducting materials having dimensions in the range of a few nanometers (roughly 10–50 atoms). They differ from other nanomaterials on account of their distinct size-dependent photophysical properties attributed to the quantum size effect. If QDs are exposed to their band gap energy (an amount of energy sufficient to excite an electron from a lower electronic band into a higher energy level), the subsequent removal of energy leads to the relaxation of excited electrons to the ground state which is associated with residual energy emission in the form of fluorescence (Padilha et al. 2011). Semiconducting nanomaterials with a size range of 2–10 nm, similar to the bulk Bohr exciton radius, exhibit a prominent phenomenon. These nanomaterials, known as quantum dots (QDs), are widely used in biomedical applications due to their outstanding photostability, broad absorption spectra, high extinction coefficient, and adjustable emission wavelengths (Kargozar et al. 2020).

For biomedical applications, the bottom-up colloidal synthesis approach is preferred to generate highly dispersed QDs. QDs composed of cadmium (Cd)-, indium (In)-, and lead (Pb) have been synthesized using this approach. Other fabrication approaches consist of biomolecule- or polymer-templated synthesis, microwave-assisted synthesis, cation-exchange-based synthesis, and some less preferred top-down approaches (like molecular beam epitaxy and electron beam lithography) (Wagner et al. 2019). Initially, a combination of elements from groups III–V and II–VI were explored for QDs fabrication, but with time QDs with fluorescence emissions ranging from UV to red in the visible region have been designed using "non-traditional" sources like Au/Ag/Cu cluster, graphene oxide, and silicon (He and Ma 2014). More recently, luciferase-conjugated self-illuminating QDs have been reported wherein the emission can be induced by the addition of luciferin (based on bioluminescence resonance energy transfer). Water solubilization and stability can be enhanced by encapsulating QDs in hydrophilic or amphiphilic polymers (Tsuboi and Jin 2018).

QDs are excellent for labeling and imaging stem cells in regenerative medicine due to their high sensitivity compared to other methods. They can be easily modified with biomolecules through covalent and non-covalent bonds, using chemical groups like carboxyl, amine, and thiol, as well as electrostatics, metal chelation, and biotin-avidin binding. Functionalized QDs allow for precise tracking of individual particles within stem cells. To achieve targeted bioimaging, specific ligands that bind to stem cell receptors or intracellular targets can be used (Yukawa and Baba 2017). Cytotoxicity is a major concern in the clinical translation of QDs, as prolonged accumulation within cells or in the vicinity of targeted tissue can lead to heavy metal and free radical/reactive species-induced damage. To mitigate this, novel synthesis approaches are being developed to yield QDs with refined physiological properties (surface charge, hydrophilicity, material stiffness) that aid in limiting unwanted cell interactions (Winnik 2013).

11.4 1D NANOMATERIALS

11.4.1 NANOFIBERS

Nanofibers, with diameters ranging from 50 to 1000 nm, are a versatile type of one-dimensional polymeric nanomaterials. They have gained significant attention in tissue engineering and stem cell research for their ability to mimic the ECM. Nanofibers function as artificial ECMs, providing a scaffold that replicates intercellular reactions and promotes intracellular responses. Their small size enables direct contact with cells, aiding in spatial orientation and playing a crucial role in the success or failure of implanted bioengineered scaffolds (Nemati et al. 2019). Additionally, nanofibers possess a high surface-to-volume and interconnective porous structure that helps in cell adhesion and cellular signal/nutrient exchange, respectively (Gao et al. 2019).

Various biocompatible and biodegradable polymers, both natural and synthetic, have been studied for creating nanofibers. Natural polymers, including sodium alginate, hyaluronic acid, chitosan, gelatin, elastin, and fibrinogen, as well as synthetic polymers like poly(glycolic acid) (PGA), poly(L-lactide) (PLLA), polycaprolactone (PCL), and polydioxanone (PDO), have been investigated. However, natural polymers have limitations such as high viscosity, weak mechanical properties, and fast enzymatic degradation. Synthetic polymers, on the other hand, can be toxic in high doses. To overcome these challenges, a combination of different polymers, such as natural-natural, natural-synthetic, and synthetic-synthetic blends, can be utilized (Gunn and Zhang 2010). Nanofibers can be produced using electrospinning, phase separation, and self-assembly techniques. Electrospinning involves charging and ejecting a polymer melt/solution through a syringe tip under high-voltage electricity. The fibers solidify into filaments upon a collector. Coaxial electrospinning is an advanced method that incorporates a "core" polymer inside a "shell" polymer, enabling controlled release and modulation of degradation pathways (Xue et al. 2019). Phase separation comprises dissolution, gelation, and solvent-based extraction of polymeric blends, followed by their freezing and drying, which results in the formation of porous nanofibers (Shao et al. 2012). Lastly, self-assembly techniques involve the unprompted organization of smaller fiber-like fragments (not more than 100 nm) into an ordered and stable structure with pre-programmed non-covalent bonds. This method is mainly applicable to copolymers based on peptide-amphiphiles (Palmer and Stupp 2008).

Nanofibers provide numerous advantages in clinical applications, such as biocompatibility, biodegradability, efficient drug/cell entrapment, physiological stability, and convenient sterilization. They offer the remarkable ability to control their mechanical properties precisely. When nanofibers are combined with stem cells, their modulus can be adjusted to match the specific tissue being targeted for implantation. This allows for safeguarding cells against compressive forces while enabling biomechanical signaling (Xie et al. 2020a).

11.4.2 CARBON NANOTUBES

Among carbon-based nanomaterials, carbon nanotubes (CNTs) possess a unique combination of mechanical, electrical, and optical properties in addition to the possibility of loading them with different molecular compounds. Since their first report in 1991 by Iijima and coworkers, CNTs have evolved as promising nanomaterials for biomedical applications, especially in stem cell research (Saliev 2019). Also called buckytube, CNTs are nanoscale hollow cylindrical molecules that consist of rolled-up sheets of single-layer carbon atoms. Here, each atom is linked to three adjacent atoms (as in graphite), resulting in a characteristic hexagonal structure and π electrons conjugation (sp^2 bonding). CNTs can be open-ended or capped (with half of a fullerene molecule). They possess a high aspect ratio with a diameter of a few nanometers and length in the order of several micrometers. The carbon backbone can be additionally functionalized with various non-covalent and covalent modifications. Based on structure, CNTs can be classified as single-walled carbon nanotubes (SWCNTs) and multi-walled carbon nanotubes (MWCNTs) (Hasnain and Nayak 2019). SWCNTs comprise a single graphene cylinder, while MWCNTs consist of at least two coaxial cylinders that enclose a hollow core.

There are three main methods to synthesize CNT_S, namely arc discharge, laser ablation, and chemical vapor deposition. All methods use a source of carbon and energy to fabricate CNTs. In the arc discharge method, carbon electrodes are expulsed with an electric discharge (50–100 A, 20 V) in the presence of inert gas and low pressure. It produces exceptionally high temperatures that vaporize the surface of carbon electrodes and generate a rod-shaped deposit of nanotubes. Pure carbon electrodes form MWCNTs, while metal-doped electrodes form SWCNTs (Arora and Sharma 2014). The concept is exactly similar for the laser ablation approach wherein the energy source is a high-intensity laser that is pulsed repeatedly. Chemical vapor deposition is a more widely used method for the large-scale production of CNTs. Here, hydrocarbons (like CH_4, acetylene, or carbon monoxide) are decomposed at high temperatures using specific catalysts to generate CNTs with a high yield (Kumar and Ando 2010). Following synthesis, impurities can be removed via acid treatment, magnetic purification, or size-exclusion chromatography (Eatemadi et al. 2014).

The size and morphological structure of CNTs is similar to those of many ECM proteins, such as collagen and laminin. Thus, their use as scaffold promotes cell growth. When used as a material for biomedical devices or implants, the flexible yet high mechanical strength provides CNTs the ability to preserve the structural integrity of scaffolds (Prajapati et al. 2022). CNTs have a high aspect ratio, allowing them to carry multiple biomolecules. Through surface functionalization, they can target and cross cell membranes without causing significant harm. This makes them valuable for labeling and tracking cells. CNTs also possess excellent electrical conductivity, making them efficient for differentiating stem cells into electro-active cells like muscle, cardiac, and neural cells. They can conduct electrical currents when integrated into tissue engineering scaffolds. Furthermore, CNTs can modulate various cellular behaviors in electro-active cells, including alignment, differentiation, metabolic activity, and protein synthesis, through external electrical stimulation (Stout and Webster 2012).

11.5 2D NANOMATERIALS

11.5.1 Graphene (and Its Derivatives)

Graphene is a widely used crystalline carbon allotrope consisting of a monolayer of sp^2 bonded atoms organized in a tightly packed two-dimensional honeycomb-like (hexagonal) lattice nanostructure. Here, each carbon atom has three σ-bonds and an out-of-plane π-bond that allows binding with adjacent atoms, resulting in a sheet-like construct. Ever since its discovery and characterization in 2004 by Geim et al. (awarded the 2010 Nobel Prize in Physics), graphene has been at the focal point of material science research (Geim and Novoselov 2007). As a biomaterial, it possesses extraordinary mechanical, physicochemical, thermal, and optical properties, which can be attributed to its strong planar C-C bonding, aromatic structural backbone, and presence of free π electrons along with reactive sites for surface reactions (Han et al. 2019b).

Using appropriate chemical and physical modifications, graphene sheets can be transformed into materials such as bi-/multilayered graphene, graphene oxide (GO), and reduced GO (rGO). Together these materials are categorized as graphene family nanomaterials (Ghosal et al. 2021). Single, bi-, and multilayered graphene are synthesized by repeated mechanical exfoliation or extremely controlled growth on substrates like silicon carbide via chemical vapor deposition (Saeed et al. 2020). GO is a highly oxidized form of graphene, consisting of a single-atom-thick layer with carboxylic, epoxide, and hydroxyl modifications. These modifications allow for weak interactions, such as hydrogen bonding, and other surface reactions. Multilayered GO is produced from crystalline graphite through oxidation and dispersion in water. Through treatment with reducing agents like hydrazine, thermal, chemical, or UV, rGO is formed, which restores electrical conductivity and optical absorbance while reducing oxygen content, surface charge, and hydrophilicity (Tarcan et al. 2020).

Graphene's planar 2D structure and large surface area enable interactions with organic molecules. Its abundance of free π electrons results in excellent thermal and electric conductivity. Due to high light transmittance and photoluminescent capabilities, graphene-based nanomaterials are extensively studied for biomedical imaging (Qu et al. 2018). Like other carbon-based nanomaterials, graphene is non-biodegradable and leads to dose-dependent accumulation in the lung and liver over long periods of time, causing severe pulmonary and hepatotoxicity. Graphene and GO demonstrate minimal biological degradation, while rGO undergoes degradation under appropriate circumstances (Orsu and Koyyada 2020). Surface modification strategies such as PEGylation can greatly improve the biocompatibility of graphene and its derivatives. This enhances their systemic clearance and reduces organ-specific localization. Graphene family nanomaterials have the potential to serve as non-cytotoxic, transferable, and implantable platforms for stem cell culture. They promote stem cell attachment, growth, and directed differentiation. Graphene, in particular, can accelerate the differentiation of mesenchymal stem cells into bone cells, even in the absence of BMP-2. Additionally, graphene preferentially induces the differentiation of neural stem cells into functional neurons rather than glial cells (Halim et al. 2018).

11.5.2 Black Phosphorous

Black phosphorus (BP) is a thermodynamically stable allotrope of phosphorus comprising multiple layers having 2D structures, weakly bonded to each other by means of van der Waals interactions. When separated, the resulting monolayer (or few-layer) material is known as phosphorene. In monolayer form, each phosphorus atom is covalently bonded with three adjacent atoms, forming a bilayer structure along the zigzag direction and a wrinkled structure along the armchair direction (Jing et al. 2015). Attributable to its distinctive structure, BP possesses unique properties and has sparked great interest within the scientific community for utilization in biomedicine and theranostics (Choi et al. 2018).

P.W. Bridgman synthesized bulk BP in 1914 from white phosphorus under high pressure and temperature. Although its electrical and optical properties were studied in the 1980s, it wasn't until 2014 that monolayer BP gained significant attention as a potential substitute for graphene in the realm of thin-layer 2D biomaterials (Anju et al. 2019). Monolayer BP production can be divided into top-down and bottom-up methods. Mechanical and liquid-phase stripping are top-down approaches, involving exfoliation from bulk BP. Chemical vapor deposition and wet-chemical methods are bottom-up approaches, utilizing precursor deposition and vapor-solid transformation (Luo et al. 2019).

Amongst other 2D nanomaterials, BP has exceptional physicochemical, electronic, and optical properties. It exhibits a wide and tunable (layer thickness dependent) band gap that allows broad absorption/interaction with electromagnetic waves (from ultraviolet to mid-infrared regions). Such unique optical property allows BP to be effectively used in biosensing, photoacoustic imaging, photothermal therapy, and drug delivery (Chen et al. 2020a). Photothermal therapy refers to the use of electromagnetic radiation for the treatment of various medical conditions, including cancer. It requires a photosensitizer that can locally convert optical energy (from a source like a laser) to thermal energy via photoexcitation. BP has a high photothermal conversion efficiency due to its large extinction coefficient and strong nonradiative relaxation derived from phonon interactions in BP crystals (Xie et al. 2020b). Additionally, BP exhibits relatively low cytotoxicity and undergoes rapid biodegradation inside the body producing non-toxic and biocompatible intermediates like phosphates and phosphites (Ma et al. 2021).

BP has been widely explored for its ability to induce neuronal and osteogenic differentiation of stem cells. When irradiated with near infrared (NIR) light, BP can provide mild heat stimulation (even when covered by thick biological tissues), leading to an upregulation in the expressions of heat shock proteins that can promote osteogenesis. At a cellular level, BP provides an excellent microenvironment for osteogenic cell differentiation. When used as a cell scaffold, BP can promote the biological mineralization of calcium phosphate crystals through degradation products (Huang et al. 2018). Very recently, it was reported that BP could effectively improve the antioxidant capacity of stem cells and protect stem cells from oxidative stress-induced cell damage (He et al. 2023). By acting as a neuroprotective nanomedicine, BP can be explored in the management of neurodegenerative disorders as it can capture excess transition metals like copper (prevalent in Alzheimer's and Parkinson's disease) (Chen et al. 2018). Additionally, it is an excellent choice for neural stem cell research as it stimulates axon and myelin sheath formation in a calcium-dependent manner (Peng et al. 2020).

11.5.3 Transition Metal Dichalcogenides

Transition metal dichalcogenides (TMDCs) are an emerging category of 2D nanomaterials with fascinating properties appropriate for various biomedical applications. They can be described as layered materials made up of stacked planar crystals with a chemical representation of MX_2, where M and X represent the central transition metal atom (belonging to group 4–7; e.g., Mo, W, Nb, Ti) and the surrounding chalcogen atoms (e.g., Mo, S, Se), respectively (Manzeli et al. 2017). Chalcogens bind to the central transition metal atom in a characteristic trigonal prismatic arrangement. Covalent bonding forms a stable hexagonal lattice within a layer, while weak van der Waals forces hold adjacent layers together. Prominent TMDCs under active study include MoS_2, WS_2, WSe_2, and $MoSe_2$. The electrical characteristics of these materials are influenced by the occupancy of outer d-orbitals. Partially filled d-orbitals lead to metal-like conductivity, while fully occupied d-orbitals result in semiconducting properties (Kuc et al. 2015). TMDCs are notable for their adjustable electronic band gap, which increases as their thickness decreases. Unlike graphene, TMDC-based biosensors offer greater sensitivity by achieving a larger current on-off ratio, thus enhancing the signal-to-noise ratio (Presutti et al. 2022). In monolayer form, TMDCs display impressive photoluminescence. Moreover, their electronic band structure and extinction coefficient enable strong optical absorption in the NIR region, making them effective for photothermal applications (Zhu et al. 2020).

MoS_2 and WS_2 are the only naturally occurring TMDCs. Synthesis of TMDCs can be achieved through both top-down and bottom-up methods. Bottom-up methods, such as chemical vapor transport with suitable precursors, are used to grow bulk crystals. Top-down methods involve exfoliation techniques like micromechanical exfoliation, hydrothermal liquid exfoliation, and etching, which aim to create monolayer derivatives by disrupting the weak inter-planar Van der Waals forces binding adjacent layers (Anju and Mohanan 2021).

TMDCs have a high surface-to-volume ratio, making it easy to functionalize them through non-covalent or covalent methods. Polymers such as PEG, PLGA, PLA, and pluronic can be used to improve colloidal stability and circulation time in the body. Thiol groups can be added to enhance antimicrobial properties. TMDCs can also be combined with active biomolecules like peptides, nucleic acids, antibodies, and aptamers to enhance their inherent properties (Li and Wong 2017). Although TMDCs display negligible cytotoxicity, they can adhere to serum proteins in circulation and provoke oxidative stress-mediated through redox reactions inside cells; hence appropriate measures must be taken before biological use (Teo et al. 2014).

TMDC nanosheets, such as MoS_2 and WS_2, can be transformed into quantum dots for cell imaging without requiring external fluorescent labels. These quantum dots emit durable fluorescence with minimal photobleaching. TMDCs are commonly utilized as 2D scaffolds, either alone or in conjunction with polymers or ceramics, to control stem cell fate in tissue engineering studies (Liu et al. 2021). In bone regeneration, they serve as ideal reinforcing agents (due to their excellent mechanical and chemical properties) while also displaying positive biological effects on bone stem cells. Studies have reported that MoS_2 scaffolds, due to their surface roughness, can promote the cell attachment, proliferation, and differentiation of pre-osteoblasts and rat bone marrow mesenchymal stem cells (Zhang et al. 2018a). Whole-transcriptome sequencing of human stem cells showed that MoS_2-induced heat under NIR irradiation could upregulate the expression of cellular adhesion-related genes. Considering this, TMDCs-based materials have been utilized for the osteogenic as well as cardiac differentiation of human induced pluripotent stem cells (Nazari et al. 2020).

11.5.4 Laponite

The use of phyllosilicate clay minerals as bioactive agents by humans dates back to prehistoric times (roughly 2500 BC). Owing to their favorable physicochemical properties, low cost, and wide availability, phyllosilicate clay minerals are often used in numerous pharmaceutical and cosmetic products. The smectite group of phyllosilicate clay minerals grants them high sorption capability, swelling capacity, and functional surface area (Khurana et al. 2015). Natural clay minerals differ in composition and properties due to their geological source, making them unsuitable for advanced biomedical applications like regenerative medicine and tissue engineering. To overcome this, synthetic mineral clays have been created with customizable purity, composition, and crystal dimensions as an alternative (Peña-Parás et al. 2018).

One such synthetic smectite alternative is Laponite (commercially available as a trademark product of BYK Additives & Instruments). It can be described as 2D nanosized crystals of synthetic clay mineral that belongs to the smectite group of phyllosilicates, characterized by 2:1 crystal composed of layered units in which two tetrahedral silica sheets sandwich one Mg^{2+}-containing octahedral sheet (Dawson and Oreffo 2013). The nanocrystals are disk-shaped and have an empirical formula of $Na^{+}_{0.7}[(Si_8Mg_{5.5}Li_{0.3})O_{20}(OH)_4]^{-0.7}$. The face of the Laponite disks bears a net negative charge which is counter-balanced by the positive charge of the sodium ions and adjacent water molecules. This results in a unique electrostatic interaction that promotes the formation of crystal stacks that exist in dry powder form (Das et al. 2019). Laponite, with its large surface area and charged properties, interacts with biomolecules and shows potential for sustained drug delivery. By combining with various polymers, both synthetic and natural, it forms biomaterials with shear-thinning properties (Gaharwar et al. 2019).

Degradation and cytotoxic behavior are key characteristics of Laponite for its usage in biomedical applications. It undergoes deterioration at the physiological acidic pH by releasing degradation products such as aqueous silica, sodium, magnesium, and lithium ions (Jatav and Joshi 2014). Laponite, at concentrations below 1 mg/mL, shows no toxicity to human mesenchymal stem cells. It can be functionalized using covalent or non-covalent methods. Covalent functionalization involves reacting with alkoxysilanes, utilizing silanol groups on Laponite edges. This method enables the creation of inorganic/organic hybrid nanomaterials and Laponite fringed structures by grafting polymer chains onto the clay's edges. Additionally, employing organic bridges to connect adjacent particles covalently can produce gel-like materials (Massaro et al. 2020). On the other hand, electrostatic interactions are primarily the basis for non-covalent functionalization. Laponite gels with superior mechanical strength have been reported using non-covalent functionalization (Wang et al. 2010).

Laponite induces osteogenic differentiation in various types of cells, including pre-osteoblasts, hMSCs, and human adipose-derived stem cells, without the need for growth factors. It enhances alkaline phosphatase activity, stimulates the expression of osteo-specific genes, and promotes the production of Collagen I (Mihaila et al. 2014). It is hypothesized that osteogenic differentiation results from the dissociation of Laponite into its individual ions (Li^{+}, Mg^{2+}, and $Si(OH)_4$). In a comprehensive transcriptomic analysis-based study, it was observed that Laponite could upregulate the expression of chondro-related genes such as cartilage oligomeric matrix protein and aggrecan. Significant production of chondrogenic ECM, such as glycosaminoglycans, was also observed (Carrow et al. 2018).

11.6 3D NANOSTRUCTURES

From the perspective of stem cell research, 3D nanostructures refer to the 3D spatial constructs of 1D (such as nanofibers and nanotubes) and/or 2D nanomaterials (such as nanosheets). These 3D nanostructures are characterized by hierarchical material arrangement and nanoscale features such as nano-porosity and surface nano-topology (e.g., groves, roughness) (Kerativitayanan et al. 2015). The physicochemical attributes of 3D nanostructures can be thought of as an extension of their individual constituents. This facilitates researchers to design unique combinations of 1D/2D nanomaterials that provide synergistic advantages of constituents. Examples under this category include but are not limited to polymeric hydrogels, bioactive ceramic-based implants, organic-inorganic hybrid constructs, and multi-component nanocomposites.

11.7 STEM CELLS AND NANO-BIOMATERIAL INTERACTIONS

Recent developments in nanotechnology have enabled the scientists to artificially create the native microenvironment for studying the stem cell-nanomaterial interactions. The interactions of stem cells with nanostructured materials are being explored from past two decades. Many studies have reported the effect of nanomaterials on stem cell differentiation, proliferation and behavior. This section discusses the effects of surface and structural properties of the nanomaterials on the stem cell behavior.

11.8 SURFACE PROPERTIES

To mimic the native stem cell niche, myriad of nanomaterials has been employed to direct the stem cell proliferation (Chueng et al. 2016). The surface properties of nanomaterials, along with the functionalizing molecules present, impact how nanomaterials will work with stem cells. Many studies have proved that stem cells can sense and respond to the microenvironment using their nanolength projections called nanopodia. Stem cells recognize and respond to the chemical and physical surface characteristics, as well. In biomaterial design for cell culture, tissue engineering, and regenerative medicine, the regulation of cell adhesion behavior affects cell migration, proliferation, and differentiation (Wei et al. 2017). Academically, any study of cell–surface interactions would be simplified if only a single parameter at any given time point was examined, but this may not be readily apparent in the literature due to the tendency toward more complex system designs of biomaterials.

The properties of nanomaterials that affect stem cell behavior are surface chemistry, topography, surface pattern, surface charge, and hydrophilicity. Surface modifications can be achieved by introducing new chemical groups like amine, acetyl, or silane, or by modifying existing groups through oxidation and reduction reactions. Surface chemistry of nanomaterials is mainly controlled by the chemical groups present. Chemical modification techniques, such as self-assembled monolayers (SAMs), can create a one molecule thick layer to study the molecular interactions between stem cells and nanomaterials (Stenger et al. 1992). A recent study introduced a hydrogel embedded with poly(lactic-co-glycolic acid) microspheres, coated with carbon nanotubes (CNTs) to imitate the neural stem cell environment. The presence of PLGA-CNT microspheres increased the porosity of the hydrogel and significantly improved stem cell adhesion on their surface (Shafiee et al. 2021).

Cell adhesion, proliferation, and osteogenic differentiation of mesenchymal stem cells were promoted when the surface was functionalized with an amino group (Kuddannaya et al. 2013).

Ma et al. improved polymeric nanofibers (PLGA-PCL) by adding hematite nanoparticles. This enhanced the adhesion and osteoconductivity of adipose-derived stem cells (ADSCs). The nanoparticles increased surface roughness, making the nanofibers more hydrophilic and promoting cell adhesion through protein adsorption. Additionally, the coated nanofibers improved stiffness and regulated mechanical properties, enhancing their osteoinductive properties (Ma et al. 2019). Another way to modify the surface of the nanomaterials is coating the nanomaterial with extracellular proteins like fibronectin, laminin, collagen, fibrin and elastin to enhance cell attachment and proliferation (Hung et al. 2021; Kam et al. 2009). Integrin is an important cell membrane protein which senses the changes in the physical micro-environment of the cells (Lv et al. 2015; Cary et al. 1999). The cytoskeletal fibers attach the nuclear membrane of the stem cells to integrin-based focal adhesion which helps in sensing and transmitting the physical stimuli that directs the shape of the nucleus and expression n of proteins on nuclear membrane. These integrin-mediated interactions impact the cell shape, morphology, proliferation, cytoskeletal reorganization, motility, and differentiation of the stem cells (Park et al. 2012; Dalby et al. 2014).

Gold nanoparticles (AuNPs) alter the morphology of MSCs, causing them to flatten and adhere through filopodia formation. Additionally, AuNPs serve as activating molecules for the MAP kinase pathway, which responds to mechanical stress and alters the phenotype of MSCs. Previous studies investigating the mechanism of osteoblastic differentiation by AuNPs have found that unmodified AuNPs induce differentiation through the p38 MAPK pathway, while chitosan-conjugated AuNPs activate the Wnt/β-catenin pathway and inhibit adipogenic differentiation in hADMSCs (Choi et al. 2015; Yi et al. 2010).

Li et al. performed a study to investigate the effect of surface charge of chemically functionalized AuNPs on osteogenic differentiating ability of mesenchymal stem cells. The surface of AuNPs was modified with amine (AuNP-NH_2), carboxyl (AuNP-COOH), and hydroxyl (AuNP-OH) groups. AuNP-NH_2 group showed positive surface charge (36.74 ± 2.35) while other functional groups showed negative surface charge but showed no important difference in osteogenic differentiation. The cellular uptake of AuNP-COOH was highest than other modified nanoparticles but inhibited the activity of alkaline phosphatase and matrix mineralization which indicated that AuNP-COOH group was more suitable for chondrogenic differentiation (Figure 11.2A) (Kawazoe and Chen 2015). The interaction of low oxygen functionalized graphene with MSCs resulted in increased expression of integrin heterodiamers without any chemical induction. These MSCs showed higher concentrations of bone specific ECM proteins such as collagen I, fibronectin and others, which assisted in faster osteogenic differentiation (Newby et al. 2020).

Topography influences stem cell behavior and fate, while changes in nanomaterial surface chemistry and topography affect wetting and interactions with ECM proteins. By modifying hydrogels made with poly(2-hydroxyethyl methacrylate), different surface wrinkle sizes and shapes were achieved, altering the morphology of cultured hMSCs. Cells on a lamellar surface adopted a similar shape and differentiated into osteogenic cells, while cells on a hexagonal surface exhibited a rounded morphology and underwent adipogenic differentiation (Figure 11.2B) (Guvendiren and Burdick 2010). Yu et al. conducted an experiment comparing chemically modified surfaces' effects on adhesion and differentiation of human dental pulp stem cells (hDPSCs). They created nanofilms of titanium (10 nm) and gold (40 nm) and incorporated amino, carboxyl, hydroxyl, and methyl groups using a self-assembling monolayer technique. The presence of different chemical groups influenced cell viability, morphology, and density. Amino group surfaces resulted in flattened, multilayered cell growth resembling osteoblast-like morphology. Methyl and hydroxyl group surfaces promoted clustered, elongated cell growth. Carboxyl group surfaces exhibited rounded cells with enhanced attachment. The highest expression levels of odontoblast-specific markers DSPP and DMP-1, confirming hDPSCs' differentiation into odontogenic cells, were observed on surfaces modified with amino groups (Figure 11.2C)

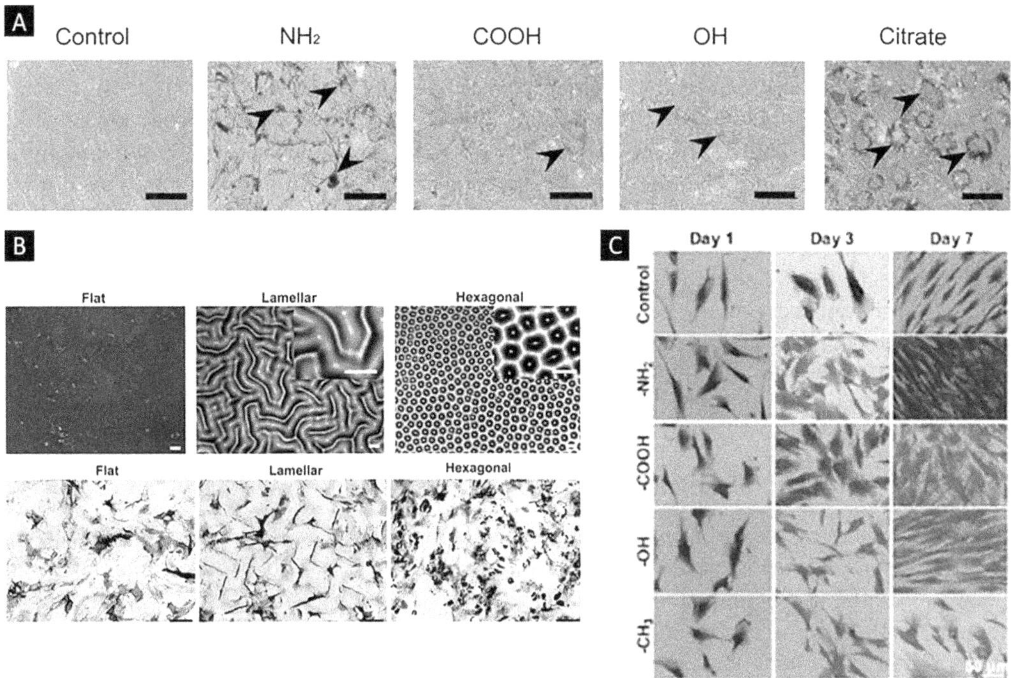

FIGURE 11.2 Effect of surface properties of nanomaterials on stem cell proliferation and differentiation. A) shows the effect of chemically modified gold nanoparticles on osteogenic differentiation of hMSCs at day 21 after culturing in osteogenic induction medium (Scale bar = 100 μm). Reprinted with permission from (Kawazoe and Chen 2015) (Copyright 2015, Elsevier). B) shows the effect of different hydrogel surface topography on stem cell morphology. The osteogenic differentiation was found to be ~91% on lamellar surface, ~74% on flat surface, and ~61% on hexagonal surface hydrogel (Scale bars = 50 and 100 μm). Reprinted with permission from (Guvendiren and Burdick 2010) (Copyright 2010, Elsevier). C) shows the microscopic images indicating the morphological effect on different chemical functional groups of self-assembled layers on hDPSCs. Reprinted with permission from (Yu et al. 2017) (Copyright 2017, American Chemical Society).

(Yu et al. 2017). The behavior of stem cells can be controlled by providing specific biophysical cues. When MSCs were cultured on a surface with random circular nanostructures instead of a smooth surface, using a non-differentiating media, their osteogenic differentiation significantly increased (Dalby et al. 2007). An experiment was conducted to analyze how surface wettability affects the proliferation of BMSCs. A gradient wettable surface was created using corona discharge on a polyethylene sheet. The water contact angle varied across different points of the modified surface, ranging from 45 to 98°. This variation influenced the attachment, morphology, and viability of the stem cells cultured on the surface. Cells on the hydrophobic surface exhibited a rounded morphology, while those on the hydrophilic surface appeared more flattened. The study found that BMSCs showed higher rates of proliferation and viability when cultured on intermediate hydrophilic surfaces (Shin et al. 2008).

Mechanical pressures can have both positive and negative effects on stem cell proliferation and destruction. Positive effects include activating signaling pathways and transcriptional regulation, leading to increased stem cell proliferation and tissue healing. Studies have shown that mechanical stretching promotes stem cell multiplication in cardiac cells, bone marrow, skin, and cartilage (Han et al. 2020). Excessive mechanical stress can trigger programmed cell death (apoptosis) in stem cells and generate reactive oxygen species that damage DNA and other cellular components, resulting in cell death. The impact of mechanical forces on stem cells is complex and varies depending on the type and intensity of the forces, as well as the specific stem cell type and tissue context (Mayr et al. 2002). With the exception of neurons, most cell types adhere and spread better on stiff surfaces (Georges et al. 2006). The process by which stem cells respond to the rigidity and stiffness of the biomaterial is called mechanotransduction (Discher et al. 2005; Engler et al. 2006). The cytoskeletal molecules which are involved in the mechanosensation are microtubules, integrins, fibronectin, f-actin, microfilaments, focal adhesion proteins, nucleoskeletal related proteins, and cadherins (Caille et al. 2002; Naqvi et al. 2020). To comprehend the mechanotransduction pathways through which cortical stiffness of the cells is regulated which in turn affects the cell proliferation and differentiation, hMSCs were cultured on the polyacrylamide gel of different rigidity. When cultured on a soft gel (rigidity: 1–2 kPa) hMSCs could not spread and adhere and also displayed a rounded morphology while on a stiffer gel (rigidity: >5 kPa) the cells were spread in a larger area with more number of stress fibers (Tee et al. 2011).

11.9 STRUCTURAL PROPERTIES

In tissue engineering, the architecture of the biomaterials plays an essential role in regulating the function of stem cells. It is possible that the appropriate selection of designed scaffolds cannot only offer a biocompatible environment in which cells can grow, but also provide an efficient method for the regulation of the function of stem cells (Cha et al. 2012). The incorporation of growth factors into scaffolds is one method for achieving this regulation. However, their use can lead to systemic side effects, especially at high doses. For instance, bone morphogenetic proteins (BMPs) at large doses may cause adverse outcomes like immunogenic response, soft-tissue edema, and abnormal bone growth. Consequently, researchers have focused on studying the biomechanical and physical characteristics of biomaterial scaffolds, including elasticity, porosity, degradation, and size. Nanostructured materials have been extensively researched for their ability to regulate stem cell differentiation and proliferation. The transmembrane receptors on stem cells can detect and respond to changes in the extracellular matrix (Argentati et al. 2019). The elastic modulus of the matrix developed to induce differentiation of the stem cells should match the native values of elastic modulus of the targeted cells (Flanagan et al. 2002; Engler et al. 2004). When MSCs were cultured on a gel system with tunable elastic modulus similar to brain tissue (0.1–1 kPa), muscle tissues (8–17 kPa), and bone tissues (25–40 kPa), were differentiated into neurons, myoblasts, and osteoblasts, respectively (Figure 11.3A) (Engler et al. 2006). Nanopillar array of different designs and polymeric compositions are studied by researchers for their effect on stem cell differentiation, especially osteogenic differentiation (Dalby et al. 2007; Wang et al. 2013a; Bucaro et al. 2012; Ahn et al. 2014). Despite of these experiments it still remains unclear that how only nanotopography will affect the stem cell differentiation and to study this, Zhang et al. developed a nanoarray of PLA pillars coated with anodic aluminum oxide. hADSCs were cultured on the surface of these nanopillars with different diameters (100, 200, and 300 nm) where the cells cultured on the array of pillars having diameter 200 nm showed the morphological change as well as enhanced osteogenesis (Figure 11.3B) (Zhang et al. 2018b). A study examined how the diameter of TiO_2 nanotubes affected the attachment of rat MSCs. The cells exhibited different growth patterns on surfaces formed by self-assembled nanotubes with diameters ranging from 15–100 nm. Tubes with a diameter of 15 nm showed increased integrin clustering, while cells seeded on tubes larger than 50 nm displayed reduced cell activity and higher programmed cell death. The researchers also investigated the response of MSCs on super-hydrophobic surfaces and found that surface wettability influenced cell attachment, which could be controlled by changing the nanotube diameter. Initially, cell attachment was significant, but cells were unable to remain attached for more than 3 days (Figure 11.3C) (Bauer et al. 2008; Park et al. 2007).

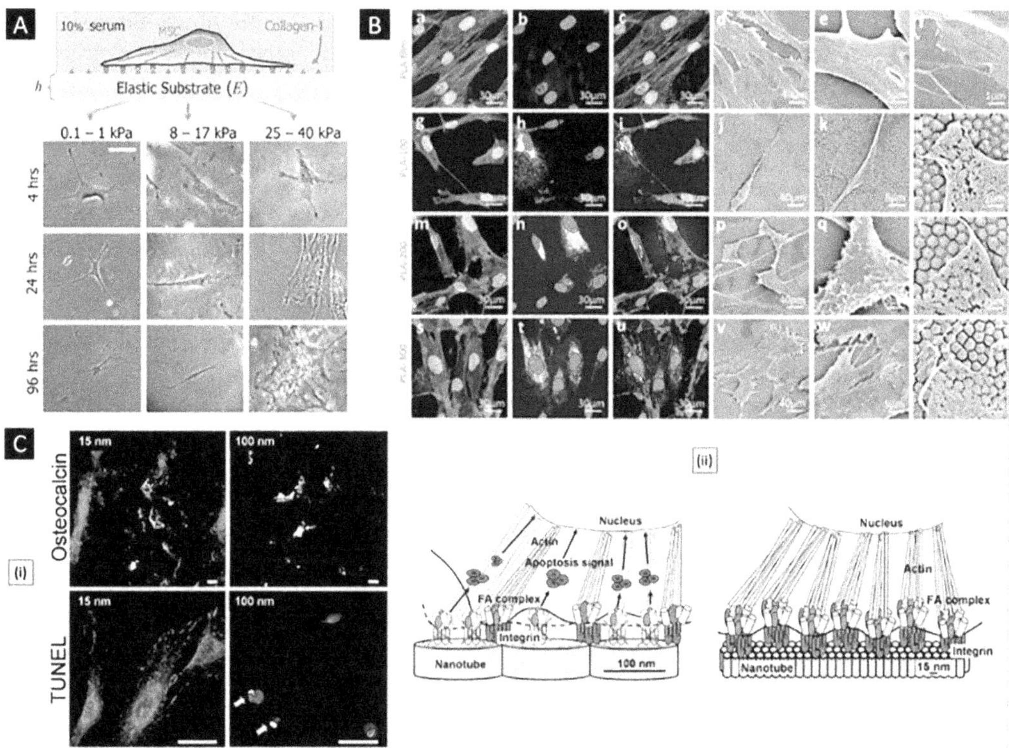

FIGURE 11.3 Effect of nanotopographical properties on stem cell differentiation. A) shows the stiffness dependent differentiation of stem cells into neuronal cells (0.1–1 kPa), myoblasts (8–17 kPa), and osteogenic cells (25–40 kPa) (Scale bar = 20 μm). Reprinted with permission from (Engler et al. 2006) (Copyright 2006, Elsevier). B) shows the morphological features of the stem cells cultured on the surfaces of different nanotopographical array of PLA nanopillars (Scale bar = 30 and 40 μm). Reprinted with permission from (Zhang et al. 2018b) (Copyright 2018, American Chemical Society). C) (i) shows the formation of focal adhesion on the TiO_2 nanotubes and microscopic images denoting the formation of actin filament of the stem cells (Scale bar = 20 and 100 μm). (ii) Hypothetical model showing the lateral spacing of focal contacts on nanotubes of different diameters. Reprinted with permission from (Park et al. 2007) (Copyright 2007, American Chemical Society).

The physical properties of materials used greatly influence the control of stem cell destiny. Porous biomaterials play a vital role in regenerative biomaterials by facilitating revascularization and allowing blood vessels to infiltrate without extensive matrix breakdown (Xiao et al. 2015; Tsou et al. 2016). To investigate the influence of pore size and porosity on stem cell differentiation, using a scaffold with a gradient pore size is advantageous. In particular, PCL scaffold with a porosity ranging from 80% to 97% along the cylindrical axis exhibited distinct effects on the chondrogenic differentiation of adipose stem cells. Notably, scaffolds with 96% porosity demonstrated the highest levels of chondrogenic differentiation and extracellular matrix synthesis (Oh et al. 2010). A study conducted to determine the optimal pore sizes required in tissue regeneration showed that the optimal pore sizes in regeneration correlate as follows: 5 μm for neovascularization, 20–125 μm for regeneration of adult mammalian skin, 100–350 μm for regeneration of bone, and 20 μm for the ingrowth of hepatocytes. In addition, while working with central nervous system tissue engineering, the usage of a macroporous scaffold that has pores larger than 40 μm is required (Murphy et al. 2014; Li et al. 2012).

The size of nanoparticles is considered as an important parameter which affects stem cell differentiation. Smaller nanoparticles (30–70 nm) tend to be more effective at inducing stem cell differentiation, as they can more easily enter cells and interact with intracellular signaling pathways (Li et al. 2016; Ko et al. 2015). On the other hand, larger nanoparticles are less cytotoxic but not be as effective at inducing stem cell differentiation, as they may have difficulty entering cells or may be less able to interact with intracellular signaling pathways (Wei et al. 2017; Nel et al. 2006; Rivera-Gil et al. 2013). There is growing evidence that cells respond to their mechanical environment. On a more rigid matrix, cells proliferate and migrate toward the region with the highest modulus. An experiment conducted to evaluate the effect of stiffness on stem cell proliferation and differentiation when cultured on RGD nanopatterned hydrogel with two different stiffness values (130 and 3170 kPa), concluded that the stiff hydrogel promoted osteogenesis while the soft hydrogel favored the adipogenesis (Ye et al. 2016). The substrate stiffness also controls the adhesion, tyrosine signaling, and proliferation of fibroblasts, smooth muscle cells, and chondrocytes.

The degradation of polymers and the byproducts of the degradation can have a significant effect on stem cell proliferation (Peng et al. 2018). Various factors affect polymer degradation, such as polymer type, size, shape, and processing conditions. Certain polymers are designed to degrade slowly for better biocompatibility, while others are engineered to degrade rapidly. Controlling the degradation rate enables synchronization with tissue repair and regeneration (Zhang et al. 2014). Local degradability of the polymer matrix is required for the cytoskeletal rearrangement of the cells which will direct the stem cells to undergo osteogenesis while adipogenesis is favored when the cytoskeletal rearrangements are restricted with non-degradable or slow degrading polymers (Khetan et al. 2013).

11.10 APPLICATIONS OF NANOMATERIALS IN STEM CELL RESEARCH

11.10.1 Stem Cell Labeling and Tracking

Stem cell therapies show promise as alternative treatments for various diseases and infections. Factors such as integration with host cells, microenvironment, migration, proliferation, and differentiation influence their success (Srijaya et al. 2014). Preclinical studies and clinical trials are assessing the safety and potential of stem cell-based therapies (Trounson and McDonald 2015; Trounson et al. 2011; Tompkins et al. 2018; Lopes et al. 2018). Histological techniques are used in preclinical studies to evaluate the behavior of transplanted stem cells, but they are ineffective for monitoring stem cell activities in clinical trials. Non-invasive methods are needed to track transplanted stem cells and understand their distribution and delivery at the targeted site after implantation in the host (Saw et al. 2015).

Various strategies for labeling stem cells can be categorized into indirect and direct techniques. Indirect labeling methods encompass receptor-based and reporter gene labeling, while direct labeling involves the use of fluorescent probes, superparamagnetic iron oxide, contrast agents, cell penetrating peptides, and microencapsulation with contrast agents (Fu and Kraitchman 2010). When receptor-labeled stem cells are transplanted, monitoring their activities becomes challenging due to the loss of markers after phenotypical changes and differentiation. Reporter gene labeling, although effective for longer tracking, has limitations such as immunogenicity, malignancy, and cellular dysfunction. On the other hand, direct labeling techniques involve introducing labeling agents into the stem cells before transplantation. The ideal agent should possess specificity, biocompatibility, and enable real-time tracking and imaging without disturbing the cells' normal environment. This can be achieved using fluorescent probes, polymeric nanoparticles, metallic nanoparticles, quantum dots, or magnetic nanoparticles (Schäfer and Hemotherapy 2010).

Nanoparticles possess a crucial characteristic of small size, enabling them to enter cells and engage with internal molecules and structures. This quality renders nanoparticles highly suitable for labeling and tracking stem cells, providing a non-invasive and biocompatible approach. Quantum dots, gold nanoparticles, and magnetic nanoparticles are among the diverse types utilized for this purpose (Hachani et al. 2013; Bengel 2011). Quantum dots and gold nanoparticles are fluorescent semiconductor and metal particles, respectively, with applications in stem cell tracking. Quantum dots emit stable fluorescence when illuminated, while gold nanoparticles can be modified with biomolecules for targeted cell tracking. Magnetic nanoparticles, detectable through MRI, are also valuable for in vivo stem cell tracking (Liang et al. 2013).

The behavior of stem cells can be studied by using labeled nanoparticles to track their movement and differentiation in real-time. This provides valuable insights into how stem cells migrate, proliferate, and differentiate in different environments. Understanding these mechanisms is crucial for advancing stem cell-based therapies for tissue repair and regeneration. A new type of nanoparticle called quaternized carbon dots was developed to label mesenchymal stem cells (MSCs). Lee et al. developed bicyclo[6.1.0]nonyne-modified glycol chitosan nanoparticles (BCN-CNPs) as imageable nanoparticles. These nanoparticles can carry fluorescent molecules like Cy5.5 and be used for in vivo imaging of transplanted cells. Animals pretreated with Ac4ManNA retained fluorescence longer and showed a 15-fold increase in fluorescence compared to the untreated group. In Figure 11.4A, the near infrared fluorescence images after 15 days of sub-cutaneous delivery of tagged stem cells are presented (Lee et al. 2017). In Figure 11.4B, fluorescence images of mice and their organs after 24 hours of implantation demonstrate the effectiveness of this labeling system. Unlabeled stem cells could not be tracked using the imaging techniques (Malina et al. 2019). Stem cell labeling and tracking with nanoparticles is crucial for monitoring stem cell therapies in clinical trials. It

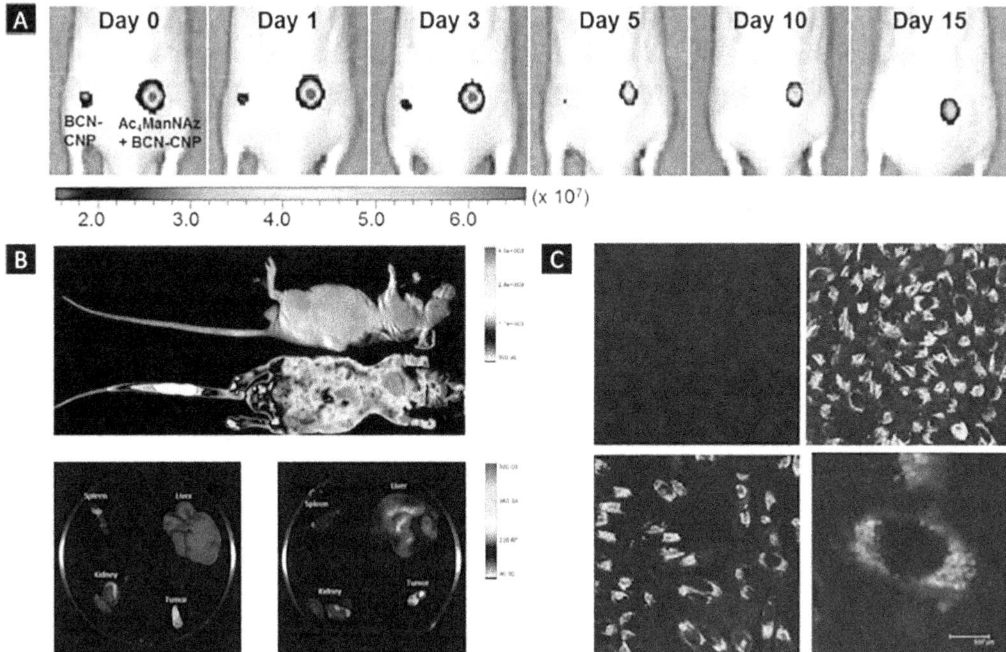

FIGURE 11.4 A) Shows the in vivo near infrared images indicating the retention of subcutaneously injected labeled stem cells. Reprinted with permission from (Lee et al. 2017) (Copyright 2017, Elsevier). B) Images indicating the labeled and unlabeled stem cells and their distribution in animal organs. Reprinted with permission from (Malina et al. 2019) (Copyright 2019, Elsevier). C) Shows the uptake of fluorescent quantum dots by the stem cells. Reprinted with permission from (Liu et al. 2015) (Copyright 2015, American Chemical Society).

allows researchers to observe the behavior and outcomes of transplanted stem cells, ensuring their safety and effectiveness. This information is vital for evaluating therapeutic effects and detecting any possible side effects.

Fluorescent probes containing chloromethyl and bromomethyl chromophores are utilized for stem cell labeling. FDA-approved fluorescent dyes such as Near Infrared fluorophores, Indocyanine green, Carboxyfluorescein Diacetate, Succinimidyl Ester (CM-DiI), and DAPI are commonly used for stem cell labeling and imaging. Quantum dots, carbon-based colloidal nanocrystals synthesized in the 1980s, have gained attention due to their biocompatibility, easy synthesis, surface modifications, and lack of photobleaching. Smaller quantum dots exhibit longer fluorescence. Figure 11.4C presents a fluorescence image of cells labeled with quantum dots (Liu et al. 2015). Ohyabu et al. labeled MSCs with an anti-mortalin peptide antibody attached to quantum dots. The antibody-quantum dot complex was effectively taken up and maintained by MSCs during their transformation into adipogenic, chondrogenic, and osteogenic cells (Ohyabu et al. 2009). Iron oxide nanoparticles, known as SPIONs, align themselves with an external magnetic field and return to their random motion when the field is removed. SPIONs are effective at low concentrations, reducing side effects. They act as negative contrast agents, producing dark images when the transverse relaxation time is shorter. Labeling stem cells with SPIONs allows non-invasive monitoring of migration patterns, as shown in a study where SPION-labeled stem cells were analyzed after 60 days of transplantation, demonstrating long-term efficacy. Researchers improved SPION internalization by modifying them with peptides, achieving up to 10–30 pg of iron per cell. Additionally, silica modification enabled SPIONs to function as both optical and magnetic contrast agents, facilitating bioconjugation and easy loading of dye molecules for optical imaging (Lu et al. 2007). Feridex and ferumoxytol are other examples of clinically approved iron oxide nanoparticles used for cell labeling and tracking. Ferumoxytol, when used to label the cardiac progenitor cells derived from embryonic stem cells showed the fluorescence up to 40 days after implanting it in animal model (Van de Walle et al. 2020). Gold nanoparticles (<100 nm) were used to label MSCs and tracked by using micro-CT, were able to track the stem cells up to 30 days of transplantation in sub-retinal layer of rat (Mok et al. 2017). Fluorescent polystyrene nanoparticles are widely used for stem cell labeling, and their surface can be modified with sulfate, aldehyde, carboxylate, or amine groups. However, these nanoparticles have limitations such as dye leaching and low dye loading. Firefli (Thermo Fisher) and FluoSpheres (Invitrogen) have addressed photobleaching issues by incorporating color internally. For MRI imaging, gadolinium ion complexed with diethylenetriaminepentacetate (DTPA) is commonly used as a contrast agent. However, the impact of gadolinium nanoparticles on cell viability and proliferation should be considered. BioPal has developed GadoCellTrack, a commercially available negative contrast agent (Wang et al. 2013b).

In summary, stem cell labeling and tracking using nanoparticles is a powerful tool for studying stem cell behavior and fate in vivo, and for monitoring the safety and efficacy of stem cell therapies. While there are still many challenges to be overcome, such as improving the stability and biocompatibility of nanoparticles, the potential of this approach is vast and holds promise for advancing the field of stem cell research and therapy.

11.11 STEM CELL-BASED REGENERATIVE MEDICINE

Tissue engineering has rapidly grown, offering solutions for tissue repair and regeneration. Researchers use the body's stem cells, which have the unique ability to differentiate and renew themselves, for this purpose. Integrating nanotechnology with stem cells has revolutionized regenerative medicine, providing potential cures for challenging disorders. This section explores the latest strategies that combine nanotechnology and stem cells to restore damaged tissues both structurally and functionally.

11.11.1 Bone Tissue

Bone is made up of four different cell types, namely, osteoblasts, osteocytes, osteoclasts, and bone lining cells. Its functions include providing structure, regulating mineral balance, protecting organs, enabling movement, and bearing weight. Additionally, bone tissues play a role in endocrine functions by controlling osteocalcin and fibroblast growth factor 23 levels. When subjected to minor injuries, bone tissues regenerate partially without forming scars in the healing process (Marsell and Einhorn 2011); but in case of injuries due to infection, tumor resection, fractures, and oral-maxillofacial surgeries, external medications are required to guide the regeneration of injured tissues (Alghazali et al. 2015). Current strategies for treating bone injuries involve surgical grafting and the use of scaffolds. However, surgical grafting carries risks such as hematoma, infections, and morbidity at the injury site. In order to address these drawbacks, extensive research has been conducted over the past two decades to develop long-term treatments for bone tissue regeneration (Dimitriou et al. 2011; Winkler et al. 2018). The main strategies involved in bone tissue repair and regeneration include biomaterials based scaffolds loaded with biochemical cues, stem cells, or both to guide the regeneration.

The ideal material for bone tissue regeneration should have excellent mechanical properties, biocompatibility, and the ability to promote cell adhesion. When selecting a scaffold material, important considerations are osteoconduction and osteoinduction. Osteoinduction refers to the material's ability to stimulate the differentiation of stem cells into chondrocytes and osteoblasts, while osteoconductivity refers to its ability to support cell proliferation and maintain vasculature. Osteoinductive materials are preferred because they attract progenitor cells, enhance their proliferation and differentiation, and serve as structural building blocks for new tissue formation (García-Gareta et al. 2015).

The scaffolds for bone regeneration can be made using natural polymers like chitosan, cellulose, collagen, hyaluronic acid, fibroin, alginate, gelatin, and silk, as well as synthetic polymers such as PCL, Poly(lactic acid) (PLA), PGA, Polypropylene fumarate (PPF), and Polyhydroxyalkanoates. Natural polymers are enzymatically degradable and biocompatible but have lower mechanical strength. Synthetic polymers can overcome these disadvantages by offering tailored structures, improved mechanical strength, controlled degradation rate, and reduced risk of microbial contamination (Sabir et al. 2009).

Collagen, being the major component of bone (90%) is one of the most studied natural polymers for its application in bone tissue regeneration. Being a polymer of natural origin, the cell adhesion is more favored with regulated cell behavior. The different types of biomaterials implied in tissue engineering include hydrogel, microspheres, porous scaffolds. It provides a 3D matrix for the cells which mimics the native extracellular microenvironment (Ferreira et al. 2012).

Gelatin, derived from porcine skin or bones, is a natural polymer with reversible solubility in water. It contains an amino acid sequence called arginine-glycine-aspartic acid (RGD), which helps cells interact with their surroundings. The mechanical strength of gelatin-based scaffolds depends on polymer concentration, crosslinking, and temperature. However, using gelatin alone as a biomaterial has drawbacks such as the need for cytotoxic chemical crosslinking and its unsuitability as a cell carrier due to its melting at body temperature. These issues can be addressed by modifying gelatin with methacrylate groups, resulting in GelMA, which has higher mechanical strength and mimics the native bone environment. Researchers have used microfluidics technology to create GelMA hydrogel microspheres, which provide a 3D matrix for cell growth. BMSCs and BMP-2 were loaded into these microspheres, enabling optimal support for the growth, proliferation, migration, and differentiation of the BMSCs into osteoblasts (Jaipan et al. 2017).

Hyaluronic acid, another polysaccharide polymer, mimics the original extracellular matrix and can be employed alone or with additional polymers. Hyaluronic acid hydrogels containing inorganic minerals like calcium phosphate, bioactive glass, and hydroxyapatite have improved mechanical characteristics. Methacrylate modification of hyaluronic acid (MeHA) tunes the degradation properties. This can also be used as bio-ink to synthesize the MSC-seeded 3D scaffolds (Ding et al. 2022).

Graphene-based nanomaterials have shown promise in bone regeneration due to their biocompatibility, osteogenic support, and low toxicity. Graphene can be modified by adding oxygen, resulting in oxygen-modified graphene nanoparticles (LOG). In a study by Elkhenany et al., LOG nanoparticles were combined with MSCs and implanted into rat tibial damage sites. After 45 days, the groups treated with MSCs and LOG nanoparticles exhibited improved mineralization and new bone formation for damage reconstruction (Figure 11.5A) (Elkhenany et al. 2017). Nanofiber membranes created through electrospinning technique have significant

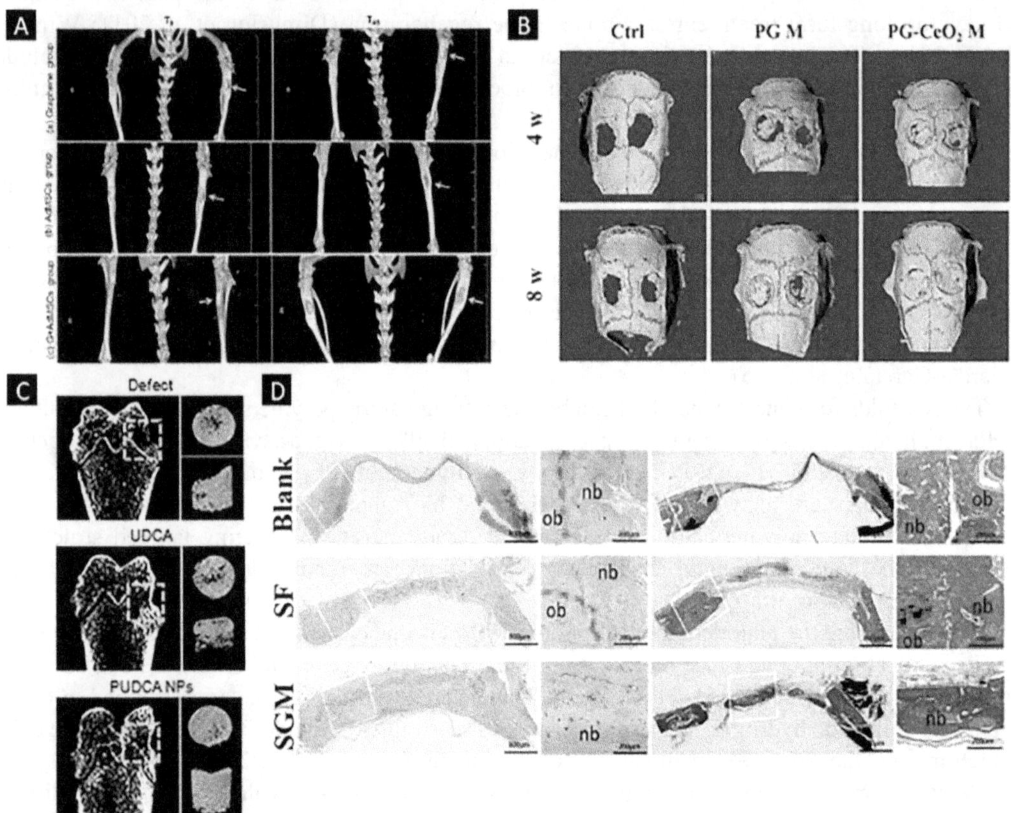

FIGURE 11.5 Indicates the new bone formation after using the stem cell-based nanotechnology. A) shows the micro-CT images of the long bone defects regenerated (at 45 days) after the treatment of graphene oxide and MSCs. Reprinted with permission from (Elkhenany et al. 2017) (Copyright 2017, Elsevier). B) shows the formation of new bone after 8 weeks of treatment in rat calvarial defect. The group treated with cerium oxide loaded nanofibers showed almost complete formation of new bone with enhanced mineralization. Reprinted with permission from (Ren et al. 2022) (Copyright 2022, Elsevier). C) shows the micro-CT images of the healing of long bone defects of rats which was confirmed by enhanced trabecular thickness. The antioxidant effect of UDCA nanoparticles helped in enhanced mineralization as well as fastened the new bone formation. Reprinted with permission from (Arai et al. 2020) (Copyright 2020, Elsevier). D) shows the H&E and Masson staining of the bone tissues regenerated after the nanofiber patch (silk fibroin patch and sandwich-nanofiber patch) implantation. The group treated with sandwich-like patch (SGM) showed the highest mineralization (Scale bar = 800 and 200 μm). Reprinted with permission from (Xiang et al. 2022) (Copyright 2022, Elsevier).

applications in delivering biomaterials for treating damaged areas. This technique allows the incorporation of biomolecules, nanoparticles, and clays, aiding tissue regeneration and enhancing mechanical strength. A study on bone regeneration utilized cerium oxide nanoparticles-loaded PCL-gelatin nanofibers. Stem cells cultured on these membranes demonstrated increased cell proliferation compared to polymeric nanofibers without nanoparticles. In an experimental calvarial defect model, the nanoparticle-loaded nanofiber membrane facilitated substantial recovery and formation of new bone within 8 weeks of implantation, as observed through micro-CT and Bone Volume/Tissue Volume analysis (Figure 11.5B) (Ren et al. 2022). Osteogenic differentiation of MSCs is negatively affected by the generation of reactive oxidative species (ROS) and the inflammatory cytokines present at the site of damage. To prevent this, ROS-responsive ursodeoxycholic acid (UDCA) nanoparticles were prepared and cultured with MSCs which prominently showed osteogenic differentiation than adipogenic differentiation. The collage sponges loaded with these nanoparticles and MSCs were surgically implanted at the defect site in long bones of rats. After 4 weeks of the implantation, the 1.42 times more trabecular thickness with improved mineralization and the formation of new bone was observed (Figure 11.5C) (Arai et al. 2020). A study found that a sandwich-like electrospun membrane, consisting of zinc oxide and hydroxyapatite, exhibited both osteogenic (bone-forming) and antibiotic properties. The membrane's nanofibers had a hydrophilic surface that facilitated cell attachment and promoted mineralization at the damaged site. H&E and Masson staining images demonstrated that the sandwich-like membrane significantly enhanced periodontal bone formation compared to membranes made solely from silk fibroin or a blank control group. This membrane structure created a sealed environment for bone regeneration, preventing the growth of epithelial connective tissues at the defect site (Figure 11.5D) (Xiang et al. 2022).

It's important to note that these are just a few examples of biomaterials that have been explored for use in stem cell-based bone tissue regeneration. There are many other polymers that have also been researched for this purpose, and the potential applications of these materials may depend on the specific properties of the polymer and the desired tissue regeneration.

11.11.2 Cardiac Tissue

Heart failure resulting from myocardial infarction (MI) and coronary artery disease is responsible for 29% of global deaths. MI, or heart attack, occurs when a coronary artery is blocked, leading to cell death and the formation of an infarcted area that impairs heart function. As individuals age, their heart tissues lose the ability to repair and regenerate themselves. Although new drugs and surgical interventions have enhanced patient outcomes, they cannot fully restore the function of damaged heart tissue (Sridhar et al. 2015). Among future therapeutic strategies for treating infarcted myocardium, tissue engineering shows the most promise. Key biological events for cardiac tissue regeneration include cardiac stem cell (CSC) recruitment, endogenous cardiomyocyte proliferation, neovascularization, and the presence of anti-inflammatory, anti-apoptotic, and anti-remodeling cytokines. The immune microenvironment, along with biochemical and biophysical cues, plays a crucial role in cell recruitment, proliferation, and differentiation for natural tissue restoration. Stem cells used in heart regeneration therapies include cardiac progenitor cells, pluripotent stem cells, mesenchymal stem cells, hematopoietic stem cells, bone marrow stromal cells, skeletal myoblasts, and others (Emmert 2017).

Injectable nanomaterials and nanostructured cardiac patches have reported the increase in cell retention after *in vivo* transplantation. The SPIONs can be used to deliver the stem cells at ischemic site using magnetic field which demonstrated the increased cell retention after implantation. Ferumoxytol, FDA-approved SPIONs, is used as a stem cell labeling agent as well as for the targeted delivery of cardiosphere-derived stem cells to repair the infarcted myocardium. A group of researchers investigated the effect of iron oxide nanoparticles modified with dimercaptosuccinic acid for their cardio protective effect. The developed nanoparticles

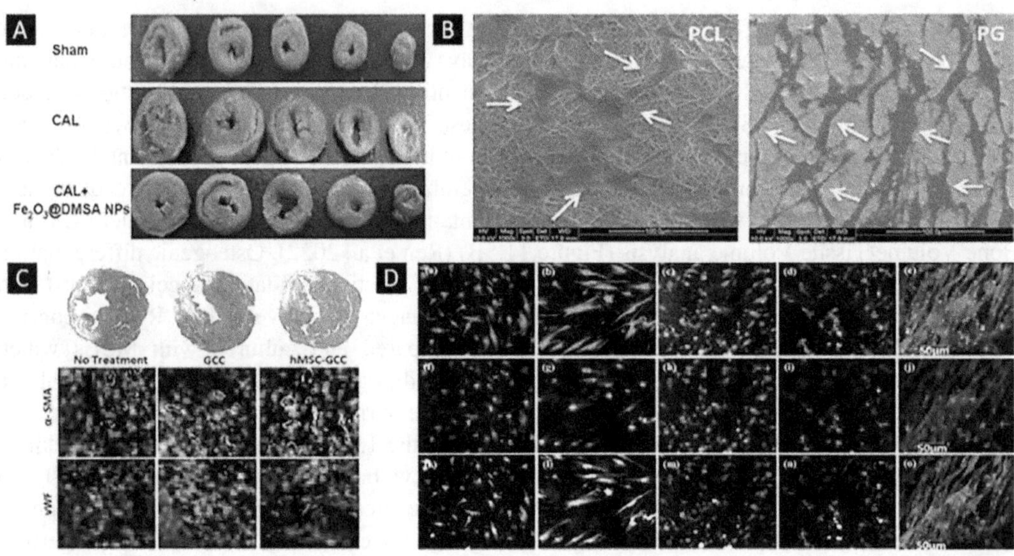

FIGURE 11.6 A) shows the reduced size of infarct in the heart after receiving the treatment of the iron oxide nanoparticles. The sections of heart were cut after 30 min of the induction of the infarct. Reprinted with permission from (Xiong et al. 2015) (Copyright 2015, Springer Nature). B) shows the SEM images of morphology and cardiomyocyte differentiation of the MSCs cultured on the nanofiber patches of PCL and PCL-Gelatin. The PCL-gelatin nanofiber patch offered the mechanical and conductive properties which guided the stem cell differentiation (Scale bar = 100 μm). Reprinted with permission from (Kai et al. 2014) (Copyright 2014, Elsevier). C) shows the reduced size of infarcts (sections taken on 28 days of implantation), the graphs indicate the decrease in collagen content, reduced area of infarct, ratio of heart weight and body weight, and the capillary formation after receiving the treatment. Reprinted with permission from (Si et al. 2020) (Copyright 2020, Elsevier). D) Shows the confocal microscopy images, (a-e) indicates the presence of MSC marker CD105(green) and (f-j) shows the presence of actinin (red), (k-o) shows the merged images, after 10 days of culture on the gold nanoparticles and vitamin B12 loaded PCL/Coll nanofibers (Scale bar = 50 μm). Reprinted with permission from (Sridhar et al. 2015) (Copyright 2015, Elsevier).

were administered to rats before induction of an ischemic injury. The group treated with nanoparticles showed the reduced ischemic lesion after 30 minutes of induction of injury (Figure 11.6A) (Xiong et al. 2015). The nanofiber patch prepared from PCL-gelatin acts as an excellent adherent surface which supported the angiogenesis as well as growth and differentiation of the MSCs in cardiomyocytes (Figure 11.6B). This patch of nanofibers was then loaded with rat MSCs and transplanted in myocardial infarction rat model. The cell-loaded patch shown the reduced size of infarct and promoted angiogenesis. The expansion of left ventricular wall of infarcted heart was decelerated after transplantation which eventually aided in cardiac repair (Kai et al. 2014).

Graphene oxide quantum dots offer several advantageous properties, including higher surface area, lower cytotoxicity, antioxidant activity, biocompatibility, optical activity, and biodegradability. In this study, stem cell-loaded graphene quantum were incorporated into a biodegradable hydrogel made of chitosan and collagen. The resulting injectable hydrogel was evaluated for its ability to protect the heart in a rat model of myocardial infarction. Figure 11.6C depicts sections of the infarcted heart after treatment with the hydrogel containing stem cells (hMSCs-GCC) and without stem cells (GCC). The group treated with hMSCs-GCC exhibited an increased cell density at the infarct site, along with evidence of capillary formation, indicating angiogenesis (Si et al. 2020).In a study on myocardial infarction, researchers implanted a bioengineered patch consisting of PCL-gelatin nanofibers loaded with MSCs. The implanted MSCs exhibited increased levels of CD31, indicating angiogenesis. The researchers developed a nanofiber patch using PCL and included gold nanoparticles (15–30 nm) to match the mechanical properties of the native myocardium. They also loaded vitamin B12 onto the PCL nanofibers to enhance hydrophilicity, regulate homocysteine levels, and promote the differentiation of stem cells into cardiomyocytes. Confocal microscopy images on day 10 of culturing showed the expression of CD105 marker (green) and cardiac-specific marker actinin (red) in the MSCs (Figure 11.6D) (Sridhar et al. 2015).

Researchers developed PCL nanofibers loaded with cardiac progenitor cells and CNTs to mimic human myocardium tissue. The incorporation of CNTs, chemically conjugated with thiophene group, improved the elasticity, mechanical strength, and mesh forming ability of PCL nanofibers. This prevented CNT aggregation and increased solubility. These modifications guided cardiac progenitor cells to form myocytes, strengthening the dilated heart wall and promoting new myocyte formation (Wickham et al. 2014). Ravichandran et al. conducted a study using albumin-coated gold nanoparticles in nanofibers to regenerate cardiac tissue. They used polyvinyl alcohol instead of organic solvents to prevent cytotoxicity. The nanofibers successfully induced cardiac differentiation in MSCs, as evidenced by the expression of cardiac proteins (Troponin-T, Actin, and Cx43 (Ravichandran et al. 2014).

11.11.3 Nervous Tissue

Human nervous system is divided in two parts, central nervous system (CNS) and peripheral nervous system (PNS). CNS mainly comprises of brain and spinal cord which are contained in the cranial cavity and vertebral column, respectively. The PNS is mainly composed of motor and sensory neurons which maintain the communication with CNS (Noback et al. 2005). Neuronal injury can result from neurodegenerative diseases, traumatic injury, and iatrogenic causes, leading to inflammation, cell death, protein degradation, and nervous system dysfunction. The central nervous system (CNS) has limited regenerative capacity, whereas the peripheral nervous system (PNS) can regenerate to some extent. Current therapies for CNS and PNS injuries mainly focus on addressing demyelination, inflammation, and neuronal apoptosis, but they do not effectively support neuronal regeneration (Rakhit et al. 2021). The discovery of stem cells revolutionized the previously impossible task of repairing the nervous system. Scientists have developed scaffolds using biomaterials and growth factors to guide the development of brain stem cells into specific cell types. By using oligodendrocytes derived from embryonic stem cells, researchers have successfully restored damaged myelin in the adult central nervous system. Transplanting oligodendrocyte progenitors from human embryonic stem cells into rats with spinal cord injuries proved to be a safe treatment and improved their ability to move. Optimal conditions for growing and differentiating embryonic stem cells in fibrin-based scaffolds have been identified for neural tissue engineering purposes (Subramanian et al. 2009).

Neurological tissue engineering relies on cytocompatibility, mechanical properties, and electrical characteristics. Inadequate cytocompatibility can hinder neuron growth, causing inflammation or infection. Insufficient mechanical characteristics can impede brain tissue regeneration. Conductive materials promote neuron proliferation, synapse formation, and neural regeneration. Improving electrical properties is crucial for initiating and regulating neuron function during electrical stimulation, aiding brain tissue recovery. Natural and synthetic polymers have been used as nerve grafts to bridge gaps and guide neuron sprouting (Zhang and Webster 2009). Because of the biocompatible, mechanical, and electrical conductive properties of the carbon nanotubes (CNTs), a group of researchers have developed the thin films using single-walled and multiple walled CNTs. The iPSCs showed enhanced cell adhesion, proliferation and neuronal regeneration when cultured on the thin films of the CNTs. The expression of the adhesion molecules was upregulated while of NES marker was reportedly downregulated indicating the formation of mature and healthy neurons (Figure 11.7A) (Mondal et al. 2022).

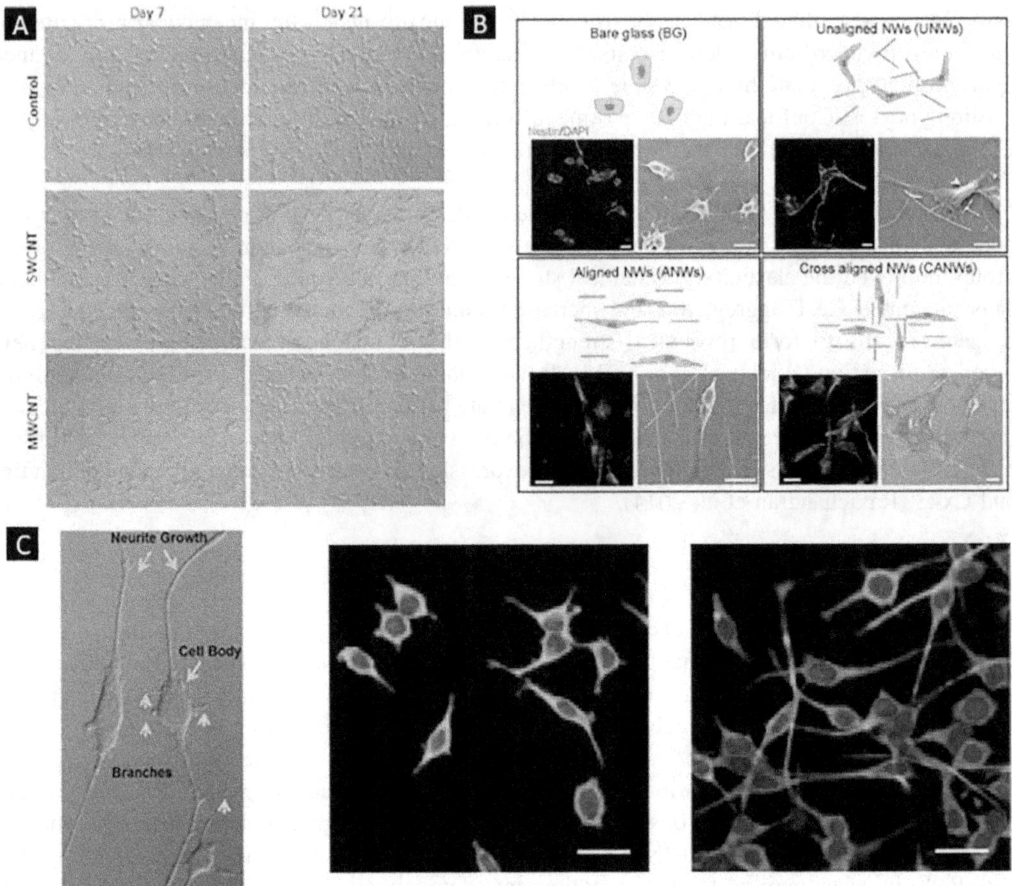

FIGURE 11.7 A) shows the differentiation of stem cells into neuronal cells on culturing with films of single-walled and multi-walled carbon nanotubes. The multi-walled carbon nanotubes showed the higher amount of neuron formation than the former one. Reprinted with permission from (Mondal et al. 2022) (Copyright 2022, Elsevier). B) shows the direction dependent differentiation of stem cells into neurons where the neurons grown on aligned nanowires demonstrated the longer microtubule formation as compared to the neurons differentiated on bare glass surface and unaligned nanowires (Fluorescence images: nestin (green) and DAPI (blue), scale bar = 20 μm; SEM Images, scale bar = 10 μm) Reprinted with permission from (Li et al. 2022) (Copyright 2022, Elsevier). C) shows the graphs indicating the increase in the number branch formation, neurite length, and area of cell body. Other two images show the PC-12 derived neurons stained with α-tubulin (Scale bar = 20 μm). Reprinted with permission from (Pradhan et al. 2018) (Copyright 2018, American Chemical Society).

Nanowires prepared by using brush shearing technology are used to check their application in neuronal regeneration by Li et al. The neuronal stem cells cultured on randomly oriented and aligned nanowires showed the direction dependent growth (Figure 11.7B). The polarity of neurons differentiated on such surface can easily be controlled along with the guided directional orientation of the nerve tissue (Li et al. 2022). A recent study utilized PCL nanowires coated with polypyrrole to enhance neuronal differentiation. These nanowires demonstrated optimal nanotopographical features and electrical conductivity. The differentiation of C17.2 cells into neurons was confirmed through increased expression of brain-derived neurotrophic factor, glial cell-derived neurotrophic factor, and nerve growth factor (Bechara et al. 2011). Owing to the properties such as electrical stimulation and photo stimulation, graphene-based nanomaterials are highly explored for their potential use in neuronal tissue regeneration. A study conducted by Akhavan et al. reported that the nanogrid of reduced graphene oxide (rGO) provided a surface for better neuronal stem cell attachment as compared to quartz. The morphology of the neuronal stem cells cultured on rGO surface was elongated and neurite outgrowth was observed (Akhavan and Ghaderi 2013). Pradhan et al. developed an injectable graphene oxide hydrogel functionalized with choline. Maintaining choline balance is crucial for restoring brain damage. The hydrogel was tested by culturing neuronal stem cells and evaluating the expression of the neuronal marker GAP43. Increased expression of neuronal markers at day 7 confirmed successful neuronal differentiation. Microtubule formation, important for neuronal regeneration and maintaining neuron structure, was confirmed by staining with anti-tubulin antibodies, showing significant growth of neurons with microtubules on day 7 of the culture (Figure 11.7C) (Pradhan et al. 2018; Dent and Baas 2014).Stem cell-based nanotechnology therapy holds potential, yet regulating cell growth and tissue formation remains challenging. Concerns such as carcinogenicity and ethical dilemmas surrounding embryonic stem cells hinder their widespread use. Additionally, a study revealed allodynia in the forepaws of rats treated with stem cell therapy to repair spinal cord injuries (Hofstetter et al. 2005). Therefore, to revolutionize the treatment of neurological disorders, comprehensive research is required to optimize the conditions for controlled and appropriate development of transplanted stem cells.

11.11.4 Skeletal Tissue

Skeletal muscles possess natural regenerative capabilities to repair damage caused by physical injuries and wear and tear (Bursac et al. 2015). Surgery and transplantation are the main treatments for severe muscle injuries, but they have drawbacks like immune rejection and cancer risk, limiting their use for large injuries. To overcome these challenges, innovative strategies in biomimetic engineering and stem cell therapy have emerged (Koike et al. 2022). Skeletal tissues consist of nanosized functional units that play a vital role in regenerating skeletal muscle tissues. Effective communication between stem cells is necessary for this process. Nanomaterials have significant potential in drug delivery and enhancing therapeutic applications. Various nano biomaterials, ranging from zero to three dimensions, are employed to maintain the anisotropic structure and microenvironment of skeletal tissue, overcoming biological challenges in muscle regeneration.

Nanoparticles promote skeletal tissue regeneration by activating p38 MAPK signaling pathways, leading to increased myoblast differentiation and proliferation. In an in vivo study, complete skeletal muscle regeneration was observed in a muscle defect. Another study showed that incorporating therapeutic proteins (IL-4 and IL-10) into nanoparticles improved muscle functions in muscular dystrophy (Lee et al. 2020). Muscle rupture leads to the generation of reactive oxygen species, resulting in increased oxygen consumption and tissue damage due to cellular imbalance. To reduce inflammation in chronic muscle injury, sodium diclofenac, an NSAID, is combined with gold nanoparticles. Animal research shows that this approach decreases pro-inflammatory cytokines (TNF-α, IL-1β, and IL-6), stimulates muscle repair by releasing growth factors (IL4 and IL10), and encourages macrophage (M2) polarization (dos Santos Haupenthal et al. 2020).

Nanomaterial composites, such as graphene oxide nanofibers and nanohybrid hydrogels, have been utilized to enhance therapeutic functions through a range of physical, chemical, and biological properties. A recent study successfully incorporated reduced graphene oxide (rGO) nanosheet into a nanohybrid hydrogel composed of alginate and PCL, cross-linked with Ca^{++} ions. This combination addressed the limitations of alginate, such as insufficient cell adhesion and structural stability. The conductive nature of graphene oxide enabled the resulting hydrogel to be suitable for electro-active skeletal tissues. In vitro experiments with C2C12 myoblasts demonstrated that the nanohybrid hydrogel exhibited non-toxic properties while promoting

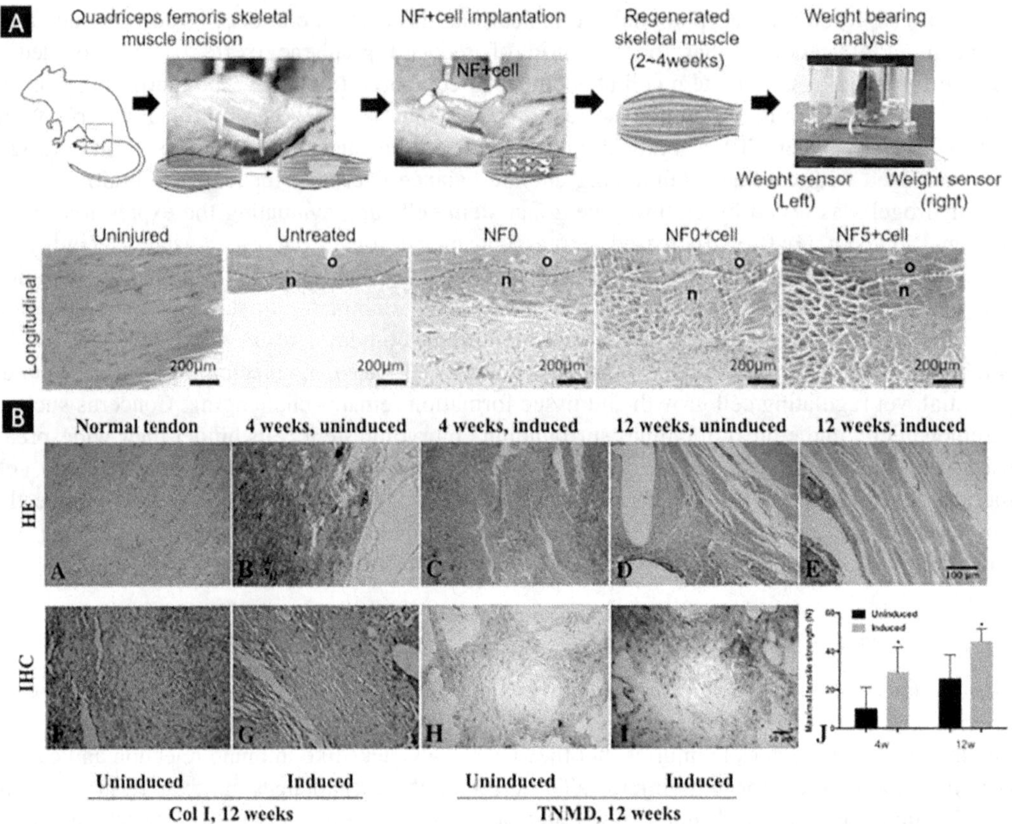

FIGURE 11.8 A) Indicates xenograft cell sheet that contains nanofibers and C2C12 in quadriceps femoris muscle. Image (a) shows the schematic demonstration of surgical process of recovery of muscle function and quadriceps femoris muscle incision was done on VML mouse model, to implant the nanofibers (PCL and the gelatin/PVA) +cells. After 4 week of implantation, skeletal tissue assessed by weight-bearing study, [(o) indicates the original tissue and (n) indicates the regenerated] (Scale bar = 200 μm). Reprinted with permission from (Pham-Nguyen et al. 2021) (Copyright 2021, wiley-VCH). B) H&E staining representation of tendon tissue in which (a) image is of normal tendon tissue, (b, c) after implantation at week 4 shows regenerative tissue in induced and uninduced (d,e) after implantation at week 12 shows the regenerative areas (f,g) Immunohistochemistry staining show collagen type I was observed in an induced group than in uninduced group, (h,i) Immunohistochemistry data of 12 weeks show maximum TNMD expressions in induced than in uninduced, (j) image show the maximum tensile strength. (Scale bar =) Reprinted with permission from (Chen et al. 2021) (Copyright 2021, Elsevier).

myoblast differentiation and proliferation (Aparicio-Collado et al. 2022). In a recent experiment, nanowires were combined with collagen-based bio-ink to 3D print C2C12 cells. This technique created an electric field on the scaffold, resulting in parallel alignment and enhanced differentiation of the cells. Furthermore, implanting the hydrogel into the VML model reduced fibrotic area and promoted muscle regeneration (Jeong et al. 2022). A xenograft cell sheet made of PCL/gelatin nanofibers containing C2C12 cells was implanted on the quadricep femoris skeletal muscle model through an incision. After 4 weeks, a weight-bearing analysis was conducted on the regenerated tissue. Histology images were used to distinguish between the original (o) and regenerated (n) areas, which were divided by a red dashed line. The myotubes in the regenerated area (n) showed better alignment of nuclei compared to the original area (o) (Figure 11.8A) (Pham-Nguyen et al. 2021).

Researchers have developed biomimetic nanomaterials using a PCL polymer and electrospinning technique to enhance tendon repair. These nanofiber yarns, combined with tendon stem/progenitor cells (TSPCs), demonstrated increased TSPC numbers on day 1. The nanofibers provided a favorable microenvironment for TSPCs, promoting their alignment and potentially increasing TNMD gene expressions. In a rat Achilles tendon model, the placement of 50 nanofiber yarns led to improved cell alignment, neo-collagen orientation, and enhanced tendon regeneration (Yang et al. 2022). In a study, adipose-derived stem cells were combined with growth differentiation factor-5 to create a sheet mixed with P(LLA-CA)/silk fibroin nanofibers, aiming to regenerate tendon tissue. A rabbit tendon defect model was used to evaluate the histology and biochemical properties. The results showed that the implanted tissue after 12 weeks had a higher extracellular matrix and fewer cells compared to 4 weeks in both induced and uninduced groups. The induced group exhibited more spindle-shaped cells and a tendon-like structure compared to the uninduced group. Immunohistochemistry staining revealed higher expression of TNMD gene, collagen type I, and tenomodulin in the induced group compared to the uninduced group. Additionally, the tensile strength of the adipose stem cell sheet was greater in the induced group than in the uninduced group. These findings demonstrate promising results for tendon regeneration (Figure 11.8B) (Chen et al. 2021). The integration of nanomaterials and bioengineering technologies is a promising approach for treating skeletal disorders.

11.11.5 Skin Tissue

The skin, the body's largest organ, acts as a protective barrier against chemicals, physical forces, and mechanical substances. However, it is vulnerable to severe injuries and diseases such as chronic wounds, diabetic wounds, and skin cancer. Wounds can be classified into two types based on their healing time: acute wounds and chronic wounds (Dreifke et al. 2015). Acute injuries such as abrasions, bruises, and lacerations generally take 8–12 weeks to heal, while chronic injuries like surgical wounds or diabetes type II require more than 12 weeks or months. Various methods and treatments are available for different types of wounds, such as fat grafting, leech therapy, shock wave therapy, intralesional cryotherapy, platelet-rich plasma gels, and wound dressings. However, delayed wound healing can lead to chronicity and increased risk of infection, causing various health complications (Sethuram and Thomas 2023).

Angiogenesis, the formation of blood vessels, is crucial in wound healing as it provides blood flow, oxygen, and nutrients necessary for the development of granulation tissues. Insufficient blood vessel formation can lead to chronic wounds. The four stages of wound repair include hemostasis, inflammation, proliferation, and remodeling of the extracellular matrix (Goonoo and Bhaw-Luximon 2020). The main aim of wound healing is to mimic the healing process with less scar formation at injured site.

Biomaterials and stem cell therapy are vital in tissue engineering to address challenges. Stem cell transplantation is an efficient method for skin regeneration, where adipose-derived and bone marrow-derived stem cells are commonly used. Hair follicle bulge stem cells aid tissue regeneration and promote hair growth after injury. These multipotent cells have high proliferation efficiency, modulate immune response, and reduce pro-inflammatory cytokine release. Mesenchymal stem cells (MSCs) secrete TGF beta, which enhances burn wound healing and cell signaling pathways. Recent research demonstrates that local injection of MSCs in a mouse burn wound model improves neoangiogenesis and burn wound healing (Xue et al. 2013). Fibrinogen nanofibers show great potential for wound healing by promoting blood clot formation, fibroblast migration, and cell adhesion. When 3T3 fibroblasts were incorporated into nanofibrous fibrinogen, they exhibited different morphologies within 24–72 hours. Both fibrinogen scaffolds showed strong proliferation with higher cell density. Nanofibrous fibrinogen facilitated fibroblast growth and close contact, evident from autofluorescence, while planar samples did not exhibit

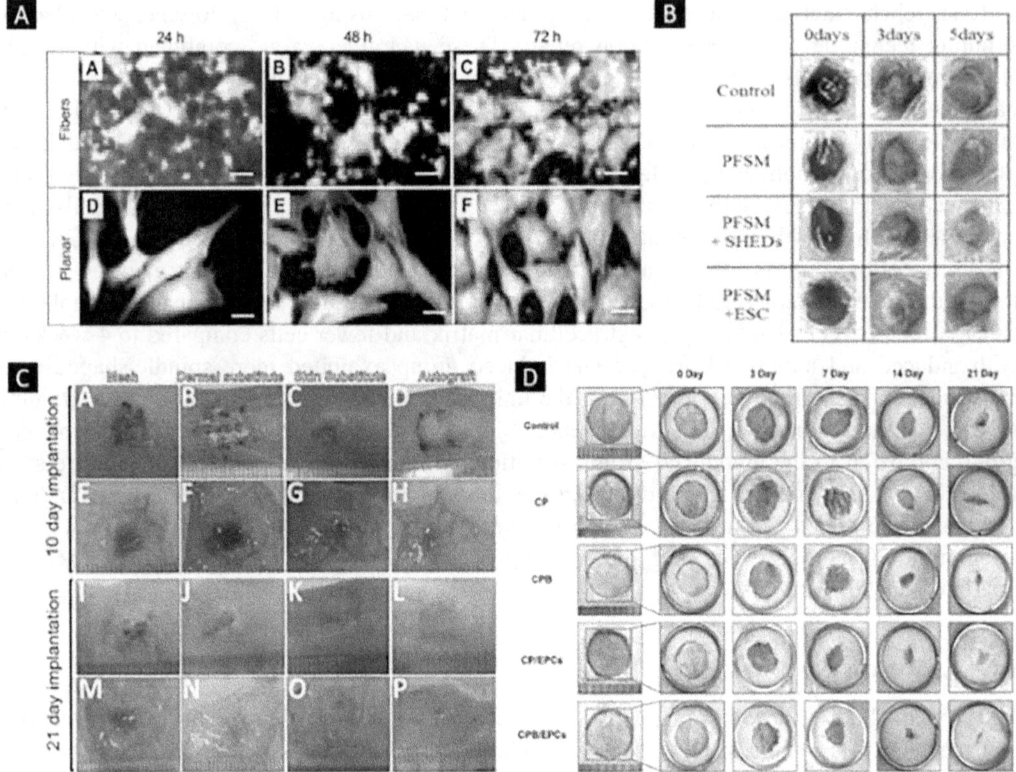

FIGURE 11.9 A) Images of confocal microscopy showed different morphologies after 24 h, 48 h and 72 h, (A-C) shows the 3T3 fibroblasts on fibrinogen scaffolds, (D-F) planar fibrinogen; phalloidin (red) indicates actin filament and DAPI (blue) indicates cell nuclei (Scale bar = 20 μm). Reprinted with permission from (Suter et al. 2021) (Copyright 2021, Elsevier). B) Images of wounds treated with Methanol-treated PVA-SF dressing (PFSM), PFSM with Stem cells from human exfoliated deciduous teeth (SHED) and PFSM with ESCs which shows the greater wound closure. Reprinted with permission from (Huang et al. 2019a) (Copyright 2019, Elsevier). (C) Macroscopic overview of wound treated with different grafts including nanofibers meshes for 10 and 21 days, autografts show best healing with minimal contraction than the others. Reprinted with permission from (Mahjour et al. 2015) (Copyright 2015, Elsevier). (D) Representative images of wound repair at day 0,3, 7, 14 and 21 by taking saline as control, CP scaffold, Collagen-PCL-Bioglass nanoparticles (CPB) scaffold, and CPB with endothelial progenitor cells (EPCs). Reprinted with permission from (Wang et al. 2018) (Copyright 2018, Elsevier).

autofluorescence. Nanofibrous fibrinogen resulted in smaller fibroblast cell size and fewer actin stress fibers, indicating enhanced fibroblast proliferation compared to planar fibrinogen (Figure 11.9A) (Suter et al. 2021).

Stem cell therapy typically takes 7–28 days, whereas combining biomaterials with stem cell therapy can reduce the duration to 7–14 days for wound repair. Utilizing suitable biomaterials that are well-designed can improve cell engraftment by controlling the rate of biodegradation (Riha et al. 2021). Natural biomaterials employed in skin regeneration include hyaluronic acid, collagen, chitosan, and gelatin. Hyaluronic acid, the complex of N-acetyl-D-glucosamine and D-glucuronic acid connected via alternating β-(1 → 4) and β-(1 → 3) glycosidic bonds, is the most widely used polymer is (Bouten et al. 2014). Hyaluronic acid (HA) coating on biodegradable electrospun nanofibers of PHBV enhanced wound healing and skin tissue regeneration. HA is released after 4 hours, promoting wound closure in 8 hours and stimulating keratinocyte activity (Kaniuk et al. 2022). Researchers utilized electrostatic spinning to create PVA/silk fibroin nanofibers integrated

TABLE 11.1
Outlining the Intersection of Biomaterials and Stem Cell Applications in Skin Regeneration

Type of Stem Cells	Biomaterials Used		Healing Time (in days)	Application	Therapeutic Outcome	Ref.
MSC	Chitosan and Arginine Based Polyester Amide	Hydrogel	7	3rd degree burn wounds in a mouse mode	Promote wound healing, re-epithelialization, regeneration of blood vessels and granulation tissue development.	(Alapure et al. 2018)
	Cerium dioxide nanoparticles	Nanoparticles in Hydrogel	14	Rat skin wound healing model	epithelialization of wounds, faster regeneration, anti-inflammatory effect.	(Silina et al. 2021)
	Simvastatin-Lipid nano carrier	Lipid nano carrier	14	Diabetic rat skin wound model	Endothelial maturation, proliferation, vascularization, improved healing	(Örgül et al. 2021)
BMSCs	PCL and Gelatin	3-D Nanofibrous scaffold	7	Full-thickness excisional wound in diabetic mouse	Improved granulation tissue formation, angiogenesis, stimulate pro-regenerative response to enhance wound healing and ECM deposition.	(Chen et al. 2020b)
	Silk	Nanofiber hydrogel	14	Rat skin wound model	Scarless skin regeneration including hair follicles, stimulate wound healing.	(Zheng et al. 2020)
	Collagen/poly(L-lactic acid)–co-poly(3-caprolactone) (Coll/PLLCL)	Nanofibers	–	Cell proliferation assay	Differentiation, wound healing, epidermal cell maturation.	(Jin et al. 2011)
	Gelatin	Hydrogel	14	Murine burn model	Highest wound healing rate, lessened discoloration rating, less roughness score and scab formation	(Lu et al. 2020)

(Continued)

TABLE 11.1 ***(Continued)***
Outlining the Intersection of Biomaterials and Stem Cell Applications in Skin Regeneration

Type of Stem Cells	Biomaterials Used		Healing Time (in days)	Application	Therapeutic Outcome	Ref.
ADSCs	PCL and Gelatin	Nanofibers	14	Rat skin wound healing models	Tissue regeneration, wound repair, enhance the production of collagen type I and formation of CD31-positive vessels in dermis layer	(Liu et al. 2023)
	Graphene oxide (GO) and Genipine	Nano-graphene oxide composite sponge	28	Xenogenic animal (rat) model	lower inflammatory reaction, skin or soft-tissue regeneration	(Nyambat et al. 2018)
	PLGA	Nanofibers	7	Full-thickness excisional wound in mouse model	higher cell activity in terms of cell adhesion, cell survival and proliferation with improved wound healing	(Tang et al. 2019)
Human endometrial stem cells (hEnSCs)	Collagen and PCL	Nanofibers	–	–	Improve angiogenesis in wound repair, hEnSCs adhesion and proliferation	(Sharif et al. 2018)
Umbilical Cord Mesenchymal Stem Cells (UC-MSCs)	Gelatin methacryloyl (GelMA)	Hydrogel	14	Epidermal rat skin model	increase the integration of host tissues and exogenous skin substitutes, proliferation and epidermis maturation.	(Zhang et al. 2017)
ESCs	PVA-Silk	Nanofibers	5	Wound repair	Faster rate of wound closure	(Huang et al. 2019b)

with epidermal stem cells and stem cells derived from human exfoliated deciduous teeth (SHEDS). In an in vivo study, it was observed that wounds treated with nanofibers containing epidermal stem cells exhibited a 1.86-fold faster closure rate, while those treated with nanofibers containing SHEDS showed a 1.35-fold faster closure rate, compared to the control group and nanofiber biomaterial alone indicating the application of nanofiber dressing for wound healing (Figure 11.9B) (Huang et al. 2019a). The nanofiber mesh prepared using blend of two different polymers (PCL and Collagen type I), were loaded with the human keratinocytes to act as skin substitutes. The cell-loaded nanofibers showed excellent re-epithelization (highest wound healing in the group treated with autografts). The incorporation of cells in the polymeric nanofiber mesh helped in integration with existing epithelial cells which accelerated the wound healing (Figure 11.9C) (Mahjour et al. 2015).

Additionally, in vitro experiments demonstrated that endothelial progenitor cells (EPCs) combined with nanofibrous CPB (collagen-PCL-bioactive glass nanoparticles) promoted angiogenesis. When treating rat wounds with various substances (saline, CP, CPB, CP/EPCs, and CPB/EPCs), the CPB/EPCs group showed an accelerated wound closure, significantly closing the wound by day 7 and fully healing it by day 14 (Figure 11.9D) (Wang et al. 2018). The table provides several additional instances of nanomaterials utilized in stem cell therapy for wound healing and epithelial regeneration. (See Table 11.1).

11.12 CONCLUSION

In conclusion, the integration of nanomaterials with stem cell research has been a rapidly growing area of investigation in recent years, with numerous exciting advancements and breakthroughs being made. Ongoing research into stem cell treatment aims to identify the best methods for isolating stem cells, creating the best possible culture conditions for them, and identifying the elements that can influence their proliferation and differentiation. The fundamental barriers to the efficacy of stem cell therapy are administering the stem cells as well as managing the homing mechanisms and tracking the condition of stem cell growth. Nanomaterials offer a unique tool to study stem cells and control their behavior, and their unique physical and chemical properties provide new avenues for manipulating stem cells in ways that were previously not possible. By using nanomaterials to regulate the microenvironment of stem cells, researchers are able to improve cell proliferation, differentiation, and functionalization, and to develop new strategies for regenerative medicine and tissue engineering. Labeling and tracking the stem cells with nanoparticles such as quantum dots, fluorescent nanoparticles, polymeric nanoparticles, and superparamagnetic nanoparticles allows for real-time evaluation of stem cell activities. In addition to this, stem cells can be transported in nanomaterials that act as cell transporters. Nanomaterials are being extensively researched for application in tissue regeneration due to their ability to mimic the cells' natural extracellular milieu. Stem cell-based therapies are also quite popular in the field of regenerative medicine due to their ability to repair and rejuvenate. A synergistic benefit of both nanotechnology and stem cell-based therapeutics is offered in medical field by their convergence. Nanomaterials can help direct stem cell fate because of their unique qualities, such as their ease of preparation, high surface area, high uptake by stem cells, biocompatibility, and the ability to fine-tune structural and surface properties. When developing stem cell-based nanomaterial therapy for the treatment of various pathological disorders, it is important to keep in mind the interactions occurring at the stem cell-nanomaterial interface as well as the cytotoxicity of the nanomaterials themselves.

However, despite the many advances that have been made in this field, there is still much to be learned about the safety and toxicity of nanomaterials in stem cell research (Rajani et al. 2022). Further studies are needed to fully understand the interactions between nanomaterials and stem cells, and to optimize the use of nanomaterials for stem cell-based applications. Additionally, more research is needed to determine the long-term effects of nanomaterials on stem cells and the tissues they form, as well as to ensure that the materials used in stem cell research are safe and biocompatible.

REFERENCES

Abdal Dayem, A., S. B. Lee, and S.-G. Cho. 2018. The impact of metallic nanoparticles on stem cell proliferation and differentiation. *Nanomaterials* 8(10): 761.

Ahmadi, A. and S. Arami. 2014. Potential applications of nanoshells in biomedical sciences. *Journal of Drug Targeting* 22(3): 175–190.

Ahn, E. H., et al. 2014. Spatial control of adult stem cell fate using nanotopographic cues 35(8): 2401–2410.

Akhavan, O. and E.J.J.o.M.C.B. Ghaderi. 2013. Differentiation of human neural stem cells into neural networks on graphene nanogrids 1(45): 6291–6301.

Alapure, B. V., et al. 2018. Accelerate healing of severe burn wounds by mouse bone marrow mesenchymal stem cell-seeded biodegradable hydrogel scaffold synthesized from arginine-based poly(ester amide) and chitosan. *Stem Cells Dev* 27(23): 1605–1620.

Alghazali, K. M., et al. 2015. Bone-tissue engineering: complex tunable structural and biological responses to injury, drug delivery, and cell-based therapies 47(4): 431–454.

Aly, R. M. 2020. Current state of stem cell-based therapies: an overview. *Stem Cell Investigation* 7.

Anju, S. and P. Mohanan. 2021. Biomedical applications of transition metal dichalcogenides (TMDCs). *Synthetic Metals* 271: 116610.

Anju, S., J. Ashtami, and P. Mohanan. 2019. Black phosphorus, a prospective graphene substitute for biomedical applications. *Materials Science and Engineering: C* 97: 978–993.

Aparicio-Collado, J. L., et al. 2022. Electroactive calcium-alginate/polycaprolactone/reduced graphene oxide nanohybrid hydrogels for skeletal muscle tissue engineering. *Colloids and Surfaces B: Biointerfaces* 214: 112455.

Apolinario, A. C., et al. 2021. Lipid nanovesicles for biomedical applications:'What is in a name'? *Progress in Lipid Research* 82: 101096.

Arai, Y., et al. 2020. Bile acid-based dual-functional prodrug nanoparticles for bone regeneration through hydrogen peroxide scavenging and osteogenic differentiation of mesenchymal stem cells 328: 596–607.

Argentati, C., et al. 2019. Insight into mechanobiology: how stem cells feel mechanical forces and orchestrate biological functions 20(21): 5337.

Arora, N. and N. Sharma. 2014. Arc discharge synthesis of carbon nanotubes: Comprehensive review. *Diamond and Related Materials* 50: 135–150.

Basu, S. M., et al. 2021. Lipid nanocapsules co-encapsulating paclitaxel and salinomycin for eradicating breast cancer and cancer stem cells. *Colloids and Surfaces B: Biointerfaces* 204: 111775.

Bauer, S., et al. 2008. Improved attachment of mesenchymal stem cells on super-hydrophobic TiO2 nanotubes 4(5): 1576–1582.

Bechara, S., L. Wadman, and K.C.J.A.b. Popat. 2011. Electroconductive polymeric nanowire templates facilitates in vitro C17. 2 neural stem cell line adhesion, proliferation and differentiation 7(7): 2892–2901.

Begines, B., et al. 2020. Polymeric nanoparticles for drug delivery: Recent developments and future prospects. *Nanomaterials* 10(7): 1403.

Bengel, F.M.J.J.o.N.C. 2011. Noninvasive stem cell tracking 18(5): 966–973.

Bouten, P. J. M., et al. 2014. The chemistry of tissue adhesive materials. *Progress in Polymer Science* 39(7): 1375–1405.

Bucaro, M. A., et al. 2012. Fine-tuning the degree of stem cell polarization and alignment on ordered arrays of high-aspect-ratio nanopillars 6(7): 6222–6230.

Bursac, N., M. Juhas, and T. A. Rando. 2015. Synergizing Engineering and Biology to Treat and Model Skeletal Muscle Injury and Disease. *Annu Rev Biomed Eng* 17: 217–242.

Caille, N., et al. 2002. Contribution of the nucleus to the mechanical properties of endothelial cells 35(2): 177–187.

Carrow, J. K., et al. 2018. Widespread changes in transcriptome profile of human mesenchymal stem cells induced by two-dimensional nanosilicates. *Proceedings of the National Academy of Sciences* 115(17): E3905–E3913.

Cary, L., et al. 1999. Invited Reviews-Integrin-mediated signal transduction pathways 14(3): 1001.

Cha, B. G. and J. Kim. 2019. Functional mesoporous silica nanoparticles for bio-imaging applications. *Wiley Interdisciplinary Reviews: Nanomedicine and Nanobiotechnology* 11(1): e1515.

Cha, C., et al 2012. Designing biomaterials to direct stem cell fate 6(11): 9353–9358.

Chen, S., et al. 2019. Recent advances in non-ionic surfactant vesicles (niosomes): Fabrication, characterization, pharmaceutical and cosmetic applications. *European Journal of Pharmaceutics and Biopharmaceutics* 144: 18–39.

Chen, S., et al. 2020b. Mesenchymal stem cell-laden, personalized 3D scaffolds with controlled structure and fiber alignment promote diabetic wound healing. *Acta Biomaterialia* 108: 153–167.

Chen, S., et al. 2021. Tenogenic adipose-derived stem cell sheets with nanoyarn scaffolds for tendon regeneration. *Mater Sci Eng C Mater Biol Appl* 119: 111506.

Chen, W., et al. 2018. Black phosphorus nanosheets as a neuroprotective nanomedicine for neurodegenerative disorder therapy. *Advanced Materials* 30(3): 1703458.

Chen, X., et al. 2020a. An overview of the optical properties and applications of black phosphorus. *Nanoscale* 12(6): 3513–3534.

Choi, J. R., et al. 2018. Black phosphorus and its biomedical applications. *Theranostics* 8(4): 1005.

Choi, S. Y., et al. 2015. Gold nanoparticles promote osteogenic differentiation in human adipose-derived mesenchymal stem cells through the Wnt/β-catenin signaling pathway 10: 4383.

Chueng, S.-T.D., et al. 2016. Multidimensional nanomaterials for the control of stem cell fate 3(1): 1–15.

Dalby, M. J., et al. 2007. The control of human mesenchymal cell differentiation using nanoscale symmetry and disorder 6(12): 997–1003.

Dalby, M. J., N. Gadegaard, and R.O.J.N.m. Oreffo. 2014. Harnessing nanotopography and integrin–matrix interactions to influence stem cell fate 13(6): 558–569.

Das, S. S., et al. 2019. Laponite-based nanomaterials for biomedical applications: a review. *Current Pharmaceutical Design* 25(4): 424–443.

Dawson, J. I. and R. O. Oreffo. 2013. Clay: new opportunities for tissue regeneration and biomaterial design. *Advanced Materials* 25(30): 4069–4086.

Dent, E. W. and P.W.J.J.o.n. Baas. 2014. Microtubules in neurons as information carriers 129(2): 235–239.

Desai, N., et al. 2021. Nanomedicine in the treatment of diabetic nephropathy 13(07): 663–686.

Desai, N., et al. 2023. Chitosan: a potential biopolymer in drug delivery and biomedical applications 15(4): 1313.

Dimitriou, R., et al. 2011. Bone regeneration: current concepts and future directions 9(1): 1–10.

Ding, Y.-W., et al. 2022. Recent advances in hyaluronic acid-based hydrogels for 3D bioprinting in tissue engineering applications.

Discher, D. E., P. Janmey, and Y.-I.J.S. Wang 2005. Tissue cells feel and respond to the stiffness of their substrate 310(5751): 1139–1143.

dos Santos Haupenthal, D. P., et al. 2020. Effects of phonophoresis with diclofenac linked gold nanoparticles in model of traumatic muscle injury. *Materials Science and Engineering: C* 110: 110681.

Dreifke, M. B., A. A. Jayasuriya, and A. C. Jayasuriya. 2015. Current wound healing procedures and potential care. *Mater Sci Eng C Mater Biol Appl* 48: 651–662.

Duan, Y., et al. 2020. A brief review on solid lipid nanoparticles: Part and parcel of contemporary drug delivery systems. *RSC Advances* 10(45): 26777–26791.

Dulińska-Litewka, J., et al. 2019. Superparamagnetic iron oxide nanoparticles—Current and prospective medical applications. *Materials* 12(4): 617.

Eatemadi, A., et al. 2014. Carbon nanotubes: properties, synthesis, purification, and medical applications. *Nanoscale Research Letters* 9(1): 1–13.

Elkhenany, H., et al. 2017. Graphene nanoparticles as osteoinductive and osteoconductive platform for stem cell and bone regeneration 13(7): 2117–2126.

Emmert, M. Y. 2017. Cell-based cardiac regeneration. *European Heart Journal* 38(15): 1095–1098.

Engler, A. J., et al. 2004. Myotubes differentiate optimally on substrates with tissue-like stiffness: pathological implications for soft or stiff microenvironments 166(6): 877–887.

Engler, A. J., et al. 2006. Matrix elasticity directs stem cell lineage specification 126(4): 677–689.

Ferreira, A. M., et al. 2012. Collagen for bone tissue regeneration 8(9): 3191–3200.

Flanagan, L. A., et al. 2002. Neurite branching on deformable substrates 13(18): 2411.

Fu, Y. and D.L.J.E.r.o.c.t. Kraitchman. 2010. Stem cell labeling for noninvasive delivery and tracking in cardiovascular regenerative therapy 8(8): 1149–1160.

Gaharwar, A. K., et al. 2019. 2D nanoclay for biomedical applications: regenerative medicine, therapeutic delivery, and additive manufacturing. *Advanced Materials* 31(23): 1900332.

Gao, X., et al. 2019. Progress in electrospun composite nanofibers: composition, performance and applications for tissue engineering. *Journal of Materials Chemistry B* 7(45): 7075–7089.

García-Gareta, E., M. J. Coathup, and G. W. J. B. Blunn. 2015. Osteoinduction of bone grafting materials for bone repair and regeneration 81: 112–121.

Geim, A. K. and K. S. Novoselov. 2007. The rise of graphene. *Nature Materials* 6(3): 183–191.

Georges, P. C., et al. 2006. Matrices with compliance comparable to that of brain tissue select neuronal over glial growth in mixed cortical cultures 90(8): 3012–3018.

Ghosal, K., et al. 2021. Graphene family nanomaterials-opportunities and challenges in tissue engineering applications. *FlatChem* 30: 100315.

Goonoo, N. and A. Bhaw-Luximon. 2020. 7 - Nanomaterials combination for wound healing and skin regeneration. In *Advanced 3D-Printed Systems and Nanosystems for Drug Delivery and Tissue Engineering*, Editor by L.C.d. Toit, et al. 159–217. Elsevier.

Gunn, J. and M. Zhang. 2010. Polyblend nanofibers for biomedical applications: perspectives and challenges. *Trends in Biotechnology* 28(4): 189–197.

Guvendiren, M. and J. A. J. B. Burdick. 2010. The control of stem cell morphology and differentiation by hydrogel surface wrinkles 31(25): 6511–6518.

Hachani, R., et al. 2013. Tracking stem cells in tissue-engineered organs using magnetic nanoparticles 5(23): 11362–11373.

Halim, A., et al. 2018. A mini review focused on the recent applications of graphene oxide in stem cell growth and differentiation. *Nanomaterials* 8(9): 736.

Han, S., et al. 2019b. The application of graphene-based biomaterials in biomedicine. *American Journal of Translational Research* 11(6): 3246.

Han, S. B., et al. 2020. Mechanical properties of materials for stem cell differentiation 4(11): 2000247.

Han, X., et al. 2019a. Applications of nanoparticles in biomedical imaging. *Nanoscale* 11(3): 799–819.

Hasnain, M. S. and A. K. Nayak. 2019. Background: carbon nanotubes for targeted drug delivery. In *Carbon Nanotubes for Targeted Drug Delivery*, 1–9. Springer.

He, F., et al. 2023. Black phosphorus nanosheets suppress oxidative damage of stem cells for improved neurological recovery. *Chemical Engineering Journal* 451: 138737.

He, X. and N. Ma. 2014. An overview of recent advances in quantum dots for biomedical applications. *Colloids and Surfaces B: Biointerfaces* 124: 118–131.

Hiew, V. V., S. F. B. Simat, and P. L. Teoh. 2018. The advancement of biomaterials in regulating stem cell fate. *Stem Cell Reviews and Reports* 14(1): 43–57.

Hofstetter, C. P., et al. 2005. Allodynia limits the usefulness of intraspinal neural stem cell grafts; directed differentiation improves outcome 8(3): 346–353.

Huang, D.-M., et al. 2008. Internalization of mesoporous silica nanoparticles induces transient but not sufficient osteogenic signals in human mesenchymal stem cells. *Toxicology and Applied Pharmacology* 231(2): 208–215.

Huang, K., J. Wu, and Z. Gu. 2018. Black phosphorus hydrogel scaffolds enhance bone regeneration via a sustained supply of calcium-free phosphorus. *ACS Applied Materials & Interfaces* 11(3): 2908–2916.

Huang, T.-Y., et al. 2019a. Epidermal cells differentiated from stem cells from human exfoliated deciduous teeth and seeded onto polyvinyl alcohol/silk fibroin nanofiber dressings accelerate wound repair 104: 109986.

Huang, T.-Y., et al. 2019b. Epidermal cells differentiated from stem cells from human exfoliated deciduous teeth and seeded onto polyvinyl alcohol/silk fibroin nanofiber dressings accelerate wound repair. *Materials Science and Engineering: C* 104: 109986.

Hung, H.-S., et al. 2021. Anti-inflammatory fibronectin-AgNP for regulation of biological performance and endothelial differentiation ability of mesenchymal stem cells 22(17): 9262.

Illath, K., et al. 2021. Metallic nanoparticles for biomedical applications. In *Nanomaterials and Their Biomedical Applications*, 29–81. Springer.

Jaipan, P., A. Nguyen, and R. J. J. M. C. Narayan. 2017. Gelatin-based hydrogels for biomedical applications 7(3): 416–426.

Jamkhande, P. G., et al. 2019. Metal nanoparticles synthesis: An overview on methods of preparation, advantages and disadvantages, and applications. *Journal of Drug Delivery Science and Technology* 53: 101174.

Jarai, B. M., et al. 2020. Polymeric nanoparticles, In *Nanoparticles for Biomedical Applications*, 303–324. Elsevier.

Jatav, S. and Y. M. Joshi. 2014. Chemical stability of Laponite in aqueous media. *Applied Clay Science* 97: 72–77.

Jeong, G.-J., et al. 2022. Nanomaterial for Skeletal Muscle Regeneration. *Tissue Engineering and Regenerative Medicine* 19.

Jin, G., M. P. Prabhakaran, and S. Ramakrishna. 2011. Stem cell differentiation to epidermal lineages on electrospun nanofibrous substrates for skin tissue engineering. *Acta Biomaterialia* 7(8): 3113–3122.

Jing, Y., et al. 2015. Small molecules make big differences: molecular doping effects on electronic and optical properties of phosphorene. *Nanotechnology* 26(9): 095201.

Kai, D., et al. 2014. Stem cell-loaded nanofibrous patch promotes the regeneration of infarcted myocardium with functional improvement in rat model 10(6): 2727–2738.

Kam, N. W. S., E. Jan, and N.A.J.N.l. Kotov. 2009. Electrical stimulation of neural stem cells mediated by humanized carbon nanotube composite made with extracellular matrix protein 9(1): 273–278.

Kaniuk, Ł., et al. 2022. Accelerated wound closure rate by hyaluronic acid release from coated PHBV electrospun fiber scaffolds. *Journal of Drug Delivery Science and Technology* 77: 103855.

Kargozar, S., et al. 2020. Quantum dots: a review from concept to clinic. *Biotechnology Journal* 15(12): 2000117.

Kawazoe, N. and G. J. B. Chen. 2015. Gold nanoparticles with different charge and moiety induce differential cell response on mesenchymal stem cell osteogenesis 54: 226–236.

Kerativitayanan, P., J. K. Carrow, and A. K. Gaharwar. 2015. Nanomaterials for engineering stem cell responses. *Advanced Healthcare Materials* 4(11): 1600–1627.

Khetan, S., et al. 2013. Degradation-mediated cellular traction directs stem cell fate in covalently crosslinked three-dimensional hydrogels 12(5): 458–465.

Khurana, I. S., et al. 2015. Multifaceted role of clay minerals in pharmaceuticals. *Future Science OA* 1(3).

Ko, W.-K., et al. 2015. The effect of gold nanoparticle size on osteogenic differentiation of adipose-derived stem cells 438: 68–76.

Koike, H., I. Manabe, and Y. Oishi. 2022. Mechanisms of cooperative cell-cell interactions in skeletal muscle regeneration. *Inflammation and Regeneration* 42(1): 48.

Kuc, A., T. Heine, and A. Kis. 2015. Electronic properties of transition-metal dichalcogenides. *MRS Bulletin* 40(7): 577–584.

Kuddannaya, S., et al. 2013. Surface chemical modification of poly (dimethylsiloxane) for the enhanced adhesion and proliferation of mesenchymal stem cells 5(19): 9777–9784.

Kumar, M. and Y. Ando. 2010. Chemical vapor deposition of carbon nanotubes: a review on growth mechanism and mass production. *Journal of Nanoscience and Nanotechnology* 10(6): 3739–3758.

Kumar, R. 2019. Lipid-based nanoparticles for drug-delivery systems. In *Nanocarriers for Drug Delivery*, 249–284. Elsevier.

Kwon, S. G., et al. 2018. Recent advances in stem cell therapeutics and tissue engineering strategies. *Biomaterials Research* 22(1): 1 8.

Lagarce, F. and C. Passirani. 2016. Nucleic-acid delivery using lipid nanocapsules. *Current Pharmaceutical Biotechnology* 17(8): 723–727.

Lee, J. R., et al. 2020. Nanovesicles derived from iron oxide nanoparticles-incorporated mesenchymal stem cells for cardiac repair. *Sci Adv* 6(18): eaaz0952.

Lee, S., et al. 2017. In vivo stem cell tracking with imageable nanoparticles that bind bioorthogonal chemical receptors on the stem cell surface 139: 12–29.

Li, H., A. Wijekoon, and N.D.J.P.o. Leipzig. 2012. 3D differentiation of neural stem cells in macroporous photopolymerizable hydrogel scaffolds 7(11): e48824.

Li, J., et al. 2016. Gold nanoparticle size and shape influence on osteogenesis of mesenchymal stem cells 8(15): 7992–8007.

Li, L., et al. 2022. A nanobrush-shearing strategy enabling the alignment of 1D nanomaterials for synchronous electrochromic actuators and controlled growth of neural stem cells 20: 100256.

Li, Z. and S. L. Wong. 2017. Functionalization of 2D transition metal dichalcogenides for biomedical applications. *Materials Science and Engineering: C* 70: 1095–1106.

Liang, C., C. Wang, and Z. Liu. 2013. Stem cell labeling and tracking with nanoparticles. *Particle & Particle Systems Characterization* 30(12): 1006–1017.

Lin, Y.-S., K. R. Hurley, and C. L. Haynes. 2012. Critical considerations in the biomedical use of mesoporous silica nanoparticles. *The Journal of Physical Chemistry Letters* 3(3): 364–374.

Liu, C., et al. 2020. Barriers and strategies of cationic liposomes for cancer gene therapy. *Molecular Therapy-Methods & Clinical Development* 18: 751–764.

Liu, J.-H., et al. 2015. Carbon "quantum" dots for fluorescence labeling of cells 7(34): 19439–19445.

Liu, N., et al. 2023. Enhancing the paracrine effects of adipose stem cells using nanofiber-based meshes prepared by light-welding for accelerating wound healing. *Materials & Design* 225: 111582.

Liu, Y., et al 2021. A bibliometric analysis: Research progress and prospects on transition metal dichalcogenides in the biomedical field. *Chinese Chemical Letters* 32(12): 3762–3770.

Lopes, L., et al. 2018. Stem cell therapy for diabetic foot ulcers: a review of preclinical and clinical research 9(1): 1–16.

Lu, C.-W., et al. 2007. Bifunctional magnetic silica nanoparticles for highly efficient human stem cell labeling 7(1): 149–154.

Lu, T.-Y., et al. 2020. Enzyme-Crosslinked Gelatin Hydrogel with Adipose-Derived Stem Cell Spheroid Facilitating Wound Repair in the Murine Burn Model 12(12): 2997.

Luo, M., et al. 2019. 2D black phosphorus–based biomedical applications 29(13): 1808306.

Lv, H., et al. 2015. Mechanism of regulation of stem cell differentiation by matrix stiffness 6(1): 1–11.

Ma, S., et al. 2019. Enhanced osteoinduction of electrospun scaffolds with assemblies of hematite nanoparticles as a bioactive interface 14: 1051.

Ma, X., et al. 2021. Molecular mechanisms underlying the role of the puckered surface in the biocompatibility of black phosphorus. *Nanoscale* 13(6): 3790–3799.

Mahjour, S. B., et al. 2015. Rapid creation of skin substitutes from human skin cells and biomimetic nanofibers for acute full-thickness wound repair. *Burns* 41(8): 1764–1774.

Malina, T., et al. 2019. Carbon dots for in vivo fluorescence imaging of adipose tissue-derived mesenchymal stromal cells 152: 434–443.

Manzeli, S., et al. 2017. 2D transition metal dichalcogenides. *Nature Reviews Materials* 2(8): 1–15.

Marsell, R. and T. A. J. I. Einhorn. 2011. The biology of fracture healing 42(6): 551–555.

Massaro, M., et al. 2020. Covalently modified nanoclays: synthesis, properties and applications, In *Clay Nanoparticles*, 305–333. Elsevier.

Mayr, M., et al. 2002. Mechanical stress-induced DNA damage and rac-p38MAPK signal pathways mediate p53-dependent apoptosis in vascular smooth muscle cells 16(11): 1423–1425.

Mihaila, S. M., et al. 2014. The osteogenic differentiation of SSEA-4 sub-population of human adipose derived stem cells using silicate nanoplatelets. *Biomaterials* 35(33): 9087–9099.

Mohamed, M., et al. 2019. PEGylated liposomes: immunological responses. *Science and Technology of Advanced Materials* 20(1): 710–724.

Mok, P. L., et al. 2017. Micro-computed tomography detection of gold nanoparticle-labelled mesenchymal stem cells in the rat subretinal layer 18(2): 345.

Mondal, T., et al. 2022. Thin films of functionalized carbon nanotubes support long-term maintenance and cardio-neuronal differentiation of canine induced pluripotent stem cells 40: 102487.

Mu, H. and R. Holm. 2018. Solid lipid nanocarriers in drug delivery: characterization and design. *Expert Opinion on Drug Delivery* 15(8): 771–785.

Murphy, W. L., T. C. McDevitt, and A.J.J.N.m. Engler. 2014. Materials as stem cell regulators 13(6): 547–557.

Naqvi, S., L.J.F.i.b. McNamara, and biotechnology. 2020. Stem cell mechanobiology and the role of biomaterials in governing mechanotransduction and matrix production for tissue regeneration 8: 597661.

Narayan, R., et al. 2018. Mesoporous silica nanoparticles: A comprehensive review on synthesis and recent advances. *Pharmaceutics* 10(3): 118.

Nazari, H., et al. 2020. Incorporation of two-dimensional nanomaterials into silk fibroin nanofibers for cardiac tissue engineering. *Polymers for Advanced Technologies* 31(2): 248–259.

Nel, A., et al. 2006. Toxic potential of materials at the nanolevel 311(5761): 622–627.

Nemati, S., et al. 2019. Current progress in application of polymeric nanofibers to tissue engineering. *Nano Convergence* 6(1): 1–16.

Newby, S. D., et al. 2020. Functionalized graphene nanoparticles induce human mesenchymal stem cells to express distinct extracellular matrix proteins mediating osteogenesis 15: 2501.

Nikzamir, M., et al. 2021. Applications of dendrimers in nanomedicine and drug delivery: A review. *Journal of Inorganic and Organometallic Polymers and Materials* 31(6): 2246–2261.

Noback, C. R., et al. 2005. *The human nervous system: structure and function*. Springer Science & Business Media.

Nyambat, B., et al. 2018. Genipin-crosslinked adipose stem cell derived extracellular matrix-nano graphene oxide composite sponge for skin tissue engineering 6(6): 979–990.

Oh, S. H., et al. 2010. Investigation of pore size effect on chondrogenic differentiation of adipose stem cells using a pore size gradient scaffold 11(8): 1948–1955.

Ohyabu, Y., et al. 2009. Stable and nondisruptive in vitro/in vivo labeling of mesenchymal stem cells by internalizing quantum dots 20(3): 217–224.

Örgül, D., et al. 2021. In-vivo evaluation of tissue scaffolds containing simvastatin loaded nanostructured lipid carriers and mesenchymal stem cells in diabetic wound healing. *Journal of Drug Delivery Science and Technology* 61: 102140.

Opatha, S. A. T., V. Titapiwatanakun, and R. Chutoprapat. 2020. Transfersomes: A promising nanoencapsulation technique for transdermal drug delivery. *Pharmaceutics* 12(9): 855.

Orsu, P. and A. Koyyada. 2020. Recent progresses and challenges in graphene based nano materials for advanced therapeutical applications: a comprehensive review. *Materials Today Communications* 22: 100823.

Padilha, L. A., et al. 2011. Optimization of band structure and quantum-size-effect tuning for two-photon absorption enhancement in quantum dots. *Nano Letters* 11(3): 1227–1231.

Palmer, L. C. and S. I. Stupp. 2008. Molecular self-assembly into one-dimensional nanostructures. *Accounts of Chemical Research* 41(12): 1674–1684.

Park, J., et al. 2007. Nanosize and vitality: TiO2 nanotube diameter directs cell fate 7(6): 1686–1691.

Park, J., et al. 2012. Control of stem cell fate and function by engineering physical microenvironments 4(9): 1008–1018.

Park, J., et al. 2021. Bioactive lipids and their derivatives in biomedical applications. *Biomolecules & Therapeutics* 29(5): 465.

Patel, V., et al. 2020. Dendrimers as novel drug-delivery system and its applications, In *Drug delivery systems*, 333–392. Elsevier.

Pawar, P. V., et al. 2013. Functionalized polymersomes for biomedical applications. *Polymer Chemistry* 4(11): 3160–3176.

Peng, L., et al. 2020. Black phosphorus: Degradation mechanism, passivation method, and application for in situ tissue regeneration. *Advanced Materials Interfaces* 7(23): 2001538.

Peng, Y., et al. 2018. Degradation rate affords a dynamic cue to regulate stem cells beyond varied matrix stiffness 178: 467–480.

Peña-Parás, L., J. A. Sánchez-Fernández, and R. Vidaltamayo. 2018. Nanoclays for biomedical applications. *Handbook of Ecomaterials* 5: 3453–3471.

Pham-Nguyen, O. V., et al. 2021. Preparation of stretchable nanofibrous sheets with sacrificial coaxial electrospinning for treatment of traumatic muscle injury 10(8): 2002228.

Pradhan, K., et al. 2018. Neuro-regenerative choline-functionalized injectable graphene oxide hydrogel repairs focal brain injury 10(3): 1535–1543.

Prajapati, S. K., et al 2022. Biomedical applications and toxicities of carbon nanotubes. *Drug and Chemical Toxicology* 45(1): 435–450.

Presutti, D., et al. 2022. Transition metal dichalcogenides (TMDC)-based nanozymes for biosensing and therapeutic applications. *Materials* 15(1): 337.

Qu, Y., et al. 2018. Advances on graphene-based nanomaterials for biomedical applications. *Materials Science and Engineering: C* 90: 764–780.

Rajabzadeh, N., E. Fathi, and R. Farahzadi. 2019. Stem cell-based regenerative medicine. *Stem Cell Investigation* 6.

Rajani, C., et al. 2022. Developmental toxicity of nanomaterials used in drug delivery: Understanding molecular biomechanics and potential remedial measures. In *Pharmacokinetics and Toxicokinetic Considerations*, 685–725. Elsevier.

Rakhit, S., et al. 2021. Management and challenges of severe traumatic brain injury. In *Seminars in Respiratory and Critical Care Medicine*. Thieme Medical Publishers, Inc.

Ramalingam, M. and D. Rana. 2015. Impact of nanotechnology in induced pluripotent stem cells-driven tissue engineering and regenerative medicine. *Journal of Bionanoscience* 9(1): 13–21.

Ravichandran, R., et al. 2014. Gold nanoparticle loaded hybrid nanofibers for cardiogenic differentiation of stem cells for infarcted myocardium regeneration 14(4): 515–525.

Ren, S., et al. 2022. Cerium oxide nanoparticles loaded nanofibrous membranes promote bone regeneration for periodontal tissue engineering 7: 242–253.

Riha, S. M., M. Maarof, and M. B. Fauzi. 2021. Synergistic Effect of Biomaterial and Stem Cell for Skin Tissue Engineering in Cutaneous Wound Healing: A Concise Review 13(10): 1546.

Rivera-Gil, P., et al. 2013. The challenge to relate the physicochemical properties of colloidal nanoparticles to their cytotoxicity 46(3): 743–749.

Sabir, M. I., X. Xu, and L.J.J.o.m.s. Li. 2009. A review on biodegradable polymeric materials for bone tissue engineering applications 44(21): 5713–5724.

Saeed, M., et al. 2020. Chemical vapour deposition of graphene—Synthesis, characterisation, and applications: A review. *Molecules* 25(17): 3856.

Saliev, T. 2019. The advances in biomedical applications of carbon nanotubes. *C* 5(2): 29.

Saw, K.-Y., et al. 2015. High tibial osteotomy in combination with chondrogenesis after stem cell therapy: a histologic report of 8 cases 31(10): 1909–1920.

Schäfer, R. J. T. M. and Hemotherapy. 2010. Labeling and imaging of stem cells–promises and concerns 37(2): 85–89.

Sethuram, L. and J. Thomas. 2023. Therapeutic applications of electrospun nanofibers impregnated with various biological macromolecules for effective wound healing strategy – A review. *Biomedicine & Pharmacotherapy* 157: 113996.

Shafiee, A., et al. 2021. An in situ hydrogel-forming scaffold loaded by PLGA microspheres containing carbon nanotube as a suitable niche for neural differentiation 120: 111739.

Shao, J., et al. 2012. Early stage evolution of structure and nanoscale property of nanofibers in thermally induced phase separation process. *Reactive and Functional Polymers* 72(10): 765–772.

Sharif, S., et al. 2018. Collagen-coated nano-electrospun PCL seeded with human endometrial stem cells for skin tissue engineering applications 106(4): 1578–1586.

Shin, Y. N., et al. 2008. Adhesion comparison of human bone marrow stem cells on a gradient wettable surface prepared by corona treatment 255(2): 293–296.

Si, R., et al. 2020. Human mesenchymal stem cells encapsulated-coacervated photoluminescent nanodots layered bioactive chitosan/collagen hydrogel matrices to endorse cardiac healing after acute myocardial infarction 206: 111789.

Silina, E. V., et al. 2021. Application of polymer drugs with cerium dioxide nanomolecules and mesenchymal stem cells for the treatment of skin wounds in aged rats 13(9): 1467.

Singla, R., et al. 2019. Nanomaterials as potential and versatile platform for next generation tissue engineering applications. *Journal of Biomedical Materials Research Part B: Applied Biomaterials* 107(7): 2433–2449.

Sridhar, S., et al. 2015. Cardiogenic differentiation of mesenchymal stem cells with gold nanoparticle loaded functionalized nanofibers 134: 346–354.

Srijaya, T. C., T. S. Ramasamy, and N.H.A.J.J.o.t.m. Kasim. 2014. Advancing stem cell therapy from bench to bedside: lessons from drug therapies 12(1): 1–17.

Stenger, D. A., et al. 1992. Coplanar molecular assemblies of amino-and perfluorinated alkylsilanes: characterization and geometric definition of mammalian cell adhesion and growth 114(22): 8435–8442.

Stout, D. A. and T. J. Webster. 2012. Carbon nanotubes for stem cell control. *Materials Today* 15(7-8): 312–318.

Subramanian, A., U. M. Krishnan, and S.J.J.o.b.s. Sethuraman. 2009. Development of biomaterial scaffold for nerve tissue engineering: Biomaterial mediated neural regeneration 16(1): 1–11.

Suter, N., et al. 2021. Self-assembled fibrinogen nanofibers support fibroblast adhesion and prevent E. coli infiltration. *Materials Science and Engineering: C* 126: 112156.

Tang, K.-C., et al. 2019. Human Adipose-Derived Stem Cell Secreted Extracellular Matrix Incorporated into Electrospun Poly(Lactic-co-Glycolic Acid) Nanofibrous Dressing for Enhancing Wound Healing 11(10): 1609.

Tarcan, R., et al. 2020. Reduced graphene oxide today. *Journal of Materials Chemistry C* 8(4): 1198–1224.

Tee, S.-Y., et al. 2011. Cell shape and substrate rigidity both regulate cell stiffness 100(5): L25–L27.

Teo, W. Z., et al. 2014. Cytotoxicity of exfoliated transition-metal dichalcogenides (MoS2, WS2, and WSe2) is lower than that of graphene and its analogues. *Chemistry–A European Journal* 20(31): 9627–9632.

Tompkins, B. A., et al. 2018. Preclinical studies of stem cell therapy for heart disease 122(7): 1006–1020.

Trounson, A. and C.J.C.s.c. McDonald. 2015. Stem cell therapies in clinical trials: progress and challenges 17(1): 11–22.

Trounson, A., et al. 2011. Clinical trials for stem cell therapies 9(1): 1–7.

Tsou, Y.-H., et al. 2016. Hydrogel as a bioactive material to regulate stem cell fate 1(1): 39–55.

Tsuboi, S. and T. Jin. 2018. Recombinant protein (luciferase-IgG binding domain) conjugated quantum dots for BRET-coupled near-infrared imaging of epidermal growth factor receptors. *Bioconjugate Chemistry* 29(4): 1466–1474.

Tu, J., et al. 2016. Mesoporous silica nanoparticles with large pores for the encapsulation and release of proteins. *ACS Applied Materials & Interfaces* 8(47): 32211–32219.

Uchida, S. and K. Kataoka. 2019. Design concepts of polyplex micelles for in vivo therapeutic delivery of plasmid DNA and messenger RNA. *Journal of Biomedical Materials Research Part A* 107(5): 978–990.

Van de Walle, A., et al. 2020. Magnetic nanoparticles in regenerative medicine: what of their fate and impact in stem cells? 11: 100084.

Vrignaud, S., et al. 2013. Aqueous core nanocapsules: a new solution for encapsulating doxorubicin hydrochloride. *Drug Development and Industrial Pharmacy* 39(11): 1706–1711.

Wagner, A. M., et al. 2019. Quantum dots in biomedical applications. *Acta Biomaterialia* 94: 44–63.

Wang, C., et al. 2018. Highly efficient local delivery of endothelial progenitor cells significantly potentiates angiogenesis and full-thickness wound healing. *Acta Biomaterialia* 69: 156–169.

Wang, Q., et al. 2010. High-water-content mouldable hydrogels by mixing clay and a dendritic molecular binder. *Nature* 463(7279): 339–343.

Wang, X., et al. 2013a. Effect of RGD nanospacing on differentiation of stem cells 34(12): 2865–2874.

Wang, X., et al. 2019. Advanced functional biomaterials for stem cell delivery in regenerative engineering and medicine. *Advanced Functional Materials* 29(23): 1809009.

Wang, Y., C. Xu, and H. Ow. 2013b. Commercial nanoparticles for stem cell labeling and tracking. *Theranostics* 3(8): 544.

Wei, M., S. Li, and W.J.J.o.N. Le. 2017. Nanomaterials modulate stem cell differentiation: biological interaction and underlying mechanisms 15(1): 1–13.

Wickham, A. M., et al. 2014. Polycaprolactone–thiophene-conjugated carbon nanotube meshes as scaffolds for cardiac progenitor cells 102(7): 1553–1561.

Winkler, T., et al. 2018. A review of biomaterials in bone defect healing, remaining shortcomings and future opportunities for bone tissue engineering: The unsolved challenge 7(3): 232–243.

Winnik, F. M. and D. Maysinger. 2013. Quantum dot cytotoxicity and ways to reduce it. *Accounts of Chemical Research* 46(3): 672–680.

Xiang, J., et al. 2022. Sandwich-like nanocomposite electrospun silk fibroin membrane to promote osteogenesis and antibacterial activities 26: 101273.

Xiao, X., et al. 2015. The promotion of angiogenesis induced by three-dimensional porous beta-tricalcium phosphate scaffold with different interconnection sizes via activation of PI3K/Akt pathways 5(1): 1–11.

Xie, X., et al. 2020a. Electrospinning nanofiber scaffolds for soft and hard tissue regeneration. *Journal of Materials Science & Technology* 59: 243–261.

Xie, Z., et al. 2020b. Black phosphorus-based photothermal therapy with aCD47-mediated immune checkpoint blockade for enhanced cancer immunotherapy. *Light: Science & Applications* 9(1): 1–15.

Xiong, F., et al. 2015. Cardioprotective activity of iron oxide nanoparticles 5(1): 1–8.

Xue, J., et al. 2019. Electrospinning and electrospun nanofibers: Methods, materials, and applications. *Chemical Reviews* 119(8): 5298–5415.

Xue, L., et al. 2013. Effects of human bone marrow mesenchymal stem cells on burn injury healing in a mouse model. *International Journal of Clinical and Experimental Pathology* 6(7): 1327–1336.

Yadava, S. K., et al. 2020. Low temperature, easy scaling up method for development of smart nanostructure hybrid lipid capsules for drug delivery application. *Colloids and Surfaces B: Biointerfaces* 190: 110927.

Yang, Q., et al. 2022. Electrospun aligned poly(ε-caprolactone) nanofiber yarns guiding 3D organization of tendon stem/progenitor cells in tenogenic differentiation and tendon repair. *Front Bioeng Biotechnol* 10: 960694.

Ye, K., et al. 2016. Interplay of matrix stiffness and cell–cell contact in regulating differentiation of stem cells 8(34): 21903–21913.

Yi, C., et al. 2010. Gold nanoparticles promote osteogenic differentiation of mesenchymal stem cells through p38 MAPK pathway 4(11): 6439–6448.

Yu, T.-T., et al. 2017. Influence of surface chemistry on adhesion and osteo/odontogenic differentiation of dental pulp stem cells 3(6): 1119–1128.

Yukawa, H. and Y. Baba. 2017. In vivo fluorescence imaging and the diagnosis of stem cells using quantum dots for regenerative medicine. *Analytical Chemistry* 89(5): 2671–2681.

Zakrzewski, W., et al. 2019. Stem cells: past, present, and future. *Stem Cell Research & Therapy* 10(1): 1–22.

Zhang, H., L. Zhou, and W. J. T. E. P. B. R. Zhang. 2014. Control of scaffold degradation in tissue engineering: a review 20(5): 492–502.

Zhang, L. and E. Wang. 2014. Metal nanoclusters: new fluorescent probes for sensors and bioimaging. *Nano Today* 9(1): 132–157.

Zhang, L. and T.J.J.N.t. Webster. 2009. Nanotechnology and nanomaterials: promises for improved tissue regeneration 4(1): 66–80.

Zhang, S., et al. 2018b. Polylactic acid nanopillar array-driven osteogenic differentiation of human adipose-derived stem cells determined by pillar diameter 18(4): 2243–2253.

Zhang, X., et al. 2017. Coculture of mesenchymal stem cells and endothelial cells enhances host tissue integration and epidermis maturation through AKT activation in gelatin methacryloyl hydrogel-based skin model. *Acta Biomaterialia* 59: 317–326.

Zhang, X., et al. 2018a. Nanostructured molybdenum disulfide biointerface for adhesion and osteogenic differentiation of mesenchymal stem cells. *Applied Materials Today* 10: 164–172.

Zhao, Y., et al. 2021. Advanced bioactive nanomaterials for biomedical applications. In *Exploration.* Wiley Online Library.

Zheng, X., et al. 2020. Microskin-Inspired Injectable MSC-Laden Hydrogels for Scarless Wound Healing with Hair Follicles 9(10): 2000041.

Zhu, Y., et al. 2020. Multicomponent transition metal dichalcogenide nanosheets for imaging-guided photothermal and chemodynamic therapy. *Advanced Science* 7(23): 2000272.

12 Conjugated Nanoparticles for Biotechnology and Bio-Analysis

Aparajita Ghosh, Ambati Himaja, Swati Biswas, Onkar Kulkarni, and Balaram Ghosh

12.1 INTRODUCTION

Nanoparticles (NPs) offer enormous advantages to be used as drug carriers due to their dimension and physicochemical properties. Their size ranges from 1 nm to 100 nm, allowing them to readily pass-through cell membranes and remain in circulation within the body for extended periods. Because they are encapsulated and biologically unavailable while in the systemic circulation, their high surface/volume ratio enables the incorporation of significant amounts of two or more drugs for combination therapy or for the simultaneous delivery of two or more therapeutic modalities, such as radiation and drugs with lower systemic toxicity when compared to free drugs in solution (Babu 2014, 709). Additionally, they can prevent drug degradation and improve the stability of biomolecular drugs like peptides, oligonucleotides, DNA and RNA molecules, and other substances by making hydrophobic compounds more soluble in water (Xiao 2020, 2001048). Another benefit is the ease with which drugs may be delivered across several barriers, the blood-brain barrier (BBB) being the most significant (Ceña and Játiva 2018, 1513). Liposomes (Akbarzadeh 2013, 1), polymeric nanoparticles (PNP) (Zielińska 2020, 1), polymeric micelles (PM) (Vinchurkar and Kuchekar 2021, 629), dendrimers, gold NPs, quantum dots, carbon nanotubes, and nanofibers are examples of NPs now in use.

The combination of targeted delivery and controlled drug release technology offers numerous advantages in chemotherapy with greater efficacy because the optimal concentration of active drug can be maintained on the desired local site thus decreasing the toxic side effects, and greater convenience owing to the need for fewer applications or treatments (ud Din 2017, 7291). Therapeutic NPs that are specifically targeted for a cell, tissue, or illness offer a potentially powerful technology. Targeted NPs must be engineered to have the desired properties to overcome the various physiological barriers and conditions amongst their passage to the target, as well as the ability to selectively recognize the target cells, and deliver the drug being internalized. Investigating the physiological changes in the diseased cells and tissues may lead to precise identification preferably by unique moieties, such as specific antigens, carbohydrates, or surface receptors, that are exclusively produced or overexpressed by disease cells (Zhao 2020, 153). Fabricating NPs to further increase their specificity and effectiveness, by conjugating specific units such as antibodies, lectins, peptides, proteins, DNA or RNA aptamers, and small molecules able to recognize those entities and deliver the drug to the target cells, called "active targeting" (Ranganathan 2012, 1044). Furthermore, intensive research is going on for the development of NPs as effective drug delivery systems for medical practice, especially for chemotherapy and gene delivery to which progress in NP technology, material science, and cellular and molecular physiology have contributed.

DOI: 10.1201/9781003305583-12

Bioconjugation is the chemical or biological bonding of biomolecules to nanoparticles, suitable for therapeutic applications. It also comprises the conjugation of biologically active substances to nanoparticles. The development of hybrid nanomaterials, which combine the highly selective catalytic and recognition properties of biomaterials like proteins/enzymes and DNA with the distinctive electronic, photonic, and catalytic features of nanoparticles, is the outcome of the convergence of biotechnology and nanotechnology. Within nanobiotechnology, the conjugation of NPs with biomolecules is a promising research field. Given their special capabilities for identification, transport, and catalysis, biomolecules are interesting macromolecular structures. The creation of innovative biosensors may benefit from the electrical or optical transmission of biological events provided by the conjugation of NPs with biomolecules. Nanoparticles and biomolecules have structural compatibility because the dimensions of biomolecules are similar to those of nanoparticles. The end result is a blend of favorable characteristics. Functionalizing nanoparticles by conjugating them with distinct groups will help in increasing stability, biocompatibility, and functionality and it also helps them in having novel and improved characteristics and increase their application range. A range of moieties can be conjugated to the nanoparticles including low molecular weight ligands (folic acid, thiamine, dimercaptosuccinic acid), peptides (RGD, LHRD, antigenic peptides, internalization peptides), proteins (BSA, transferrin, antibodies, lectins, cytokines, fibrinogen, thrombin), polysaccharides (hyaluronic acid, chitosan, dextran, oligosaccharides, heparin), polyunsaturated fatty acids (palmitic acid, phospholipids), DNA, plasmids, siRNA, and so forth (Arruebo 2009, 1). In this Chapter, we discuss different conjugation strategies of bioconjugation, factors involved in efficient and robust bioconjugation, and available bioconjugated nanoparticles with their diverse applications.

12.2 CONJUGATION STRATEGIES

12.2.1 Noncovalent Interaction – Physical Conjugation Strategy

Electrostatic interaction – Electrostatic adsorption is a physical method for binding biomolecules to the surface of nanoparticles (Figure 12.1). The nanoparticle-biomolecule system is made up of nanoparticles and biomolecules bearing opposite charges (Shen 2003, 389; Zhang 2012, 239). At the isoelectric point, maximum adsorption will take place because there will be less repulsive forces between the biomolecules so that they will pack on the surface of the nanoparticle more efficiently (Saallah and Lenggoro 2018, 95).

For example, the effective binding of Human Serum Albumin on the surface of silver nanoparticles which is produced by the reduction of citrate, at higher pH values than the HSA isoelectric point can be achieved by the attraction between the negatively charged silver nanoparticles and the positively charged HSA. To develop biomolecule-nanoparticle conjugates, electronic adsorption is one of the simple and convenient methods, but it has some limitations. Conjugates obtained by this method are unstable and changes in the conformation of biomolecules are also induced (Zhang 2010, 240).

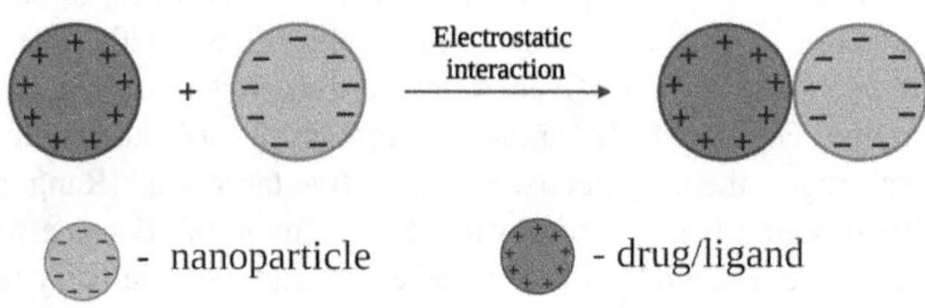

FIGURE 12.1 Electrostatic Interaction between drug and nanoparticle.

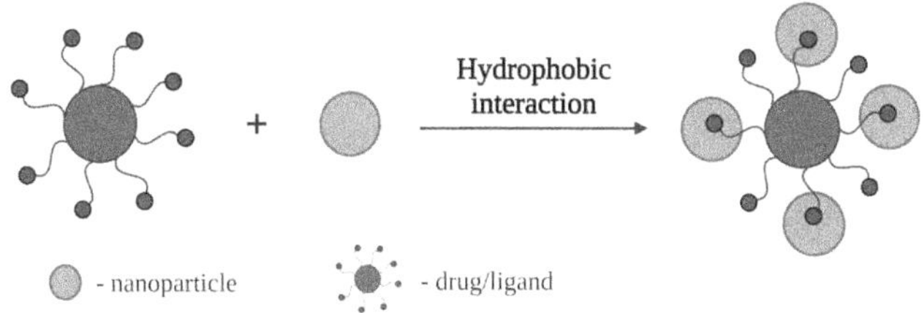

FIGURE 12.2 Hydrophobic interaction between drug and nanoparticle.

12.2.2 Hydrophobic Interaction

Bioconjugation of nanoparticles with hydrophobic biomolecules is favored when the surface charge of nanoparticles is close to zero. Hydrophobic interaction is helpful in recognizing and adsorbing biomolecules on nanoparticles of countercharge at variable pH conditions (Figure 12.2). We can also modulate the biomolecule uptake onto the nanoparticle surface by tuning the surface properties and environmental conditions (Puddu 2012, 6356). But these interactions sometimes induce significant conformational changes in biomolecules which may cause denaturation and loss of biological activity (Shemetov 2012, 4588; Marques 2020, 183).

12.2.3 Streptavidin-Biotin Affinity Binding (Bio-Affinity Interaction)

The introduction of specific functional groups onto the surface of the nanoparticle will help in the efficient binding of biomolecules to nanoparticles via affinity interactions (Figure 12.3). Avidin, a glycoprotein has high specificity and affinity to biotin, the result is the interaction with a stronger affinity (Saallah and Lenggoro 2018, 98; Zhang 2010, 240). Biotin-functionalized nanoparticles have been frequently employed to link with streptavidin, which may be used as a linker for further conjugation with biomolecules (Jain 2017, 28; Zhu 2008, 193; Haes 2002, 10598; Nath 2002, 508). Compared to non-biotin conjugated nanoparticles, biotin-decorated PEG-PCL nanoparticles co-delivering DOX and quercetin to DOX-resistant breast cancer were shown to increase in vivo antitumor drug efficacy (Marques 2020, 188; Lv 2016, 32185; Mehdizadeh 2017, 496). However, this system has some disadvantages like the high possibility of immunogenicity, and glycosylation, it also has a high isoelectric point and nonspecific binding *in vivo,* and also due to stronger avidin-biotin interaction, there will be a hindrance to the release of biomolecules tagged to avidin or biotin (Zhang 2010, 240). To overcome these disadvantages, we should have to choose non-glycosylated and near-neutral forms of avidin. Streptavidin, an avidin analog was discovered after extensive research effort. This resulted in a decrease in nonspecific electrostatic interactions. The interaction of streptavidin/biotin is based on five important hydrogen bonds between the ureido ring of biotin and the indole ring of tryptophan at the hydrophobic binding pocket of streptavidin. This hydrogen bonding makes the complex more stable/resistant to proteolytic enzymes, denaturants, pH, and temperature under extreme conditions (Marques 2020, 188; Liu 2016, 1; Huberman 2001, 32031; Liu W 2017, 2392).

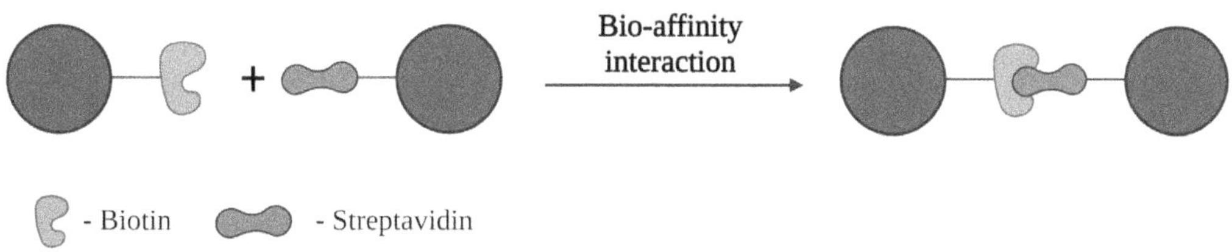

FIGURE 12.3 Binding of drug/biomolecule to nanoparticle via affinity interactions.

12.2.4 Encapsulation

Encapsulation of biomolecules inside nanoparticles is a unique approach that can provide better protection than surface binding and can regulate substrate diffusion, product release, or keep inhibitors out. Wrapping biomolecules entirely in NPs provide sturdy protection for the fragile molecules against the extreme conditions encountered in many applications (Liu Y 2017, 2). The biomolecules can be encapsulated within the nanoparticle (nanocapsule), or adsorbed on the surface of the nanoparticles (nanoshells). Encapsulation helps in the controlled release of biomolecules at the targeted sites. β-Carotene encapsulation into whey protein resulted in the formation of nanocapsules with high stability against photooxidation and high encapsulation efficiency (Zhang 2010, 241).

12.2.5 Covalent Interactions

Covalent binding is chosen over adsorption because of its great stability and repeatability. Covalent bonds are less vulnerable to disassembly, resulting in better conjugation with biomolecules. The frequently used covalent binding techniques for nanoparticle-biomolecule conjugation are carbodiimide chemistry and maleimide chemistry.

12.2.5.1 Carbodiimide Chemistry

It is a crosslinking method, in which carboxylic acids are crosslinked to primary amines using the carbodiimide compound 1-ethyl-3-(3-dimethyl aminopropyl) carbodiimide (EDC) (Figure 12.4) (Lin 2021, 43; Conde 2014, 14). Generally, primary amines are available abundantly on the surface of biomolecules and they are also more reactive with a variety of functional groups, such as aldehydes, and cyanogen bromide present on the surface of

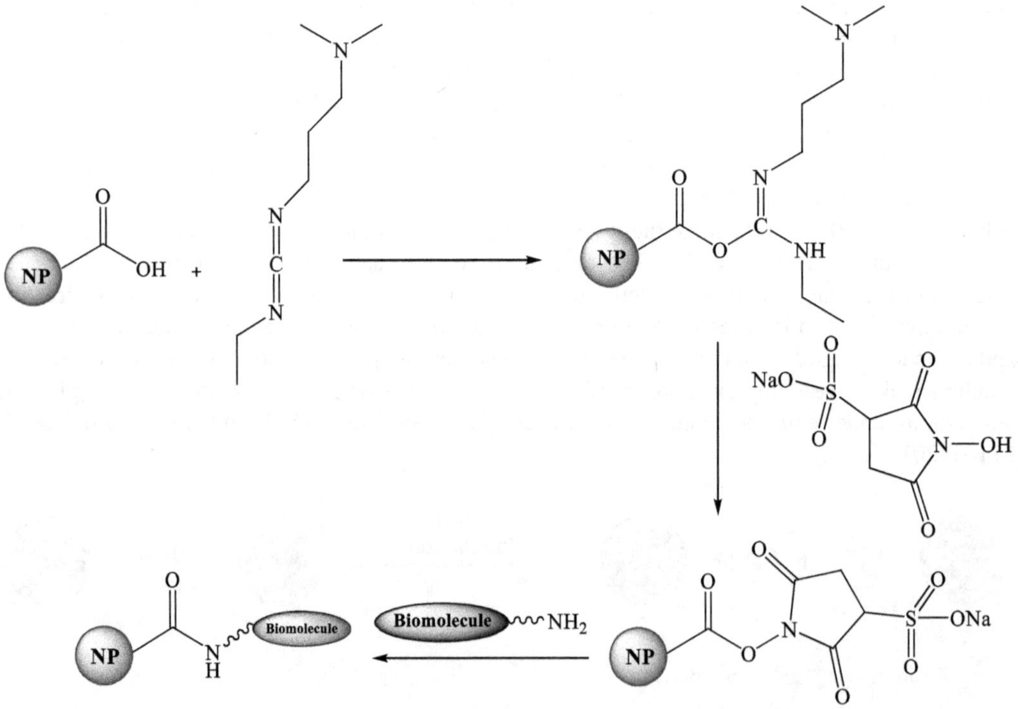

FIGURE 12.4 Crosslinking of drug/biomolecule with nanoparticles using carbodiimide chemistry.

nanoparticles without further chemical modification. For better coupling, EDC is often used in combination with *N*-hydroxy succinimide (NHS) or *N*-hydroxysulfosuccinimide (sulfo-NHS). The presence of NHS/sulfo-NHS prevents intra and intermolecular crosslinking of biomolecules (Conde 2014, 14; Yao 2016, 276). During bioconjugation of carboxyl moiety with amine, EDC reacts with carboxyl groups on the surface of the nanoparticle forming an unstable reactive amine intermediate, *o*-acylisourea ester. If this intermediate fails to react with the amine in an aqueous solution, then the ester will get hydrolyzed and regenerate acid. So, to stabilize this intermediate NHS/sulfo-NHS modify this reactive amine ester to a semi-stable active ester. These semi-stable states are reactive toward amines present in biomolecules and produce stable amide linkages releasing NHS/sulfo-NHS group.

For example, Docetaxel (DOC)-loaded poly (ethylene glycol)-b-poly (-caprolactone) (PEG-PCL) nanoparticles are functionalized with a programmed death-ligand 1 (PD-L1) antibody. Using EDC and sulfo-NHS, DOC-PEG-PCL nanoparticles were effectively attached to antiPD-L1, resulting in an amide bond. The cellular absorption, cytotoxic effects, and cellular apoptosis of PD-L1-conjugated nanoparticles were then investigated in three gastric cancer (GC) cell lines (HGC27, MGC803, and MKN45). In comparison to non-functionalized DOC-PEG-PCL nanoparticles, antibody-conjugated nanoparticles were more effective in the inhibition of PD-L1 expression in GC cells (Marques 2020, 185; Xu 2019, 19). The major drawback of carbodiimide chemistry is that biomolecules contain both carboxyl and amino groups, so this may lead to the activation of carboxyl groups in biomolecules and there is a possibility of self-polymerization. Additionally, there is a lack of selectivity between functional groups during coupling as we cannot predict the orientation of biomolecules (Marques 2020, 185; Kocbek 2007, 20; Li 2015, 4277).

12.2.5.2 Maleimide-thiol Chemistry

Maleimides have emerged as a prominent method for producing site-specific cysteine modifications in biomolecules (Ravasco 2019, 45; Lu 2021, 10). Because of its high conjugation efficiency and the stability of the conjugates, maleimide–thiol chemistry is preferred for conjugation (Figure 12.5). Maleimides form a thioether bond specifically by reacting with thiols. If thiols are not available or not exposed enough on the surface, thiolation is required. The most commonly used thiolation reagent is 2-iminothiolane (Liu 2017, 2584; Yanase 2014, 5919). We can also functionalize biomolecules using N-succinimidyl 3-(2-pyridyldithio) propionate), which can be reduced to thiol by a reducing agent dithiothreitol or β-mercaptoethanol. The reduction results in producing reactive sulfhydryls so that biomolecules can be selectively immobilized onto the surface of the nanoparticles. Due to their greater selectivity for the sulfhydryl side chain of cysteine, maleimide-activated crosslinkers stand out among all sulfhydryl-reactive crosslinkers. The reactivity of maleimide toward the sulfhydryl group is much faster compared to reactivity with primary amines (Marques 2020, 185; Borges 2014, 3; Akkapeddi 2016, 2956).

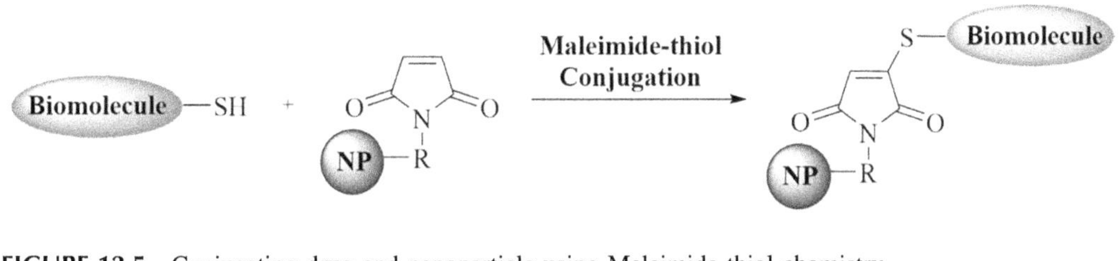

FIGURE 12.5 Conjugating drug and nanoparticle using Maleimide-thiol chemistry.

12.2.5.3 Click Chemistry

"Click chemistry" refers to a class of chemical reactions that are orthogonal, site-specific, and have a high reaction rate (Figure 12.6). The reactions are simple and occur with great efficiency at room temperature. As a result, irreversible chemical bonds are formed, and harmful wastes are eliminated. The main reactions of click chemistry include cycloadditions, Staudinger ligation, and thiol-ene reactions (Marques 2020, 187; Yi 2018, 3; Hein 2008, 2218). In biological systems, azides and alkynes are non-existent and they are also very inert to most functional groups and biomolecules. These characteristics assure that conjugation occurs exclusively at the specific site (s). EDC/NHS coupling or maleimide–thiol conjugation can be used to functionalize nanoparticles with azide or alkyne molecules. Each heterobifunctional linker in this context has an NHS ester or a maleimide at one end and an azide or alkyne at the other end (Marques 2020, 187–188; Pickens 2017, 687).

Generally, click chemistry reactions are accelerated significantly in the presence of a copper catalyst (Copper-catalyzed cycloaddition). However, the toxicity associated with copper catalysts limits their applications in living systems. With significant efforts, a strain-promoted azide-alkyne cycloaddition (SPAAC) has been developed to overcome the limitations (Smyth 2014, 1778; Haldón 2015, 9530; Sun 2015, 17073; Hatit 2018, 3). To create stable triazoles, SPAAC is one of the spontaneous and stable cycloaddition reactions. Cyclooctyne (OCT) is the smallest stable cycloalkyne with low activation energy that allows cycloaddition without a catalyst. Therefore, it was the first strain-promoted reactive alkyne, though comparatively slower than copper-catalyzed cycloaddition. Therefore, optimization of second-order rate constants of SPAAC was done with different OCT-based derivatives like bicyclo [6.1.0] nonyne (BCN), dibenzocyclooctynol (DIBO), and dibenzocyclooctyne (DBCO) (Pickens 2017, 687; Dommerholt 2016, 4; Dommerholt 2014, 3; Ning 2008, 2285; Dommerholt 2010, 9423; Debets 2010, 98).

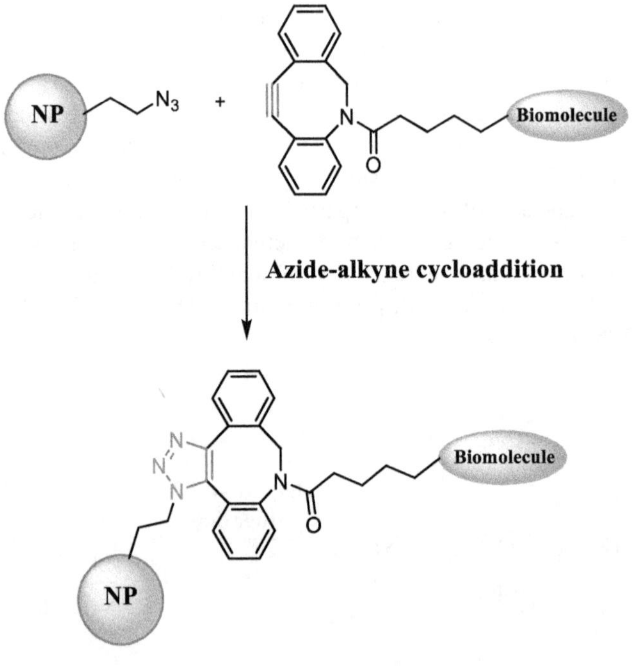

FIGURE 12.6 Conjugation using click chemistry.

Jeong et al. synthesized silica-encapsulated nanoparticles (SiNP) which had been functionalized with a human epidermal growth factor receptor 2 (HER2) antibody (anti-HER2-SiNP) using the coupling methods SPAAC and EDC/sulfo-NHS. Surface of SiNP was modified with azide and the cycloalkyne DIBO was introduced onto the antibody using aza-dibenzocyclooctyne-maleimide with the help of free thiol groups which are obtained by reduction. Then the modified nanoparticles and antibody are mixed at room temperature which resulted in the antibody-conjugated SiNP. Despite the efforts to fasten kinetics, SPAAC remains significantly slower than copper-catalyzed cycloaddition. Furthermore, the hydrophobicity of the cyclooctyne ring reduces the water solubility of OCT-containing reagents. Further, the side reaction between strained alkynes, particularly BCN, and thiol groups is defying biorthogonality (Jeong 2017, 2550).

Another catalyst-free click chemistry reaction, an alternative to SPAAC, is inverse electron demand Diels-Alder Reactions (iEDDA). It is a reaction between a diene and dienophile which allows orthogonal bioconjugation with a faster rate compared to copper-catalyzed cycloaddition and SPAAC (Marques 2020, 187; Kim 2019, 7836; Devaraj 2011, 818). Covalent interactions between biomolecules and nanoparticles can also be achieved through chemisorption and also by introducing biofunctional linkers.

- A simple chemical reaction that occurs between the nanoparticles and the thiol groups of biomolecules through residues of cysteine which are present on the surface of the biomolecules is called chemisorption (Figure 12.7). Gold nanoparticles are highly reactive toward the thiol group and form an Au-S bond very easily and with a stronger affinity. By taking the advantage of this strong binding affinity, many biomolecules, such as DNA, peptides, proteins, and antibodies, are conjugated onto the surface of the nanoparticles. If thiol residues are not available on the surface of the biomolecules, we need to introduce them onto the surface of biomolecules for interaction with nanoparticles. Generally, cysteine is used to introduce thiol groups on the surface of biomolecules, or else we can also use 2-iminothiolane (Traut's reagent) (De 2008, 4227; Saallah and Lenggoro 2018, 98).
- Covalent binding through bifunctional linkers (Figure 12.8) uses the low molecular weight bifunctional linkers for the conjugation of biomolecules with noble metal nanoparticles. These bifunctional linkers possess both the anchor groups, such as thiols, disulfides, and phosphine ligands, which help in attaching to the nanoparticle surface and the functional groups which help in covalent binding to the target molecule (Haes 2004, 1030; Riboh 2003, 1773; Nath 2002, 505; Haes 2002, 10599). For example, the mixed

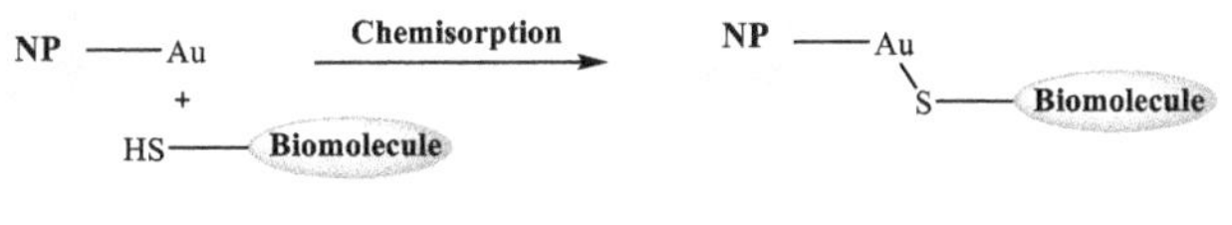

FIGURE 12.7 Conjugation of drug and nanoparticle via chemisorption.

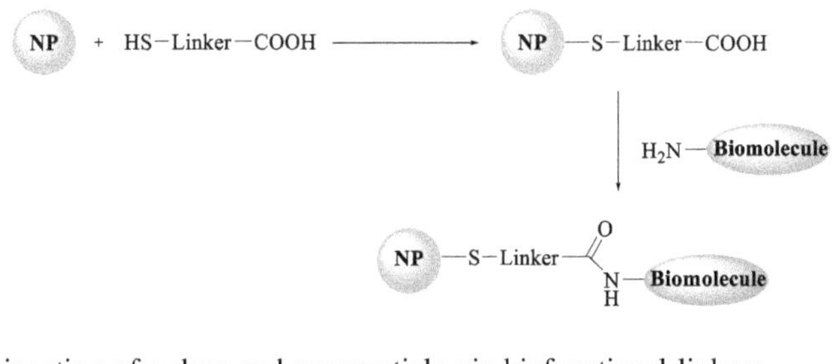

FIGURE 12.8 Conjugation of a drug and nanoparticle via biofunctional linkers.

monolayer of mercaptoundecanoic acid/6-mercapto-1-hexanol is used as a bifunctional linker, which is combined with the surface of silver nanoparticles using thiol anchor groups and further conjugated with IgG protein by forming a peptide bond. Anchor groups in bifunctional linkers help replace the molecules which are weakly adsorbed to nanoparticles. By using bifunctional linkers, we can overcome the problem of instability and inactivation. As the approach maintains the biomolecules' stability and activity, it allows subsequent conjugation with additional biomolecules (Huang 2007, 7710).

12.3 FACTORS AFFECTING CONJUGATED SYSTEM EFFICIENCY

A range of chemotherapies has been evaluated for encapsulation in nanoparticle systems. Whilst there are various forms of nanoparticles, from liposomes to dendrimers, the choice of nanoparticle carrier is usually dependent on the nature and solubility of the cargo drug to ensure a high drug entrapment. For example, polymeric nanoparticles consisting of commonly employed poly-lactide-based polymers are frequently used for the entrapment of hydrophobic agents, whereas liposomes can entrap hydrophobic or hydrophilic drugs in either the phospholipid bilayer of the liposome or its aqueous lumen respectively. The combination of chemotherapies in nanoparticles has also been explored. This has advantages as it offers the opportunity to overcome pharmacokinetic differences in drug agents to ensure their delivery at the disease site in the required proportions/ ratios to provide synergistic therapeutic effects.

12.3.1 Drug Load Onto Nanoparticle

Based on diverse nanoparticle-based drug delivery technologies, several nanomedicines have been created. However, only limited number of nanomedicines, including Abraxane® and Doxil, have been approved for clinical use by the US Food and Drug Administration (US-FDA) despite extensive research efforts and significant investments over the past decade. Drug-loading is one of the major issues that prevents the clinical application of nanomedicine. One of the major difficulties is inadequate drug-loading with uncontrolled drug release (Liu 2020, 590). Due to difficulties in scale-up in manufacturing with repeatable characteristics, high production costs, and potentially harmful side effects, clinical translation of low drug-loading nanomedicines is difficult. Additionally, a very high particle concentration is necessary to obtain a pharmacological therapeutic window, yet the highly viscous solution of such a high NP concentration makes intravenous administration challenging. Therefore, it is crucial to reduce these side effects to enhance drug-loading. The number of nanoparticles required to attain the therapeutic level is much lower when using high drug-loading nanoparticles, which can both lessen the risk of side effects from overdosing and lower the cost of producing the nanomedicine. To deliver the high drug dosage with a small amount of carrier material, high drug-loading nanoparticles would be desired (Yang 2019, 14357; Shen 2017, 1369; Zhao 2020, 152).

12.3.2 Immune Evasion

Fabricated nanoparticles are readily internalized by phagocytes which is a primary step of innate immune system activation. The immune system has the astonishing ability to recognize and produce antibodies against almost any chemical species, natural or manufactured as it identifies repeating patterns of any resemblance to those found on (nano-sized) viruses (Fadeel 2019, 2). Interestingly, researchers have found specific antibodies to be produced against structured nanosystems, such as single-walled carbon nanotubes (SWCNTs) (Erlanger 2001, 465). In another study, an antibody fragment with high affinity and selectivity for gold surfaces was identified. Therefore, anit-gold human antibodies have been used in various types of research to bio-functionalize NPs to increase affinity toward specific targets. Moreover, it was observed through intensive study that NP conjugation to certain key protein carriers is necessary for successful antibody induction. Hereby, the NPs attached to small proteins incline to behave as haptens and prompt an immune response once enter the systemic circulation. Nevertheless, as nanomaterials rapidly associate with proteins when they enter the body, close attention to the potential immunogenicity of nanomaterials is necessary. Various factors drive the intensity of the immune response generated, such as physicochemical factors, size and shape of NPs, and most importantly the surface modifications of designed nanosystems (Liu 2017, 2393). In this context, antibodies against the surface coating of nanomaterials, including polyethylene glycol (PEG), are also important to consider. To add to the complexity, metal/metal oxide NPs may undergo dissolution with the release of metal ions, and though this is recognized as one potential mechanism of nanotoxicity, there are few studies on the immunogenic role of the released ions. NPs can be designed to either specifically interact with or avoid recognition by the immune system and to generate modification to immune response for therapeutic goals as in novel immunotherapy approaches (Liu 2017, 2394).

12.3.3 Drug Release Factors

Liposomes and nanoparticles are two examples of colloidal drug delivery technologies for intravenous injection that are highly appealing for achieving controlled release, preventing drug degradation, and avoiding adverse consequences. However, two factors – a short blood half-life (rapid removal from the bloodstream) and nonspecific targeting – may impair effective drug delivery with these devices. The mononuclear phagocyte system (MPS), which is recognized by macrophages because of the adsorption of blood proteins (opsonins) onto the surface of colloidal carriers, is what causes the fast removal from the bloodstream. As a result, the liver and spleen, which are MPS organs, begin to accumulate particulate drug carriers. It is feasible to change the surface of colloidal drug carriers with hydrophilic, flexible, and non-ionic polymers, such as poly (ethylene glycol), to produce long-circulating colloidal drug carriers (PEG). Some authors have sought to boost the tissue specificity of colloidal drug carriers by coupling targeting agents, such as monoclonal antibodies, to address the issue of site-specific targeting. However, this method has at least two drawbacks: the immunogenicity of the antibodies and their overall size, which make it difficult for particles to pass across biological barriers.

The targeted release of the chemotherapy at the site of the disease is crucial with conjugated nanosystem technologies. This is mainly owing to the possibility of adverse effects if healthy tissues are exposed to the very effective payload in the case of bioconjugates like antibody conjugates. Conjugates are often built in the ways so that the drug will only be released upon internalization into the target cell to prevent the side effects. Complex linker chemistry using either cleavable or non-cleavable linkers, is frequently needed for this. A change in the physical environment, such as a lysosomal pH that drops or the presence of a cathepsin-labile motif, typically causes cleavable linkers to break (Sheyi 2022, 2). With non-cleavable linkers, the conjugates general breakdown/metabolize in the endo-lysosomal lumen that determines how quickly the payload is released (Tsuchikama 2018, 31).

The linker utilized to conjugate the antibody in the developed nanoparticles can also be polymers or lipids, in which case the drug release is unaffected. Drug release in these formulations is, therefore, simply a consequence of both drug diffusion and particle degradation. The use of lower potency chemotherapies, as previously stated, may make it possible to tolerate a tolerable premature release of the medication, even though it is nearly impossible to completely prevent some leaching of the drugs from the nanoparticles unless covalently linked. The use of ion-pairing, which helps to hold the drug within the nanoparticle longer through electrostatic interactions, has been investigated as a strategy to maintain the drug inside polymeric nanoparticles until uptake. This limits early release and provides a longer therapeutic window once they are at the disease site.

12.3.4 Toxicity

The enormous benefits that nanoparticles provide are largely known to everyone, but less is known about their toxicity, nonspecific protein interactions, translocation to secondary target organs, etc. Due to their widespread usage in cancer therapy, carbon nanotubes have received a lot of attention among the many nanocarriers. However, the use of carbon nanotubes in non-anti-cancer medications has also been extended, either as a carrier or as an adjuvant. When utilized in drug delivery systems, the advanced physical, mechanical, and chemical features of CNT cause significant concern among researchers. This is because CNTs may transport the carrier biomolecule while moving across cell membranes. Therefore, it is necessary to understand the cytotoxicity of CNTs. According to studies, CNTs can have negative effects on the nervous system, the lungs, the immunological system, the embryo, the genitalia, the liver, the heart, and the blood vessels (Gholamine 2017, 341; Park 2016, 258; Lee 2015, 7109).

One of the FDA-approved superparamagnetic iron oxide nanoparticles (SPIONs), is widely employed in both medicinal and diagnostic applications. The use of SPIONs as prospective drug delivery methods to treat cancer and other disorders has, however, increased in recent years. However, the safety issues with SPIONs have in some way restricted their usage. Some SPIONs that had previously received FDA approval for use in medical imaging have been withdrawn due to toxicity and allergic response issues. SPIONs are also known to affect the complement system, which can have negative effects on innate immunity. One such SPION-based medication that is known to affect the complement system is FeridexTM, which the FDA has asked to stop selling. Recent investigations have shown that the application of coating agents can lessen the toxicity of SPIONs. Other metal nanoparticles, including as titanium, gold, silver, and zinc, have been utilized for drug delivery but have shown considerable toxicity, similar to SPIONs (Bakhtiary 2016, 289; Veranth 2007, 4; Benasutti 2017, 2750).

For a variety of medications used in treatments, lipid-based nanoparticles function as amphipathic nanocarriers. Liposomes interact with a variety of biomolecules, including LDL, HDL, and opsonins, when they enter a live system. Opsonins trigger the activation of RES recognition, which then works to remove it from the system. The stability of liposomes within the biological system is known to be decreased by other lipoproteins in the blood by causing a rearrangement of surface lipids, that is one of the main issues with employing liposomes as a method of medication administration is this. PfzerBioNTech used two lipid nanoparticle-based mRNA COVID vaccines, which have been shown to be 95% effective, have just received FDA approval for emergency use. However, vaccine volatility has been shown to be a significant problem. To improve the vaccine stability, lipid nanoparticle structural modifications are required. For disorders brought on by changed TTR (Transthyretin) protein, the FDA and EC authorized Alnylam Pharmaceuticals' ONPATTRO in 2018. However, respiratory problems, headaches, and changes in blood pressure are a few of the frequent side effects seen after using medications (Daraee 2016, 410; Alavi 2017, 3).

Due to its advantages, silica nanoparticles have been utilized to transport cargos of such anti-cancer medications (such as camptothecin and doxorubicin in the treatment of breast cancer), nucleic acids, and proteins (such as RGD peptide and aptamers) through controlled releasing technique. Assessment of possible toxicity issues linked with nanoparticles is frequently required for the expanding applications. According to certain report, silica nanoparticle toxicity affects the immune system, causes neurotoxicity and pulmonary illnesses (Li 2012, 5380; Lu 2010, 1798; Bancos 2014, 4).

Smart medication delivery uses polymeric nanoparticles, which can be created from synthetic or natural polymers. In-depth research is being done on them as carriers for controlled/sustained release in drug delivery systems. The limited action of these nanopharmaceuticals is caused by safety concerns, toxicity concerns, poor biocompatibility, and physiological difficulties. Hazardous monomer aggregation, leftover material associated with them, and toxic breakdown processes are drawbacks of these nanoparticles. Quantum size effects, which are associated to oxidative stress, cytotoxicity, and genotoxicity, have an impact on the toxicity of polymeric NPs (Anagnostou 2020, 332; Singh 2017, 1517). Thus, in order to identify possible problems and develop preventative measures, the safety evaluation is essential for the usage of nanoparticles in living things. It is critical to evaluate the risk to benefit ratio assessment before introducing nanoparticles into the therapeutic market. In order to give a more thorough investigation of the nanoparticles utilized in medication delivery, toxicological studies are also required. Researchers will be able to further improve the characteristics of nanoparticles to lessen their toxicity and will develop effective yet non-toxic nano formulations for therapeutic/diagnostic usage.

12.4 AVAILABLE CONJUGATED NANOPARTICLES

12.4.1 Antibody-Conjugated Nanoparticles

Since the beginning of the 20th century, cancer research has placed a lot of emphasis on the targeted administration of chemotherapy to tumors. One approach of improving therapeutic targeting to the tumor location is the use of antibody-drug conjugates (ADCs). Modern ADCs are monoclonal antibodies that have been covalently coupled to extremely powerful medicinal compounds. These “Trojan horse” treatments are intended to bind to a tumor-specific receptor, internalize, and undergo lysosomal metabolization, releasing one to eight drug molecules that can then trigger their deadly mechanism(s) of action. ADCs were initially tested in the late 1950s employing leukemia cell-targeting antibodies with methotrexate conjugation. More than 40 years down the line, in 1983, positive findings from the first human clinical study employing an ADC were announced. Antibody and ADC medication development, however, being stalled for a while. To solve the problems of single antibody species manufacturing and immunogenicity, it needed discoveries like hybridoma fusion and humanization, which made it possible to create antibody-based medicines.

The US-FDA granted expedited clearance to gemtuzumab ozogamicin (Mylotarg1) in 2000 as the first ADC medication for the treatment of acute myeloid leukemia (AML). Mylotarg1 had a DNA-fracturing payload and was directed at CD33 (calicheamicin). It became the first formulation to receive rapid approval before being later withdrawn after receiving poor findings from a post-approval trial. The industry’s interest in ADCs was somewhat stifled by Mylotarg1’s under-whelming performance, but in 2011 and 2013, respectively, the FDA-approved the anti-tubulin payload-based ADCs Brentuximab vedotin (Adcetris1) which targets CD30 positive lymphoma, and trastuzumab emtansine (Kadcyla1) which targets HER2 positive breast cancer. The most recent ADC to receive approval was Besponsa1 in 2017. It targets CD22-positive leukemias and, like Mylotarg1, employs the DNA-targeting calicheamicin payload. About 60 ADCs are now undergoing clinical studies for different malignancies.

A technique that draws on the success and potential of both ADCs and nanotechnology is antibody-conjugated nanoparticles (Figure 12.9(A)). In that the antibodies may be utilized to precisely target sick cells and deliver encapsulated cargo medications, conceptually antibody-conjugated nanoparticles (ACNPs) and ADCs are comparable. 1980 saw the publication of the first targeted nanoparticles, while 2011 saw the first ones enter clinical trials.

12.4.2 Folic Acid Conjugated Nanoparticles

Folic acid is one kind of man-made form of folate, a B vitamin, naturally found in certain fruits, vegetables, and nuts. It was originally extracted by Lucy Willis in 1931 from a type of yeast, as the nutrient able to prevent anemia during pregnancy. Biologically, folic acid is one of the key elements essential for transferring one-carbon units involved in phospholipid, DNA, protein, and neurotransmitter syntheses. It has been investigated to have multifactorial benefits in birth defects,

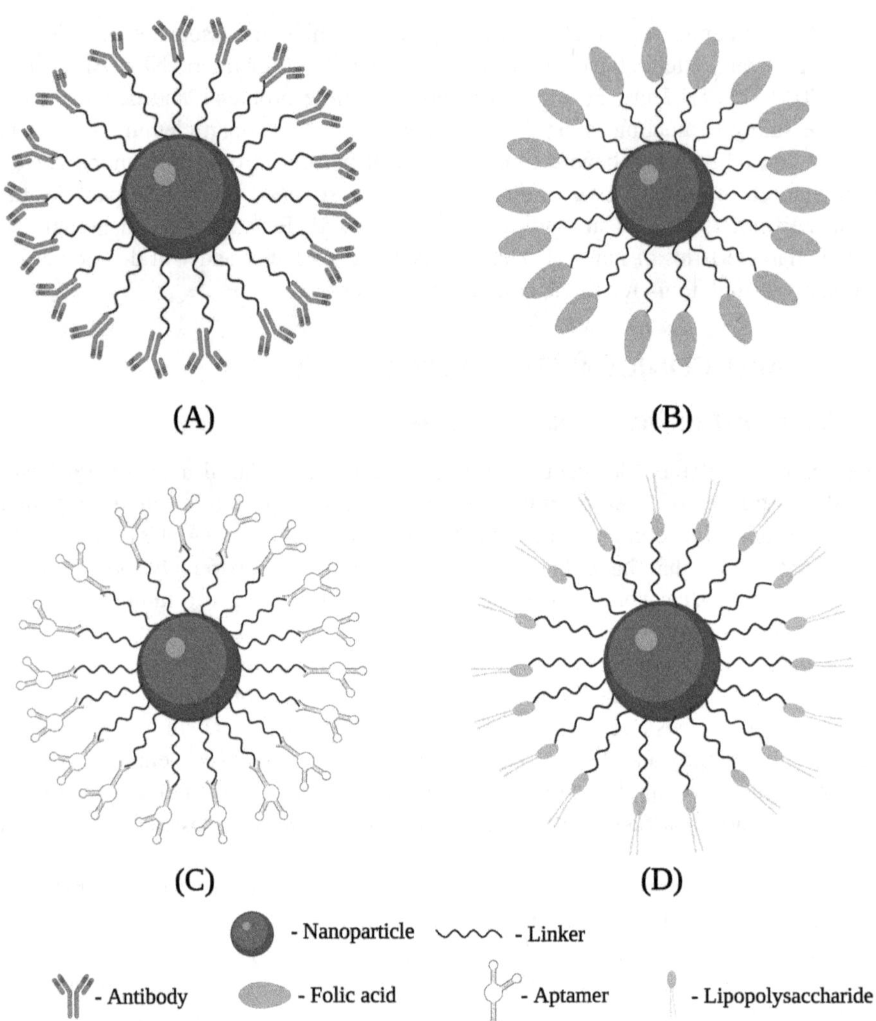

FIGURE 12.9 Different types of Conjugated nanoparticles.

fertility, heart disease, stroke, and neurological disorders. Moreover, folate intake appears to be inversely related to the risk of colorectal cancer (Ho 2009, 5672).

The role of folic acid as a biological ligand has been explored widely in the past few decades. It is useful for targeting cell membrane and enhancing endocytosis of nanoparticles selectively into folate receptor overexpressed cells. The efficiency of this biomolecule relies on the factors such as high stability, inexpensive, and poorly immunogenic with high affinity for its cell surface receptor (Kd ~1nM)(Cheung 2016, 52554). The alpha isoform of the folate receptors has been reported to be overexpressed on the surface of cancer cells often of epithelial origin, including the carcinoma of the ovary, lungs, breast, kidney, uterus, and colon (Lutz 2015, 119; O'Shannessy 2012, 2). The overexpression of folate receptor has been correlated with promoting the growth and metastasis of cancer and therefore has been explored as one of the tumor markers. The cellular uptake mechanism of Folic acid conjugates, which are covalently derivatized via folate's γ-carboxyl moiety, is generally uncompromised, as the bioligand helps entry of the conjugated nanoparticles by folate receptor-mediated endocytosis and move through many organelles by vesicular trafficking (Yang 2010, 679).

A diverse range of folic acid conjugated systems (Figure 12.9(B)) have been exploited in the case of cancer imaging and therapeutics. Folic acid derivatives conjugated to a variety of metal chelates have allowed non-invasive detection and offer greater tissue selectivity, especially in FRα-positive tumors (Stallivieri 2017, 2–4). One of the best studied of these probes, 111In-DPTA-folate, showed excellent on-target to off-tumor contrast in animal models, resulting to get into a phase I/II study (Radioisotopes and No 2009). In preclinical models of ovarian cancer, a different folic acid peptide derivative (EC20) conjugated with 99mTechnetium (99mTc-EC20), which is less expensive and has a shorter half-life, has also shown encouraging results (Cheung 2016, 52554). Etarfolatide, commonly known as FolateScan, is a 99mTc-based imaging agent that has been the focus of several trials in patients with ovarian, kidney, lung, and other resistant solid tumors (Yamada 2015, 794). In contrast, mice models of ovarian cancer that were imaged with magnetic resonance imaging (MRI) and macromolecular dendrimer polychelates attached to folic acid revealed the enhanced tumor to non-target contrast, with selective accumulation in the tumor. FR-specific ultrasmall superparamagnetic iron oxides, which were specifically maintained in FR-positive breast tumors in MRI of a rat model of breast cancer, showed potential as a non-invasive diagnostic tool in a more recent investigation.

The first folate-conjugated cytotoxic agent to be evaluated in tumor therapy was a maytansinoid conjugate. Since then, a number of chemotherapeutic drugs have been successfully targeted by being conjugated to folate. Vintafolide, also known as EC145, is a folate conjugation of the microtubule-destabilizing compound vinblastine's derivative desacetylvin-blastinemonohydrazide (DAVLBH) (Vergote and Leamon 2015, 207). The self-immolative disulfide-linker system permits the release of DAVLBH into the tumor cell endosome following receptor-mediated endocytosis, which inhibits cell division and induces cell death. The action was restricted to just FR-positive tumors in several phase II and phase III clinical studies for lung and ovarian cancer. Other folate conjugates include: a) EC0225, which is folate-conjugated to both a vinca alkaloid and mitomycin (Marchetti 2014, 1225); b) BMS-753493 (Epofolate), which is a folate conjugate of epothilone A (Gokhale 2013, 1316); c) EC0489, which is an analog of vintafolide (Leamon 2011, 338); and d) EC1456, which is a folic acid-tubulysin small-molecule drug conjugate (Reddy 2018, 1) are currently undergoing phase I trial in patients. While numerous other microtubule poisons have demonstrated some promise in *in vitro* and *in vivo* models, folaneutralizeded carboplatin was not regarded as a potential therapeutic since it was neutralized by folate receptor-mediated endocytosis (Scaranti 2020, 350).

PEG-coated biodegradable nanoparticles conjugated to folic acid, a novel colloidal drug carrier were designed by chemical coupling of folic acid for the specific recognition of the soluble form of the folate receptor expressed at the surface of cancer cells (Stella 2000, 1453).

The binding of drugs that include folate is potentially hindered by an excess of free folate in the blood. The exposure of FR expressed in the proximal renal tubules‘ apical membrane as a result of the filtration of low molecular weight (LMW) folate conjugates is another possible problem. The clinical use of LMW folate conjugate-based radiopharmaceuticals for targeted radiotherapy may be impacted, even though this has not negatively impacted the clinical translation of folate conjugates.

12.4.3 Aptamar Conjugated Nanoparticles

Advances in DNA synthesis technologies have permitted the creation of enormous populations of degenerate oligodeoxynucleotides, while PCR allows researchers to amplify modest quantities of molecules into levels that may be easily controlled. In vitro selection of functional nucleic acids (systematic evolution of ligands by exponential enrichment, or SELEX) was formed when these two advancements were repeatedly combined with the capacity to partition oligonucleotides based on their binding or catalytic activity. Aptamers based on nucleic acids were initially disclosed in 1990, when they were in vitro chosen using SELEX and were termed "aptamers," an invented

Latin term that meaning "to fit." Like antibodies, aptamers are one of a select few kinds of molecules that can be designed to attach to a variety of diverse targets (Keefe 2010, 540).

Single-stranded RNA or DNA oligonucleotides called aptamers fold up into a characteristic 3D shape that allows them to attach to tiny molecules up to whole organisms with great affinity and selectivity, with dissociation constants in the nanomolar range. A variety of biological applications, such as therapeutic, aptasensors, biosensors, diagnostic, and imaging systems, are now covered by the aptamer sector. Due to their tiny size, aptamers are vulnerable to renal filtration and require stabilization for usage in vivo against nuclease destruction effect. Different methods of chemical modification can be used to stabilize aptamers. Additionally, adding chemical changes to nucleic acid libraries expands the target spectrum of aptamers' interaction potential (Odeh 2019, 2).

Aptamers that bind to the cell surface can specifically cause therapeutics (such as drugs, toxins or siRNA) to persist in the vicinity of a specific cell or tissue type. This may also speed up the process of receptor-mediated endocytosis, which is the internalization of substances into cells. Prostate-specific membrane antigen (PSMA) (Chang 2004, S14), a protein that is extensively expressed but almost ever present on the cell surface, is one common epitope that has been targeted for treatment. Some prostate cancer cells have PSMA on their surfaces, hence two aptamers, designated as A9 and A10, were chosen to target this receptor and showed low nanomolar IC50 values (Gourni 2017, 2). The PSMA-specific aptamer was a great choice for an escort aptamer that may promote transport via endocytosis since PSMA is constitutively internalized. Aptamers have been used to deliver traditional small-molecule medications. A10, a PSMA-specific aptamer, has been used to deliver the anthracycline medication doxorubicin directly to the target cells (Zhang 2010, 1).

The creation of phototoxic aptamers for the targeted treatment of certain cancer cells is a more creative use. DNA aptamers were internalized into epithelial tumor cells after being changed at their 5′ ends using the photodynamic chemical chlorin e6. These aptamers were chosen to target short O-glycan peptides that are only expressed on the surface of cancer cells. The destruction of cancer cells via light-activated cytotoxicity was shown to be 500-fold more effective than the medication alone.

The delivery of biopolymer therapeutics has also been examined. Because gelonin lacks a translocation domain in the absence of conjugation, it has a low level of natural cytotoxicity. Gelonin is a toxin that has been conjugated to antibodies or other proteins for delivery to tumor cells. Prostate cancer cells that overexpress PSMA can be precisely targeted and eliminated by gelonin coupled to the PSMA-specific aptamer A9. The conjugates are at least 600 times more potent than cells that do not produce PSMA. It is interesting to note that research using both gelonin and medications have demonstrated that conjugation can lessen the free therapeutic's spontaneous absorption into non-cancerous cells, potentially reducing side effects.

The use of aptamers to deliver other oligonucleotide therapies, such as siRNAs, is an intriguing recent development. Streptavidin was used to bind biotin-labeled aptamers to biotin-labeled siRNAs, or aptamers were hybridized to siRNAs. A tat/rev-specific siRNA and an aptamer specific to HIV gp120 have also been combined to create aptamer-siRNA chimaeras. Through the use of both the aptamer and the siRNA components, this construct can prevent HIV replication in cells that are HIV-1-infected. Combination aptamer-siRNA delivery therapy may be possible if more aptamers are produced against gp120 as well as other HIV proteins like gp160.

Finally, aptamers can be used to direct the delivery of supramolecular structures. The application of PSMA-specific aptamer A10 for the delivery of nanoparticles to tumor cells was reported in a recent study. The study also administered docetaxel, an encapsulated cancer medication, to LNCaP xenografts in nude mice. The tumor volume was reduced in all seven mice on given injection treatment, and all seven animals lived. Additionally, some effectiveness was shown when cisplatin, a chemotherapy medication, was administered to tumor cells using PLGA-PEG nanoparticle conjugates that had been functionalized with aptamers. Similar to this, a liposome containing the strong chemotherapy cisplatin was coupled to a nucleolin-specific aptamer, which enabled the medicine to be delivered to tumors. Furthermore, an antisense oligonucleotide was used to induce regulated release of the conjugate to tumors. Multiple applications have made use of aptamer-conjugated nanoparticles (Figure 12.9(C)), notably dual nanoparticles for fluorescent labeling and magnetic extraction. In order to identify and extract specific cells from a range of matrixes, silica-coated magnetic and fluorophore-doped silica nanoparticles have recently been coupled to highly selective aptamers. To make this two-particle test effective in a clinical environment, significant enhancements are needed also to boost the selectivity and sensitivity (Medley 2011, 728). By effectively using aptamer-based technologies, these tailored encapsulation techniques may reduce the systemic toxicity often associated with chemotherapies.

12.4.4 Lipopolysaccharide Conjugated Nanoparticles

Complex, lipid-linked, negatively charged molecules of carbohydrates are known as lipopolysaccharides (LPS). Typically, they consist of three distinct regions: a short oligosaccharide, the core region, an O-antigen portion made up of a polymer of repeating oligosaccharide units, the composition of which varies greatly among Gram-negative bacteria, and a fatty-acylated, highly conserved region known as lipid A. The active moiety of LPS is lipid A, which is in charge of many of the pathophysiological outcomes related to Gram-negative bacterial infection. A hydrophilic, negatively charged bisphosphorylated glucosamine disaccharide makes up Lipid A, which is covalently joined to a hydrophobic, 12- to 16-carbon acyl chain domain via amide and ester linkages (Molinaro 2015, 504). It consistently demonstrates an amphiphilic nature and the capacity to form aggregates. Despite the fact that LPS is chemically inert, its presence in the blood (endotoxemia) triggers a series of host reactions that alter the structure and operation of organs and cells, alter metabolic processes, increase body temperature, alter hemodynamics, and result in septic shock (Wassenaar 2018, 64).

Through the activation with cyanogen bromide, an effective technique for conjugating LPS with probes carrying primary amino and hydrazine moieties was developed (Figure 12.9(D)) (Battaglini 2011, 162). The endotoxic activity of LPS is maintained, good conjugate ratios are produced, and a variety of probes with the aforementioned functional groups can be combined with LPS. Another method for activating the hydroxyl groups in carbohydrates is to use 1-cyano-4-dimethylaminopyridinium tetrafluoroborate (CDAP) (Pallarola 2009, 3825). A fluorescent probe was used to investigate the conjugation of LPS in an aqueous solution in the presence of amphiphilic substances such sodium dodecyl sulfate (SDS), triton X-100, and sodium deoxycholate (NaDC), at various concentrations (dansyl hydrazine) (Pallarola 2009, 3826). From these experiments, NaDC was found to be the best suited. LPS from S. minnesota is coupled with the fluorescent probe at good labeling ratios (110 nmol mg1), maintaining 70% of its endotoxic action. Diamino polyethylene glycol, or DAPEG), was utilized as a spacer for the conjugation with HRP (Xue 2011, 297). The spacer interacted with the CDAP-activated LPS after being attached to a periodate-oxidized HRP (oHRP). The compound LPSDAPEGoHRP maintained its endotoxic and enzymatic activity while exhibiting an excellent conjugation ratio. The endotoxin neutralizing protein (ENP) was then immobilized on the new matrix and employed in conjunction with a gold electrode array modified by this conjugation (Priano 2007, 109).

B. pseudomallei has a variety of surface polysaccharides, and lipopolysaccharide (LPS) and capsular polysaccharide have been investigated as vaccine candidates. It has been suggested that the O-polysaccharide of the lipopolysaccharide (LPS), an unbranched heteropolymer made up of varying amounts of acetylated and methylation -3)-β-D-glucopyranose-(1-3)-6-deoxy-α-L-talopyranose-(1-, contributes to virulence and resistance to serum death. In some animal models, monoclonal and polyclonal antibodies produced against LPS provide passive defense against B. pseudomallei infection. Previously, a conjugate vaccine made from the *O*-antigen of B.

pseudomallei was utilized, and the sera produced as a result of these vaccines offered passive protection to diabetic rats.

There are several approved polysaccharide vaccines available today, including those to prevent infections from *Streptococcus pneumoniae*, *Neisseria meningitidis*, and *Haemophilus influenzae*. IgM and IgG3 antibodies are often produced in response to polysaccharides, with a widespread inability to convert to IgG synthesis. Polysaccharides can be attached to proteins to generate T-cell memory and change the response into a more benevolent T-dependent response. This is true for meningococcal type C vaccines and the *H. influenzae* type b vaccine, both of which have clinical use licenses. Furthermore, researches are going on for the development and use of an LPS-protein conjugate vaccine that, in a murine infection model, induces balanced immune responses and effectively protects against melioidosis and other infections (Scott 2014, 2).

12.5 APPLICATIONS

The ultimate objective of personalized medicine is to provide the most efficient course of therapy for each patient based on their unique ailment, while reducing any potential negative effects. Significant molecular variations across related tumors have been found to significantly affect patient outcomes. Cancer may respond better or worse to therapies because of the molecular variations, which can also signal the course of the condition. Characterizing disease states requires an understanding of these molecular variations, particularly in tumors linked to certain biomarkers. The proteins that are expressed differently by various tumors can serve as the foundation for screening methods and therapeutic choices. However, the dual clinical requirements of diagnosis and therapy have been constrained by the relative scarcity of certain biomarkers. The systematic isolation and identification of biological molecules from complicated body fluids or tissues is a laborious procedure involved in the development of biomarkers. Molecular techniques that not only distinguish between cancer subtypes to disclose the most effective treatment choices but also enable early identification to guarantee that the therapy's efficacy is maximized must be developed to realize the paradigm of personalized medicine.

Polyclonal and monoclonal antibodies have long been the workhorses for biomolecular identification due to their selectivity and high binding affinities. Certain limits of antibodies have come to light because of the incorporation of nanotechnology into biosensor, assay, and diagnostic tool applications. Large proteins that are immobilized on the surface of nanoparticles (NPs) may aggregate and reduce the effectiveness of the nanomaterials. The final utility of antibodies combined with nanomaterials is also influenced by the complexity, cost, and challenging bioconjugation processes. A number of these problems are addressed by the development of aptamers for molecular recognition, which have a low molecular weight, easy and repeatable synthesis, simple modification, quick tissue penetration, low toxicity or immunogenicity, simple storage, and high binding affinity and selectivity.

12.6 CONCLUSIONS

Nano-delivery systems are of utmost importance for precisely targeting the desired location in the treatment of several diseases, and they have been the subject of intensive study over the past decade in the creation of nanoparticle-based therapeutic agents. Currently, polymers or lipids make up the bulk of the nanoparticles employed in the targeted delivery strategy. Although polymeric nanoparticles show significant benefits in the treatment of disease, they also have associated drawbacks, including scaling-up, the use of organic solvents in their manufacturing process, biocompatibility, cytotoxicity, and immunogenicity. However, due to their resemblance to cell membrane, lipid-based nanoparticles show the capacity to penetrate difficult-to-reach locations even without any surface functionalization. Therapeutic nanoparticles are now mostly designed for the treatment or prevention of a single illness. However, when researchers began combining different therapeutic molecules and different kinds of nanoparticles, the development of multi-therapeutic nanoparticles, which can be used to treat several diseases, is the direction that therapeutic nanoparticles are heading in the future. The cost of nanomedicine and large-scale production are other crucial problems that have to be resolved. Therapeutics cannot advance from their conception to the market due to a lack of funding or a poor cost-benefit ratio. Therefore, by comprehending the properties of nanoparticles and their interactions with their biological surroundings, such as their targeting receptors or mechanisms of action in disease pathophysiology, we will be able to get around obstacles and create novel approaches for the treatment, prevention, and diagnosis of many diseases, especially those that are difficult to cure.

ABBREVIATION

ACNPs (Antibody-conjugated nanoparticles)
ADCs (Antibody-drug conjugates)
AML (Acute myeloid leukemia)
BBB (Blood-brain barrier)
BCN (Bicylco nonyne)
BSA (Bovine serum albumin)
CD22 (Cluster of differentiate 20)
CD30 (Cluster of differentiate 30)
CD33 (Cluster of differentiate 33)
CDAP (1-cyano-4-dimethylaminopyridinium tetrafluoroborate)DAPEG (Diamino polyethylene glycol)
DAVLBH (Microtubule-destabilizing compound vinblastine's derivative desacetylvin-blastinemonohydrazide)
DBCO (Dibenzocyclooctyne)
DIBO (Dibenzocyclooctynol)
DOC (Docetaxel)
DOX (Doxicillin)
DNA (Deoxyribonucleic acid)
EDC (1-ethyl-3-(3-dimethyl aminopropyl) carbodiimide)
ENP (Endotoxin neutralizing protein)
FRα (Folate receptor α)
GC (Gastric cancer)
HDL (High-density lipid)
HER2 (Human epidermal growth factor receptor 2)
HIV (Human immunodeficiency virus)
HSA (Human serum albumin)
iEDDA (Inverse electron demand Diels-Alder reactions)
IgG (Immunoglobulin G)
IgM (Immunoglobulin M)
LDL (Low-density lipid)
LMW (Low molecular weight)
LPS (Lipopolysaccharide)
MPS (Mononuclear phagocytes system)
MRI (Magnetic resonance imaging)
mRNA (Messenger RNA)
NaDC (Sodium deoxycholate)
NHS (*N*-hydroxy succinimide)

NP	(Nanoparticle)
OCT	(Cyclooctyne)
oHRP	(Oxidized-horse radish peroxide)
PCR	(Polymeric chain reaction)
PD-L1	(Programmed death-ligand 1)
PEG-PCL	(Polyethylene glycol-b-poly-caprolactone)
PLGA	(Polylactic-co-glycolic acid)
PM	(Polymeric micelles)
PNP	(Polymeric nanoparticles)
PSMA	(Prostate-specific membrane antigen)
RGD	(Arginyl-gylcyl-aspartic acid)
RNA	(Ribonucleic acid)
SDS	(Sodium dodecyl sulfate)
SELEX	(Systematic evolution of ligands by exponential enrichment)
SiNP	(Silica-encapsulated nanoparticle)
siRNA	(Silencing RNA)
SPAAC	(Strain-promoted azide-alkyne cycloaddition)
SPIONs	(Superparamagnetic iron oxide)
SWCNTs	(Single-walled carbon nanotubes)
TTR	(Transthyretin)
US-FDA	(United States Food and Drug Administration)

REFERENCES

Akbarzadeh, Abolfazl, Rogaie Rezaei-Sadabady, Soodabeh Davaran, Sang Woo Joo, Nosratollah Zarghami, Younes Hanifehpour, Mohammad Samiei, Mohammad Kouhi, and Kazem Nejati-Koshki. 2013. "Liposome: classification, preparation, and applications." *Nanoscale Research Letters* 8 (1):1–9.

Akkapeddi, Padma, Saara-Anne Azizi, Allyson M. Freedy, Pedro M. S. D. Cal, Pedro M. P. Gois, and Gonçalo JL Bernardes. 2016. "Construction of homogeneous antibody–drug conjugates using site-selective protein chemistry." *Chemical Science* 7 (5):2954–2963.

Alavi, Mehran, Naser Karimi, and Mohsen Safaei. 2017. "Application of various types of liposomes in drug delivery systems." *Advanced Pharmaceutical Bulletin* 7 (1):3.

Anagnostou, Katerina, Minas M. Stylianakis, Grigoris Atsalakis, Dimitrios M. Kosmidis, Athanasios Skouras, Ioannis J. Stavrou, Konstantinos Petridis, and Emmanuel Kymakis. 2020. "An extensive case study on the dispersion parameters of HI-assisted reduced graphene oxide and its graphene oxide precursor." *Journal of Colloid and Interface Science* 580:332–344.

Arruebo, Manuel, Mónica Valladares, and África González-Fernández. 2009. "Antibody-conjugated nanoparticles for biomedical applications." *Journal of Nanomaterials* 2009.

Babu, Anish, Amanda K. Templeton, Anupama Munshi, and Rajagopal Ramesh. 2014. "Nanodrug delivery systems: a promising technology for detection, diagnosis, and treatment of cancer." *Aaps Pharmscitech* 15 (3):709–721.

Bakhtiary, Zahra, Amir Ata Saei, Mohammad J. Hajipour, Mohammad Raoufi, Ophir Vermesh, and Morteza Mahmoudi. 2016. "Targeted superparamagnetic iron oxide nanoparticles for early detection of cancer: Possibilities and challenges." *Nanomedicine: Nanotechnology, Biology and Medicine* 12 (2):287–307.

Bancos, Simona, and Katherine M. Tyner. 2014. "Evaluating the effect of assay preparation on the uptake of gold nanoparticles by RAW264. 7 cells." *Journal of nanobiotechnology* 12:1–11.

Battaglini, Fernando, and Diego Pallarola. 2011. "Two efficient methods for the conjugation of smooth-form lipopolysaccharides with probes bearing hydrazine or amino groups. I. LPS activation with cyanogen bromide." In *Microbial Toxins*, 147–160. Springer.

Benasutti, Halli, Guankui Wang, Vivian P. Vu, Robert Scheinman, Ernest Groman, Laura Saba, and Dmitri Simberg. 2017. "Variability of complement response toward preclinical and clinical nanocarriers in the general population." *Bioconjugate Chemistry* 28 (11):2747–2755.

Borges, Chad R., and Nisha D. Sherma. 2014. "Techniques for the analysis of cysteine sulfhydryls and oxidative protein folding." *Antioxidants & Redox Signaling* 21 (3):511–531.

Ceña, Valentín, and Pablo Játiva. 2018. "Nanoparticle crossing of blood–brain barrier: A road to new therapeutic approaches to central nervous system diseases." Future Medicine.

Chang, Sam S. 2004. "Overview of prostate-specific membrane antigen." *Reviews in Urology* 6 (Suppl 10): S13.

Cheung, Anthony, Heather J. Bax, Debra H. Josephs, Kristina M. Ilieva, Giulia Pellizzari, James Opzoomer, Jacinta Bloomfield, Matthew Fittall, Anita Grigoriadis, and Mariangela Figini. 2016. "Targeting folate receptor alpha for cancer treatment." *Oncotarget* 7 (32):52553.

Conde, João, Jorge T. Dias, Valeria Grazú, Maria Moros, Pedro V. Baptista, and Jesus M. de la Fuente. 2014. "Revisiting 30 years of biofunctionalization and surface chemistry of inorganic nanoparticles for nanomedicine." *Frontiers in Chemistry* 2:48.

Daraee, Hadis, Ali Eatemadi, Elham Abbasi, Sedigheh Fekri Aval, Mohammad Kouhi, and Abolfazl Akbarzadeh. 2016. "Application of gold nanoparticles in biomedical and drug delivery." *Artificial Cells, Nanomedicine, and Biotechnology* 44 (1):410–422.

De, Mrinmoy, Partha S. Ghosh, and Vincent M. Rotello. 2008. "Applications of nanoparticles in biology." *Advanced Materials* 20 (22):4225–4241.

Debets, Marjoke F., Sander S. Van Berkel, Sanne Schoffelen, Floris P. J. T. Rutjes, Jan C. M. Van Hest, and Floris L. Van Delft. 2010. "Aza-dibenzocyclooctynes for fast and efficient enzyme PEGylation via copper-free (3+ 2) cycloaddition." *Chemical Communications* 46 (1):97–99.

Devaraj, Neal K., and Ralph Weissleder. 2011. "Biomedical applications of tetrazine cycloadditions." *Accounts of chemical research* 44 (9):816–827.

Dommerholt, Jan, Floris P. J. T. Rutjes, and Floris L. van Delft. 2016. "Strain-promoted 1, 3-dipolar cycloaddition of cycloalkynes and organic azides." *Cycloadditions in Bioorthogonal Chemistry* 57–76.

Dommerholt, Jan, Samuel Schmidt, Rinske Temming, Linda J. A. Hendriks, Floris P. J. T. Rutjes, Jan C. M. Van Hest, Dirk J. Lefeber, Peter Friedl, and Floris L. Van Delft. 2010. "Readily accessible bicyclononynes for bioorthogonal labeling and three-dimensional imaging of living cells." *Angewandte Chemie International Edition* 49 (49):9422–9425.

Dommerholt, Jan, Olivia Van Rooijen, Annika Borrmann, Célia Fonseca Guerra, F Matthias Bickelhaupt, and Floris L. Van Delft. 2014. "Highly accelerated inverse electron-demand cycloaddition of electron-deficient azides with aliphatic cyclooctynes." *Nature Communications* 5 (1):1–7.

Erlanger, Bernard F., Bi-Xing Chen, Min Zhu, and Louis Brus. 2001. "Binding of an anti-fullerene IgG monoclonal antibody to single wall carbon nanotubes." *Nano Letters* 1 (9):465–467.

Fadeel, Bengt. 2019. "Hide and seek: Nanomaterial interactions with the immune system." *Frontiers in Immunology* 10:133.

Gholamine, Babak, Isaac Karimi, Amir Salimi, Parisa Mazdarani, and Lora A. Becker. 2017. "Neurobehavioral toxicity of carbon nanotubes in mice: Focus on brain-derived neurotrophic factor messenger RNA and protein." *Toxicology and Industrial Health* 33 (4):340–350.

Gokhale, Madhushree, Ajit Thakur, and Frank Rinaldi. 2013. "Degradation of BMS-753493, a novel epothilone folate conjugate anticancer agent." *Drug Development and Industrial Pharmacy* 39 (9): 1315–1327.

Gourni, Eleni, and Gjermund Henriksen. 2017. "Metal-based PSMA radioligands." *Molecules* 22 (4):523.

Haes, Amanda J., W Paige Hall, Lei Chang, William L. Klein, and Richard P. Van Duyne. 2004. "A localized surface plasmon resonance biosensor: First steps toward an assay for Alzheimer's disease." *Nano Letters* 4 (6):1029–1034.

Haes, Amanda J., and Richard P. Van Duyne. 2002. "A nanoscale optical biosensor: sensitivity and selectivity of an approach based on the localized surface plasmon resonance spectroscopy of triangular silver nanoparticles." *Journal of the American Chemical Society* 124 (35):10596–10604.

Haldón, Estela, M Carmen Nicasio, and Pedro J. Pérez. 2015. "Copper-catalysed azide–alkyne cycloadditions (CuAAC): an update." *Organic & Biomolecular Chemistry* 13 (37):9528–9550.

Hatit, Marine Z. C., Linus F. Reichenbach, John M. Tobin, Filipe Vilela, Glenn A. Burley, and Allan J. B. Watson. 2018. "A flow platform for degradation-free CuAAC bioconjugation." *Nature Communications* 9 (1):4021.

Hein, Christopher D., Xin-Ming Liu, and Dong Wang. 2008. "Click chemistry, a powerful tool for pharmaceutical sciences." *Pharmaceutical Research* 25 (10):2216–2230.

Ho, Ja-an Annie, Chi-Hsiang Hung, Li-Chen Wu, and Ming-Yuan Liao. 2009. "Folic acid-anchored PEGgylated phospholipid bioconjugate and its application in a liposomal immunodiagnostic assay for folic acid." *Analytical Chemistry* 81 (14):5671–5677.

Huang, Tao, Prakash D. Nallathamby, Daniel Gillet, and Xiao-Hong Nancy Xu. 2007. "Design and synthesis of single-nanoparticle optical biosensors for imaging and characterization of single receptor molecules on single living cells." *Analytical Chemistry* 79 (20):7708–7718.

Huberman, Tamir, Yael Eisenberg-Domovich, Gerry Gitlin, Tikva Kulik, Edward A. Bayer, Meir Wilchek, and Oded Livnah. 2001. "Chicken avidin exhibits pseudo-catalytic properties: biochemical, structural, and electrostatic consequences." *Journal of Biological Chemistry* 276 (34): 32031–32039.

Jain, Akshay, and Kun Cheng. 2017. "The principles and applications of avidin-based nanoparticles in drug delivery and diagnosis." *Journal of Controlled Release* 245:27–40.

Jeong, Sinyoung, Ji Yong Park, Myeong Geun Cha, Hyejin Chang, Yong-il Kim, Hyung-Mo Kim, Bong-Hyun Jun, Dong Soo Lee, Yoon-Sik Lee, and Jae Min Jeong. 2017. "Highly robust and optimized conjugation of antibodies to nanoparticles using quantitatively validated protocols." *Nanoscale* 9 (7):2548–2555.

Keefe, Anthony D., Supriya Pai, and Andrew Ellington. 2010. "Aptamers as therapeutics." *Nature Reviews Drug Discovery* 9 (7):537–550.

Kim, Eunha, and Heebeom Koo. 2019. "Biomedical applications of copper-free click chemistry: in vitro, in vivo, and ex vivo." *Chemical Science* 10 (34):7835–7851.

Kocbek, Petra, Nataša Obermajer, Mateja Cegnar, Janko Kos, and Julijana Kristl. 2007. "Targeting cancer cells using PLGA nanoparticles surface modified with monoclonal antibody." *Journal of Controlled Release* 120 (1-2):18–26.

Leamon, Christopher P., Joseph A. Reddy, Patrick J. Klein, Iontcho R. Vlahov, Ryan Dorton, Alicia Bloomfield, Melissa Nelson, Elaine Westrick, Nikki Parker, and Kristen Bruna. 2011. "Reducing undesirable hepatic clearance of a tumor-targeted vinca alkaloid via novel saccharopeptidic modifications." *Journal of Pharmacology and Experimental Therapeutics* 336 (2):336–343.

Lee, Byeongho, Youngbin Baek, Minwoo Lee, Dae Hong Jeong, Hong H. Lee, Jeyong Yoon, and Yong Hyup Kim. 2015. "A carbon nanotube wall membrane for water treatment." *Nature Communications* 6 (1):7109.

Li, Shi-Yong, Rong Sun, Hong-Xia Wang, Song Shen, Yang Liu, Xiao-Jiao Du, Yan-Hua Zhu, and Wang Jun. 2015. "Combination therapy with epigenetic-targeted and chemotherapeutic drugs delivered by nanoparticles to enhance the chemotherapy response and overcome resistance by breast cancer stem cells." *Journal of Controlled Release* 205:7–14.

Li, Xuan, and John J. Lenhart. 2012. "Aggregation and dissolution of silver nanoparticles in natural surface water." *Environmental Science & Technology* 46 (10):5378–5386.

Lin, Li-Sen, Zhong-Xiao Cong, Jian-Bo Cao, Kai-Mei Ke, Qiao-Li Peng, Jinhao Gao, Huang-Hao Yang, Gang Liu, and Xiaoyuan Chen. 2014. "Multifunctional Fe3O4@ polydopamine core–shell nanocomposites for intracellular mRNA detection and imaging-guided photothermal therapy." *ACS Nano* 8 (4):3876–3883.

Lin, Tzu-Tang, Li-Yen Yang, I-Hsuan Lu, Wen-Chih Cheng, Zhe-Ren Hsu, Shu-Hwa Chen, and Chung-Yen Lin. 2021. "AI4AMP: an antimicrobial Peptide predictor using physicochemical property-Based encoding method and deep learning." *Msystems* 6 (6):e00299-21.

Liu, Bei, Chunxia Li, Ziyong Cheng, Zhiyao Hou, Shanshan Huang, and Jun Lin. 2016. "Functional nanomaterials for near-infrared-triggered cancer therapy." *Biomaterials Science* 4 (6):890–909.

Liu, Rui, Yang An, Wenfeng Jia, Yushan Wang, Yue Wu, Yonghuan Zhen, Jun Cao, and Huile Gao. 2020. "Macrophage-mimic shape changeable nanomedicine retained in tumor for multimodal therapy of breast cancer." *Journal of Controlled Release* 321:589–601.

Liu, Yuanchang, Joseph Hardie, Xianzhi Zhang, and Vincent M. Rotello. 2017. "Effects of engineered nanoparticles on the innate immune system." Seminars in immunology.

Lu, Jie, Monty Liong, Zongxi Li, Jeffrey I. Zink, and Fuyuhiko Tamanoi. 2010. "Biocompatibility, biodistribution, and drug-delivery efficiency of mesoporous silica nanoparticles for cancer therapy in animals." *Small* 6 (16):1794–1805.

Lu, Lantian, Viet Tram Duong, Ahmed O. Shalash, Mariusz Skwarczynski, and Istvan Toth. 2021. "Chemical conjugation strategies for the development of protein-based subunit nanovaccines." *Vaccines* 9 (6):563.

Lutz, Robert J. 2015. "Targeting the folate receptor for the treatment of ovarian cancer." *Transl Cancer Res* 4 (1):118–126.

Lv, Li, Chunxia Liu, Chuxiong Chen, Xiaoxia Yu, Guanghui Chen, Yonghui Shi, Fengchao Qin, Jiebin Ou, Kaifeng Qiu, and Guocheng Li. 2016. "Quercetin and doxorubicin co-encapsulated biotin receptor-targeting nanoparticles for minimizing drug resistance in breast cancer." *Oncotarget* 7 (22):32184.

Marchetti, Claudia, Innocenza Palaia, Margherita Giorgini, Caterina De Medici, Roberta Iadarola, Laura Vertechy, Lavinia Domenici, Violante Di Donato, Federica Tomao, and Ludovico Muzii. 2014. "Targeted drug delivery via folate receptors in recurrent ovarian cancer: a review." *OncoTargets and Therapy* 7:1223.

Marques, A. C., P. J. Costa, S. Velho, and M. H. Amaral. 2020. "Functionalizing nanoparticles with cancer-targeting antibodies: A comparison of strategies." *Journal of Controlled Release* 320: 180–200.

Medley, Colin D., Suwussa Bamrungsap, Weihong Tan, and Joshua E. Smith. 2011. "Aptamer-conjugated nanoparticles for cancer cell detection." *Analytical Chemistry* 83 (3):727–734.

Mehdizadeh, Mozhdeh, Hasti Rouhani, Nima Sepehri, Reyhaneh Varshochian, Mohammad Hossein Ghahremani, Mohsen Amini, Mehdi Gharghabi, Seyed Nasser Ostad, Fatemeh Atyabi, and Azin Baharian. 2017. "Biotin decorated PLGA nanoparticles containing SN-38 designed for cancer therapy." *Artificial Cells, Nanomedicine, and Biotechnology* 45 (3):495–504.

Molinaro, Antonio, Otto Holst, Flaviana Di Lorenzo, Maire Callaghan, Alessandra Nurisso, Gerardino D'Errico, Alla Zamyatina, Francesco Peri, Rita Berisio, and Roman Jerala. 2015. "Chemistry of lipid A: at the heart of innate immunity." *Chemistry–A European Journal* 21 (2):500–519.

Nath, Nidhi, and Ashutosh Chilkoti. 2002. "A colorimetric gold nanoparticle sensor to interrogate biomolecular interactions in real time on a surface." *Analytical Chemistry* 74 (3):504–509.

Ning, Xinghai, Jun Guo, Margreet A. Wolfert, and Geert-Jan Boons. 2008. "Visualizing metabolically labeled glycoconjugates of living cells by copper-free and fast Huisgen cycloadditions." *Angewandte Chemie* 120 (12):2285–2287.

O'Shannessy, Daniel J., Elizabeth B. Somers, Julia Maltzman, Robert Smale, and Yao-Shi Fu. 2012. "Folate receptor alpha (FRA) expression in breast cancer: identification of a new molecular subtype and association with triple negative disease." *Springerplus* 1 (1):1–9.

Odeh, Fadwa, Hamdi Nsairat, Walhan Alshaer, Mohammad A. Ismail, Ezaldeen Esawi, Baraa Qaqish, Abeer Al Bawab, and Said I. Ismail. 2019. "Aptamers chemistry: Chemical modifications and conjugation strategies." *Molecules* 25 (1):3.

Pallarola, Diego, and Fernando Battaglini. 2009. "Surfactant-assisted lipopolysaccharide conjugation employing a cyanopyridinium agent and its application to a competitive assay." *Analytical Chemistry* 81 (10):3824–3829.

Park, Sujin, and Yeh-Jin Ahn. 2016. "Multi-walled carbon nanotubes and silver nanoparticles differentially affect seed germination, chlorophyll content, and hydrogen peroxide accumulation in carrot (Daucus carota L.)." *Biocatalysis and Agricultural Biotechnology* 8:257–262.

Pickens, Chad J., Stephanie N. Johnson, Melissa M. Pressnall, Martin A. Leon, and Cory J. Berkland. 2017. "Practical considerations, challenges, and limitations of bioconjugation via azide–alkyne cycloaddition." *Bioconjugate Chemistry* 29 (3):686–701.

Priano, Graciela, Diego Pallarola, and Fernando Battaglini. 2007. "Endotoxin detection in a competitive electrochemical assay: synthesis of a suitable endotoxin conjugate." *Analytical Biochemistry* 362 (1): 108–116.

Puddu, Valeria, and Carole C. Perry. 2012. "Peptide adsorption on silica nanoparticles: evidence of hydrophobic interactions." *ACS Nano* 6 (7):6356–6363.

Radioisotopes, IAEA, and Radiopharmaceuticals Series No. 2009. 1: Technetium-99m radiopharmaceuticals: status and trends. Vienna.

Ranganathan, Ramya, Shruthilaya Madanmohan, Akila Kesavan, Ganga Baskar, Yoganathan Ramia Krishnamoorthy, Roy Santosham, D. Ponraju, Suresh Kumar Rayala, and Ganesh Venkatraman. 2012. "Nanomedicine: towards development of patient-friendly drug-delivery systems for oncological applications." *International Journal of Nanomedicine* 7:1043.

Ravasco, João MJM, Hélio Faustino, Alexandre Trindade, and Pedro M. P. Gois. 2019. "Bioconjugation with maleimides: A useful tool for chemical biology." *Chemistry–A European Journal* 25 (1): 43–59.

Reddy, Joseph A., Ryan Dorton, Alicia Bloomfield, Melissa Nelson, Christina Dircksen, Marilynn Vetzel, Paul Kleindl, Hari Santhapuram, Iontcho R. Vlahov, and Christopher P. Leamon. 2018. "Pre-clinical evaluation of EC1456, a folate-tubulysin anti-cancer therapeutic." *Scientific Reports* 8 (1): 1–10.

Riboh, Jonathan C., Amanda J. Haes, Adam D. McFarland, Chanda Ranjit Yonzon, and Richard P. Van Duyne. 2003. "A nanoscale optical biosensor: real-time immunoassay in physiological buffer enabled by improved nanoparticle adhesion." *The Journal of Physical Chemistry B* 107 (8): 1772–1780.

Saallah, Suryani, and I Wuled Lenggoro. 2018. "Nanoparticles carrying biological molecules: Recent advances and applications." *KONA Powder and Particle Journal* 35:89–111.

Scaranti, Mariana, Elena Cojocaru, Susana Banerjee, and Udai Banerji. 2020. "Exploiting the folate receptor α in oncology." *Nature Reviews Clinical Oncology* 17 (6):349–359.

Scott, Andrew E., Sarah A. Ngugi, Thomas R. Laws, David Corser, Claire L. Lonsdale, Riccardo V. D'Elia, Richard W. Titball, E Diane Williamson, Timothy P. Atkins, and Joann L. Prior. 2014. "Protection against experimental melioidosis following immunisation with a lipopolysaccharide-protein conjugate." *Journal of Immunology Research* 2014.

Shemetov, Anton A., Igor Nabiev, and Alyona Sukhanova. 2012. "Molecular interaction of proteins and peptides with nanoparticles." *ACS nano* 6 (6):4585–4602.

Shen, Shiyang, Meng Liu, Teng Li, Shiqi Lin, and Ran Mo. 2017. "Recent progress in nanomedicine-based combination cancer therapy using a site-specific co-delivery strategy." *Biomaterials science* 5 (8):1367–1381.

Shen, Xingcan, Qi Yuan, Hong Liang, Haigang Yan, and Xiwen He. 2003. "Hysteresis effects of the interaction between serum albumins and silver nanoparticles." *Science in China Series B: Chemistry* 46 (4):387–398.

Sheyi, Rotimi, Beatriz G. de la Torre, and Fernando Albericio. 2022. "Linkers: An Assurance for Controlled Delivery of antibody-drug conjugate." *Pharmaceutics* 14 (2):396.

Singh, Bijay, Yoonjeong Jang, Sushila Maharjan, Hyeon-Jeong Kim, Ah Young Lee, Sanghwa Kim, Nomundelger Gankhuyag, Myeon-Sik Yang, Yun-Jaie Choi, and Myung-Haing Cho. 2017. "Combination therapy with doxorubicin-loaded galactosylated poly (ethyleneglycol)-lithocholic acid to suppress the tumor growth in an orthotopic mouse model of liver cancer." *Biomaterials* 116:130–144.

Smyth, Tyson, Krastina Petrova, Nicole M. Payton, Indushekhar Persaud, Jasmina S. Redzic, Michael W. Graner, Peter Smith-Jones, and Thomas J. Anchordoquy. 2014. "Surface functionalization of exosomes using click chemistry." *Bioconjugate Chemistry* 25 (10):1777–1784.

Stallivieri, Aurélie, Ludovic Colombeau, Gulim Jetpisbayeva, Albert Moussaron, Bauyrzhan Myrzakhmetov, Philippe Arnoux, Samir Acherar, Régis Vanderesse, and Céline Frochot. 2017. "Folic acid conjugates with photosensitizers for cancer targeting in photodynamic therapy: Synthesis and photophysical properties." *Bioorganic & Medicinal Chemistry* 25 (1):1–10.

Stella, Barbara, Silvia Arpicco, Maria Teresa Peracchia, Didier Desmaële, Johan Hoebeke, Michel Renoir, Jean D'Angelo, Luigi Cattel, and Patrick Couvreur. 2000. "Design of folic acid-conjugated nanoparticles for drug targeting." *Journal of Pharmaceutical Sciences* 89 (11):1452–1464.

Sun, Lingyi, Yongkang Gai, Carolyn J. Anderson, and Dexing Zeng. 2015. "Highly-efficient and versatile fluorous-tagged Cu (i)-catalyzed azide–alkyne cycloaddition ligand for preparing bioconjugates." *Chemical Communications* 51 (96):17072–17075.

Tsuchikama, Kyoji, and Zhiqiang An. 2018. "Antibody-drug conjugates: recent advances in conjugation and linker chemistries." *Protein & Cell* 9 (1):33–46.

ud Din, Fakhar, Waqar Aman, Izhar Ullah, Omer Salman Qureshi, Omer Mustapha, Shumaila Shafique, and Alam Zeb. 2017. "Effective use of nanocarriers as drug delivery systems for the treatment of selected tumors." *International Journal of Nanomedicine* 12:7291.

Veranth, John M., Erin G. Kaser, Martha M. Veranth, Michael Koch, and Garold S. Yost. 2007. "Cytokine responses of human lung cells (BEAS-2B) treated with micron-sized and nanoparticles of metal oxides compared to soil dusts." *Particle and Fibre Toxicology* 4 (1):1–18.

Vergote, Ignace, and Christopher P. Leamon. 2015. "Vintafolide: a novel targeted therapy for the treatment of folate receptor expressing tumors." *Therapeutic Advances in Medical Oncology* 7 (4): 206–218.

Vinchurkar, Rutuja Hemant, and Ashwin Bhanudas Kuchekar. 2021. "Polymeric Micelles: A Novel Approach towards Nano-Drug Delivery System." *Biosciences Biotechnology Research Asia* 18 (4):629–649.

Wassenaar, Trudy M., and Kurt Zimmermann. 2018. "Lipopolysaccharides in food, food supplements, and probiotics: should we be worried?" *European Journal of Microbiology and Immunology* 8 (3): 63–69.

Xiao, Fan, Zhe Chen, Zixiang Wei, and Leilei Tian. 2020. "Hydrophobic interaction: a promising driving force for the biomedical applications of nucleic acids." *Advanced Science* 7 (16):2001048.

Xu, Shijie, Fangbo Cui, Dafu Huang, Dinghu Zhang, Anqing Zhu, Xia Sun, Yiming Cao, Sheng Ding, Yao Wang, and Eryun Gao. 2019. "PD-L1 monoclonal antibody-conjugated nanoparticles enhance drug delivery level and chemotherapy efficacy in gastric cancer cells." *International Journal of Nanomedicine* 14:17.

Xue, Ying, Megan L. O'Mara, Peter P. T. Surawski, Matt Trau, and Alan E. Mark. 2011. "Effect of poly (ethylene glycol)(PEG) spacers on the conformational properties of small peptides: a molecular dynamics study." *Langmuir* 27 (1):296–303.

Yamada, Yoshinori, Hiroshi Nakatani, Hisashi Yanaihara, and Masahiro Omote. 2015. "Phase I clinical trial of 99mTc-etarfolatide, an imaging agent for folate receptor in healthy Japanese adults." *Annals of Nuclear Medicine* 29 (9):792–798.

Yanase, Noriko, Hiroko Toyota, Kikumi Hata, Seina Yagyu, Takahiro Seki, Mitsunori Harada, Yasuki Kato, and Junichiro Mizuguchi. 2014. "OVA-bound nanoparticles induce OVA-specific IgG1, IgG2a, and IgG2b responses with low IgE synthesis." *Vaccine* 32 (45):5918–5924.

Yang, Guangze, Yun Liu, Haofei Wang, Russell Wilson, Yue Hui, Lei Yu, David Wibowo, Cheng Zhang, Andrew K. Whittaker, and Anton P. J. Middelberg. 2019. "Bioinspired core–shell nanoparticles for hydrophobic drug delivery." *Angewandte Chemie International Edition* 58 (40): 14357–14364.

Yang, Shu-Jyuan, Feng-Huei Lin, Kun-Che Tsai, Ming-Feng Wei, Han-Min Tsai, Jau-Min Wong, and Ming-Jium Shieh. 2010. "Folic acid-conjugated chitosan nanoparticles enhanced protoporphyrin IX accumulation in colorectal cancer cells." *Bioconjugate Chemistry* 21 (4):679–689.

Yao, Virginia J., Sara D'Angelo, Kimberly S. Butler, Christophe Theron, Tracey L. Smith, Serena Marchiò, Juri G. Gelovani, Richard L. Sidman, Andrey S. Dobroff, and C Jeffrey Brinker. 2016. "Ligand-targeted theranostic nanomedicines against cancer." *Journal of Controlled Release* 240:267–286.

Yi, Gawon, Jihwan Son, Jihye Yoo, Changhee Park, and Heebeom Koo. 2018. "Application of click chemistry in nanoparticle modification and its targeted delivery." *Biomaterials Research* 22 (1):1–8.

Zhang, Huiming. 2010. "Quantum dot-A10 RNA aptamer-doxorubicin conjugate."

Zhang, Yu Juan, Rao Huang, Xian Fang Zhu, Lian Zhou Wang, and Chen Xu Wu. 2012a. "Synthesis, properties, and optical applications of noble metal nanoparticle-biomolecule conjugates." *Chinese Science Bulletin* 57 (2):238–246.

Zhang, Yu Juan, Rao Huang, Xian Fang Zhu, Lian Zhou Wang, and Chen Xu Wu. 2012b. "Synthesis, properties, and optical applications of noble metal nanoparticle-biomolecule conjugates." *Chinese Science Bulletin* 57:238–246.

Zhao, Zongmin, Anvay Ukidve, Jayoung Kim, and Samir Mitragotri. 2020. "Targeting strategies for tissue-specific drug delivery." *Cell* 181 (1):151–167.

Zhu, Shaoli, Fei Li, Chunlei Du, and Yongqi Fu. 2008. "A localized surface plasmon resonance nanosensor based on rhombic Ag nanoparticle array." *Sensors and Actuators B: Chemical* 134 (1):193–198.

Zielińska, Aleksandra, Filipa Carreiró, Ana M. Oliveira, Andreia Neves, Bárbara Pires, D Nagasamy Venkatesh, Alessandra Durazzo, Massimo Lucarini, Piotr Eder, and Amélia M Silva. 2020. "Polymeric nanoparticles: production, characterization, toxicology and ecotoxicology." *Molecules* 25 (16):3731

13 Quantum Dots for Bioimaging and Sensing Application

Manisha Patel, Vineet Kumar Jain, Harvinder Popli, Parth Patel, and Keerti Jain

13.1 INTRODUCTION

Quantum dots (QDs) are luminescent nanomaterials with size ranging from 2–10 nm with unique optical and surface properties. QDs are semiconductors made up of either organic materials like citric acid, amino acids, proteins, etc., or inorganic materials belonging to elemental groups II-VI and III-V. QDs were first discovered in the 1980s by scientist Alexei Ekimov, who focused on semiconductor research. The extensive research in the field of life sciences inspired scientists to explore luminescent materials for both detection and cellular imaging. QDs have remarkable properties like, high quantum yield, large stoke shift, tunable emission, and excellent photostability. Their optical properties, such as excellent luminescence, confined emission wavelength, and wide range of excitation, provide them an obligatory tool in biomedical applications like imaging, targeted drug delivery, and sensing (Bajwa et al. 2016). Consequently, various types of QDs have been produced with emissions ranging from ultraviolet to near-infrared (NIR) region. In comparison to materials that emit light at ultraviolet or visible wavelength, NIR is preferred because long-wavelength excitation and emission could greatly minimize the background signal-to-noise and increase detection sensitivity (Dhas et al. 2022; Li et al. 2019; Bajwa et al. 2015).

In contrast to fluorescent materials emitting visible light, nanomaterials having emissions in the NIR I band (650–950 nm) can penetrate biological tissues more deeply. Still it experience the noise of autofluorescence in deep tissue but upto the depth of only 1–2 cm. The signal-to-noise ratio should be significantly enhanced by producing a fluorophore with an emission region from 1000–1450 nm (Michalet et al. 2005; Xu et al. 2016). Therefore, various types of QDs have been synthesized with emission wavelengths ranging from ultraviolet to NIR region. Several NIR imaging agents, including gold nanorods, single-walled carbon nanotubes, core/shell rare earth nanocrystals, and other types of QDs that exhibit bioimaging activity, have been reported so far (Xu et al. 2016). In biomedical applications, nanomaterials showing emission in the NIR-II range ($1000 \leq \lambda \leq 1700$ nm) are well accepted because of their ability to overcome the absorption and scattering of photon-induced by tissue. Generally, NIR-II emission depends upon the QDs' band gap and particle size. A large band gap is smaller in size and emits blue light, while QDs having red emission showed a smaller band gap and larger size, as mentioned in Figure 13.1 (Li et al. 2019; Bajwa et al. 2015).

The presence of narrow emission and broad absorption spectrum provides stable and intense fluorescence. Due to its superior fluorescence properties, it can also be used in multiplexing, which includes the simultaneous detection of multiple targets (Park et al. 2017; Alivisatos et al. 2005; Michalet et al. 2005; Martynenko et al. 2017). QDs have distinctive surface properties due to the presence of various functional groups that allows the binding of biomolecules like peptides and antibodies and improve the selectivity of QDs.

 DOI: 10.1201/9781003305583-13

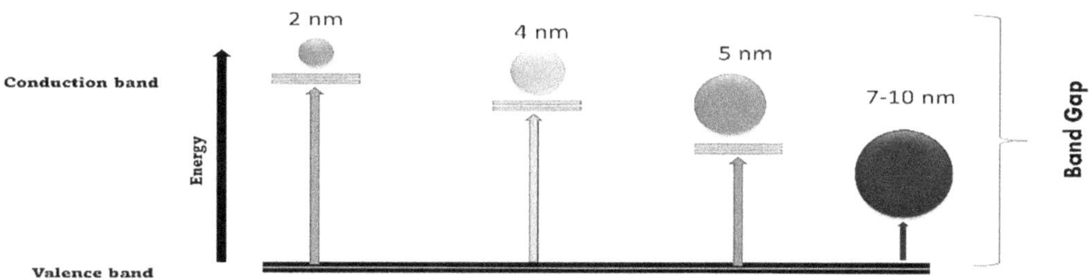

FIGURE 13.1 Pictorial representation of relation between size of QDs and band gap. *Adapted with permission from reference Bajwa et al. 2015.*

A drug delivery system is a formulation or device for the introduction of therapeutic moieties into the body in a controlled manner. To improve the therapeutic efficacy of any drug, it is important to deliver the right amount of dose at a diseased site in a controlled way. The complete drug delivery process includes administration of formulation, release of active ingredients into blood stream, and transport of the therapeutics to the diseased site. Nowadays, drug delivery systems are more focused on the delivery of therapeutics to target sites using nanotechnology (Jain et al. 2022; Ahmad et al. 2022; Patel et al. 2023). Nanotechnology includes the formation of nano-sized particles containing active pharmaceutical agents. Due to its inherent large surface area associated with small particle size, it poses some benefits like improved solubility, enhanced absorption by the biological membrane, provides better bioavailability, and improved *in vivo* stability (Juneja et al. 2022). Further modification in the nanoparticles enables them to target specific receptors or organs and provide immense benefits compared to conventional formulations. QDs alone or in combination with other nanocarriers like dendrimers, liposomes, polymeric or lipid nanoparticles have been used in biosensing or bioimaging as an effective fluorescence probe due to their good cytocompatibility, intense fluorescence, and stability in the aqueous medium. In the current chapter, we tried to cover the biosensing and bioimaging applications of QDs along with their properties, different types, method of preparations, and associated delivery systems.

13.2 QUANTUM DOTS BASIC PROPERTIES

13.2.1 Functionalization and Solubility

The surface functionalization controls the dispersibility of QDs. Although ultrasmall sizes of QDs are ideal for use in living systems, but aggregation in biological fluids or high ionic strength buffers are the drawbacks associated with biological applications of QDs. The sterically stabilized QDs are significantly larger and unable to enter inside the cells, but they are less sensitive to aggregation associated with ionic strength. There are numerous ways to produce solubilization, including surface functionalization, which also provides the target recognition ability to QDs without changing the properties or roles of such molecules (Dhas et al. 2022; Michalet et al. 2005).

13.2.2 Optical Properties

There are several types of cancer, and some of them are diagnosed at stage III or later, which is difficult to treat. The early detection of disease provides a better opportunity to eliminate tumors. There are several imaging techniques involved in detecting the tumor such as computed tomography, x-ray imaging, magnetic resonance imaging, fluorescence imaging, etc. In cancer imaging, QDs utilized as a tagging agent to mark/label the biological targets. Its optical properties are influenced by several variables, including the composition of QDs, particle size, surface chemistry, dispersity, etc. The necessary optical qualities are often considered while choosing the primary ingredient. QDs show maximum absorption toward shorter wavelengths and asymmetric narrow emission spectra. The position of absorption and emission spectra is adjusted or controlled by changing the particle size. The emission of QDs varies from ultraviolet to near-infrared region, depending on particle sizes, shape, layer, and functional group. In *in vivo* imaging of cancer and photodynamic therapy, generally, imaging agents with an emission maximum in the NIR range is preferable over tagging agents having emission in the visible range due to deep tissue penetration ability. Some examples of QDs having emission in the NIR region are cadmium telluride/ cadmium selenide CdTe/CdSe (core/shell) and CdTe (core only) (Dhas et al. 2022; Xu et al. 2016).

13.2.3 Stability

In comparison to other imaging agents, QDs have superior photochemical and thermal stability. Sometime in CNTs photobleaching occures, which is the results from the breakage of covalent bonds or interactions between fluorophores and the environment's molecules. Although, QDs are less vulnerable to photobleaching phenomena and hence suitable for fluorescence imaging (Dhas et al. 2022).

13.3 CLASSIFICATION OF QUANTUM DOTS

QDs are classified into two categories based on the starting material: organic and inorganic QDs. Organic QDs are mostly prepared from carbon-based materials, while inorganic QDs are prepared from metals like Cd, Te, Se, etc. The detailed characteristics and their applications are discussed below.

13.3.1 Organic Quantum Dots

Organic QDs contain carbon and its allotropes, which are derived from the same element but structurally different from each other. Carbon-based QDs (CQDs) were discovered in 2004 during the purification of single-wall nanotubes. It has been identified as discrete, non-toxic, highly fluorescent, water-soluble nanomaterials with size ranges between 2–8 nm. The surface of CQDs contains polar groups, such as carboxylic (-COOH), amine (NH_2), and hydroxy (OH), based on the groups present in raw materials (Dhas et al. 2022; Sk et al. 2020). The presence of -COOH moieties on the surface of CQDs makes them biocompatible and provides water solubility (Lim et al. 2015; Namdari et al. 2017). Their unique physicochemical properties make them suitable candidates for biological application, while their small size and biocompatibility established them as drug delivery carriers (Eatemadi et al. 2014; Valizadeh et al. 2012). There are several methods used to prepare CQDs, such as thermal oxidation of suitable molecule precursors, laser ablation, electrochemical oxidation of graphite, and microwave-assisted techniques (Li et al. 2010).

13.3.2 Inorganic Quantum Dots

Inorganic QDs were discovered in 1981, and they are made up of inorganic elements like cadmium, selenium, lead, etc. Due to their unique photoluminescence properties, good photostability, and colloidal stability, inorganic QDs being widely explored by the researchers (Gidwani et al. 2021). Some widely explored inorganic QDs are discussed in the following subsections.

13.3.2.1 Cadmium quantum dots

Cadmium quantum dots (CdQDs) is an inorganic QDs that are prepared from semiconductor materials such as cadmium selenium (CdSe) or cadmium telluride (CdTe). The CdQDs are prepared by an aqueous or organometallic method. The aqueous method uses water-soluble thio-compound as a ligand, while the organometallic procedure is performed at elevated temperature by using tri-n-octylphosphine oxide (TOPO) as a solvent (Zhang et al. 2014). The major disadvantage of CdQDs is cadmium leaching, which leads to cell death associated with DNA damage. Their toxicity can be minimized by utilizing various strategies, such as passivation of QDs core using polymeric shell coating, which prevents the leaching of cadmium from CdQDs and provides stability. Some other techniques are utilized to reduce cellular toxicity, such as surface modification with peptides, DNA, and carbohydrates (Zhu et al. 2019). The high-quality CdSe/CdTe QDs can be produced by hydrothermal, microwave, or ultrasound methods. CdSe-based nanomaterials have been shown to reduce amyloid-beta accumulation (amyloid-beta-induced cytotoxicity) in Alzheimer's disease at the neuronal site, which is the pathological hallmark of Alzheimer's disease (Bu et al. 2013; Saini et al. 2022). The surface modification, charge, and chirality influence the QD's cellular uptake. For example, the quantitative analysis of cellular uptake showed higher uptake in the case of surface-modified CdSe and CdTe QDs (Zhang et al. 2009).

13.3.2.2 Silver quantum dots

Silver-QDs have unique electro-optical properties, which make them suitable for various chemical, medicinal, and microbiological applications. Silver-QDs emit NIR wavelength, which makes them promising imaging agents for *in vivo* and *in vitro* applications. They are sub-categorized as Ag_2S, Ag_2Se, Ag_2Te, and $AgInS_2$ QDs (Borovaya et al. 2021). Researchers have used different techniques to formulate silver-QDs. For instance, Bao et al. prepared NIR-II emitting AgS QDs *via* the microwave-assisted synthesis technique. Breifly 0.0191 g of D-penicillamine was dissolved into 38.4 mL of water, after that 1.6 mL (0.1 M) of silver nitrate was mixed with above solution under continuous stirring for 20 min at room temperature. The results demonstrated that color was changed from colorless to red-brown color, an indication of successful synthesis of AgS QDs. Further, to get emission in the IR region, 4 mL of red-brown solution was continuously stirred along with silver nitrate (40 μL, 0.1 M) solution for 7 h. Further, scientists performed fluorescence quenching study for these AgS QDs in the presence of different biological compounds, such folic acid, dopamine, riboflavin, glutathione, cysteine, glucose, and ascorbic acid. To analyses the dopamine concentration, an ultrasensitive fluorescent probe with a detection limit of 0.77 μM was used which showed a fast response because of fluorescence quenching of AgSQDs (Bao et al. 2022).

13.3.2.3 Silicon QDs

Silicon QDs (SiQDs) are extremely promising for a variety of applications, as they can emit and amplify light with high fluorescence intensity. The advancement in SiQDs has opened new technological frontiers, such as optoelectronics and photonics. SiQDs are non-toxic, biodegradable, spectrally tunable, and have bright emissions. These properties provide new opportunities to further explored the SiQDs in drug delivery, bioimaging, lab-on-chip-sensing, cosmetics, and in phototherapeutic applications (Dohnalová et al. 2014). In the last few years, different techniques have been used to synthesize the SiQDs, such as sono chemical synthesis, laser-induced pyrolysis of silane (laser ablation), electrochemical and plasma-assisted aerosol precipitation method to get the good quality SiQDs for biomedical applications (Chinnathambi et al. 2014).

13.4 QUANTUM DOTS SYNTHESIS

QDs can be prepared by using various methods, which may be broadly subcategorized into two approaches: bottom-up and top-down approaches. Each method has its own merits and demerits.

13.4.1 Top-Down Approach

In the top-down technique, the bulk material is converted into nanomaterial. To obtain QDs desired size, different processes were available, such as the arc-discharge method, electrochemical oxidation method, and laser ablation method (Bera et al. 2010), which are described in the following subsection.

13.4.1.1 Arc-discharge technique

In 2004, a team in the US was working on a method for making single-walled nanotubes and accidentally created some new kinds of nanomaterial. This led to the discovery of the arc discharge method for making QDs. One of the freshly created nanoparticles displayed fluorescence in the visible spectrum and had some other properties like good water solubility, but their major limitation was large particle size. In an instrumental setup, anode and cathode electrodes are positioned either horizontally or vertically in a chamber that is part of the arc-discharge process for the creation of QDs. The carbon precursor, macromolecule, and catalyst are positioned on the anode, and a graphite rod serves as the cathode. When the power is turned on, an arc is created, which causes the plasma to form at a very high temperature (4000K–6000 K). The carbon precursor on the anode is sublimated as a result of plasma. Due to the temperature gradient present, the gaseous particle travels towards the cathode, causing carbon vapors to deposit on the cathode (Dhas et al. 2022). Using this technique, one can create QDs from crude carbon nanotubes. In order to add carboxyl groups, the raw material z(sediment) was oxidized with 3.3 M HNO_3, and the resulting material was then extracted with NaOH/basic solution of pH 8.4 to produce a stable dark-colored suspension. The collected material was cleaned using gel electrophoresis (Singh et al. 2018). These QDs showed excellent potential in the medical field as imaging agents in targeted delivery, diagnostic purpose, and biosensing.

13.4.1.2 Electrochemical oxidation

In electrochemical oxidation, high-purity graphite rods were installed into an ionic fluid/water solution and employed pyrolytic graphite with a 2 cm partition as the anode and platinum wire as the counter electrode (Lu et al. 2009). Static potentials were used to start the exfoliation of carbon particles. Exfoliation took place because of a complicated interaction between the anodic oxidative cleavage of water and anionic intercalation from the ionic fluid. Exfoliating products were rinsed with ethanol and water until the pH was neutral. CQDs with a quantum yield of 2.8–5.2% were produced after separation by filtration and ultracentrifugation at 15,000 rpm at 20°C (Singh et al. 2018).

13.4.1.3 Laser ablation method

In order to create a thermodynamic state with higher pressure and temperature, which causes heating and evaporation of the carbon precursor and the production of some vapors that crystallize to form nanoparticles, is what laser ablation entails using a laser (high energy pulse of laser) on a surface. In addition to a carbon source, the laser excision procedure also depends on the parameters of the surrounding environment and requires a liquid (usually water) that serves as a solvent. This method of QDs generation has several benefits, like (i) laser ablation is a very straightforward and clean process that produces few byproducts without the use of complicated materials and catalysts; (ii) this technique does not require extreme conditions, such as high temperatures and pressures; and (iii) it can be carried out in mild ambient conditions. This method is performed by using a variety of substances and solvents. By altering the synthetic parameters and conditions, laser ablation can be used to produce the nanostructure with different shapes and sizes (Cho et al. 2008; Dhas et al. 2022).

13.4.2 Bottom-up Approaches

The bottom-up technique of QDs synthesis entails the condensation reaction, which produces larger compounds from smaller precursors like benzene derivatives. QDs synthesis is a multi-step procedure that uses extremely complicated methods to self-assemble the precursor molecules. The microwave method, ultrasonic, hydrothermal/solvothermal, and combustion method comes under this category, which is further discussed in detail in the following subsections (Bera et al. 2010).

13.4.2.1 Microwave/ultrasonic method

Heating organic molecules, such as proteins, amino acids, carbohydrates, etc., as precursors in the presence of a solvent is the first step in the microwave-assisted technique. It is a suitable technique, and has greater potential for biomedical applications because it doesn't require harsh environmental conditions and is simple to use, also exhibits good features (Guo et al. 2018). To create CQDs that contain phosphorus in a water solvent, Wang et al. described a simple one-stage microwave-assisted technique. In this procedure, 25 mL of triple distilled water was used to make a mixture of 2 mL phytic acid (70%) and 1 mL ethylenediamine. The turbid mixture was subsequently heated in a 700 W microwave for approximately 8 minutes. Phosphorus-containing CQDs were created because of the crude substance's purification; additionally, the aromatic structures of these CQDs were covalently joined to phosphorus groups. When triggered at low wavelengths, the phosphorus containing CQDs showed two peak emissions. However, when activated at high wavelengths (360–460 nm), the nanomaterials showed a single peak at 525 nm (green fluorescence). The quantum yield of the following CQDs, which included phosphorus, was 21.65% (Wang et al. 2014). While in the case of graphene QDs (GQDs) and nitrogen-doped GQDs synthesis through this method, citric acid and urea can be used as precursors (Singh et al. 2018).

13.4.2.2 Combustion/thermal method

The CQDs synthesis through the combustion method was reported in 2007. This technique uses oxidative acid treatments to condense tiny carbon resources into CQDs, which regulate the fluorescence characteristics and improve water solubility. Scientists observed that candle ash is produced when a candle is partially burned along with aluminum foil, and obtained powder ash is further refluxed in nitric acid solution. The pure CQDs were produced after dissolving candle ashes in a neutral medium, and the resulting solution was further centrifuged and dialyzed. The obtained QDs showed excellent fluorescence and low quantum yield (Bera et al. 2010; Singh et al. 2018).

13.4.2.3 Hydrothermal/ solvothermal method

The QDs prepared through the hydrothermal method provide excellent photostability along with high quantum yield. This methodology is industrially viable, cost-effective, and advantageous (Fatima et al. 2021). The hydrothermal synthetic process is performed in a Teflon-coated stainless-steel autoclave at 160°C temperature for 8 h by adding a 1% buffer solution (bis(2-hydroxyethyl) amine, 2-(N-morpholino) ethane sulfonic acid and (4-(2-hydroxyethyl)-1-piperazineethanesulfonic acid) (BES, MES, and HEPES) (Samantara et al. 2016). While the solvothermal synthesis process has been utilized to manage the morphology of nanomaterials for around the last 20 years. Instead of using water, the solvothermal approach uses alternative organic solvents with higher boiling points, such as benzene, dimethyl sulfoxide, and N, N-Dimethylformamide. Usually, these solvents are used to dissolve carbon sources, which are subsequently extracted from the solution and concentrated on it. Mitra et al. prepared self-passivated CQDs using a green synthesis solvothermal procedure, in which polyethylene glycol-200 and NaOH were mixed and vigorously shaken to create a homogenous solution. This solution was formerly allowed to sonicate for one hour in a sonicator bath. After being left undisturbed for an entire night, a separate, yellow-colored solution was produced, which gave CQDs with a distinctive yellowish-brown color. In this procedure, polyethylene glycol-200 works as the surface passivating agent for QDs. The obtained CQDs showed negligible toxicity with good biocompatibility and the ability for bioimaging applications (Mitra et al. 2013; Jha et al. 2018).

13.5 APPLICATION

QDs have gained significant attention in biomedical research due to their unique optical properties and potential for various diagnostic and therapeutic purposes. Here are some notable biomedical applications of QDs:

13.5.1 Bioimaging

QDs can be used as fluorescent probes for high-resolution cellular and molecular imaging. Their narrow emission spectra, high brightness, and photostability make them ideal for visualizing biological structures and processes. QDs can be functionalized with specific ligands or antibodies to target specific cells or molecules, enabling targeted imaging in areas such as cancer research. QDs possess unique optical properties that make them highly suitable for bioimaging applications. Currently, several researchers are working on CQDs to utilize in biomedical fields, for the first time, Yang et al. investigated the viability of CQDs as a fluorescence-contrasting agent (Bhunia et al. 2013). Like this, Sun et al. employed PEGylated CQDs to perform *in vivo* optical imaging of a variety of organs, including the liver, kidney, and bladder. The image obtained from mice after receiving subcutaneous injections of PEGylated CQDs (440 g in 200 L) showed an adequate amount of contrast (Yang et al. 2009). One of the most alluring technologies is multi-imaging, in which magnetic resonance imaging (MRI) and optical imaging combinedly performed, MRI has the potential to produce simultaneous information about physiology and anatomy and have great spatial resolution, although optical imaging allows for quick screening. Recently, it was revealed that thermal breakdown might be used to create ultrafine, water-dispersible gadolinium (III) -dopped CQDs (Gd (III)-doped CQDs) which is bimodal imaging nanoprobes with a dual feature. In comparison with commercial Gadovist® (MRI contrast agent), the produced Gd (III)-doped CQDs displayed vivid fluorescence in the visible region and strong T1-weighted image (in which water signal suppressed and fatty tissue signal enhanced) with minimal cytotoxicity (Bourlinos et al. 2012).

13.5.1.1 Dual modal imaging

In recent times, dual-modal imaging gained much attention due to its accuracy and high efficiency. Dual modal imaging combines two different imaging techniques like, fluorescence and MRI or fluorescence and computed tomography, to obtain complementary information about the sample being studied. In this imaging technique, a comprehensive and accurate knowledge of biological systems is obtained. The purpose of discovered the multi or dual modal imaging because of the limitation in simple imaging technique like limited target specificity, photobleaching and stability problems and less penetration in depth of biological tissue (Hasegawa et al. 2002). There are several research has been done on dual modal imaging such as, for cancer therapy, Huang et al., developed a paramagnetic foalte Gd-graphene QDs complex, in which folic acid and diethylene triamine penta acetic acid gadolinium used as a conjugating agent, and doxycycline (DOX) drug was loaded on this nanocarrier. The developed nanocarrier serves as a versatile chemotherapeutic agent, functioning as both a targeting ligand and a drug delivery vehicle. The results demonstrated that folate Gd-graphene QDs have good targeting ability in case of non-invasive cancer diagnosis, excellent biocompatibility in desired concentration, and showed excellent therapeutic activity

(Huang et al. 2015). Similarly, Zhao et al. synthesized dual model procaine-derived CQDs for anticancer activity and imaging application. The *in vitro* and *in vivo* results declared that procaine-derived CQDs reduced the proliferation rate in colon cancer cells by inducing apoptosis and provided desired bioimaging (Zhao et al. 2020). For antitumor activity and bioimaging purposes, Lu et al. synthesized dual-functionalized gallic-acid-based CQDs because gallic acid has antitumor activity. A simple microwave-assisted technique is used to synthesize gallic-acid-based CQDs and found that they have tremendous potential for usage in future clinical applications as an imaging material as well as an anticancer agent. The functionalized gallic-acid-based CQDs showed excellent photoluminescence behavior, and potential to be used as a bioimaging material. (Lu et al. 2019).

13.5.1.2 Magnetic resonance imaging

MRI is an advanced imaging technique used in both clinical diagnosis and scientific research. In MRI, generally, T1 and T2 contrasting agents are used for a clear diagnosis of T1 and T2-weighted images. These agents shorten the longitudinal and transverse relaxation time by increasing or decreasing signal intensity and improve the resolution as well as the sensitivity of MRI. The major advantage of MRI is deep tissue penetration and high spatiotemporal resolution in soft tissue. It is a non-invasive and extremely sensitive technique used in biomedical fields. For assessments of biocompatibility and biotoxicity, Huang et al. synthesized biocompatible iron-doped CQDs (FeCQDs), which work as a fluorescent imaging probe as well as the magnetic resonance contrast agent. The one-pot synthesis procedure is followed, in which 1.45 g ethylenediaminetetraacetic acid (EDTA), and 3 g glutathione along with 0.54 g $FeSO_4{\cdot}7H_2O$ are dissolved into 500 mL water. The mixture solution was heated in stainless steel autoclave at 180°C for 6 h, a red-brown color solution was obtained, cooled down at 25°C, and purification was done through the dialysis bag. To confirm the internalization of Fe-II into CQDs, different characterizations were performed, such as Energy Dispersive X-Ray Spectroscopy, X-ray photoelectron spectroscopy, and Fourier transform infrared. Their assessment result in A549 cell line demonstrated that FeCQDs showed excellent biocompatibility during biotoxicity analysis (Huang et al. 2019).

In current times, MRI is used for different biomedical applications. One of them is cardio magnetic resonance which is a rapidly growing field in cardio-oncology. It is used to measure the cardiovascular effect during cancer therapy (Harries et al. 2020). In tissue imaging, QDs showed some limitations. For example, the effectiveness of QDs is significantly constrained because of tissue absorption, autofluorescence, and photon scattering. The NIR-II element-based QDs have optimal wavelengths up to 1300 nm. So during *in vivo* imaging, they showed deep tissue penetration. In transplanted malignancies, to recognize the tissue autofluorescence from the QDs signal, a spectrum scattering method must be developed (Dong et al. 2013).

Wang et al. synthesized boron-dopped magnetic GQDs and found that it contains paramagnetic properties, which is a suitable MRI contrast agent for imaging. These QDs were injected in mice models to evaluate the biodistribution, and the fluorescent intensities were measured in different organs after 72 h. The QDs distribution in targeted organs found a significant proportion, and the assay of serum markers in the liver and kidney showed no systemic cytotoxicity. The subcutaneous administration of boron-doped QDs increases the longitudinal relaxation time, and *in vivo,* substantial time-dependent contrast signal also enhances. Their results showed that boron-doped-QDs are safer contrast agents and have enormous potential for clinical MRI application (Wang et al. 2017).

13.5.1.3 Dynamic cellular imaging

Dynamic cellular imaging includes the conjugation of QDs with antibodies or ligands, which are directly attached to membrane receptors and provide information about their complex working with extreme sensitivity and high resolution. Neurons are known to have very complex plasma membranes with different kinds of microdomains that form synapses. QDs help in the study of neurotransmitter diffusion in and out of the synapse by attaching to the glycine neurotransmitters and α-amino-3-hydroxy-5-methyl-4-isoxazolepropionic acid (AMPA) glutamate receptors. Certain receptors, like neuronal growth factor receptors, are internalized inside the cell after binding with specific ligands, and the process was studied in detail. To induce the endocytosis of the receptor-ligand pair inside vesicular structures, QDs conjugated with the ligand of neuronal growth factor, which attaches to the neuronal growth factor receptor at the terminals of neuronal axons. According to QDs imaging, these vesicles usually have only one neuronal growth factor receptor and provide great potential to the cell through numerous molecular pathways (Ruan et al. 2007; Kairdolf et al. 2013). Ruan and their colleague formulated peptide-conjugated QDs to study the intracellular transport mechanism. Dynamic confocal imaging study demonstrated that the peptide-conjugated QDs were internalized through micropinocytosis, a fluid-phase endocytosis process induced upon QDs attachment to cell membranes. The internalized QDs were entrapped in cytoplasmic organelles and attached to the inner vesicle surfaces (Ruan et al. 2007). Due to insufficient delivery techniques, live cell monitoring is limited to lab-scale studies. Recently developed derivatives of QDs can easily track the cytoplasmic process (Kim et al. 2008).

13.5.2 Biosensing

Biosensing through QDs is a promising field of nanotechnology, which involves the use of QDs as optical probes for detecting and analyzing biological molecules and processes (Chern et al. 2019; Ding et al. 2022). In biosensing applications, the surface of the QDs is functionalized with specific biomolecules such as antibodies, peptides, or DNA probes, which can selectively bind the target biomolecules such as proteins, nucleic acids, or small molecules. These QDs are exposed to the target molecules, and they undergo a change in their fluorescence properties, allowing for the detection and quantification of the target analytes (Singh et al. 2018; Sapsford et al. 2006). QDs have high photostability that enables long-term imaging and monitoring of the activity of desired molecules (Chern et al. 2019; Şahin et al. 2021). Furthermore, QDs can be integrated with other nanomaterials to develop advanced biosensing platforms. The applications of QDs biosensing are diverse and spread to numerous fields, including medical diagnostics, environmental monitoring, food safety, and drug discovery (Ding et al. 2022).

The outstanding optical properties of QDs make them excellent fluorophore sensors. In the case of cancer diagnosis, surface functionalized QDs are widely explored in sensing as well as targeting biomarkers in patients. MicroRNA is a noncoding endogenous RNA biomarker that helps in the early detection of cancer disease. For rapid detection of microRNA, QDs strip-based biosensor was developed by Deng and coworkers. They prepared hairpin probe by utilizing nonenzymatic amplification strategy for sensitive and rapid detection of the endogenous RNA biomarkers (Deng et al. 2017). Similarly, Hou and their colleague synthesized CdSe QDs for the detection of acetylcholinesterase enzyme activity (AchE). Scientists have performed bioanalytical assay to study the acetylcholine activity via photochemical sensing strategy. The results of this study demonstrated the inhibition of acetylcholinesterase, and this activity depends upon the biphasic reaction mode, as mentioned in Figure 13.2 (Hou et al. 2016). However, it is important to note that QDs biosensing is still an unexplored area of research, and there is an immense need to overcome challenges such as biocompatibility, potential cytotoxicity, and stability concerns. Researchers are actively working on addressing these issues and optimizing QDs-based biosensing platforms for practical applications.

13.5.3 Theranostic

QDs can easily be incorporated inside the various theranostic nanocarriers, which have therapeutics and diagnostics capabilities in a single system (Jain et al. 2022). Nanotechnology

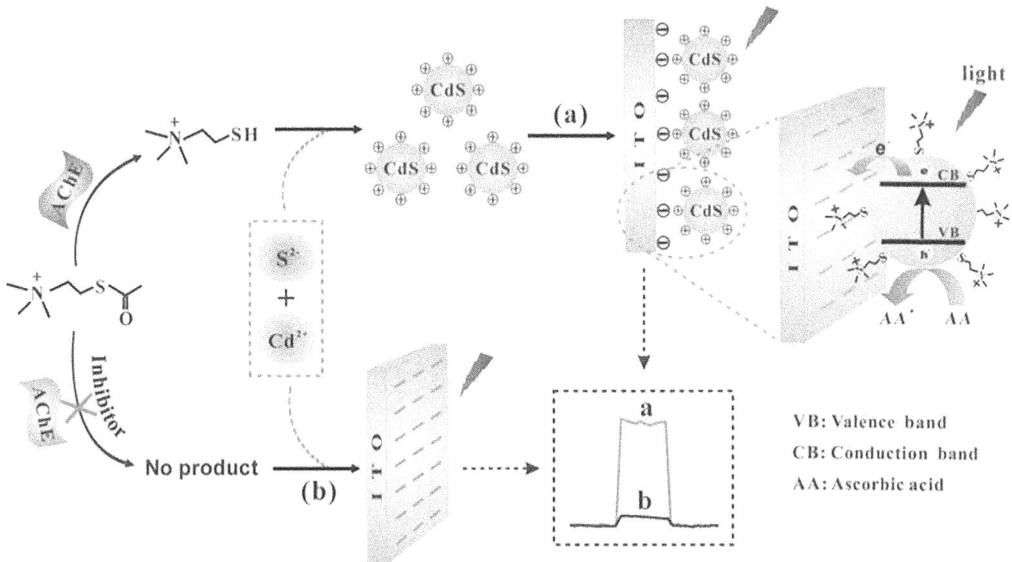

FIGURE 13.2 Biphasic PEC sensing CdS-based QDs for the detection of AChE activity and inhibition. Reproduced with permission from (Hou et al., 2016).

is currently moving towards the concept of theranostic, and it is employed as a coordinate strategy for simultaneous therapy and disease regression analysis. Theranostic tools or devices are biocompatible, lesser side effects, and are non-toxic to healthy cells (Tandale et al. 2021). There are several QDs conjugated theranostic nanocarriers that were developed for the therapy and diagnosis of disease. One of them is drug-loaded RNA-aptamer conjugate, developed by Bagalkot and co-workers. RNA-aptamer has excellent targeting properties, while QDs have fluorescence properties. The conjugation of aptamer with QDs made a theranostic tool that can easily identify the prostate-specific membrane antigen present in prostate cancer cells, and doxorubicin provides therapeutic activity (Bagalkot et al. 2007). In current times, several QDs-based nanotheranostics formulations were developed by the scientific community, as described in detail in Table 13.1.

13.6 CONCLUSIONS

The QDs have tremendous potential for applications as sensors, bio-imaging tools, and drug delivery systems. Many challenges still need to be resolved before its successful clinical translation, including overall toxicity, clearance, scalability of the synthesis technique, environmental impact, manufacturing costs, and others. Theranostic platforms are continuously being explored by administring QDs with different kinds of nanoparticles and/or pharmacologically active compounds. Apart from bioimaging applications, biosensing applications of QDs is gaining more attention recently which may open up the new doors of research.. Conclusively, the QDs is a emerging as potent and futuristic imaging agents with wide scope for further exploration as bioimaging and biosensing agent in health care and biomedication applications.

ACKNOWLEDGMENTS

We acknowledge the Department of Pharmaceuticals (DOP), Ministry of Chemicals and Fertilizers, Government of India, for providing facilities for writing this book chapter. The NIPER-R communication number for this chapter is NIPER-R/Communication/457.

TABLE 13.1
Summary of Various Theranostic Nanocarriers

S. No.	Type of QDs	Therapeutic Moiety	Synthesis Method	Type of Disease	Outcome	Ref.
1.	GQDs@Psi	Graphene along with porous silica	One-step pyrolysis	Diabetic wound healing	*In vitro* release study showed stimuli-responsive drug release pattern in slightly acidic and high oxidative stress condition. It leads to the enhanced cell proliferation and fasten the wound healing process.	(Cui et al. 2021)
2.	NO-CQDs	Carbon material and Nitric oxide (NO)	Hydrothermal method	Cancer	It was found that both NO payloads and surface functionalization were necessary for the clinical efficacy of the NO-releasing CQDs against prostate cancer, lung cancer, and colon cancer cell lines. The most effective CQDs for treating cancers were complexes with primary amine and NO payloads.	(Jin et al. 2021)
3.	Black phosphorus based QDs	Black phosphorous, cysteine-based poly-disulfide amide polymer and Paclitaxel	Simple exfoliation technique and nanoprecipitation technique.	Cancer	The produced NPs have outstanding near-infrared radiation-activated photothermal transduction efficiency and drug transport capability, making them a unique treatment platform with minimal side effects, biodegradability, and significantly reduced chemo dose to get the same therapeutic action.	(Chen et al. 2021)
4.	GO-PEI-GQDs	Graphene-oxide-graphene and methotrexate	Hydrothermal method	Cancer	It acts as potential nanotheranostic system. It is effective at very low concentration and low power density of incident light and showed biocompatibility and cellular imaging capability.	(Kumawat et al. 2019)
5.	Iron oxide/bismuth oxide-GQDs (GQDs-Fe/Bi)	Graphene, bismuth oxide and superparamagnetic iron oxide	One-step pyrolysis	Cancer	Attenuating characteristics of bismuth in GQDs complex provide extreme contrast in computed tomography imaging.	(Badrigilan et al. 2019)
6.	Polyamidoamine (PAMAM) dendrimer functionalized CQDs (CQDs-PAMAM)	PAMAM dendrimer and carbon-based material	Carbodiimide coupling reaction	Cancer and gene therapy	The prepared CQDs and PAMAM conjugate showed selective fluorescence quenching in presence of Cu^{+2} ions with quenching efficiency of 93%. By using this characteristic scientist have shown the triple negative breast cancer targeting ability of nanocarrier because Cu ions largely present over here.	(Ghosh et al. 2019)

7.	GQDs	Graphene, cyclodextrin and Herceptin	One-step pyrolysis	Breast cancer	Herceptin decorated GQDs was found to be potential in targeting the breast cancer and suitable intracellular uptake and desired drug release.	(Ko et al. 2017)
8.	Liposome loaded QDs	D-alpha-tocopheryl polyethylene glycol 1000 succinate monoester (TGPS), docetaxel and folic acid	Solvent injection method	Cancer	D-alpha-tocopheryl polyethylene glycol 1000 succinate monoester enhanced the *in vivo* circulation time, while folic acid conjugation provides targeting ability to nanocarrier.	(Muthu et al. 2012)
9.	CdTe/CdS QDs	Ganciclovir	N/A	Cancer gene therapy	Thymidine kinase conjugated QDs prepared through using sulfo-N-hydroxy sulfosuccunimide coupling reagent. This conjugate showed higher toxicity in Hela cells and QDs luminescence helped in tracking the intracellular traffic.	(Shao et al. 2012)
10.	Mesoporous carbon@ silicon-silica	Carbon, mesoporous silica, polyethylene glycol and hyaluronic acid	One-step carbonization	Cancer	Polyethylene glycol associated phospholipid compound and hyaluronic acid conjugated with mesoporous silica nanoparticles imparted great targeting towards CD44 cells and simultaneous imaging properties.	(He et al. 2012)
11.	Bi_2O_3@PVA QDs	poly (vinyl alcohol), Bi_2O_3 and temozolomide	N/A	Cancer	Temperature-responsive polymer (Bi_2O_3@PVA) provides highly sensitive fluorescent signals. QDs internalized in melanoma cells provide dual modal imaging and enhanced the drug release at target site.	(Zhu et al. 2012)
12.	CQDs	Carbon, folic acid, and methotrexate	N/A	Cancer	CQDs actively targeting the breast cancer cells and desired biodistribution minimize the tumor growth. A significant amount of drug is release into tumor tissue and potential fluorescence properties is better for imaging.	(Pandey et al. 2017)
13.	Polyethylene glycol coated QDs	Curcumin, poly (lactic-co-glycolic acid) and polyethylene glycol	N/A	Primary effusicn lymphoma	Poly (lactic-co-glycolic acid) based theranostic nanocarrier found to be effective in localization of curcumin and tumor imaging associated with QDs in primary effusion lymphoma.	(Belletti et al. 2017)

INDEX WORDS

A

AChE	Acetylcholinesterase
AgQDs	Silver-quantum dots
AMPA	α-amino-3-hydroxy-5-methyl-4-isoxazolepropionic acid

B

BES	N, N-Bis(2-hydroxyethyl)-2-aminoethanesulfonic acid

C

CdQDs	Cadmium quantum dots
CdTe/CdSe	Cadmium telluride/Cadmium selenide
CQDs	Carbon-based QDs

D

DOX	Doxycycline

E

EDTA	Ethylenediaminetetraacetic acid
EGFR	Epidermal growth factor receptor

F

FeCQDs	Iron-doped carbon QDs

G

Gd (III)-doped CQDs	Gadolinium (III) -dopped CQDs
GQDs	Graphene quantum dots

H

HCC-15	Hepatocellular carcinoma
HEPES	(4-(2-hydroxyethyl)-1-piperazineethanesulfonic acid)

I

InP-QDs	Indium-based QDs

M

MES	2-Morpholinoethanesulfonic acid monohydrate
MIP@QDs	Molecularly imprinted polymer QDs
MRI	Magnetic resonance imaging
MSN@QDs	Mn-doped ZnSe QDs

N

NC-Si QDs	Colloidal nanocrystalline silicon QDs
NIR	Near-infrared

P

PET	Positron emission tomography
PL	Photoluminescence
PQDs	Procaine-derivative carbon QDs

Q

QDs	Quantum dots

R

RLE-6NT	Rat lung epithelial-T-antigen negative
ROS	Reactive oxygen species

S

SiQDs	Silicon-based quantum dots

T

TOPO	Tri-n-octylphosphine oxide

Z

Zn@BPQDs	Zin-ion-coordinated black phosphorus QDs

REFERENCES

Ahmad, Javed, Hassan A. Albarqi, Mohammad Zaki Ahmad, Rizwanullah Md., Teeja Suthar, Keerti Jain, Parameswara Rao Vuddanda, and Mohammad Ahmed Khan. 2022. "Receptor-Targeted Surface-Engineered Nanomaterials for Breast Cancer Imaging and Theranostic Applications." *Critical Reviews™ in Therapeutic Drug Carrier Systems* 39 (6). Begel House Inc.: 1–44.

Alivisatos, A. Paul, Weiwei Gu, and Carolyn Larabell. 2005. "Quantum Dots as Cellular Probes." *Annual Review of Biomedical Engineering* 7: 55–76.

Badrigilan, Samireh, Behrouz Shaabani, Nahideh Gharehaghaji, and Asghar Mesbahi. 2019. "Iron oxide/bismuth oxide nanocomposites coated by graphene quantum dots: "Three-in-one" theranostic agents for simultaneous CT/MR imaging-guided in vitro photothermal therapy." *Photodiagnosis and photodynamic therapy* 25: 504–514.

Bagalkot, Vaishali, Liangfang Zhang, Etgar Levy-Nissenbaum, Sangyong Jon, Philip W. Kantoff, Robert Langer, and Omid C. Farokhzad. 2007. "Quantum dot– aptamer conjugates for synchronous cancer imaging, therapy, and sensing of drug delivery based on bi-fluorescence resonance energy transfer." *Nano letters* 10: 3065–3070.

Bajwa, Neha, Neelesh Kumar Mehra, Keerti Jain, and Narendra Kumar Jain. 2015. "Targeted anticancer drug delivery through anthracycline antibiotic bearing functionalized quantum dots." *Artificial cells, nanomedicine, and biotechnology* 7: 1774–1782.

Bajwa, Neha, Neelesh K. Mehra, Keerti Jain, and Narendra K. Jain. 2016. "Pharmaceutical and biomedical applications of quantum dots." *Artificial cells, nanomedicine, and biotechnology* 3: 758–768.

Bao, Wenhui, Lu Ga, Ruiguo Zhao, and Jun Ai. 2022. "Microwave synthesis of silver sulfide near-infrared fluorescent quantum dots and their detection of dopamine." *Biosensors and Bioelectronics* 10: 100112.

Belletti, Daniela, Giovanni Riva, Mario Luppi, Giovanni Tosi, Flavio Forni, Maria Angela Vandelli, Barbara Ruozi, and F. R. A. N. C. E. S. C. A. Pederzoli. 2017. "Anticancer drug-loaded quantum dots engineered polymeric nanoparticles: Diagnosis/therapy combined approach." *European Journal of Pharmaceutical Sciences* 107: 230–239.

Bera, Debasis, Lei Qian, Teng-Kuan Tseng, and Paul H. Holloway. 2010. "Quantum dots and their multimodal applications: a review." *Materials* 4: 2260–2345.

Bhunia, Susanta Kumar, Arindam Saha, Amit Ranjan Maity, Sekhar C. Ray, and Nikhil R. Jana. 2013. "Carbon nanoparticle-based fluorescent bioimaging probes." *Scientific reports* 1: 1473.

Borovaya, Mariya, Inna Horiunova, Svitlana Plokhovska, Nadia Pushkarova, Yaroslav Blume, and Alla Yemets. 2021. "Synthesis, properties and bioimaging applications of silver-based quantum dots." *International Journal of Molecular Sciences* 22: 12202.

Bourlinos, Athanasios B., Aristides Bakandritsos, Antonios Kouloumpis, Dimitrios Gournis, Marta Krysmann, Emmanuel P. Giannelis, Katerina Polakova, Klara Safarova, Katerina Hola, and Radek Zboril. 2012. "Gd (III)-doped carbon dots as a dual fluorescent-MRI probe." *Journal of Materials Chemistry* 44: 23327–23330.

Bu, Hang-Beom, Hayato Kikunaga, Kunio Shimura, Kohji Takahasi, Taichi Taniguchi, and DaeGwi Kim. 2013. "Hydrothermal synthesis of thiol capped CdTe nanoparticles and their optical properties." *Physical Chemistry Chemical Physics* 8: 2903–2911.

Chen, Haolin, Zhiming Liu, Bo Wei, Jun Huang, Xinru You, Jingyang Zhang, Zhiling Yuan, Zhilie Tang, Zhouyi Guo, and Jun Wu. 2021. "Redox responsive nanoparticle encapsulating black phosphorus quantum dots for cancer theranostics." *Bioactive materials* 3: 655–665.

Chern, Margaret, Joshua C. Kays, Shashi Bhuckory, and Allison M. Dennis. 2019. "Sensing with photoluminescent semiconductor quantum dots." *Methods and applications in fluorescence* 1: 012005.

Chinnathambi, Shanmugavel, Song Chen, Singaravelu Ganesan, and Nobutaka Hanagata. 2014. "Silicon quantum dots for biological applications." *Advanced healthcare materials* 1: 10–29.

Cho, Kwangjae, X. U. Wang, Shuming Nie, Zhuo Chen, and Dong M. Shin. 2008. "Therapeutic nanoparticles for drug delivery in cancer." *Clinical cancer research* 5: 1310–1316.

Cui, Yaoxuan, Wei Duan, Yao Jin, Fangjie Wo, Fengna Xi, and Jianmin Wu. 2021. "Graphene quantum dot-decorated luminescent porous silicon dressing for theranostics of diabetic wounds." *Acta Biomaterialia* 131: 544–554.

Deng, Huaping, Qianwen Liu, Xin Wang, Ru Huang, Hongxing Liu, Qiumei Lin, Xiaoming Zhou, and Da Xing. 2017. "Quantum dots-labeled strip biosensor for rapid and sensitive detection of microRNA based on target-recycled nonenzymatic amplification strategy." *Biosensors and Bioelectronics* 87: 931–940.

Dhas, Namdev, Monarch Pastagia, Akanksha Sharma, Alisha Khera, Ritu Kudarha, Sanjay Kulkarni, Soji Soman et al. 2022. "Organic quantum dots: An ultrasmall nanoplatform for cancer theranostics." *Journal of Controlled Release* 348: 798–824.

Ding, Rui, Yue Chen, Qiusu Wang, Zhengzhang Wu, Xing Zhang, Bingzhi Li, and Lei Lin. 2022. "Recent advances in quantum dots-based biosensors for antibiotics detection." *Journal of Pharmaceutical Analysis* 3: 355–364.

Dohnalová, K., T. Gregorkiewicz, and K. Kůsová. 2014. "Silicon quantum dots: surface matters." *Journal of Physics: Condensed Matter* 17: 173201.

Dong, Bohua, Chunyan Li, Guangcun Chen, Yejun Zhang, Yan Zhang, Manjiao Deng, and Qiangbin Wang. 2013. "Facile synthesis of highly photoluminescent Ag2Se quantum dots as a new fluorescent probe in the second near-infrared window for in vivo imaging." *Chemistry of Materials* 12: 2503–2509.

Eatemadi, Ali, Hadis Daraee, Hamzeh Karimkhanloo, Mohammad Kouhi, Nosratollah Zarghami, Abolfazl Akbarzadeh, Mozhgan Abasi, Younes Hanifehpour, and Sang Woo Joo. 2014. "Carbon nanotubes: properties, synthesis, purification, and medical applications." *Nanoscale research letters* 1: 1–13.

Fatima, Iqra, Abbas Rahdar, Saman Sargazi, Mahmood Barani, Mohadeseh Hassanisaadi, and Vijay Kumar Thakur. 2021. "Quantum dots: Synthesis, antibody conjugation, and HER2-receptor targeting for breast cancer therapy." *Journal of functional biomaterials* 4: 75.

Ghosh, Santanu, Krishanu Ghosal, Sk Arif Mohammad, and Kishor Sarkar. 2019. "Dendrimer functionalized carbon quantum dot for selective detection of breast cancer and gene therapy." *Chemical Engineering Journal* 373: 468–484.

Gidwani, Bina, Varsha Sahu, Shiv Shankar Shukla, Ravindra Pandey, Veenu Joshi, Vikas Kumar Jain, and Amber Vyas. 2021. "Quantum dots: Prospectives, toxicity, advances and applications." *Journal of Drug Delivery Science and Technology* 61: 102308.

Guo, Lili, Lin Li, Meiying Liu, Qing Wan, Jianwen Tian, Qiang Huang, Yuanqing Wen, Shangdong Liang, Xiaoyong Zhang, and Yen Wei. 2018. "Bottom-up preparation of nitrogen doped carbon quantum dots with green emission under microwave-assisted hydrothermal treatment and their biological imaging." *Materials Science and Engineering*: 60–66.

Harries, Iwan, Kate Liang, Matthew Williams, Bostjan Berlot, Giovanni Biglino, Patrizio Lancellotti, Juan Carlos Plana, and Chiara Bucciarelli-Ducci. 2020. "Magnetic resonance imaging to detect cardiovascular effects of cancer therapy: JACC CardioOncology state-of-the-art review." *Cardio Oncology* 2: 270–292.

Hasegawa, Bruce H., Koji Iwata, Kenneth H. Wong, Max C. Wu, Angela J. Da Silva, H. Roger Tang, William C. Barber, Andrew H. Hwang, and Anne E. Sakdinawat. 2002. "Dual-modality imaging of function and physiology." *Academic radiology* 11: 1305–1321.

He, Qianjun, Ming Ma, Chenyang Wei, and Jianlin Shi. 2012. "Mesoporous carbon@ silicon-silica nanotheranostics for synchronous delivery of insoluble drugs and luminescence imaging." *Biomaterials* 17: 4392–4402.

Hou, Ting, Lianfang Zhang, Xinzhi Sun, and Feng Li. 2016. "Biphasic photoelectrochemical sensing strategy based on in situ formation of CdS quantum dots for highly sensitive detection of acetylcholinesterase activity and inhibition." *Biosensors and Bioelectronics* 75: 359–364.

Huang, Chun-Lin, Chih-Ching Huang, Fu-Der Mai, Chia-Liang Yen, Shin-Hwa Tzing, Hsiao-Ting Hsieh, Yong-Chien Ling, and Jia-Yaw Chang. 2015. "Application of paramagnetic graphene quantum dots as a platform for simultaneous dual-modality bioimaging and tumor-targeted drug delivery." *Journal of Materials Chemistry* 4: 651–664.

Huang, Qing, Yue Liu, Linling Zheng, Liping Wu, Zhengyu Zhou, Jiafei Chen, Wei Chen, and Huawen Zhao. 2019. "Biocompatible iron (II)-doped carbon dots as T 1-weighted magnetic resonance contrast agents and fluorescence imaging probes." *Microchimica Acta* 186: 1–10.

Jain, Keerti, and Javed Ahmad, eds. 2022. *Nanotheranostics for Treatment and Diagnosis of Infectious Diseases*. Academic Press.

Jain, Keerti, and Jian Zhong. 2022. "Theranostic Applications of Nanomaterials." *Current Pharmaceutical Design* 2: 77–77.

Jha, Swati, Prateek Mathur, Suman Ramteke, and Narendra Kumar Jain. 2018. "Pharmaceutical potential of quantum dots." *Artificial cells, nanomedicine, and biotechnology* 46 (no. supl): 57–65.

Jin, Haibao, Evan S. Feura, and Mark H. Schoenfisch. 2021. "Theranostic activity of nitric oxide-releasing carbon quantum dots." *Bioconjugate Chemistry* 2: 367–375.

Juneja, Mehak, Teeja Suthar, Vishwas P. Pardhi, Javed Ahmad, and Keerti Jain. 2022. "Emerging trends and promises of nanoemulsions in therapeutics of infectious diseases." *Nanomedicine* 11: 793–812.

Kairdolf, Brad A., Andrew M. Smith, Todd H. Stokes, May D. Wang, Andrew N. Young, and Shuming Nie. 2013. "Semiconductor quantum dots for bioimaging and biodiagnostic applications." *Annual review of analytical chemistry* 6: 143–162.

Kim, Betty Y. S., Wen Jiang, John Oreopoulos, Christopher M. Yip, James T. Rutka, and Warren C. W. Chan. 2008. "Biodegradable quantum dot nanocomposites enable live cell labeling and imaging of cytoplasmic targets." *Nano letters* 11: 3887–3892.

Ko, N. R., M. Nafiujjaman, J. S. Lee, H-N. Lim, Y-K. Lee, and I. K. Kwon. 2017. "Graphene quantum dot-based theranostic agents for active targeting of breast cancer." *Rsc Advances* 19: 11420–11427.

Kumawat, Mukesh Kumar, Mukeshchand Thakur, Rohan Bahadur, Tanvi Kaku, R. S. Prabhuraj, Aakansha Suchitta, and Rohit Srivastava.2019. "Preparation of graphene oxide-graphene quantum dots hybrid and its application in cancer theranostics." *Materials Science and Engineering* 103: 109774.

Li, Haitao, Xiaodie He, Zhenhui Kang, Hui Huang, Yang Liu, Jinglin Liu, Suoyuan Lian, Chi Him A. Tsang, Xiaobao Yang, and Shuit-Tong Lee.2010. "Water-soluble fluorescent carbon quantum dots and photocatalyst design." *Angewandte Chemie International Edition* 26: 4430–4434.

Li, Meixiu, Tao Chen, J. Justin Gooding, and Jingquan Liu. 2019. "Review of carbon and graphene quantum dots for sensing." *ACS sensors* 7: 1732–1748.

Lim, Shi Ying, Wei Shen, and Zhiqiang Gao. 2015. "Carbon quantum dots and their applications." *Chemical Society Reviews* 1: 362–381.

Lu, Jiong, Jia-xiang Yang, Junzhong Wang, Ailian Lim, Shuai Wang, and Kian Ping Loh. 2009. "One-pot synthesis of fluorescent carbon nanoribbons, nanoparticles, and graphene by the exfoliation of graphite in ionic liquids." *ACS nano* 8: 2367–2375.

Lu, Shuting, Liping Liu, Henan Wang, Wancheng Zhao, Zeyu Li, Zheng Qu, Jialu Li, Tiedong Sun, Ting Wang, and Guangchao Sui. 2019. "Synthesis of dual functional gallic-acid-based carbon dots for bioimaging and antitumor therapy." *Biomaterials science* 8: 3258–3265.

Martynenko, I. V., A. P. Litvin, F. Purcell-Milton, A. V. Baranov, A. V. Fedorov, and Y. K. Gun'Ko. 2017. "Application of semiconductor quantum dots in bioimaging and biosensing." *Journal of Materials Chemistry* 33: 6701–6727.

Michalet, Xavier, Fabien F. Pinaud, Laurent A. Bentolila, James M. Tsay, S. J. J. L. Doose, Jack J. Li, G. Sundaresan, A. M. Wu, S. S. Gambhir, and S. Weiss. 2005. "Quantum dots for live cells, in vivo imaging, and diagnostics." *Science* 5709: 538–544.

Mitra, Shouvik, Sourov Chandra, Shaheen H. Pathan, Narattam Sikdar, Panchanan Pramanik, and Arunava Goswami. 2013. "Room temperature and solvothermal green synthesis of self-passivated carbon quantum dots." *RSC advances* 10: 3189–3193.

Muthu, Madaswamy S., Sneha A. Kulkarni, Anandhkumar Raju, and Si-Shen Feng. 2012. "Theranostic liposomes of TPGS coating for targeted co-delivery of docetaxel and quantum dots." *Biomaterials* 12: 3494–3501.

Namdari, Pooria, Babak Negahdari, and Ali Eatemadi. 2017. "Synthesis, properties and biomedical applications of carbon-based quantum dots: An updated review." *Biomedicine & pharmacotherapy* 87: 209–222.

Pandey, Sunil, Ganga Raju Gedda, Mukeshchand Thakur, Mukesh Lavkush Bhaisare, Abou Talib, M. Shahnawaz Khan, Shou-Mei Wu, and Hui-Fen Wu. 2017. "Theranostic carbon dots 'clathrate-like' nanostructures for targeted photo-chemotherapy and bioimaging of cancer." *Journal of industrial and engineering chemistry* 56: 62–73.

Park, Youngrong, Sanghwa Jeong, and Sungjee Kim. 2017. "Medically translatable quantum dots for biosensing and imaging." *Journal of Photochemistry and Photobiology C: Photochemistry Reviews* 30: 51–70.

Patel, Parth, Kishore Kumar, Vineet K. Jain, Harvinder Popli, Awesh K. Yadav, and Keerti Jain. 2023. "Nanotheranostics for Diagnosis and Treatment of Breast Cancer." *Current Pharmaceutical Design.*

Ruan, Gang, Amit Agrawal, Adam I. Marcus, and Shuming Nie. 2007. "Imaging and tracking of tat peptide-conjugated quantum dots in living cells: new insights into nanoparticle uptake, intracellular transport, and vesicle shedding." *Journal of the American Chemical Society* 47: 14759–14766.

Şahin, Sultan, Caner Ünlü, and Levent Trabzon. 2021. "Affinity biosensors developed with quantum dots in microfluidic systems." *Emergent Materials* 4: 187–209.

Saini, Vanshul, Ajit Singh, Rahul Shukla, Keerti Jain, and A. K. Yadav. 2022. "Silymarin-Encapsulated Xanthan Gum–Stabilized Selenium Nanocarriers for Enhanced Activity against Amyloid Fibril Cytotoxicity." *AAPS PharmSciTech* 5: 125.

Samantara, Aneeya K., Santanu Maji, Arnab Ghosh, Bamaprasad Bag, Rupesh Dash, and Bikash Kumar Jena. 2016. "Good's buffer derived highly emissive carbon quantum dots: excellent biocompatible anticancer drug carrier." *Journal of Materials Chemistry* 14: 2412–2420.

Sapsford, Kim E., Thomas Pons, Igor L. Medintz, and Hedi Mattoussi. 2006. "Biosensing with luminescent semiconductor quantum dots." *Sensors* 8: 925–953.

Shao, Dan, Qinghui Zeng, Zheng Fan, Jing Li, Ming Zhang, Youlin Zhang, Ou Li, Li Chen, Xianggui Kong, and Hong Zhang. 2012. "Monitoring HSV-TK/ganciclovir cancer suicide gene therapy using CdTe/CdS core/shell quantum dots." *Biomaterials* 17: 4336–4344.

Singh, Inderbir, Riya Arora, Hardik Dhiman, and Rakesh Pahwa. 2018. "Carbon quantum dots: Synthesis, characterization and biomedical applications." *Turkish Journal of Pharmaceutical Sciences* 2: 219–230.

Sk, Md Palashuddin, and Anushree Dutta. 2020. "New-generation quantum dots as contrast agent in imaging." In *Nanomaterials in Diagnostic Tools and Devices*, 525–556. Elsevier.

Tandale, P., Neeraj Choudhary, Joga Singh, Akanksha Sharma, Ananya Shukla, Pavani Sriram, Udit Soni et al. 2021. "Fluorescent quantum dots: An insight on synthesis and potential biological application as drug carrier in cancer." *Biochemistry and Biophysics Reports* 26: 100962.

Valizadeh, Alireza, Haleh Mikaeili, Mohammad Samiei, Samad Mussa Farkhani, Nosratalah Zarghami, Mohammad Kouhi, Abolfazl Akbarzadeh, and Soodabeh Davaran. 2012. "Quantum dots: synthesis, bioapplications, and toxicity." *Nanoscale research letters* 7: 1–14.

Wang, Hui, Mr Richard Revia, Kui Wang, Mr Rajeev J. Kant, Qingxin Mu, Zheng Gai, Kunlun Hong, and Miqin Zhang. 2017. "Paramagnetic properties of metal-free boron-doped graphene quantum dots and their application for safe magnetic resonance imaging." *Advanced Materials (Deerfield Beach, Fla.)* 11.

Wang, Wei, Yongmao Li, Lu Cheng, Zhiqiang Cao, and Wenguang Liu. 2014. "Water-soluble and phosphorus-containing carbon dots with strong green fluorescence for cell labeling." *Journal of Materials Chemistry* 1: 46–48.

Xu, Suying, Jiabin Cui, and Leyu Wang. 2016. "Recent developments of low-toxicity NIR II quantum dots for sensing and bioimaging." *TrAC Trends in Analytical Chemistry* 80: 149–155.

Yang, Sheng-Tao, Xin Wang, Haifang Wang, Fushen Lu, Pengju G. Luo, Li Cao, Mohammed J. Meziani et al. 2009. "Carbon dots as nontoxic and high-performance fluorescence imaging agents." *The Journal of Physical Chemistry* 42: 18110–18114.

Zhang, Fan, Dongmei Yi, Haizhu Sun, and Hao Zhang. 2014. "Cadmium-based quantum dots: preparation, surface modification, and applications." *Journal of nanoscience and nanotechnology* 2: 1409–1424.

Zhang, Leshuai W., and Nancy A. Monteiro-Riviere. 2009. "Mechanisms of quantum dot nanoparticle cellular uptake." *Toxicological Sciences* 1: 138–155.

Zhao, Xiaoming, Tianyang Qi, Mingxi Yang, Wenjing Zhang, Chenfei Kong, Miao Hao, Yuqian Wang et al. 2020. "Synthesis of dual functional procaine-derived carbon dots for bioimaging and anticancer therapy." *Nanomedicine* 07: 677–689.

Zhu, Chunyan, Zhao Chen, Shuai Gao, Ban Leng Goh, Ismail Bin Samsudin, Kai Wen Lwe, Yunlong Wu, Caisheng Wu, and Xiaodi Su. 2019. "Recent advances in non-toxic quantum dots and their biomedical applications." *Progress in Natural Science: Materials International* 6: 628–640.

Zhu, Hongbo, Yaoxin Li, Runqi Qiu, Lei Shi, Weitai Wu, and Shuiqin Zhou. 2012. "Responsive fluorescent Bi2O3@ PVA hybrid nanogels for temperature-sensing, dual-modal imaging, and drug delivery." *Biomaterials* 10: 3058–3069.

14 Unraveling Neurological Enigmas with SmFRET

Exploring the Intricacies of Neuroscience with Single-Molecule Förster Resonance Energy Transfer (SmFRET)

Farhana Islam and Padmaja P. Mishra

14.1 INTRODUCTION TO FLUORESCENCE RESONANCE ENERGY TRANSFER

FRET, or Fluorescence Resonance Energy Transfer is a mechanism by which energy is transferred from a donor molecule to an acceptor molecule through non-radiative dipole-dipole interaction (Miki et al. 1992; Paul et al. 2017). It is based on the principle that when two fluorescent molecules are in close proximity to each other, the energy from one molecule (the "donor") can be transferred to the other molecule (the "acceptor") (Lakowicz 2006). This transfer of energy causes the acceptor molecule to emit fluorescence, which can be detected and used to study the interactions between the donor and acceptor molecules. FRET is commonly used in biochemistry and molecular biology to study protein-protein interactions, protein-nucleic acid interactions, and conformational changes in proteins. It is also used in cell biology to study the dynamics of protein localization and the mechanisms of signal transduction. The FRET efficiency (E) is a measure of the amount of energy transferred from the donor molecule to the acceptor molecule. The most common equation used to calculate FRET efficiency is (Ha 2001):

$$E = (I_A/I_A + I_D)$$

Where: I_A is the acceptor fluorescence intensity and I_D is the donor fluorescence intensity. The FRET efficiency can be calculated using the previous two equations using the fluorescence intensity values of the donor and acceptor.

The spectral overlap integral is a mathematical expression used to calculate the probability of energy transfer between two fluorescent molecules in FRET (Islam et al. 2022). It is calculated by multiplying the donor's emission spectrum by the acceptor's absorption spectrum and integrating the product over all wavelengths. The result is a measure of the overlap between the donor's emission spectrum and the acceptor's absorption spectrum. The greater the overlap, the higher the probability of energy transfer (Figure 14.1b). The spectral overlap integral can be represented mathematically as (Szabó et al. 2022):

$$J = \int (\text{donor emission spectrum} * \text{acceptor absorption spectrum})d\lambda$$

DOI: 10.1201/9781003305583-14

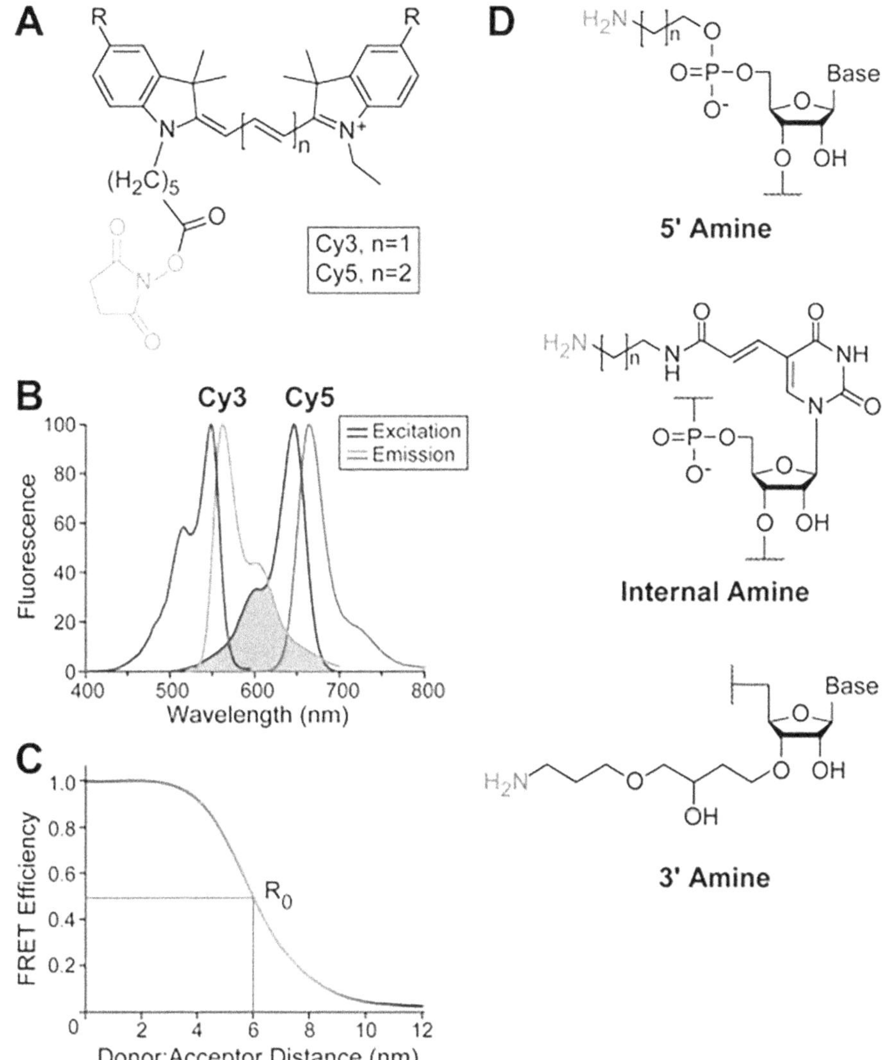

FIGURE 14.1 (A) Structure of Cy dyes that are commonly used fluorophores in FRET (Förster Resonance Energy Transfer) applications. These dyes possess NHS (N-hydroxysuccinimide) leaving groups in their structures (shown in gray), allowing for easy attachment to biomolecules for fluorescent labeling. To enhance solubility, sulfonate groups (represented as R) are often added to the Cy dyes. In FRET experiments, two fluorophores, Cy3 and Cy5, are commonly used in conjunction. (B) The excitation (Ex) and emission (Em) spectra of Cy3 and Cy5 are crucial in FRET experiments. The spectral overlap between Cy3 emission and Cy5 excitation is denoted in yellow, indicating the range of wavelengths where energy transfer between the two dyes can occur. (C) The FRET efficiency is a measure of the energy transfer efficiency between Cy3 and Cy5 fluorophores. At 0.5 FRET efficiency, the distance between the Cy3 and Cy5 fluorophores is equal to the Förster distance, R0, which is approximately 6mnm for the Cy3:Cy5 pair. (D) To facilitate fluorophore labeling in various applications, a few examples of commercially available options for Cy dyes with different properties and functional groups for attachment to biomolecules are shown. Image courtesy: https://doi.org/10.1016/j.ymeth.2017.05.011 (van der Feltz and Hoskins 2017). Creative Commons Public Domain Mark 1.0. PMC Open Access Subset. Bethesda (MD): National Library of Medicine. 2003. Available from https://www.ncbi.nlm.nih.gov/pmc/tools/openftlist/.

Or

$$J(\lambda) = \int F_D(\lambda)\varepsilon_A(\lambda)\lambda^4 d\lambda$$

Where: F_D = normalized emission spectra of the donor, ε_A = molar absorption of the acceptor, and λ = wavelength, donor emission spectrum is the spectral intensity of the donor molecule's fluorescence as a function of wavelength, acceptor absorption spectrum is the spectral intensity of the acceptor molecule's absorption as a function of wavelength, and $\int$ represents the integral symbol. The spectral overlap integral is also known as the Förster distance and is related to the efficiency of FRET. The greater the spectral overlap integral, the greater the FRET efficiency. The Förster distance is typically reported in units of nanometers (nm).

The Förster radius, also known as the Förster distance, is a measure of the distance at which energy transfer between two fluorescent molecules is most efficient. It is related to the spectral overlap integral and is used to calculate the efficiency of FRET. FRET is a powerful technique that allows scientists to study the interactions between molecules by measuring the transfer of energy from a donor molecule to an acceptor molecule. The efficiency of this energy transfer is dependent on several factors, including the distance between the donor and acceptor molecules, the spectral overlap between the donor's emission spectrum and the acceptor's absorption spectrum, and the environment in which the molecules are located. The Förster radius can be calculated using the Förster equation, which is based on the principles of quantum mechanics:

$$R_0 = (R * k_2)/(6\pi n)$$

Where: R_0 is the Förster radius, R is the distance between the donor and acceptor molecules, k_2 is the rate constant for energy transfer, n is the refractive index of the medium surrounding the molecules. The Förster radius is inversely proportional to the sixth power of the distance between the donor and acceptor molecules and is represented as:

$$(R_0)^6 = 8.79 \times 10^{-25}(\kappa^2\eta^{-4}Q_D J(\lambda)) \text{ (in cm}^6\text{)}$$
$$R_0 = 8.79 \times 10^{23}(\kappa^2\eta^{-4}Q_D J(\lambda))^{1/6} \text{(in Á)}$$
$$(R_0)^6 = 8.79 \times 10^{23}(\kappa^2\eta^{-4}Q_D J(\lambda)) \text{ (in Á}^6\text{)}$$

This means that as the distance between the molecules increases, the Förster radius decreases, and the efficiency of energy transfer decreases (Figure 14.1c). The Förster radius is a measure of the distance at which energy transfer is most efficient, and it can be used to calculate the efficiency of FRET. In addition to the Förster equation, there are other methods to calculate the Förster distance like the steady-state and time-resolved FRET methods, or by using spectroscopic techniques such as fluorescence lifetime imaging microscopy (FLIM). The Förster radius is an important parameter in FRET studies, as it can be used to determine the distance between interacting molecules, and to study the dynamics of protein-protein interactions and conformational changes in proteins. The "FPbase FRET Calculator" determines R_0 for many organic dyes and fluorescent proteins (https://www.fpbase.org/fret/).

14.2 FRET (FÖRSTER RESONANCE ENERGY TRANSFER)

FRET (Förster Resonance Energy Transfer) efficiency is a measure of the efficiency of energy transfer between two fluorescent molecules, also known as donor and acceptor (Selvin 1995). FRET occurs when a photon is absorbed by the donor molecule, exciting it to a higher energy state, and then the energy is transferred to the acceptor molecule, which then emits a photon. The efficiency of this energy transfer is determined by the distance between the two molecules and the relative orientation of their transition dipoles.

FRET efficiency is typically described as the ratio of the number of photons emitted by the acceptor to the number of photons absorbed by the donor, and is given by:

$$E_{fret} = \frac{R_0^6}{R_0^6 + r^6}$$

It is usually reported as a percentage. The maximum possible efficiency of FRET is 100%, but in practice, the efficiency is typically much lower, often on the order of 10–20%. This is because the energy transfer process is highly dependent on the distance between the donor and acceptor, and the relative orientation of their transition dipoles, so even small changes in these factors can result in large changes in the efficiency of the transfer.

14.3 SINGLE-MOLECULE FRET

Single-molecule FRET (Fluorescence Resonance Energy Transfer) is a technique used to study the dynamics and interactions of biomolecules at the single-molecule level (Roy et al. 2008). It is based on the principle that when two fluorescent dyes are in close proximity, the energy from one dye (the donor) can be transferred to the other dye (the acceptor) through non-radiative energy transfer. By measuring the fluorescence intensity of the donor and acceptor dyes separately, it is possible to determine the distance between the dyes and the dynamics of their interactions.

One of the main advantages of single-molecule FRET is that it allows for the study of biomolecular interactions and dynamics in their native environment, without the need for ensemble averaging (Sasmal et al. 2016). This is important because many biomolecular interactions and conformational changes occur on a timescale that is too fast to be observed by ensemble techniques, such as bulk fluorescence spectroscopy. Single-molecule FRET also allows for the study of rare events, such as the formation of transient complexes or the switching between different conformations of a protein. Another advantage of single-molecule FRET is that it can provide high spatial and temporal resolution. By using high-resolution microscopy techniques, such as total internal reflection fluorescence (TIRF) microscopy, it is possible to achieve sub-nanometer resolution in the distance measurements. Additionally, single-molecule FRET experiments can be performed in real-time, allowing for the study of dynamic processes at the single-molecule level. Single-molecule FRET can also be used in combination with other techniques, such as single-molecule force spectroscopy, to study the mechanical properties of biomolecules. For example, by applying force to a biomolecule using a microscope cantilever, it is possible to study the unfolding and refolding of proteins, or the stretching and bending of DNA. In summary, single-molecule FRET is a powerful technique that allows for the study of biomolecular interactions and dynamics at the single-molecule level. Its advantages include the ability to study biomolecules in their native environment, the ability to study rare events, high spatial and temporal resolution, and the ability to be combined with other techniques. With the help of single-molecule FRET, we can gain a more detailed understanding of biomolecular interactions and dynamics, which is essential for furthering our understanding of cellular processes and disease mechanisms. Single-molecule techniques have several advantages, including (Ha and Selvin 2008):

1. High sensitivity: Single-molecule techniques can detect and study individual molecules, allowing for measurements at very low concentrations.

2. High spatial resolution: Single-molecule techniques can be used to study the behavior of individual molecules at the nanoscale, providing detailed information about their structure and interactions.
3. High temporal resolution: Single-molecule techniques can be used to study the dynamics of individual molecules on a timescale of milliseconds to seconds.
4. Low sample requirements: Single-molecule techniques require very small amounts of sample (i.e., ~ pM range), making them ideal for studying rare or precious biomolecules or those susceptible to aggregation at high concentrations.
5. Label-free detection: Some single-molecule techniques do not require the use of fluorescent labels, eliminating the need for additional chemical modification of the sample and reducing the potential for artefactual results.
6. In this technique, the focus is on observing the energy transfer between a single pair of donor and acceptor molecules.
7. The study enables the identification and characterization of various hidden kinetic pathways that these molecules can undergo. Additionally, it allows for understanding the distribution of populations in each state and even detecting rare intermediate states with short lifespans.
8. By examining the energy transfer between individual molecules, researchers gain insight into dynamic heterogeneity, which refers to the differences in behavior between molecules, even in the same environment.
9. To achieve more precise imaging while minimizing issues like photobleaching and background noise, the fluorophore molecules are excited using an evanescent wave generated through total internal reflection.
10. During data analysis, post-synchronization methods are employed, eliminating the need for synchronizing measurements in real-time. This approach streamlines the experimental process and simplifies data collection and interpretation.

14.4 SELECTING THE RIGHT FLUOROPHORE PAIR FOR SmFRET

When selecting fluorophore pairs for single-molecule FRET (smFRET), there are a few key factors to consider (Lakowicz 2006):

- Spectral overlap: The emission spectrum of the donor fluorophore should overlap with the absorption spectrum of the acceptor fluorophore to ensure efficient energy transfer.
- small in size (~1 nm) so as to cause minimum perturbation to the host molecules or the system under investigation.
- Brightness: The donor and acceptor fluorophores should be bright enough to detect at the single-molecule level (high extinction coefficients >50,000 M^{-1} cm^{-1} and quantum yields).
- Photostability: The fluorophores should be stable and not degrade or bleach quickly under the excitation conditions.
- Spectral compatibility with other dyes: If other dyes are used in the experiment, their spectra should not overlap with the FRET pair's spectra.
- Large Stokes shift: A large Stokes shift between the donor and acceptor fluorophores ensures that the acceptor emission is spectrally separated from the donor emission, making it easier to detect and analyze the FRET signal.
- Distance range: The donor and acceptor should be chosen such that the distance range of interest can be probed.
- commercially available in a form that can be conjugated to biomolecules (either NHS ester modified or maleimide functionalized).

- Low background noise: The pair should have low background noise, and be able to distinguish between the FRET signal and any other signals.
- Cost-effective: The pair should be affordable and easy to obtain.

Popular FRET pairs include Cy3 and Cy5 (Figure 14.1a,d), CFP and YFP, and Alexa 488 and Alexa 647. A wide range of small molecule dyes with distinct characteristics are available along with a comprehensive review of their properties (Grimm et al. 2015, 2017). For single-molecule Förster Resonance Energy Transfer (smFRET) experiments, cyanine (Cy) dyes are commonly utilized due to their brightness and relatively high photostability. These qualities allow for longer observation of FRET signals before irreversible photobleaching occurs. Among the Cy dyes, Cy3 and Cy5 are frequently used as a donor: acceptor pair in biological FRET studies because of their well-matched spectral overlap. Additionally, their fluorescence emissions can be easily separated optically, making them suitable for such applications. Cy3 has an excitation maximum at 550 nm and an emission maximum at 570 nm, while Cy5 has excitation and emission maxima at 649 nm and 670 nm, respectively. Because Cy5's spectral properties are distinct from Cy3, the commonly used 532 nm laser excitation for Cy3 does not significantly excite Cy5. Nevertheless, this excitation is sufficient to generate robust FRET signals. To label nucleic acids or other biomolecules with these fluorophores, they can be easily linked to primary amino groups through amidation with NHS-derivatized fluorophores. For researchers conducting experiments with these dyes, Alexander et al. have compiled a list of spectroscopic properties of commonly used organic dyes, providing valuable references for their work.

14.4.1 Slide Preparation and Surface Immobilization

To prepare slides for smFRET, several steps are typically followed:

1. Sample preparation: The biomolecules of interest must be labeled with fluorescent dyes that can be used for energy transfer. This can be done by incorporating fluorescently labeled nucleotides during PCR, or by attaching fluorescent dyes to specific amino acids in proteins.
2. Slide preparation: The labeled biomolecules are then deposited onto a glass slide or coverslip. The slide should be cleaned thoroughly to minimize background fluorescence (Joo and Ha 2012a; Lamichhane et al.).
3. Immobilization: The biomolecules must be immobilized on the slide close to the surface in order to be within the evanescent field of TIRF. This can be done using a variety of methods, such as chemical crosslinking, biotin-streptavidin interaction, or microfabrication (Figure 14.2A-D).
4. TIRF Microscopy: The slide is then placed in TIRF microscope, which uses a high-angle light illumination to excite the fluorescent dyes at the surface of the slide.
5. Confocal Microscopy: The slide is then placed in a confocal microscope, which uses a pinhole aperture to reject out-of-focus fluorescence and improve the resolution.
6. Imaging: Once the slide is prepared, it is ready for imaging using TIRF microscope. The microscope should be equipped with appropriate filters for detecting the dyes used for labeling the biomolecules.

It's worth noting that the preparation of slides for smFRET will vary depending on the specific experiment and the biomolecules being studied. It is important to consult the literature and speak with experts in the field to ensure that the best practices are followed.

In single-molecule experiments, obtaining high-quality data is essential, and a key requirement is to prevent fluorescent dyes from undergoing undesired reactions in the excited state, leading to

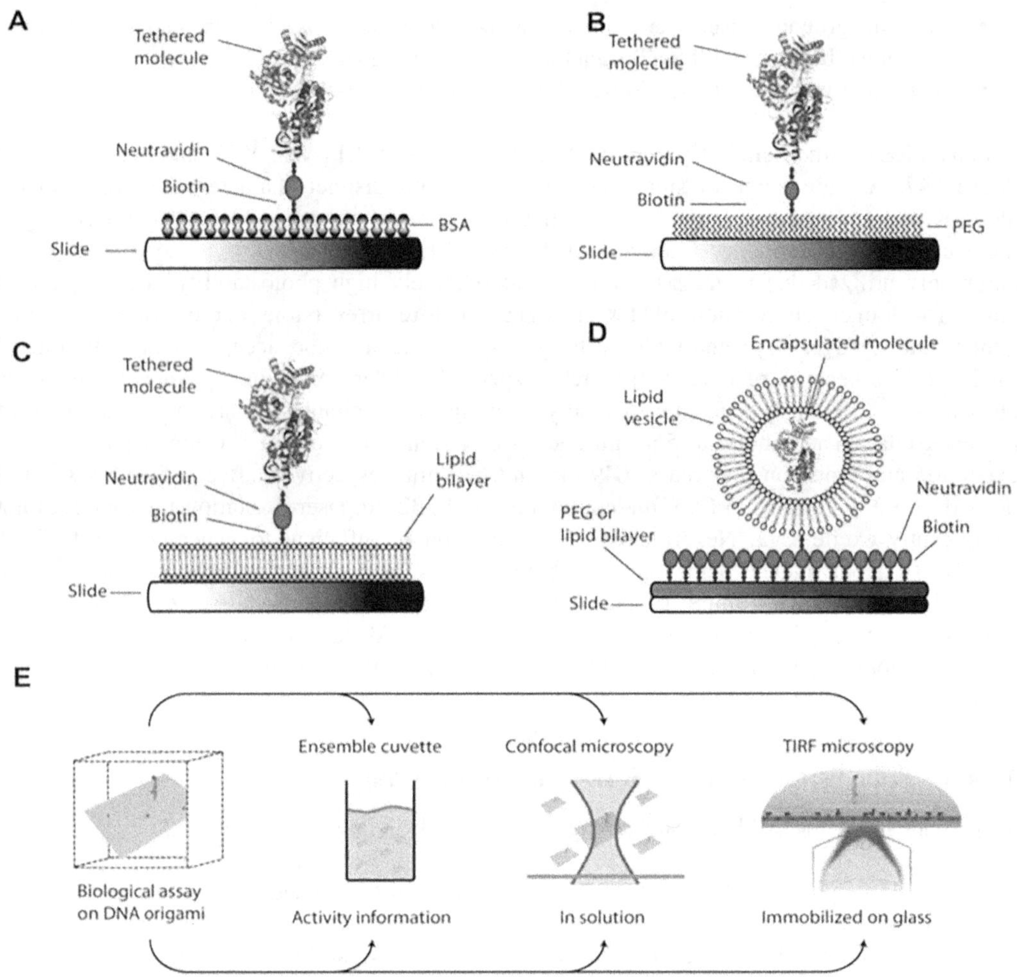

FIGURE 14.2 In single-molecule measurements involving surface-tethered molecules, immobilization and passivation strategies are crucial to ensure accurate and specific observations. Passivation of the glass slide is necessary to prevent non-specific binding of molecules of interest, and biocompatible reagents like (A) bovine serum albumin (BSA), (B) polyethylene glycol (PEG), or (C, D) lipids are commonly used for this purpose. Passivation can be achieved by mixing the passivation reagent with its biotinylated counterpart and then introducing neutravidin. This allows for the immobilization of the molecule of interest on the passivated surface through a biotin-neutravidin linkage. (E) In more advanced approaches, pseudo-surfaces such as DNA origami are utilized. DNA origami serves as a transportable and biocompatible platform that facilitates the matching of ensemble and single-molecule measurements. This enables researchers to study individual molecules within the context of an ensemble, providing valuable insights into molecular behavior and interactions. Image courtesy: doi: 10.3390/molecules191015824 (Gust et al. 2014). Attribution 4.0 International (CC BY 4.0).

photobleaching and photo blinking. Photobleaching is the permanent loss of fluorescent properties due to photon-induced chemical damage, while photo blinking refers to the intermittent loss of fluorescence caused by the dye being trapped in a non-fluorescing state, typically the triplet state. The main culprit responsible for both photobleaching and blinking is excited state oxygen. When oxygen species collide with the excited state fluorophore, they can react with it, leading to permanent damage, or temporarily push it into a triplet state, causing fluorescence intermittency that can last for milliseconds or longer. Removing oxygen molecules is crucial to prolong the residence time of the fluorophore in the dark triplet state and prevent premature signal termination. To achieve this, a triplet state quencher is employed to eliminate the effects of oxygen. The most commonly used quencher is a vitamin E analog called Trolox. Trolox acts as an excellent triplet state quencher, effectively suppressing blinking and promoting long-lasting emission of popular cyanine dyes.

To further enhance the stability of the fluorescent signal, an enzymatic oxygen-scavenging system and reducing agents are used (Joo and Ha 2012b). The system typically consists of D-glucose (at a concentration of 0.8% w/v), glucose oxidase (1 mg/ml), and catalase (0.04 mg/ml). Glucose oxidase converts glucose to hydrogen peroxide, and catalase efficiently converts hydrogen peroxide to water and oxygen. This enzymatic system helps maintain a low level of oxygen, further reducing the chances of photobleaching and blinking during the experiment. By combining triplet state quenchers like Trolox and an enzymatic oxygen-scavenging system with reducing agents, researchers can significantly improve the quality and duration of fluorescence signals in single-molecule experiments, enabling more accurate and reliable data acquisition. The resultant reactions occurring are:

$$\textbf{Glucose} + \mathbf{O_2} + \mathbf{H_2O} \xrightarrow[\text{glucose oxidase}]{} \textbf{gluconic acid} + \mathbf{H_2O_2}$$

$$\mathbf{H_2O_2} \xrightarrow[\text{catalase}]{} \tfrac{1}{2}\,\mathbf{O_2} + \mathbf{H_2O}$$

14.5 EXPERIMENTAL APPROACH: SURFACE IMMOBILIZATION OR FREE DIFFUSION

Total Internal Reflection Spectroscopy (TIRS) is a technique used to study the properties of thin films, surfaces, and interfaces. It is based on the phenomenon of total internal reflection (TIR), which occurs when light travels from a medium of higher refractive index to a medium of lower refractive index at an angle greater than the critical angle. When TIR occurs, the light is completely reflected back into the higher refractive index medium, and none of it is transmitted into the lower refractive index medium.

In TIRS, a light source is directed at a sample surface at an angle greater than the critical angle, causing TIR to occur. The light that is reflected back into the higher refractive index medium is then analyzed to study the properties of the sample surface. The intensity of the reflected light can be measured as a function of wavelength, angle, or polarization, providing information on the optical properties, composition, and structure of the sample surface.

TIRS can be used to study a wide range of samples, including thin films, surfaces, and interfaces in semiconductors, metals, polymers, and biological materials. It is a non-destructive technique that can be used to study samples in situ and in real-time, making it a powerful tool for material characterization and process control. It is important to note that the TIRS method is usually utilized with a prism that is in contact with the sample, and the prism should be chosen carefully depending on the sample and the wavelength range of the light source. Also, TIRS can be combined with other techniques such as ellipsometry, reflectometry, and microscopy for more detailed analysis.

14.5.1 Prism Type TIR and Objective-Type TIR

Prism-type Total Internal Reflection Spectroscopy (PTIRS) is a technique in which a prism is placed in contact with the sample surface, and a light source is directed at the prism-sample interface at an angle greater than the critical angle. The light that is reflected back into the prism is then analyzed to study the properties of the sample surface. In objective-type TIRS, an objective lens is used instead of a prism to direct the light onto the sample surface at an angle greater than the critical angle.

The objective lens focuses the light onto a small area of the sample surface, allowing for high spatial resolution and the ability to study localized regions of the sample. Both prism-type and objective-type TIRS can be used to study a wide range of samples, including thin films, surfaces, and interfaces in semiconductors, metals, polymers, and biological materials. The choice of using prism-type or objective-type TIRS depends on the specific application and the desired level of spatial resolution (Lerner et al. 2021). Prism-type TIRS is relatively simple, easy to implement and relatively inexpensive, but has a lower spatial resolution and is less versatile than objective-type TIRS which is more complex, more expensive, and more versatile but has a higher spatial resolution. It is also worth noting that objective-type TIRS can be combined with other techniques such as microscopy, such as confocal or scanning probe, to achieve even higher spatial resolution.

Observing fluorescence from a single fluorophore can be challenging due to high background signals arising from the excitation source. To minimize this background and improve single-molecule fluorescence observation, fluorescence microscopes often employ Total Internal Reflection Fluorescence (TIRF) microscopy. TIRF works by reflecting the laser excitation light at the interface between the glass slide and the aqueous sample (Fish 2009). This creates an evanescent field, which has a limited penetration depth (approximately 100 nm), and is used to selectively excite the donor fluorophore in Förster Resonance Energy Transfer (FRET) experiments. The evanescent field significantly reduces the background fluorescence from the bulk solution, thus enabling better signal-to-noise ratios in single-molecule observations. TIRF microscopes can be constructed using either objective or prism-based excitation schemes to achieve the high incident angle required for generating the evanescent field. In both designs, the background fluorescence signal is greatly minimized, allowing for improved detection of single fluorophores. TIRF microscopy, combined with smFRET (termed tFRET), allows simultaneous imaging of multiple surface-immobilized molecules. This capability enables the monitoring of static and dynamic heterogeneity in complex systems. To fully characterize the spatial and temporal resolution of tFRET, theoretical predictions for heterogeneity and rigorous experimental characterization of the technique using well-understood standards are necessary. While such characterization has been reported for confocal smFRET, it is essential to carry out similar studies for tFRET, as this method holds great significance for studying a wide range of biological processes at the single-molecule level. In a tFRET (time-resolved fluorescence resonance energy transfer) measurement, a series of images is captured of molecules that are immobilized on a surface. These molecules are labeled with fluorescent dyes, and the FRET data is later extracted through image analysis. The Total Internal Reflection Fluorescence (TIRF) modality allows for simultaneous imaging of hundreds to thousands of dye-labeled molecules in a single field of view (Figure 14.3). This enables the observation of individual molecules in a time-lapse manner, resembling "motion pictures," over a duration ranging from seconds to minutes until the fluorescent dyes photobleach.

Single-molecule Förster Resonance Energy Transfer (smFRET) studies can be implemented using either confocal or Total Internal Reflection Fluorescence (TIRF) microscopy (Figure 14.4A&B). The choice between the two methods depends on the specific characteristics of the biological system being investigated and the availability of appropriate instrumentation. Confocal Microscopy is mostly well-suited for Freely-Diffusing Molecules (Figure 14.2A). Confocal microscopy is ideal for studying freely-diffusing molecules as it allows the observation of single molecules as they pass through the confocal excitation volume, providing snapshots of thousands of individual molecules over hours. The short residence times and free diffusion of molecules into the observation volume enable the acquisition of many single-molecule events at rates of a few events per second. This method can offer sub-nanosecond time resolution, making it suitable for capturing rapid conformational dynamics. In the confocal modality, Time-Correlated Single Photon Counting (TCSPC) and polarization-sensitive detections, known as multiparameter fluorescence detection (MFD), are used. MFD allows for monitoring the fluorescence lifetime and anisotropy, in addition to fluorescence intensity, providing more information about the system under investigation. On the other hand, TIRF Microscopy is used for Surface-Immobilized

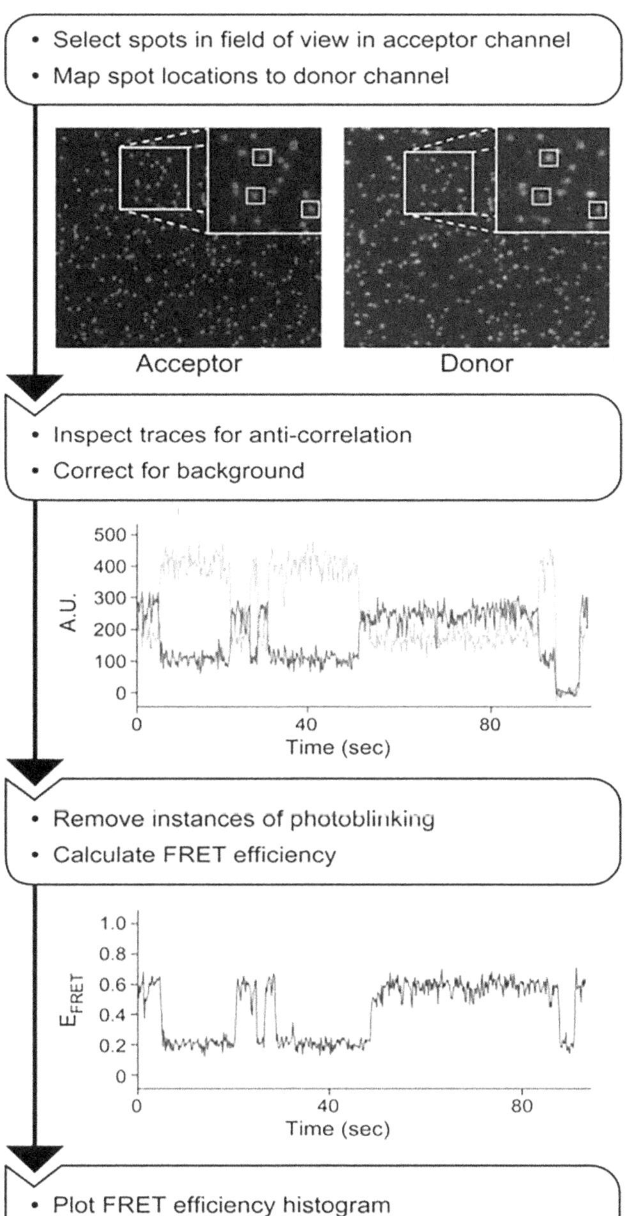

FIGURE 14.3 The workflow of smFRET data processing in TIRF mode typically involves the following steps: Image Selection: The camera images from both the donor (green) and acceptor (red) channels are examined to select molecules that are present in both channels. This is indicated by the yellow boxes. For each selected molecule, integrated fluorescence trajectories are generated by plotting the fluorescent signals from both the donor (green) and acceptor (red) channels over time. This step confirms anti-correlation between the donor and acceptor signals, which is a characteristic feature of FRET. Additionally, background correction is performed to remove any non-specific fluorescence signals or noise. The FRET efficiency (EFRET) is calculated for each time point of the trajectory. The EFRET values are then plotted as a function of time, represented by a blue line. This EFRET plot provides a dynamic profile of FRET efficiency changes over the course of the experiment. Image courtesy: https://doi.org/10.1016/j.ymeth.2017.05.011 (van der Feltz and Hoskins 2017). Creative Commons Public Domain Mark 1.0. PMC Open Access Subset. Bethesda (MD): National Library of Medicine. 2003. Available from https://www.ncbi.nlm.nih.gov/pmc/tools/openftlist/.

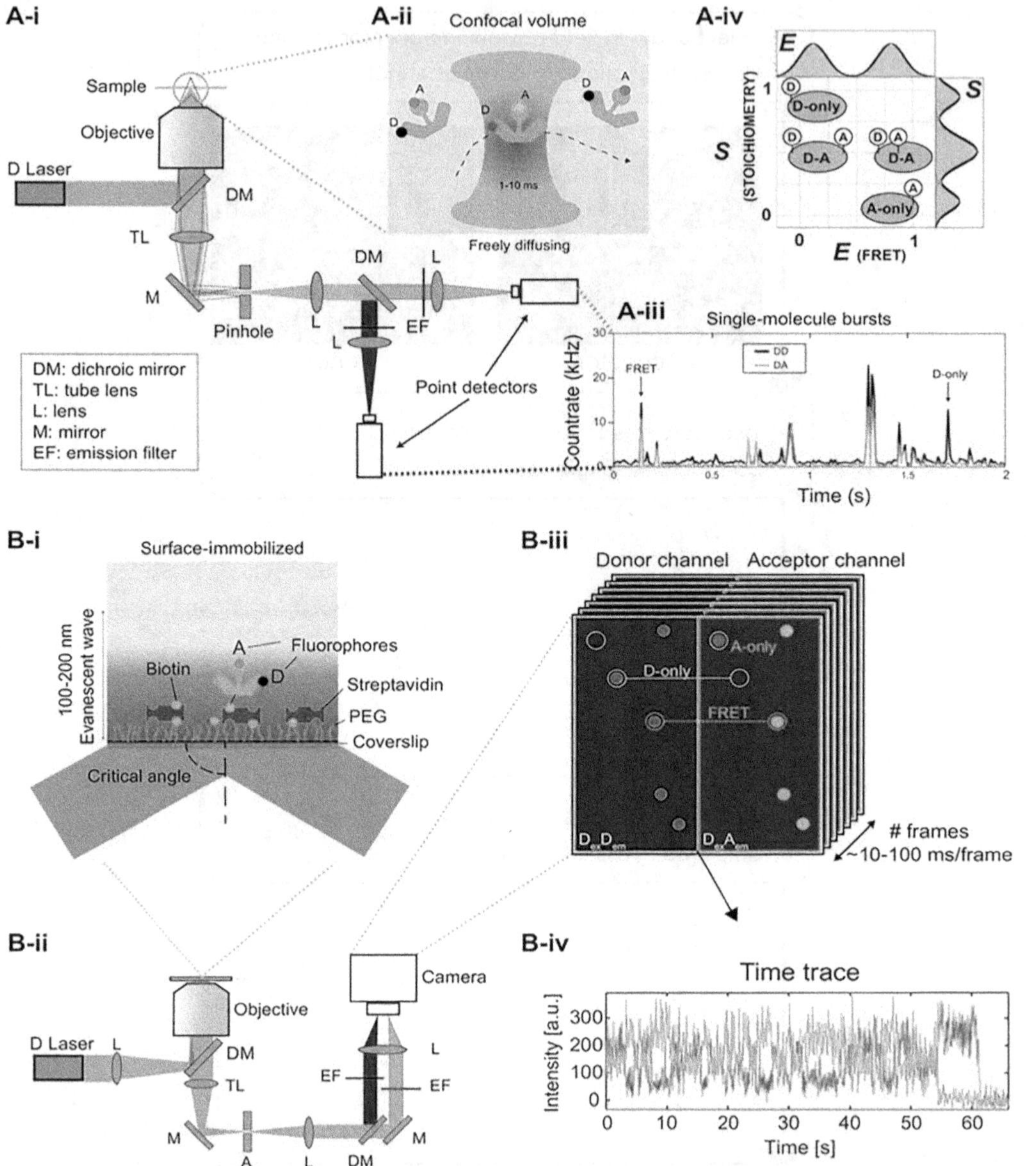

FIGURE 14.4 Two different single-molecule FRET (smFRET) modalities commonly used in research: Confocal smFRET measurements on freely-diffusing molecules and TIRF-based smFRET experiments on surface-immobilized molecules. (A) Confocal smFRET measurements on freely-diffusing molecules: 1) This method involves a single-color excitation confocal microscope with point detectors for two-color detection. The excitation light is directed toward the sample through a high-numerical aperture objective lens. Fluorescence emitted from the sample is collected through the same objective lens and then split into donor and acceptor detection channels by using dichroic mirrors and emission filters. Point detectors, such as avalanche photodiodes (APD), capture single photons emitted from individual molecules. 2) The molecules of interest, labeled with donor and acceptor fluorophores, freely diffuse through the confocal excitation spot. 3) In a confocal smFRET measurement, photon bursts arising from single molecules diffusing through the confocal volume are recorded. The bursts exhibit characteristic patterns of donor (green) and acceptor (red) emissions, enabling the identification of single-labeled and double-labeled molecules. 4) In ALEX (Alternating Laser Excitation) or PIE (Pulsed Interleaved Excitation) experiments, a two-dimensional histogram of FRET efficiency (E) and stoichiometry (S) is generated. This allows researchers to distinguish

Molecules (Figure 14.2B). TIRF microscopy is typically used for studying surface-immobilized molecules. It allows for precise observation of molecules close to the surface, minimizing background noise and increasing the signal-to-noise ratio. TIRF microscopy is well-suited for monitoring interactions and dynamics at the interface between the surface and the biomolecules. In TIRF, the excitation light penetrates only a few hundred nanometers into the sample, which is suitable for observing surface-bound molecules.

Choosing between confocal and TIRF microscopy depends on the dynamics of the biological system of interest and the availability of appropriate equipment. Some experimental setups are commercially available, but many are custom-built to suit specific research needs. For a comprehensive understanding of a biological system's dynamics, it is often beneficial to use both confocal and TIRF modalities in parallel. Each method offers unique advantages, and combining them can provide valuable insights into the conformational dynamics and interactions occurring over a wide range of timescales, from nanoseconds to seconds.

TIRF has a relatively lower temporal resolution, usually on the order of a few tens of milliseconds, though ongoing technological advancements are improving this aspect. TIRF can be accomplished by illuminating the sample through a high-numerical-aperture objective (Figure 14.4B) or using a quartz prism (Roy et al. 2008). The technique of Alternating Laser Excitation (ALEX) (Kapanidis et al. 2004) permits the distinction of molecules exhibiting fluorescence from a single dye or both dyes involved in the FRET experiment (Figure 14.4A-iv). This process also provides valuable information about the photophysics of the fluorescent dyes used. In the TIRF modality, a technique called millisecond ALEX (msALEX) is commonly utilized (Kapanidis et al. 2004, 2005; Lee et al.). Conversely, in the confocal modality, microsecond ALEX (μsALEX) or even nanosecond ALEX (nsALEX), also known as pulsed interleaved excitation (PIE), are employed (Kudryavtsev et al. 2012; Laurence et al. 2005; Mü et al.).

A few areas in the field of Neurobiology and where we saw smFRET making a big impact and answering long-standing questions have been highlighted in the following section.

14.6 F1-ATPASE CONFORMATIONAL CYCLE FROM SIMULTANEOUS SINGLE-MOLECULE FRET AND ROTATION MEASUREMENTS

The paper, "F1-ATPase conformational cycle from simultaneous single-molecule FRET and rotation measurements," describes a study of the conformational cycle of F1-ATPase, a molecular motor that uses the energy from ATP hydrolysis to drive mechanical rotation (Sugawa et al. 2016). The researchers used a technique called simultaneous single-molecule FRET (fluorescence resonance energy transfer) and rotation measurements to study the conformational changes of the motor as it rotates (Figure 14.5A-F).

The F1-ATPase is a molecular motor that converts the chemical energy of ATP hydrolysis into mechanical energy. It is a complex enzyme composed of two main domains: the F1 domain, which

between single- and double-labeled populations based on their FRET characteristics. (2005 Elsevier Ltd. All rights reserved. The figure was originally published as Figure 14.2A in Lee et al., 2005. Biophysical Journal, 88(4): 2939–2953. B) TIRF-based smFRET experiments on surface-immobilized molecules: 1) This approach involves immobilizing the molecules of interest on a surface and labeling them with donor and acceptor fluorophores. 2) In a single-color objective-type TIRF excitation two-color wide-field detection microscope, excitation light is directed through the objective lens at a specific angle to achieve Total Internal Reflection (TIR). This generates an evanescent field that selectively excites fluorophores near the surface. A is Aperture, TL is Tube lens, L is Lens, M is Mirror, DM is Dichroic mirror and EF is Emission filter. 3) The fluorescence signals from the donor and acceptor dyes are split onto two halves of a camera, and fluorescent beads or zero-mode waveguides are used for mapping the two channels and calibrating the system (Roy et al., 2008; Salem et al., 2019). 4) By observing the single-molecule fluorescence trajectories of donor and acceptor dyes, anti-correlation patterns indicative of FRET dynamics can be identified. Image courtesy: doi: 10.7554/eLife.60416 (Gust et al. 2014). CC0 1.0 Universal (CC0 1.0) Public Domain Dedication.

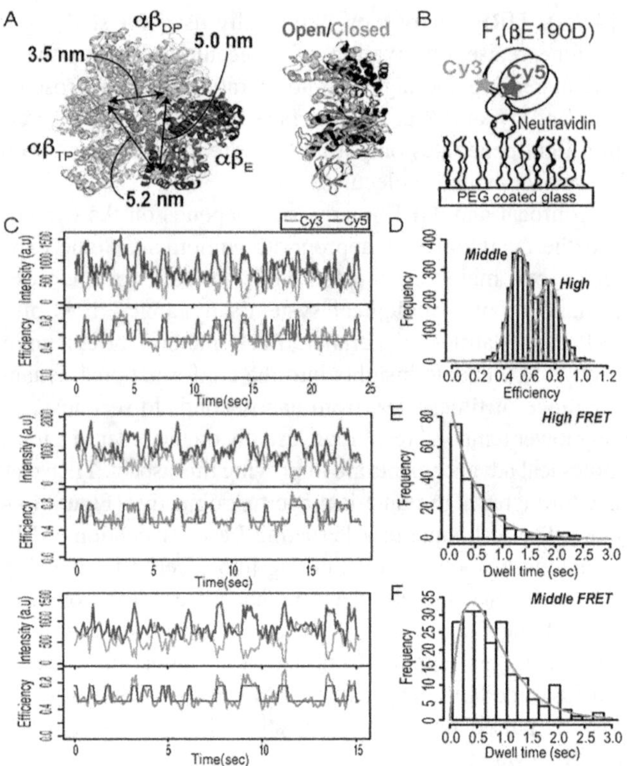

FIGURE 14.5 (A) In single-molecule FRET (Förster resonance energy transfer) measurements of F1-ATPase, specific fluorescent labeling of the protein was used to study its conformational changes and dynamics. The molecular structure that was focused on was derived from bovine F1 (an ATPase enzyme) [PDB ID code 2JDI (38)]. In the structure, they highlighted the βL402 residue, which corresponds to βL398 in thermophilic F1 as yellow spheres. The β-subunit of F1 can exist in different conformational states, represented by closed forms of βDP and βTP (shown in cyan) and an open form (shown in magenta). Additionally, the γ-subunit is shown in orange. The distances between three cysteine residues (Cβs) of βL402 in different conformations are as mentioned. (B) For the single-molecule FRET measurements, a schematic of the setup is shown where the protein was labeled with two fluorophores, Cy3 (donor, represented in green) and Cy5 (acceptor, represented in magenta). FRET occurs when the donor and acceptor fluorophores are in close proximity, and the energy from the donor is transferred to the acceptor. This FRET signal is used to monitor conformational changes in the protein. (C) Typical time trajectories of fluorescence intensities from the Cy3 and Cy5 fluorophores and the resulting FRET efficiency are shown. The FRET efficiency reflects the proximity between the labeled sites and is represented in blue. A two-state Hidden Markov Model (HMM) was performed to fit and analyze the FRET efficiency trace and infer the protein's conformational states. (D) The resulting histogram of FRET efficiencies from all the time series data provides information about the different conformational states the protein can adopt. (E,F) The dwell time histograms give insights into the stability and dynamics of these different conformational states. The two red lines in (E) & (F) are from a single exponential fit (high FRET state) and contortion of two exponentials with same rate constant (for the intermediate FRET state). Image courtesy: https://doi.org/10.1073/pnas.1524720113 (Sugawa et al. 2016). Creative Commons Public Domain Mark 1.0.

binds and hydrolyzes ATP, and the F0 domain, which drives rotation. The F1 domain contains three subunits: alpha, beta, and gamma, each with specific roles in the enzymatic cycle. The study aimed to understand the conformational changes of the F1-ATPase during the enzymatic cycle and how it relates to the rotation of the motor. The researchers used a single-molecule FRET technique in which the F1-ATPase was labeled with two different fluorescent dyes, one for the alpha subunit and one for the gamma subunit. The distance between the dyes changed as the enzyme underwent conformational changes, allowing the researchers to monitor the conformational changes in real-time. They also used a rotation assay to measure the rotation of the F0 domain in the presence of ATP. The results of the study showed that the F1-ATPase enzyme undergoes a series of conformational changes during the catalytic cycle, which can be divided into three main phases: the ATP hydrolysis phase, the rotation phase, and the ATP synthesis phase. During the ATP hydrolysis phase, the catalytic core hydrolyzes ATP and the rotor does not rotate. During the rotation phase, the rotor rotates clockwise, and during the ATP synthesis phase, the catalytic core synthesizes ATP and the rotor continues to rotate clockwise.

The study also revealed that the conformational changes in the catalytic core and rotor are tightly coupled to the rotation of the rotor. Specifically, the researchers observed that the conformational changes in the catalytic core preceded the rotation of the rotor, suggesting that the conformational changes in the catalytic core drive the rotation of the rotor.

Additionally, the study also showed that the conformational changes in the catalytic core and rotor are highly coordinated (Figure 14.6A,B). For example, they observed that the

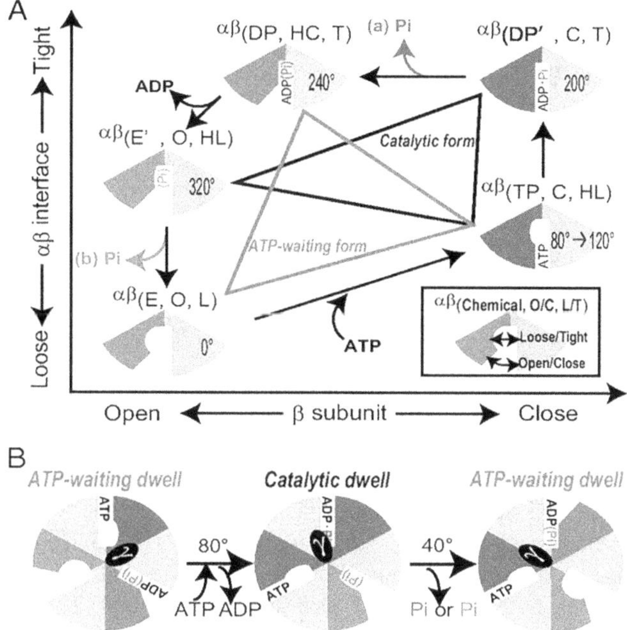

FIGURE 14.6 (A) In F1-ATPase, a tripartite coupling model exists between chemical reactions, stator conformation, and rotary angles. This model involves the αβ-dimer coupling, which can be categorized into at least five states based on three parameters: chemical states (E, TP, DP', DP, and E'), open/closed states of the β subunit (O, HC, and C), and loose/tight states at the αβ-interface (L, HL, and T). The chemical states refer to different nucleotide binding configurations in the αβ-dimer. The open/closed states of the β subunit describe the conformational changes that affect the accessibility of the active site. The loose/tight states at the αβ-interface represent the conformational changes in the interaction between α and β subunits. Pi (inorganic phosphate) is released from two specific states: (A, a) αβ(DP', C, T) or (A, b) αβ(E', O, HL). These states are associated with different steps in the ATP hydrolysis cycle. The model further extends to describe the ATP waiting form and catalytic form of F1-ATPase, with corresponding rotary angle transitions. The ATP waiting form is represented by three αβ-dimers linked together, while the catalytic form is represented by a different arrangement of three αβ-dimers, indicated by orange and blue lines, respectively. Rotary angles play a significant role in this model. The αβ(E, O, L) state is defined as 0°, and other states have specific angular relationships relative to this reference state. (B) Model depicting the transitions in rotary angle between ATP waiting and catalytic forms. Image courtesy: https://doi.org/10.1073/pnas.1524720113 (Sugawa et al. 2016). Creative Commons Public Domain Mark 1.0.

conformational changes in the catalytic core and rotor were in phase with each other, indicating that the two regions of the enzyme work together to drive the catalytic cycle. They also found that during the ATP hydrolysis phase, the catalytic core undergoes a conformation change that causes a decrease in FRET efficiency, indicating that the catalytic core changes its conformation to a state that is less favorable for FRET. The researchers also observed that during the ATP synthesis phase, the catalytic core undergoes a conformation change that causes an increase in FRET efficiency, indicating that the catalytic core changes its conformation to a state that is more favorable for FRET.

Overall, the paper provides a detailed understanding of the conformational changes that occur in F1-ATPase during its catalytic cycle, highlighting the tight coupling and coordination between the conformational changes in the catalytic core and rotor and the rotation of the rotor. These findings are important for understanding the mechanism of F1-ATPase and may have implications for the development of new therapeutic strategies for diseases that involve defects in the F1-ATPase enzyme.

14.7 THE STRUCTURAL ARRANGEMENT AND DYNAMICS OF THE HETEROMERIC GluK2/GluK5 KAINATE RECEPTOR AS DETERMINED BY SmFRET

The paper "The structural arrangement and dynamics of the heteromeric GluK2/GluK5 kainate receptor as determined by smFRET" reports the use of single-molecule fluorescence resonance energy transfer (smFRET) to study the structural arrangement and dynamics of the heteromeric GluK2/GluK5 kainate receptor. The kainate receptor is a type of ionotropic glutamate receptor (Figure 14.7) that plays a crucial role in synaptic transmission and plasticity in the nervous system. However, the structural organization and dynamics of these receptors are not well understood. The authors used smFRET to address these questions by studying the interactions between GluK2 and GluK5 subunits within the heteromeric receptor.

The smFRET experiments were performed on GluK2/GluK5 receptors expressed in HEK293 cells. The authors used a FRET pair of fluorescent proteins, where one protein was attached to the GluK2 subunit and the other to the GluK5 subunit. The FRET efficiency was used as a measure of the distance between the two subunits (Table 14.1). The study used smFRET to investigate the structural arrangement of the heteromeric GluK2/GluK5 kainate receptor, which is composed of two subunits, GluK2 and GluK5 (Figure 14.8). The study found that the GluK2 and GluK5 subunits are arranged in a "head-to-tail" configuration, with the GluK2 subunit at the N-terminal and GluK5 at the C-terminal. The study also found that the GluK2 and GluK5 subunits interact with each other through a specific interface that involves their extracellular domains. The authors found that in the absence of ligands, the GluK2 and GluK5 subunits were far apart, indicating that the receptor was in a closed conformation. However, in the presence of the ligand, the subunits were in close proximity, indicating that the receptor had undergone a conformational change to an open state. The authors studied the dynamics of the receptor by measuring the lifetime of the FRET signal. They found that the lifetime of the FRET signal was significantly longer in the presence of the ligand, indicating that the subunits were in close proximity for a longer period of time. The authors propose that this is due to the formation of a stable complex between the GluK2 and GluK5 subunits in the presence of the ligand.

The authors also found that the GluK5 subunit plays a critical role in the formation of the receptor complex and in the conformational changes that occur in response to ligand binding. The authors used a GluK2 receptor mutant that cannot interact with GluK5 and found that the FRET efficiency was low, indicating that the GluK5 subunit is necessary for the formation of the receptor complex. The authors also found that the conformational changes that occur in response to ligand binding were abolished in the GluK2 receptor mutant that cannot interact

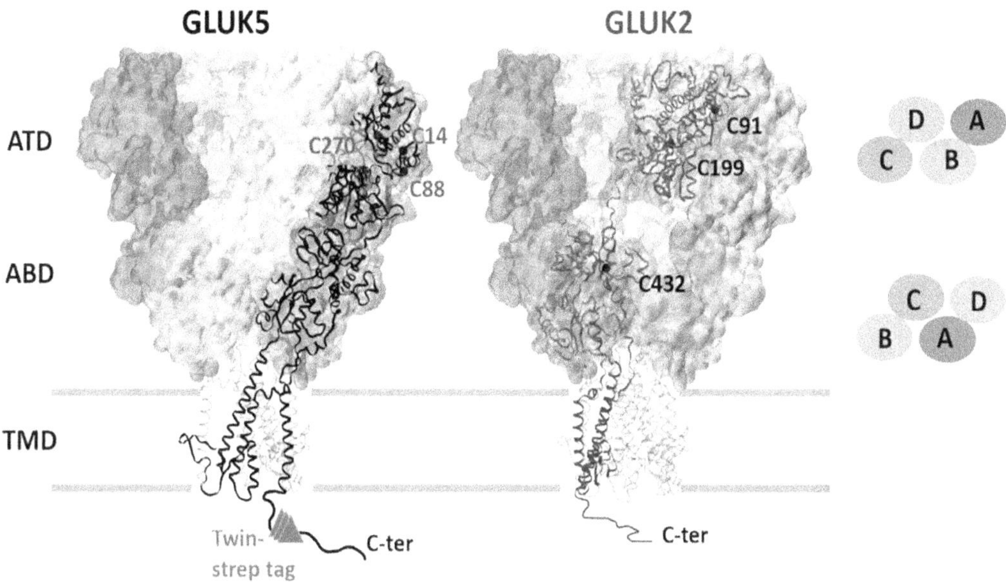

FIGURE 14.7 In the structure of GluK5 and GluK2 subunits, which are part of kainate receptors, there are important extracellular domains known as the amino-terminal domain (ATD) and the agonist-binding domain (ABD). These domains play essential roles in ligand recognition and receptor dimerization. Interestingly, the ATD and ABD form a unique dimer of dimers configuration, wherein two GluK5 subunits and two GluK2 subunits come together to create a functional unit. To achieve this configuration, domain swapping occurs between dimers, meaning that parts of one dimer interact with parts of the other dimer, forming stable interfaces. This arrangement facilitates the proper functioning of the receptor and ensures its stability. During the experiments, certain cysteine residues in the ATD and ABD regions are targeted for modification or mutation. These cysteines, represented as black spheres, are modified to create Cysless constructs. Cysless constructs are engineered versions of the receptors where cysteine residues have been replaced or modified to avoid interference with specific experimental procedures or to investigate the functional roles of these residues. The transmembrane domain (TMD) of the receptors is depicted as a cartoon structure within the cell membrane. The TMD forms the ion channel pore, and it is responsible for allowing the flow of ions in response to ligand binding in the ATD and ABD regions. To facilitate the purification and detection of the GluK5 subunit during experimental procedures, a twin-strep tag is attached to its C-terminus. The twin-strep tag is a specific peptide sequence used as an affinity tag to enable easy isolation and identification of the GluK5 subunit. Image courtesy: doi: 10.1016/j.bbamem.2019.05.023 (Litwin et al. 2019). Creative Commons Public Domain Mark 1.0.

with GluK5, indicating that GluK5 is essential for the conformational changes that occur in response to ligand binding. The authors propose that the GluK5 subunit acts as a "molecular switch" that regulates the activity of the receptor. They suggest that the GluK5 subunit binds to the GluK2 subunit in the presence of ligands, leading to the formation of a stable complex and the conformational changes that occur in response to ligand binding. This results in opening of the receptor channel and activation of the receptor.

In conclusion, the results of this study provide new insights into the structural organization and dynamics of the heteromeric GluK2/GluK5 kainate receptor. The study shows that the GluK2 and GluK5 subunits form a stable complex in the presence of ligands, and that the subunits are in close proximity to each other. Furthermore, the study reveals that the GluK5 subunit plays a critical role in the formation of the receptor complex and in the conformational changes that occur in response to ligand binding.

TABLE 14.1
Summary of All smFRET Efficiencies and Distances Calculated

SITES	APO/ANTAGONIST BOUND STATE				GLUTAMATE BOUND STATE					
	Denoised FRET Efficiency (Calculated Distance in Å)	Gaussian fit	3KG2 (AMPA) Expected FRET efficiency (Distance in Å for sites)	5KUH (Kainate)	Denoised FRET efficiency (Calculated Distance in Å)	Gaussian fit	5KUF (Kainate) Expected FRET efficiency (Distance in Å for sites)	5VHZ (AMPA)	4U4F (AMPA)	5WEO (AMPA)
GluK2*-266-GluK2*-266	0.72 (43) 0.83 (39)	0.72 ± 0.03 0.84 ± 0.01	0.70 (44)	0.62 (47)						
GluK2*-479-GluK5*-471	0.80 (40) 0.90 (35)	0.80 ± 0.01 0.90 ± 0.01	0.83 (39)	0.87 (37)	0.56 (49) 0.69 945) 0.83 (39) 0.94 (32)	0.56 ± 0.01 0.68 ± 0.01 0.83 ± 0.01 0.93 ± 0.01	0.33 (57)	0.62 (47)	0.83 (39)	0.89 (36)
GluK2*-479-GluK5*-471	Not observed		0.23 (62)	0.13 (70)	Not observed		0.20 (64)	0.22 (63)	0.25 (61)	0.20 (64)
GluK5*-471-GluK5*-471	Not observed		0.05 (82)	0.04 (87)	Not observed		0.02 (97)	0.05 (85)	0.08 (77)	0.05 (83)
GluK2*-479-GluK2*-479	Not observed		0.15 (68)	0.14 (69)	Not observed		0.11 (72)	0.12 (71)	0.15 (68)	0.24 (62)
GluK2*-523-GluK2*-523	0.96 (30) 0.83 (39)	0.96 ± 0.01 0.83 ± 0.01	0.91 (34)		0.85 (38) 0.94 (32)	0.83 ± 0.01 0.94 ± 0.01	0.85 (38)	0.87 (37)	0.90 (35)	0.71 (44)
GluK5*-515-GluK5*-515	0.94 (32) 0.81 (40)	0.94 ± 0.01 0.81 ± 0.01	0.91 (34)		0.84 (38) 0.93 (33)	0.85 ± 0.01 0.95 ± 0.01	0.85 (38)	0.90 (35)	0.91 (34)	0.87 (37)

Image courtesy: doi: 10.1016/j.bbamem.2019.05.023 (Litwin et al. 2019). Creative Commons Public Domain Mark 1.0.

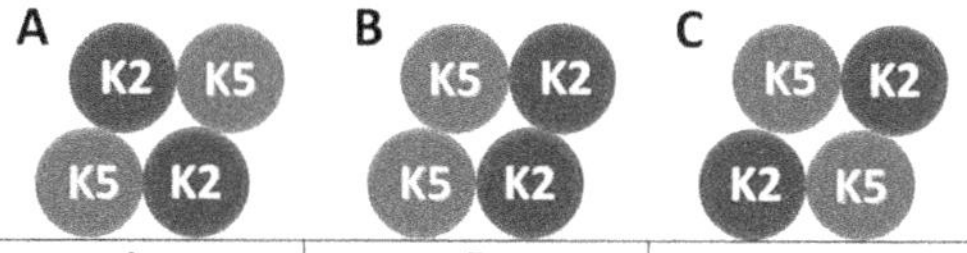

Conditions	A		B		C	
	K2-K2 (Å)	K5-K5 (Å)	K2-K2 (Å)	K5-K5 (Å)	K2-K2 (Å)	K5-K5 (Å)
Model-based distance	44	135	67	67	135	44
Experiment-based distance	42	No FRET	No FRET	No FRET	No FRET	No FRET

FIGURE 14.8 Cartoon representation of the various possible configurations of the amino-terminal domains (ATDs) of GluK2 and GluK5 subunits in the GluK2/GluK5 heterotetramer. This heterotetramer structure forms a dimer of dimers arrangement, where two GluK2 subunits and two GluK5 subunits come together to create a functional unit. However, due to the flexibility and dynamic nature of these domains, different configurations are feasible. Homology modeling predicts the structure of a protein based on the known structure of a related protein. By constructing models for the different ATD configurations in the GluK2/GluK5 heterotetramer, the distances between GluK2-GluK2 subunits and GluK5-GluK5 subunits have been calculated. This has been complemented with calculated distances from single-molecule Förster resonance energy transfer (smFRET) experiments. Image courtesy: doi: 10.1016/j.bbamem.2019.05.023 (Litwin et al. 2019). Creative Commons Public Domain Mark 1.0.

14.8 CONCLUSION

Single-molecule FRET (smFRET) has gained widespread popularity as a powerful technique for investigating the structural dynamics of biomolecules. Thanks to the collaborative efforts of numerous research groups, significant advancements have been made in sample preparation, measurement procedures, data analysis, algorithms, and documentation related to smFRET experiments. Several laboratories utilizing smFRET methods have united their expertise to streamline the process of conducting experiments and analyzing results, with the aim of obtaining accurate quantitative information on biomolecular structure and dynamics. Ensuring its reproducibility and reliability as a method, regardless of the location or conditions in which the sample is measured, is crucial for the field of structural studies. To address this, an initiative led by Thorsten Hugel brought together twenty laboratories to conduct smFRET measurements on various double-stranded DNA constructs (Hellenkamp et al.). These experiments involved studying six distinct samples with different dyes and varying distances between the dyes. Surprisingly, the mean FRET efficiencies obtained by each participating lab demonstrated a remarkably high level of agreement, with a deviation between 0.02 and 0.05, depending on the specific details of the sample. The successful quantitative assessment and reproducibility of intensity-based smFRET measurements, along with in-depth discussions about data analysis, marked a significant milestone in the field. As a result of this collaborative effort, standardized dsDNA FRET samples are now available for daily calibration, serving as valuable tools, especially for newcomers to the smFRET community.

Single-molecule Förster resonance energy transfer has emerged as a transformative technique in unraveling the mysteries of neurobiology. Neurobiology seeks to understand the complex mechanisms underlying brain function, including learning, memory, cognition, and neurological disorders. SmFRET has played a vital role in this field by providing unparalleled insights into the behavior and interactions of individual biomolecules within neurons, offering a new level of understanding that was previously unattainable. Traditional biochemical and biophysical methods often rely on ensemble averaging, masking the heterogeneity and dynamics of molecular behavior. SmFRET, on the other hand, enables researchers to examine individual molecules, revealing rare and transient events that play crucial roles in neuronal processes. A key area where SmFRET has made significant contributions is in the study of synaptic transmission. Synapses are the sites of communication between neurons, and their efficiency relies on precise protein-protein interactions. It allows researchers to monitor the dynamics of synaptic proteins, such as neurotransmitter receptors and vesicle fusion machinery, uncovering the intricacies of synaptic function. Furthermore, this technique has been instrumental in elucidating the behavior of intrinsically disordered proteins (IDPs) in neurobiology. IDPs lack stable secondary structures and are prevalent in the brain, involved in a wide range of cellular processes. SmFRET enables the characterization of IDP conformational changes upon binding to partners, shedding light on their functional relevance and how they contribute to neurological diseases. The investigation of protein conformational changes is another crucial aspect in neurobiological disorders. Proteins often undergo dynamic structural transitions to carry out their functions, and these conformational changes are central to neuronal signaling and plasticity. The real-time monitoring of these conformational changes, provides detailed mechanistic insights into how proteins modulate neuronal activities. Additionally, the technique has been applied to study the assembly and function of membrane proteins, such as ion channels and transporters, which are vital for neuronal excitability and communication.

In summary, SmFRET has revolutionized the field of neurobiology by providing unprecedented single-molecule sensitivity and high-resolution imaging capabilities. Its role in unraveling the mysteries of neurobiology lies in its capacity to unveil the dynamic behavior, interactions, and conformational changes of individual biomolecules within neurons. Through these revelations, it continues to drive groundbreaking discoveries, offering a promising future for understanding brain function and ultimately contributing to the development of therapies for neurological disorders.

REFERENCES

Fish, K. N. 2009. Total internal reflection fluorescence (TIRF) Microscopy. *Curr. Protoc. Cytom* 0 12, Unit12.18.

Grimm, J. B. *et al.* 2015. A general method to improve fluorophores for live-cell and single-molecule microscopy. doi:10.1038/nmeth.3256.

Grimm, J. B. *et al.* 2017. A general method to fine-tune fluorophores for live-cell and in vivo imaging 14.

Gust, A. *et al.* 2014. A starting point for fluorescence-based single-molecule measurements in biomolecular research. *Molecules* 19: 15824.

Ha, T. 2001. Single-molecule fluorescence resonance energy transfer. *Methods* 25: 78–86.

Ha, T., and P. R. Selvin. 2008. The new era of biology in singulo. *Cold Spring Harb. Lab. Press*: 1–36.

Hellenkamp, B. *et al.* Precision and accuracy of single-molecule FRET measurements-a multi-laboratory benchmark study Corrected: Publisher correction. *Nat. Methods* 10.

Islam, F., M. Basu, and P. P. Mishra. 2022. From ensemble FRET to single-molecule imaging: Monitoring individual cellular machinery in action. *Opt. Spectrosc. Microsc. Tech*: 113–142 doi:10.1007/978-981-16-4550-1_6.

Joo, C., and T. Ha. 2012a. Preparing Sample Chambers for Single-Molecule FRET. *Cold Spring Harb. Protoc* 2012: pdb.prot071530.

Joo, C., and T. Ha. 2012b. Imaging and identifying impurities in single-molecule FRET studies. *Cold Spring Harb. Protoc* 2012: pdb.prot071548.

Kapanidis, A. N. *et al.* 2004. Fluorescence-aided molecule sorting: Analysis of structure and interactions by alternating-laser excitation of single molecules. *Proc. Natl. Acad. Sci. U. S. A.* 101: 8936–8941.

Kapanidis, A. N. *et al.* 2005. Alternating-laser excitation of single molecules. *Acc. Chem. Res.* 38: 523–533.

Kudryavtsev, V. *et al.* 2012. Combining MFD and PIE for accurate single-pair förster resonance energy transfer measurements. *ChemPhysChem* 13: 1060–1078.

Lakowicz, J. R. 2006. Principles of fluorescence spectroscopy 954.

Lamichhane, R., A. Solem, W. Black, and D. Rueda Single molecule FRET of protein-nucleic acid and protein-protein complexes: Surface passivation and immobilization. doi:10.1016/j.ymeth.2010.06.010.

Laurence, T. A., X. Kong, M. Jäger, and S. Weiss. 2005. Probing structural heterogeneities and fluctuations of nucleic acids and denatured proteins. *Proc. Natl. Acad. Sci. U. S. A.* 102: 17348–17353.

Lee, N. K. *et al.* Accurate FRET measurements within single diffusing biomolecules using alternating-laser excitation. doi:10.1529/biophysj.104.054114.

Lerner, E. *et al.* 2021. FRET-based dynamic structural biology: Challenges, perspectives and an appeal for open-science practices. *Elife* 10.

Litwin, D. B., N. Paudyal, E. Carrillo, V. Berka, and V. Jayaraman. 2019. BBA-biomembranes. doi:10.1016/j.bbamem.2019.05.023.

Miki, M. *et al.* 1992. 6 L. Matyus. *Annu. Reo. Biophys. Biophys. Chem* 13: 39.

Mü, B. K., E. Zaychikov, C. Brä, and D. C. Lamb pulsed interleaved excitation. doi:10.1529/biophysj.105.064766.

Paul, T., S. C. Bera, and P. P. Mishra 2017. Direct observation of breathing dynamics at the mismatch induced DNA bubble with nanometre accuracy: a smFRET study †. doi:10.1039/c6nr09348e.

Roy, R., S. Hohng & T. Ha. 2008. A practical guide to single-molecule FRET. doi:10.1038/NMETH.1208.

Sasmal, D. K., L. E. Pulido, S. Kasal, and J. Huang. 2016. Single-molecule fluorescence resonance energy transfer in molecular biology. *Nanoscale Rev. Cite this Nanoscale* 8.

Selvin, P. R. 1995. Fluorescence resonance energy transfer. *Methods Enzymol* 246.

Sugawa, M. *et al.* 2016. F1 -ATPase conformational cycle from simultaneous single-molecule FRET and rotation measurements. *Proc. Natl. Acad. Sci. U. S. A.* 113: E2916–E2924.

Szabó, Á., J. Szöllősi, and P. Nagy. 2022. Principles of Resonance Energy Transfer. *Curr. Protoc.* 2.

van der Feltz, C., and A. A. Hoskins. 2017. Methodologies for studying the spliceosome's RNA dynamics with single-molecule FRET. *Methods* 125: 45.

Index

For Product Safety Concerns and Information please contact our EU representative GPSR@taylorandfrancis.com
Taylor & Francis Verlag GmbH, Kaufingerstraße 24, 80331 München, Germany

www.ingramcontent.com/pod-product-compliance
Lightning Source LLC
LaVergne TN
LVHW081314110826
845149LV00006B/1504

* 9 7 8 1 0 3 2 3 0 5 2 9 5 *